CHEMISTRY OF DRUGS

CHEMISTRY OF DRUGS

Editor

Eberhard Welter

scitus
academics

Chemistry of Drugs

Edited by **Eberhard Welter**

This Book is Printed on Acid-Free Paper. ♾

ISBN: 9781681178875

Library of Congress Control Number: 2017948914

SCITUS Academics LLC,

616, Corporate Way, Suite 2, 4766,

Valley Cottage, NY 10989

www.scitusacademics.com

Notice

Reasonable efforts have been made to publish reliable data and views articulated in the chapters are those of the individual contributors, and not necessarily those of the editors or publishers. Editors or publishers are not responsible for the accuracy of the information in the published chapters or consequences of their use. The publisher believes no responsibility for any damage or grievance to the persons or property arising out of the use of any materials, instructions, methods or thoughts in the book. The editors and the publisher have attempted to trace the copyright holders of all material reproduced in this publication and apologize to copyright holders if permission has not been obtained. If any copyright holder has not been acknowledged, please write to us so we may rectify.

Printed in United States of America.

Preface

Drugs are a way of modifying the chemistry of the body. They can be used to treat diseases and infections, correct imbalances in electrolytes and fluids, or alter mental status (such as inducing amnesia or stopping hallucinations). Drugs are used both for medical purposes and for recreation. In both cases, no drug is perfect. A perfect drug would be 100% effective while causing no side effects. Drug discovery has undergone many changes over the years but the goal has remained same: to uncover safer medicines for all diseases. Drug discovery and development is driven by the knowledge of chemistry of the molecules and their association with life process. The classical or traditional method adopted by medicinal chemists involves modifying bio-active molecules from natural products. These natural products are the source of active ingredients in most of the existing drugs. The current era has witnessed an ever changing role in modern drug discovery. The chemical methods adopted for the discovery of the molecules have also undergone changes leading to the development of technologies such as combinatorial chemistry, microwave assisted organic synthesis (MAOS) and high-throughput (HTS) biological screening. These new technologies have enabled medicinal scientists to accelerate the discovery process. The contribution of chemistry is not confined just to the discovery stage. The recent changes in synthetic chemistry is practiced in this environment center around new scientific advances in synthetic techniques and new technologies for rational drug design, combinatorial chemistry, automated synthesis, and compound purification and identification. As all drugs are chemicals, and pharmacy is mainly about the study of various aspects of drugs, including manufacture, storage, actions and toxicities, metabolisms and managements, chemistry still plays a vital role in pharmacy education. However, the extent at which chemistry used to be taught a couple of decades ago has certainly changed significantly. It has been recognized that while pharmacy students need a solid foundation in chemistry knowledge the extent cannot be the same as chemistry students may need.

This book 'Chemistry of Drugs' is an accessible introduction to organic chemistry, elementary medicinal chemistry, and biochemistry. The book offers an accessible introduction to subjects that are fundamental to pharmaceutical science. It looks at the chemical structure of drugs, and in particular to elements or fragments in these chemical structures. It covers all the key aspects of organic chemistry, elementary medicinal chemistry, and biochemistry, required by pharmacy and pharmaceutical sciences students as well as researchers.

Table of Contents

CHAPTER 1

Synthesis, Characterization and In Vitro Antitumor Evaluation of New Pyrazolo[3,4-d]Pyrimidine Derivatives

Ahmed M. El-Morsy[1], Mohamed S. El-Sayed[1], Hamada S. Abulkhair[1,2*]

[1]*Department of Organic Chemistry, College of Pharmacy, Al-Azhar University, Cairo, Egypt*

[2]*Material and Biomedical Sciences, Zewail City of Science and Technology, Cairo, Egypt*

ABSTRACT

A new series of 3-(methylthio)-1-phenyl-1H-pyrazolo[3,4-d]pyrimidine derivatives was synthesized. The structures of the new derivatives were confirmed by the spectral data and elemental analyses. The antitumor activity of this series against human breast adenocarcinoma cell line MCF7 was evaluated. Out of twenty new derivatives, ten were revealed mild to moderate activity compared with doxorubicin as a reference antitumor. Among this new series N-(2-chlorophenyl)-2-(3-(methylthio)-4-oxo-1-phenyl-1H-pyrazolo[3,4-d]pyrimidin-5(4H)-yl)acetamide (13_a) was found the most active one with IC_{50} equal to 23 μM.

Keywords: Pyrazolo[3,4-d]Pyrimidine, Antitumor, Human Breast Adenocarcinoma Cell Line MCF7

1. INTRODUCTION

Cancer is the most serious health problem and the second major cause of death in the developing countries [1] [2] . In spite of significant process in the development of novel chemotherapeutic agents in the last seven decades, success in developing targeted non-toxic drugs with minor side effects has only achieved in the last one [3] . Therefore, the discovery of new selective, potent and safe antitumor agents is a must. Pyrazolo[3,4-d]pyrimidine nucleus is the bioisostere of purine [4] [5] , which exhibits promising activity as antitumor by competitive inhibition for ATP kinase enzymes. Many pyrazolo[3,4-d]pyrimidine derivatives were reported as antitumor agents [6] [7] [8] . The cytotoxic activity of such compound may attribute to inhibition of several enzymes such as tyrosine kinase [9] , Src kinase [10] , cyclin dependent kinase (CDK) [11] , mammalian target of rapamycin (mTOR) [12] and glycogen synthase kinase (GSK) [13] [14] . In addition, the presence of methyl sulphonyl group at the 3 position of pyrazolo[3,4-d]pyrimidine nucleus was reported to potentiate the antitumor activity of such nucleus [11] [15] . For example, compound 1 and 2 (Figure 1) were exhibited excellent antitumor activity against breast adenocarcinoma cell line MCF7 with IC_{50} values of 12.0 and 7.50 μM respectively [16] . Also, compound 3 displayed superior activity as cytotoxic against A549 cell line with IC_{50} value of 5.28 μM [4] .

Based on these scientific facts and for further exploration of novel antitumor agents, we supposed that incorporation of these structural features together may result in potent antitumor agents that act on breast adenocarcinoma cell line. In this work, new 3-(methylthio)-1-phenyl-1H-pyrazolo[3,4-d]pyrimidine derivatives 10 - 16 were synthesized, incorporating the methyl sulphonyl group at the 3 position of pyrazolo[3,4-d]pyrimidine ring system and varying the substituents at the 4 and 5 positions of such ring in order to study the effect of these varying substitutions on the antitumor activity of pyrazolo[3,4-d]pyrimidine nucleus against human breast adenocarcinoma cell line MCF7.

2. RESULTS AND DISCUSSION

2.1. Chemistry

Scheme 1 shows the synthetic pathway of the starting pyrazolo[3,4-d]pyrimidin-4(5H)-one derivatives 8 and 4-chloro-3-(methylthio)-1-phenyl-1H-pyrazolo[3,4- d]pyrimidine (9) which were accomplished via reaction of malononitrile with carbon disulfide in the presence of sodium ethoxide followed by methylation of the product with dimethyl sulphate. The resulting 2 (bis(methylthio)methylene) malononitrile was then treated with phenyl hydrazine in absolute ethanol [17] . Cyclization of the 5-amino-3-(methylthio)-1-phenyl-1Hpyrazole-4-carbonitrile (7) by the action of formic acid afforded 3-(methylthio)-1-phenyl-1H-pyrazolo [3,4-d]pyrimidin-4(5H)-one [18] . Structure of the latter was confirmed by the disappearance of the $C\equiv N$ and NH characteristic absorption bands in the IR spectrum of the starting 5-amino-1Hpyrazole-4-carbonitrile 7. Chlorination of compound 8 with phosphorus oxychloride yielded the 4-chloro derivative 9 [19] [20] . The latter was allowed to react with different aliphatic and aromatic amines to afford the target pyrazolo[3,4-d]pyrimidin-4-amine derivatives 10_{a-e}, 11 and 12 (Scheme 2).

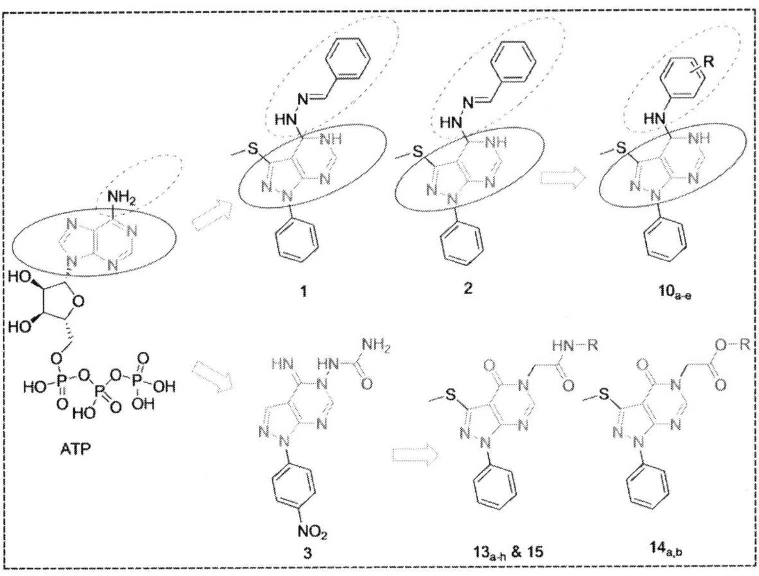

Figure 1. Structural similarities and pharmacophoric features of ATP, reported potent antitumor pyrazolo[3,4-d]pyrimidines and new designed compounds.

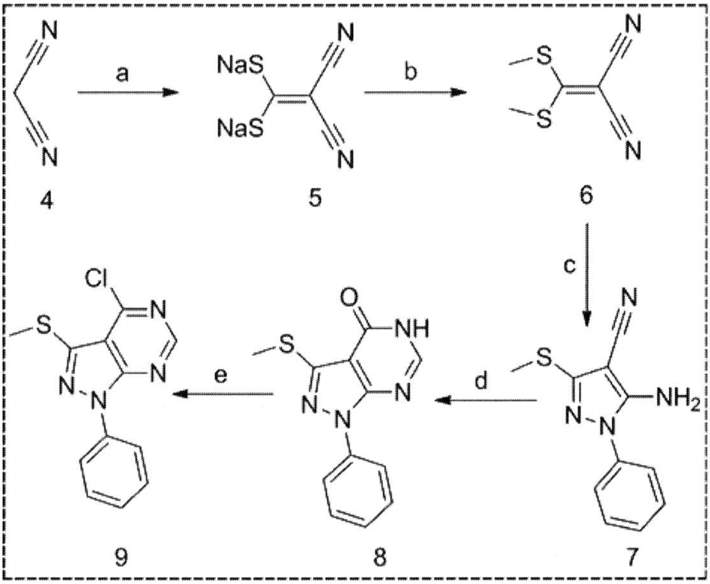

Scheme 1. Synthetic pathway of starting pyrazolo[3,4-d]pyrimidine. Reagents: (a) CS_2, C_2H_5OH; (b) $(CH_3)_2SO_4$, C_2H_5OH; (c) NH_2NH_2, C_2H_5OH; d) HCOOH; e) $POCl_3$.

Formation of compounds 10 - 12 was confirmed by spectral data and elemental analyses. The ^{1}HNMR spectra of these derivatives demonstrated the appearance of a new D_2O exchangeable singlet signals at δ 8.48 - 8.60 ppm corresponding to the NH protons. Mass spectra of these compounds showed distinctive molecular ion peaks at the right m/z values.

Scheme 3 shows the synthetic pathway of the target compounds $13_{a\text{-}h}$ through reaction of 3-(methylthio)-1-phenyl-1H-pyrazolo[3,4-d]pyrimidin-4(5H)-one (8) with 2-chloro-N-phenylacetamide or 3-chloro-N-phenylpropanamide derivatives. Structures of these amides were confirmed depending on spectral data and elemental analyses. The IR spectra of these compounds showed the characteristic NH stretching bands at the range of 3223 - 3317 cm^{-1}. In addition, the ^{1}HNMR spectra of the same derivatives showed singlet signals corresponding to NH protons at δ 9.72 - 10.78 ppm.

Ester derivatives $14_{a,b}$ were prepared via condensation of compound 8 with alkyl chloroacetate in the presence of potassium carbonate. Structures of these two ester derivatives were confirmed on the basis of their spectral data and elemental analyses. The IR spectra of these esters showed sharp elevation of the wave number of the C=O absorption compared to that of the starting amide. Condensation of 14_b with hydrazine hydrate produced the hydrazide 15. The ^{1}HNMR spectrum of this new compound revealed two D_2O exchangeable signals of the NH and NH_2 protons at 9.41 and 4.30 ppm respectively. Disappearance of the NH_2 signal in the ^{1}HNMR spectra of compounds $16_{a,b}$ confirms their structures.

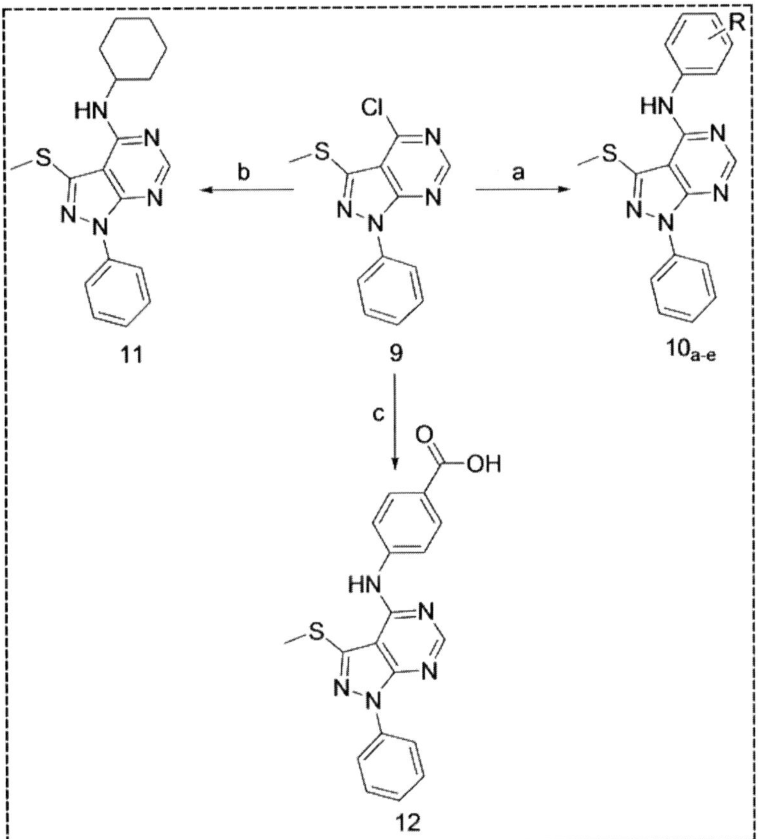

Scheme 2. Synthetic pathway of target compounds 10 - 12. R = 2-CH_3, 2-SH, 2,6-Cl_2, 2-COOH, 4-$COOC_2H_5$. Reagents: (a) $RC_6H_4NH_2$, $C_2H_5OH/(C_2H_5)_3N$; (b) Cyclohexylamine, $C_2H_5OH/(C_2H_5)_3N$; (c) 4-(Aminomethyl)benzoic acid, $C_2H_5OH/(C_2H_5)_3N$..

2.2. In Vitro Antitumor Screening

All of the newly synthesized derivatives were evaluated for antitumor activity by measuring the inhibitory effect of such compounds against human breast adenocarcinoma cell line MCF7 using MITT technique [21]

[22] . The MTT Cell Proliferation Assay measures the reduction in cancer cell viability due to apoptosis or necrosis as a response to external factor. The yellow colored tetrazolium salt of MTT is reducible by the action of metabolically active cells, through dehydrogenase enzymes that leads to generation of NADH and NADPH reducing equivalents. The produced intracellular purple formazan can be solubilized and spectrophotometrically quantified [23] . The results of in vitro antitumor activity were compared with doxorubicin as a reference antitumor agent. The parameter used herein is the IC_{50}, which represents the concentration needed for 50% inhibition of the cell viability. A relation between the IC_{50} values of the new compounds that showed more than 50% inhibition against MCF-7 and that of the reference antitumor agent is shown in Table 1 and represented graphically in Figure 2.

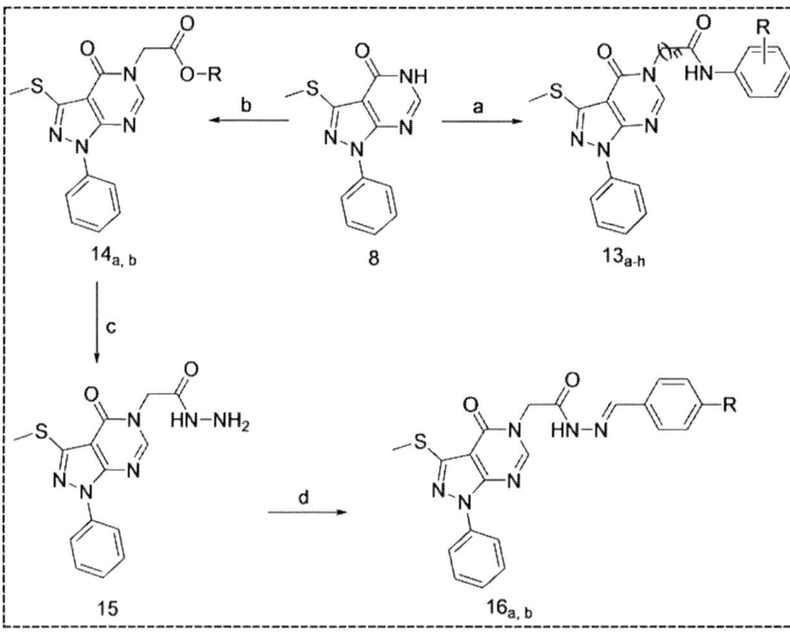

Scheme 3. Synthetic pathway of target compounds 13 - 16. For $13_{a\text{-}d}$ n = 1, R = H, 2-Cl, 3-CH_3, 4-$COOC_2H_5$; For $13_{e\text{-}h}$: n = 2, R = H, 2-Cl, 3-CH_3, 4-$COOC_2H_5$; For $14_{a\text{-}b}$: R = CH_3, C_2H_5; For $16_{a\text{-}b}$: R = H, OH. Reagents: For $13_{a\text{-}d}$: (a) $ClCH_2CONHC_6H_4R$, K_2CO_3, C_2H_5OH; For $13_{e\text{-}h}$: (a) $ClCH_2CH_2CONHC_6H_4R$, K_2CO_3, C_2H_5OH; (b) $ClCH_2CH_2COOR$, K_2CO_3, C_2H_5OH; (c) NH_2NH_2; d) RC_6H_4CHO, C_2H_5OH.

3. EXPERIMENTAL

3.1. General

All melting points were taken on electro thermal (LA9000 SERIS) digital melting point apparatus and are uncorrected. IR spectra were recorded on PyeUnicam Sp 1000 spectrophotometer and were carried out at the Pharmaceutical Analytical Unit, Faculty of Pharmacy, Al-Azhar University, Egypt. The ^{1}HNMR and ^{13}CNMR spectra were recorded in DMSO-D_6 either on Varian Mercury VXR-300 NMR spectrophotometer at the Microanalytical Unit of Cairo University or BURKER 400 MHZ spectrophotometer at the Nuclear Magnetic Resonance Lab, Faculty of Pharmacy, Zagazig University, Egypt. Chemical shifts were related to that of the solvent. TMS was used a standard. Mass spectra were recorded on Hewlett Packard 5988 spectrometer at the Regional Center for Mycology & Biotechnology, Al-Azhar University, Cairo, Egypt. Progress of the reactions was monitored by TLC pre-coated with UV fluorescent silica gel and was visualized using UV lamp and different solvent systems as mobile phases. 5- Amino-3-(methylthio)-1-phenyl-1H-pyrazole-4-carbonitrile (7) and 3-(methyl- thio)-1-phenyl-1H-pyrazolo[3,4-d]pyrimidin-4(5H)-one (8) were prepared according to published method [17] . Compound 9 was obtained following reported procedure [19] . 2-Chloro-N-arylactamide and 3-chloro-N-arylpropanamide derivatives were prepared as reported [24] .

3.2. General Procedure for Synthesis of 3-(Methylthio)-1-Phenyl- N-Aryl-1H-Pyrazolo[3,4-d]Pyrimidin-4-Amines 10_{a-e}

A mixture of compound 9 (10 mmol) and the appropriate aniline derivative (10 mmol) in absolute ethanol (35 ml) containing trimethylamine (15 mmol) was heated under reflux for 6 hours. The reaction mixture was cooled, and the separated solid was filtered, dried and finally recrystallized from ethanol.

Table 1. Results of in vitro cytotoxic activity of compounds showed more than 50% inhibition of MCF7 adenocarcinoma cell line.

Comp. No.	Conc. (μM)	Average Absorbance (%)	Viable Cells (%)	Standard Deviation	IC_{50} (μM)
10_a	0.00	0.221	100.000	0.799	128
	0.10	0.201	90.937	0.804	
	1.00	0.190	86.254	0.919	
	10.00	0.181	82.175	0.841	
	100.00	0.124	56.193	1.047	
	1000.00	0.021	9.366	0.604	
10_e	0.00	0.4220	100.000	2.052	42
	0.10	0.3893	92.259	1.241	
	1.00	0.3723	88.231	0.285	
	10.00	0.3573	84.676	0.440	
	100.00	0.3167	75.039	1.226	
	1000.00	0.1450	34.360	0.410	
11	0.00	0.221	100.000	0.799	169.8
	0.10	0.198	89.728	0.884	
	1.00	0.196	88.822	1.140	
	10.00	0.177	80.363	0.841	
	100.00	0.134	60.725	0.453	
	1000.00	0.050	22.659	0.523	
12	0.00	0.221	100.000	0.799	257
	0.10	0.194	88.066	0.218	
	1.00	0.185	83.686	0.400	
	10.00	0.156	70.846	1.180	
	100.00	0.148	67.070	0.785	
	1000.00	0.053	23.867	0.799	
13_a	0.00	0.4937	100.000	0.761	23
	0.10	0.4593	93.045	0.389	
	1.00	0.4423	89.602	0.411	
	10.00	0.3830	77.583	0.510	
	100.00	0.0417	8.440	0.294	
	1000.00	0.0187	3.781	0.243	
13_b	0.00	0.221	100.000	0.799	49
	0.10	0.188	85.045	0.371	
	1.00	0.169	76.586	1.199	
	10.00	0.167	75.529	0.545	
	100.00	0.085	38.369	0.302	
	1000.00	0.067	30.211	1.702	
13_d	0.00	0.4220	100.000	2.052	61.7
	0.10	0.3993	94.629	1.507	
	1.00	0.3777	89.494	0.553	

	10.00	0.3437	81.438	0.480	
	100.00	0.1750	41.469	0.684	
	1000.00	0.1153	27.330	0.890	
	0.00	0.221	100.000	0.799	
	0.10	0.194	88.066	1.627	
	1.00	0.188	85.196	0.785	
14_a	10.00	0.180	81.722	0.658	288.4
	100.00	0.150	67.976	1.457	
	1000.00	0.067	30.363	0.692	
	0.00	0.221	100.000	0.799	
	0.10	0.190	86.254	0.996	
	1.00	0.180	81.420	0.919	
14_b	10.00	0.171	77.341	1.089	776
	100.00	0.151	68.278	1.343	
	1000.00	0.109	49.245	2.634	
	0.00	0.4220	100.000	2.052	
	0.10	0.3907	92.575	1.139	
	1.00	0.3767	89.258	0.617	
16_b	10.00	0.3540	83.886	0.137	47.8
	100.00	0.1343	31.833	0.754	
	1000.00	0.0253	6.003	0.418	
	0.00	0.221	100.000	0.799	
	0.10	0.177	80.212	0.746	
	1.00	0.150	67.976	1.199	
Doxorubicin	10.00	0.138	62.689	0.658	4.27
	100.00	0.093	41.994	0.658	
	1000.00	0.056	25.227	1.057	

3.2.1. 3-(Methylthio)-1-Phenyl-N-o-Tolyl-1H-Pyrazolo[3,4-d]Pyrimidin- 4-Amine (10a)

White solid; Yield: 85%; m. p. 130°C - 131°C. IR (KBr) cm^{-1}: 3372 (NH), 3047 (CH aromatic), 2924 (CH aliphatic). ^{1}HNMR (DMSO-d6) δ ppm: 8.45 (s, 1H, NH, D_2O exchangeable), 8.42 (s, 1H, pyrimidine-H2), 8.19 (d, 2H, J = 1.80 Hz, phenylpyrazole-H2, H6), 7.80 (t, 2H, J = 7.80 Hz, phenylpyrazole-H3, H5), 7.59 (t, 1H, J = 2.10 Hz, phenylpyrazole-H4), 7.56 (t, 1H, J = 6.90 Hz, phenyl-H5), 7.38 (d, 1H, J = 1.50 Hz, phenyl-H3), 7.35 (d, 1H, J = 1.50 Hz, phenyl-H6), 7.29 (t, 1H, J = 6.00 Hz, phenyl-H4), 2.50 (s, 3H, SCH_3), 2.30 (s, 3H, Ar-CH_3). MS (m/z): 347 ($C_{19}H_{17}N_5S$, 53.48%, M^+), 332 ($C_{18}H_{15}N_5S$, M-

CH_3, 74.57%), 77 (C_6H_5, 100%). Anal. Calc. for: ($C_{19}H_{17}N_5S$) (M.W. = 347): C, 65.68; H, 4.39; N, 20.16%; Found: C, 65.81; H, 4.89; N, 20.31%.

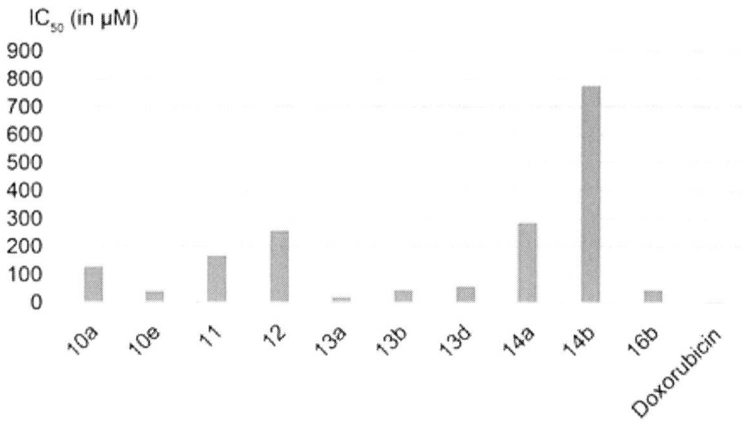

Figure 2. IC_{50} in µM of the synthesized compounds and doxorubicin against MCF7 adenocarcinoma cell line.

3.2.2. 2-(3-(Methylthio)-1-Phenyl-1H-Pyrazolo[3,4-d]Pyrimidin-4-Ylamino)Benzenethiol (10_b)

White solid; Yield: 75%; m. p. 141°C - 142°C. IR (KBr) cm^{-1}: 3291 (NH), 3053 (CH aromatic), 2918 (CH aliphatic). ^{1}HNMR (DMSO-d6) δ ppm: 12.42 (s, 1H, SH, D_2O exchangeable), 8.07 (s, 1H, pyrimidine-H2), 8.06 (d, 2H, J = 4.80 Hz, phenylpyrazole-H2, H6), 7.93 (t, 2H, J = 10.00 Hz, phenylpyrazole-H3, H5), 7.58 (t, 1H, J = 3.00 HZ, phenylpyrazole-H4), 7.49 (t, 1H, J = 10.00 HZ, phenyl-H5), 7.44 (d, 1H, J = 2.80 Hz, phenyl-H3), 7.35 (d, 1H, J = 4.00 Hz, phenyl-H6), 7.32 (t, 1H, J = 3.60 Hz, phenyl-H4), 2.60 (s, 3H, SCH_3). MS (m/z): 365 ($C_{18}H_{15}N_5S_2$, 2.33%, M^+). Anal. Calc. for: ($C_{18}H_{15}N_5S_2$) (M.W. = 365): C, 59.16; H, 4.14; N, 19.16%; Found: C, 59.21; H, 3.83; N, 19.47%.

3.2.3. N-(2,6-Dichlorophenyl)-3-(Methylthio)-1-Phenyl-1H-Pyrazolo [3,4-d]Pyrimidin-4-Amine (10_c)

White solid; Yield: 63%; m. p. 231°C - 232°C. IR (KBr) cm^{-1}: 3455 (NH), 3050 (CH aromatic), 2938 (CH aliphatic). ^{1}HNMR (DMSO-d6) δ ppm: 10.40 (s, 1H, NH, D_2O exchangeable), 8.66 (s, 1H, pyrimidine-H2), 8.18 (d, 2H, J = 4.80 Hz, phenylpyrazole-H2, H6), 8.04 (t, 2H, J = 8.10 Hz, phenylpyrazole-H3, H5), 7.59 (t, 1H, J = 6.90 Hz, phenylpyrazole-H4), 7.54 (d, 2H, J = 7.20

Hz, phenyl-H3, H5), 7.39 (t, 1H, J = 6.60 Hz, phenyl-H4), 2.60 (s, 3H, SCH_3). MS (m/z): 403 ($C_{18}H_{13}Cl_2N_5S$, 0.41%, M+2), 401 ($C_{18}H_{13}Cl_2N_5S$, 1.46%, M^+), 366 ($C_{18}H_{13}ClN_5S$, 2.42%), 331 ($C_{18}H_{13}N_5S$, 1.47%), 256 ($C_{12}H_{10}N_5S$, 3.01%), 241 ($C_{12}H_9N_4S$, 4.62%). Anal. Calc. for: ($C_{18}H_{13}Cl_2N_5S$) (M.W. = 401): C, 53.74; H, 3.26; N, 17.41%; Found: C, 53.92; H, 3.23; N, 17.68%.

3.2.4. 4-(3-(Methylthio)-1-Phenyl-1H-Pyrazolo[3,4-d]Pyrimidin-4-Ylamino)Benzoic Acid (10_d)

White solid; Yield: 85%; m. p. 162°C - 164 °C. IR (KBr) cm^{-1}: 3397 (OH), 3360 (NH), 3047 (CH aromatic), 2924 (CH aliphatic), 1689 (C=O). ^{1}HNMR (DMSO-d6) δ ppm: 13.80 (s, 1H, OH, D_2O exchangeable), 11.25 (s, 1H, NH, D_2O exchangeable), 8.52 (s, 1H, pyrimidine-H2), 8.50 - 7.30 (m, 9H, Ar-H), 2.62 (s, 3H, SCH_3). MS (m/z): 377 ($C_{19}H_{15}N_5O_2S$, 1.19%, M^+), 333 ($C_{18}H_{15}N_5S$, 2.4%). Anal. Calc. for: ($C_{19}H_{15}N_5O_2S$) (M.W. = 377): C, 60.47; H, 4.01; N, 18.56%; Found: C, 60.64; H, 4.09; N, 18.73%.

3.2.5. Ethyl-4-((3-(Methylthio)-1-Phenyl-1H-Pyrazolo[3,4-d]Pyrimidin- 4-yl)Amino)Benzoate (10_e)

White solid; Yield: 85%; m. p. 123°C - 124°C. IR (KBr) cm^{-1}: 3372 (NH), 3033 (CH aromatic), 2940 (CH aliphatic), 1710 (C=O). ^{1}HNMR (DMSO-d6) δ ppm: 8.79 (s, 1H, NH, D_2O exchangeable), 8.60 (s, 1H, pyrimidine-H2), 8.17 (d, 2H, J = 1.80 Hz, phenylpyrazole-H2, H6), 7.97 (t, 2H, J = 8.70 Hz, phenylpyrazole-H3, H5), 7.56 (d, 2H, J = 8.10 Hz, phenyl-H2, H6), 7.37 (t, 1H, J = 6.00 Hz, phenylpyrazole-H4), 7.33 (t, 2H, J = 6.00 Hz, phenyl H2, H6), 4.23 (q, 2H, J = 7.20 Hz, CH_2CH_3), 2.62 (s, 3H, S-CH_3), 1.3 (t, 3H, J = 6.6 Hz, CH_2CH_3). MS (m/z): 405 ($C_{21}H_{19}N_5O_2S$, 6.85%, M^+), 376 ($C_{19}H_{14}N_5O_2S$, 3.72%), 256 ($C_{12}H_{10}N_5S$, 2.27%), 241 ($C_{12}H_9N_4S$, 12.52%). Anal. Calc. for: ($C_{21}H_{19}N_5O_2S$) (M.W. = 405): C, 62.21; H, 4.72; N, 17.27%; Found: C, 62.47; H, 4.81; N, 17.49%.

3.3. N-Cyclohexyl-3-(Methylthio)-1-Phenyl-1H-Pyrazolo[3,4-d] Pyrimidin-4-Amine (11)

Into a solution of equimolar amounts of compound 9 and cyclohexylamine (10 mmol each) in ethanol (30 ml), trimethylamine (15 mmol) was added. The reaction mixture was heated under reflux for 6 hours then allowed to cool. The crude product was filtered out, dried and finally recrystallized from ethanol. White solid; Yield: 74%; m. p. 114°C - 116°C. IR (KBr) cm^{-1}: 3397 (NH), 3031 (CH aromatic), 2925 (CH aliphatic). ^{1}HNMR (DMSO-d6) δ ppm: 8.36 (s, 1H, pyrimidine-H2), 8.16 (d, 2H, H6, J = 8.10, phenyl-H2), 7.55 (t, 2H, H5, J = 8.40, phenyl-H3), 7.33 (t, 1H, J = 7.20, phenyl-H4), 6.32 (s, 1H, NH, D_2O exchangeable), 2.71 (s, 3H, SCH_3) 1.97-1.36 (m, 11H, cyclohexyl).

MS (m/z): 339 ($C_{18}H_{21}N_5S$, 23.10%, M^+), 257 ($C_{12}H_{11}N_5S$, M-C_6H_{11}, 100%). Anal. Calc. for: ($C_{18}H_{21}N_5S$) (M.W. = 339): C, 63.69; H, 6.24; N, 20.63%; Found: C, 63.85; H, 6.32; N, 20.86%.

3.4. 4-((3-(Methylthio)-1-Phenyl-1H-Pyrazolo[3,4-d]Pyrimidin- 4-Ylamino)Methyl)Benzoic Acid (12)

Into a solution of equimolar amounts of compound 9 and 4-(aminomethyl) benzoic acid (10 mmol each) in ethanol (30 ml), trimethylamine (15 mmol) was added. The reaction mixture was heated under reflux for 6 hours then allowed to cool. The crude product was filtered out, dried and finally recrystallized from ethanol. White solid; Yield: 70%; m. p. 170°C - 171°C. IR (KBr) cm^{-1}: 3382 (broad OH), 3027 (CH aromatic), 2985 (CH aliphatic), 1699 (C=O). ^{1}HNMR (DMSO-d6) δ ppm: 10.82 (s, 1H, OH), 8.39 (s, 1H, pyrimidine-H2), 7.88 (d, 2H, phenyl-H2, H6, J = 7.2), 7.30 (m, 5H, phenylpyrazole), 6.87 (d, 2H, phenyl-H3, H5, J = 7.2), 6.28 (s, 1H, NH), 4.32 (s, 2H, CH_2), 2.68 (s, 3H, S-CH_3). MS (m/z): 391 ($C_{20}H_{17}N_5O_2S$, M^+, 100%), 270 ($C_{19}H_{14}N_5O_2S$, M-CH_3, 46.20%), 256 ($C_{12}H_{11}N_5S$, M-COOH-$C_6H_4CH_2$, 43.23%). Anal. Calc. for: ($C_{20}H_{17}N_5O_2S$) (M.W. = 391): C, 61.37; H, 4.38; N, 17.89%; Found: C, 61.59; H, 4.43; N, 18.15%.

3.5. General Procedure for Synthesis of 2-(3-(Methylthio)-4-Oxo- 1-Phenyl-1H-Pyrazolo[3,4-d]Pyrimidin-5(4H)-yl)-N- Arylacetamide and N-Arylpropanamide $13_{a\text{-}h}$

Into a solution of compound 8 (10 mmol each) in DMF (30 ml) containing potassium carbonate (0.5 g), the appropriate 2-chloro N-arylacetamide or 3-chloro- N-arylpropanamide (10 mmol) was added. The reaction mixture was heated under reflux for 3 hours. After complete reaction, the reaction mixture was filtered while hot, concentrated, cooled and the resulting solid product was dried and finally recrystallized from ethanol.

3.5.1. 2-(3-(Methylthio)-4-Oxo-1-Phenyl-1H-Pyrazolo[3,4-d]Pyrimidin- 5(4H)-yl)-N-Phenylacetamide (13_a)

White solid; Yield: 85%; m. p. 164°C - 165°C. IR (KBr) cm^{-1}: 3307 (NH), 3053 (CH aromatic), 2925 (CH aliphatic), 1672 (C=O). ^{1}HNMR (DMSO-d6) δ ppm: 10.42 (s, 1H, NH, D_2O exchangeable), 8.47 (s, 1H, pyrimidine-H2), 8.04 (t, 2H, J = 7.20 Hz, Aniline-H3, H5), 7.59 (d, 2H, J = 2.00 Hz, Aniline-H2, H6), 7.57 (d, 2H, J = 8.40 Hz, phenylpyrazole-H3, H5), 7.41 (t, 1H, J = 7.20 Hz, 7.34 (t, 2H, J = 7.60 Hz, phenylpyrazole-H2, H6), 7.07 (t, 1H, J = 7.20 Hz, phenylpyrazole-H4), 4.87 (s, 2H, CH_2 C=O), 2.63 (s, 3H, S-CH_3). ^{13}CNMR (DMSO-d6 400 MHz) δ ppm: 13.26, 48.59, (Aliphatic CH_3 and CH_2), 104.94,

119.53, 121.76, 124.07, 127.39, 129.36, 129.72, 138.45, 139.04, 146.08, 152.96, 153.18, 156.51 (Aromatic carbons), 165.74 (C=O). MS (m/z): 391 ($C_{20}H_{17}N_5O_2S$, M, 42.33%), 299 ($C_{14}H_{11}N_4O_2S$, M-NHC_6H_5, 100%), 271 ($C_{13}H_{11}N_4OS$, 51.75%), 257 ($C_{12}H_9N_4OS$, 8.2%). Anal. Calc. for: ($C_{20}H_{17}N_5O_2S$) (M.W. = 391): C, 61.37; H, 4.38; N, 17.89%; Found: C, 61.48; H, 4.43; N, 18.12%.

3.5.2. N-(2-Chlorophenyl)-2-(3-(Methylthio)-4-Oxo-1-Phenyl-1H-Pyrazolo[3,4-d]Pyrimidin-5(4H)-yl)Acetamide (13_b)

White solid; Yield: 78%; m. p. 146°C - 147°C. IR (KBr) cm^{-1}: 3258 (NH), 3072 (CH aromatic), 2932 (CH aliphatic), 1695 (C=O). ^{1}HNMR (DMSO-d6) δ ppm: 10.09 (s, 1H, NH, D_2O exchangeable), 8.48 (s, 1H, pyrimidine-H2), 8.04 (d, 2H, J = 7.80 Hz, Phenylpyrazole-H2, H6), 7.75 (d, 1H, J = 8.10 Hz, Aniline-H6), 7.57 (d, 1H, J = 8.10 Hz, Aniline-H3), 7.55 (t, 2H, J = 8.10 Hz, Phenylpyrazole-H3, H5), 7.53 (t, 1H, J = 8.10 Hz, Phenylpyrazole-H4), 7.4 (t, 1H, J = 7.60 Hz, Aniline-H5), 7.2 (t, 1H, J = 7.60 Hz, Aniline-H4), 4.87 (s, 2H, CH_2 C=O), 2.63 (s, 3H, S-CH_3). MS (m/z): 427 ($C_{20}H_{16}ClN_5O_2S$, M+2, 1.52%), 4.82%, 425 ($C_{20}H_{16}ClN_5O_2S$, M^+, 4.82%), 299 ($C_{14}H_{11}N_4O_2S$, M-NHC_6H_4Cl, 84.7%), 271 ($C_{13}H_{11}N_4OS$, 100%). Anal. Calc. for: ($C_{20}H_{16}ClN_5O_2S$) (M.W. = 425): C, 56.40; H, 3.79; N, 16.44%; Found: C, 56.61; H, 3.76; N, 16.58%.

3.5.3. 2-(3-(Methylthio)-4-Oxo-1-Phenyl-1H-Pyrazolo[3,4-d]Pyrimidin- 5(4H)-yl)-N-m-Tolylacetamide (13_c)

White solid; Yield: 82%; m. p. 152°C - 153°C. IR (KBr) cm^{-1}: 3299 (NH), 3038 (CH aromatic), 2925 (CH aliphatic), 1672 (C=O). ^{1}HNMR (DMSO-d6) δ ppm: 10.37 (s, 1H, NH, D_2O exchangeable), 8.46 (s, 1H, pyrimidine-H2), 8.06 (d, 2H, J = 7.60 Hz, Phenylpyrazole-H2, H6), 7.59 (d, 1H, J = 7.60 Hz, Aniline-H6), 7.43 (t, 2H, J = 5.60 Hz, phenylpyrazole-H3, H5), 7.39 (t, 1H, J = 7.60 Hz, Phenylpyrazole-H4), 7.36 (s, 1H, Aniline-H2), 7.22 (t, 1H, J = 8.00 Hz, Aniline-H5), 6.90 (d, 1H, J = 7.60 Hz, Aniline-H4), 4.87 (s, 2H, CH_2 C=O), 2.63 (s, 3H, S-CH_3), 2.27 (s, 3H, Ar-CH_3). MS (m/z): 405 ($C_{21}H_{19}N_5O_2S$, M, 52.08%), 299 ($C_{14}H_{11}N_4O_2S$, M-$NHC_6H_4CH_3$, 100%), 271 ($C_{13}H_{11}N_4OS$, 67.81%). Anal. Calc. for: ($C_{21}H_{19}N_5O_2S$) (M.W. = 405): C, 62.21; H, 4.72; N, 17.27%; Found: C, 62.38; H, 4.76; N, 17.49%.

3.5.4. Ethyl 4-(2-(3-(Methylthio)-4-Oxo-1-Phenyl-1H-Pyrazolo[3,4-d] Pyrimidin-5(4H)-yl)Acetamido)Benzoate (13_d)

White solid; Yield: 65%; m. p. 72°C - 73°C. IR (KBr) cm^{-1}: 3287 (NH), 3058 (CH aromatic), 2988 (CH aliphatic), 1695 (C=O). ^{1}HNMR (DMSO-d6) δ ppm: 10.78 (s, 1H, NH), 8.4 (s, 1H, pyrimidine-H2), 8.70 (d, 2H, J = 8.80 Hz, Aniline-H3, H5), 7.80 (d, 2H, J = 6.80 Hz, Aniline-H2, H6), 7.80 (t, 2H, J = 8.40 Hz, Phenylpyrazole-H3, H5), 7.51 (t, 2H, J = 8.10 Hz, Phenylpyrazole-H2,

H6), 7.42 (t, 1H, J = 7.20 Hz, Phenylpyrazole-H4), 4.90 (s, 2H, CH_2 C=O), 4.30 (q, 2H, J = 6.80 Hz, CH_2CH_3), 2.60 (s, 3H, S-CH_3), 1.29 (t, 3H, J = 7.20 Hz, CH_2CH_3). MS (m/z): 463 ($C_{23}H_{21}N_5O_4S$, M, 5.65%), 299 ($C_{14}H_{11}N_4O_2S$, M-$NHC_6H_4COOC_2H_5$, 19.63%), 271 ($C_{13}H_{11}N_4OS$, 18.52%), 174 ($C_9H_5N_2S$, 100%). Anal. Calc. for: ($C_{23}H_{21}N_5O_4S$) (M.W. = 463): C, 59.60; H, 4.57; N, 15.11%; Found: C, 59.70; H, 4.63; N, 15.15%.

3.5.5. 3-(3-(Methylthio)-4-Oxo-1-Phenyl-1H-Pyrazolo[3,4-d]Pyrimidin- 5(4H)-yl)-N-Phenylpropanamide (13_e)

White solid; Yield: 80%; m. p. 154°C - 155°C. IR (KBr) cm^{-1}: 3317 (NH), 3010 (CH aromatic), 2924 (CH aliphatic), 1674 (C=O). ^{1}HNMR (DMSO-d6) δ ppm: 12.44 (s, 1H, NH, D_2O exchangeable), 8.44 (s, 1H, pyrimidine-H2), 8.02 (t, 2H, J = 7.20 Hz, Phenylpyrazole-H2, H6), 7.54 (d, 2H, J = 2.00 Hz, Aniline-H2, H6), 7.50 (d, 2H, J = 8.40 Hz, phenylpyrazole-H3, H5), 7.37 (t, 1H, J = 7.20 Hz), 7.33 (t, 2H, J = 7.60 Hz, phenylpyrazole-H4), 7.28 (t, 1H, J = 7.20 Hz, Aniline-H3, H5), 7.25 (t, 1H, J = 7.20 Hz, Aniline-H4), 4.26 (t, 2H, J = 7.20 Hz NCH_2), 2.87 (t, 2H, J = 7.20 Hz CH_2 C=O), 2.61 (s, 3H, S-CH_3). ^{13}CNMR (DMSO-d6 400 MHz) δ ppm: 13.27, 31.12, 31.22 (Aliphatic carbons), 105.86, 119.62, 121.66, 122.79, 129.74, 138.60, 145.89, 150.40, 152.79, 153.48, 156.58, 157.42, 162.74 (Aromatic carbons), 169.09 (C=O). MS (m/z): 405 ($C_{21}H_{19}N_5O_2S$, M, 45.17%), 313 ($C_{15}H_{13}N_4O_2S$, M-NHC_6H_5, 48.33%), 285 ($C_{14}H_{13}N_4OS$, 7.07%), 259 ($C_{12}H_9N_4OS$, 100%). Anal. Calc. for: ($C_{21}H_{19}N_5O_2S$) (M.W. = 405): C, 62.21; H, 4.72; N, 17.27%; Found: C, 62.45; H, 4.75; N, 17.53%.

3.5.6. N-(2-Chlorophenyl)-3-(3-(Methylthio)-4-Oxo-1-Phenyl-1,4-Dihydro-5H-Pyrazolo[3,4-d]Pyrimidin-5-yl)Propanamide (13_f)

White solid; Yield: 70%; m. p. 149°C - 150°C. IR (KBr) cm^{-1}: 3301 (NH), 3083 (CH aromatic), 2931 (CH aliphatic), 1680 (C=O). ^{1}HNMR (DMSO-d6) δ ppm: 9.72 (s, 1H, NH D_2O exchangeable), 8.44 (s, 1H, pyrimidine-H2), 8.03 (d, 2H, J = 7.80 Hz, Phenylpyrazole-H2, H6), 7.61 (d, 1H, J = 8.10 Hz, Aniline-H6), 7.57 (d, 1H, J = 8.10 Hz, Aniline-H3), 7.46 (t, 2H, J = 8.10 Hz, Phenylpyrazole-H3, H5), 7.39 (t, 1H, J = 8.10 Hz, Phenylpyrazole-H4), 7.34 (t, 1H, J = 7.50 Hz, Aniline-H5), 7.20 (t, 1H, J = 7.60 Hz, Aniline-H4), 4.29 (t, 2H, J = 80 Hz, CH_2CH_2), 2.8 (t, 2H, J = 10.00 Hz, CH_2-C=O), 2.50 (s, 3H, S-CH_3). MS (m/z): 441 ($C_{21}H_{18}ClN_5O_2S$, M+2, 3.78%, M^+), 439 ($C_{21}H_{18}ClN_5O_2S$, M, 11.49%), 313 ($C_{15}H_{13}N_4O_2S$, 35.54%), 285 ($C_{14}H_{13}N_4OS$, 5.89%), 257 ($C_{12}H_9N_4OS$, 8.2%). Anal. Calc. for: ($C_{21}H_{18}ClN_5O_2S$) (M.W. = 439): C, 57.34; H, 4.12; N, 15.92%; Found: C, 57.49; H, 4.19; N, 16.08%.

3.5.7. 3-(3-(Methylthio)-4-Oxo-1-Phenyl-1,4-Dihydro-5H-Pyrazolo [3,4-d]Pyrimidin-5-yl)-N-(m-Tolyl)Propanamide (13_g)

White solid. Yield: 55%; m. p. 138°C - 141°C. IR (KBr) cm^{-1}: 3223 (NH), 3049 (CH aromatic), 2980 (CH aliphatic), 1693 (C=O). ^1H NMR (DMSO-d_6) δ ppm: 9.92 (s, 1H, NH D_2O exchangeable), 8.45 (s, 1H, pyrimidine-H2), 8.02 (d, 2H, J = 7.8 Hz, Phenylpyrazole-H2, H6), 7.56 (d, 1H, J = 7.80 Hz, Aniline-H6), 7.38 (t, 2H, J = 5.70 Hz, phenylpyrazole-H3, H5), 7.32 (t, 1H, J = 8.40 Hz, Phenylpyrazole-H4), 7.17 (s, 1H, Aniline-H2), 7.14 (t, 1H, J = 8 Hz, Aniline-H5), 6.85 (d, 1H, J = 7.50 Hz, Aniline-H4), 4.2 (t, 2H, SCH_2), 2.8 (t, 2H, CH_2-C=O), 2.6 (s, 3H, Ar-CH_3), 2.2 (s, 3H, S-CH_3). MS (m/z): 419 ($C_{22}H_{21}N_5O_2S$, M, 2.06%,), 314 ($C_{15}H_{14}N_4O_2S$, 10.23%), 257 ($C_{12}H_9N_4OS$, 8.2%), 105 (C_7H_6N, 100%). Anal. Calc. for: ($C_{22}H_{21}N_5O_2S$) (M.W. = 419): C, 62.99; H, 5.05; N, 16.69%; Found: C, 63.21; H, 5.11; N, 16.87%.

3.5.8. Ethyl 4-(3-(3-(Methylthio)-4-Oxo-1-Phenyl-1,4-Dihydro-5H-Pyrazolo[3,4-d]Pyrimidin-5-yl)Propanamido)Benzoate (13_h)

White solid. Yield: 85%; m. p. 160°C. IR (KBr) cm^{-1}: 3287 (NH), 3058 (CH aromatic), 2988 (CH aliphatic), 1695 (C=O). ^1HNMR (DMSO-d_6) δ ppm: 10.78 (s, 1H, NH), 8.40 (s, 1H, pyrimidine-H2), 8.71 (d, 2H, J = 8.80 Hz, Aniline-H3, H5), 7.81 (d, 2H, J = 6.80 Hz, Aniline-H2, H6), 7.81 (t, 2H, J = 8.40 Hz, Phenylpyrazole-H3, H5), 7.53 (t, 2H, J = 8.10 Hz, Phenylpyrazole-H2, H6), 7.42 (t, 1H, J = 7.20 Hz, Phenylpyrazole-H4), 3.90 (t, 2H, J = 7.20, SCH_2), 4.35 (t, 2H, J = 7.20, CH_2 C=O), 4.3 (q, 2H, J = 6.80 Hz, CH_2CH_3), 2.6 (s, 3H, S-CH_3), 1.29 (t, 3H, J = 7.20 Hz, CH_2CH_3). MS (m/z): 477 ($C_{24}H_{23}N_5O_4S$, 27.54%, M^+), 432 ($C_{22}H_{18}N_5O_3S$, 2.51%), 313 ($C_{15}H_{13}N_4O_2S$, 55.20%), 285 ($C_{14}H_{13}N_4OS$, 8.39%). Anal. Calc. for: ($C_{24}H_{23}N_5O_4S$) (M.W. = 477): C, 60.36; H, 4.85; N, 14.65%; Found: C, 60.64; H, 4.93; N, 14.85%.

3.6. General Procedure for Synthesis of Alkyl 2-(3-(Methylthio)- 4-Oxo-1-Phenyl-1H-Pyrazolo[3,4-d]Pyrimidin-5(4H)-yl) Acetate $14_{a\text{-}b}$

Into a solution of compound 9 (10 mmol) in DMF (30 ml) containing potassium carbonate (0.5 gm), the appropriate alkyl-2-chloroacetate (10 mmol) was added. The reaction mixture was heated under reflux for 4 hours. After complete reaction (as indicated by TLC), the reaction mixture was filtered while hot, concentrated, cooled and the resulting solid product was recrystallized from ethanol.

3.6.1. Methyl -2-(3-(Methylthio)-4-Oxo-1-Phenyl-1,4-Dihydro-5H-Pyrazolo[3,4-d]Pyrimidin-5-yl)Acetate (14_a)

White solid; Yield: 70%; m. p. 78°C - 79°C. IR (KBr) cm^{-1}: 3044 (CH aromatic), 2948 (CH aliphatic), 1747 (C=O). ^{1}HNMR (DMSO-d_6) δ ppm: 8.47 (s, 1H, pyrimidine-H2), 8.02 (d, 2H, J = 7.80 Hz, phenyl-H2, H6), 7.57 (t, 2H, J = 8.10 Hz phenyl-H3, H5), 7.41 (t, 1H, J = 7.20 Hz, phenyl-H4), 4.85 (s, 2H, CH_2 C=O), 3.73 (s, 3H, O-CH_3), 2.63 (s, 3H, S-CH_3). ^{13}C NMR (DMSO-d_6. 400 MHz) δ (ppm): 13.25, 47.03, 52.95 (Aliphatic carbons), 104.77, 121.85, 127.46, 129.68, 138.33, 146.12, 152.52, 152.79, 156.30 (Aromatic carbons), 168.81 (C=O). MS (m/z): 330 ($C_{15}H_{14}N_4O_3S$, M, 100%), 315 ($C_{14}H_{11}N_4O_3S$, M-CH_3, 1.07%). Anal. Calc. for: ($C_{15}H_{14}N_4O_3S$) (M.W. = 330): C, 54.54; H, 4.27; N, 16.96%; Found: C, 54.71; H, 4.36; N, 17.21%.

3.6.2. Ethyl 2-(3-(Methylthio)-4-Oxo-1-Phenyl-1,4-Dihydro-5H-Pyrazolo [3,4-d]Pyrimidin-5-yl)Acetate (14_b)

White solid; Yield: 85%; m. p. 82°C - 83°C. IR (KBr) cm^{-1}: IR (KBr) cm^{-1}: 3044 (CH aromatic), 2948 (CH aliphatic), 1747 (C=O). ^{1}HNMR (DMSO-d_6) δ ppm: 8.47 (s, 1H, pyrimidine-H2), 8.02 (d, 2H, J = 7.80 Hz, phenyl-H2, H6), 7.57 (t, 2H, J = 8.00 Hz phenyl-H3, H5), 7.41 (t, 1H, J = 7.20 Hz, phenyl-H4), 4.85 (s, 2H, CH_2 C=O), 3.73 (s, 3H, O-CH_3), 2.63 (s, 3H, S-CH_3). ^{13}C NMR (DMSO-d_6. 400 MHz) δ (ppm): 13.25, 47.03, 52.95 (Aliphatic carbons), 104.77, 121.85, 127.46, 129.68, 138.33, 146.12, 152.52, 156.30 (Aromatic carbons), 168.81 (C=O). MS (m/z): 330 ($C_{15}H_{14}N_4O_3S$, M, 100%), 314 ($C_{14}H_{10}N_4O_3S$, M-CH_3, 100%), 299 ($C_{14}H_{11}N_4O_2S$, 4.76%), 257 ($C_{12}H_9N_4OS$, 2.35%). Anal. Calc. for: ($C_{15}H_{14}N_4O_3S$) (M.W. = 330): C, 54.54; H, 4.27; N, 16.96%; Found: C, 54.71; H, 4.36; N, 17.21%.

3.7. Synthesis of 2-(3-(Methylthio)-4-Oxo-1-Phenyl-1,4-Dihydro- 5H-Pyrazolo[3,4-d]Pyrimidin-5-yl)Acetohydrazide (15)

Into a solution of 14_b (10 mmol) in ethanol (30 ml), hydrazine-hydrate (20 mmol) was added. The reaction mixture was heated under reflux for 6 hours. After complete reaction, the reaction allowed to cool. The separated solid was and filtered out, recrystallized from ethanol. White solid; Yield: 65%; m. p. 122°C - 123°C. IR (KBr) cm^{-1}: 3307 ($NHNH_2$), 3017 (CH aromatic), 2984 (CH aliphatic), 1673 (C=O). ^{1}HNMR (DMSO-d_6) δ ppm: 9.41 (s, 1H, NH D_2O exchangeable), 8.40 (s, 1H, pyrimidine-H2), 8.05 (d, 2H, J = 8.10 Hz, phenyl-H2, H6), 7.58 (t, 2H, J = 7.20 Hz, phenyl-H3, H5), 7.39 (t, 1H, J = 1.20 Hz, phenyl-H4), 4.6 (s, 2H, CH_2C=O), 4.30 (s, 2H, NH_2 D_2O exchangeable), 2.5 (s, 3H, S-CH_3). MS (m/z): 330 ($C_{14}H_{14}N_6O_2S$, 13.32%, M^+), 299 ($C_{14}H_{11}N_4O_2S$, 100%), 271 ($C_{13}H_{11}N_4OS$, 65.87%), 257 ($C_{12}H_9N_4OS$,

1.79%). Anal. Calc. for: ($C_{14}H_{14}N_6O_2S$) (M.W. = 330): C, 50.90; H, 4.27; N, 25.44%; Found: C, 51.23; H, 4.34; N, 25.61%.

3.8. General Procedure for Synthesis of N'-Benzylidene Derivatives-2-(3-(Methylthio)-4-Oxo-1-Phenyl-1,4-Dihydro- 5H-Pyrazolo[3,4-d]Pyrimidin-5-yl)Acetohydrazide $16_{a\text{-}b}$

Into a solution of 15 (10 mmol) in glacial acetic acid (20 ml), benzaldehyde derivatives (10 mmol) was added. The mixture was then heated under reflux for 5 hours. The reaction mixture was concentrated and allowed to cool. The separated solid was filtered and finally recrystallized from ethanol.

3.8.1. (E)-N'-Benzylidene-2-(3-(Methylthio)-4-Oxo-1-Phenyl-1,4-Dihydro-5H-Pyrazolo[3,4-d]Pyrimidin-5-yl)Acetohydrazide (16_a)

White solid; Yield: 73%; m. p. 189°C - 190°C. IR (KBr) cm^{-1}: 3196 (NH), 3044 (CH aromatic), 2929 (CH aliphatic), 1680 (C=O). ^{1}HNMR (DMSO-d_6) δ ppm: 11.88 (s, 1H, NH D_2O exchangeable), 8.49 (s, 1H, pyrimidine-H2), 8.07 (t, 3H, J = 8.00 Hz, phenyl-H3, H4, H5), 7.75 (d, 2H, J = 8.00 Hz, phenyl-H2, H6), 7.58 (d, 2H, J = 8.00 Hz, phenylpyrazole-H2, H6), 7.46 (t, 2H, J = 6.00 Hz, Phenylpyrazole-H3, H5), 7.44 (t, 1H, J = 7.60 Hz, Phenylpyrazole-H4), 7.39 (s, 1H, CH=N), 5.24 (s, 2H, CH_2 C=O), 2.63 (s, 3H, S-CH_3). ^{13}C NMR (DMSO-d_6. 400 MHZ) δ (ppm): 13.26, 46.88, 47.44 (Aliphatic carbons), 104.96 121.73, 127.63, 129.93, 130.67, 134.46, 138.46, 146.06, 147.88, 152.94, 153.12, 156.51, 163.76, (Aromatic carbons), 168.64 (C=O). MS (m/z): 418 ($C_{21}H_{18}N_6O_2S$, 2.30%, M^+), 299 ($C_{14}H_{11}N_4O_2S$, 100%), 271 ($C_{13}H_{11}N_4OS$, 65.87%). Anal. Calc. for: ($C_{21}H_{18}N_6O_2S$) (M.W. = 418): C, 60.27; H, 4.34; N, 20.08%; Found: C, 51.23; H, 4.34; N, 25.61%.

3.8.2. (E)-N'-(4-Hydroxybenzylidene)-2-(3-(Methylthio)-4-Oxo-1-Phenyl-1,4-Dihydro-5H-Pyrazolo[3,4-d]Pyrimidin-5-yl) Acetohydrazide (16_b)

White solid; Yield: 78%; m. p. 205°C - 206°C. IR (KBr) cm^{-1}: 3017 (CH aromatic), 2984 (CH aliphatic), 1673 (C=O), 3307 (NH). ^{1}HNMR (DMSO-d_6) δ ppm: 11.67 (s, 1H, NH D_2O exchangeable), 9.97 (s, 1H, OH D_2O exchangeable), 8.47 (s, 1H, pyrimidine-H2), 8.05 (d, 2H, J = 8.00 Hz, phenyl-H2, H6), 7.97 (s, 2H, CH=N), 7.58 (t, 4H, J = 8.00 Hz, phenylpyrazole-H3, H5, Phenyl-H3, H5), 7.40 (t, 1H, J = 8.00 Hz, Phenylpyrazole-H4), 6.86 (d, 2H, J = 8.00 Hz, Phenylpyrazole-H2, H6), 5.19 (s, 2H, CH_2 C=O), 2.63 (s, 3H, S-CH_3). ^{13}C NMR (DMSO-d_6. 400 MHZ) δ (ppm): 13.26, 20.89, 47.37 (Aliphatic carbons), 104.96, 116.20, 121.72, 125.42, 127.33, 129.68, 138.47, 145.13,

148.16, 152.94, 156.52, 159.90, 163.33, (Aromatic carbons), 168.24 (C=O). MS (m/z): 434 ($C_{21}H_{18}N_6O_3S$, 2.66%, M^+), 328 ($C_{14}H_{12}N_6O_2S$, 19.44%), 271 ($C_{13}H_{11}N_4OS$, 4.40%), 257 ($C_{12}H_9N_4OS$, 1.79%). Anal. Calc. for: ($C_{21}H_{18}N_6O_3S$) (M.W. = 434): C, 58.08; H, 4.18; N, 19.34%; Found: C, 51.23; H, 4.34; N, 25.61%.

4. BIOLOGICAL TESTING

4.1. Materials and Method

Human breast adenocarcinoma cell line MCF7, were purchased from the American Type Cell Culture Collection (ATCC, Manassas, USA) and grown on Roswell Park Memorial Institute Medium (RPMI 1640) supplemented with 100 g/ml of streptomycin, 100 units/ml of penicillin and 10% of heat inactivated fetal bovine serum in a humidified, 5% (v/v) CO_2 atmosphere at 37°C.

4.2. Measurement of Potential Antitumor

The antitumor activity of newly synthesized pyrazolo[3,4-d]pyrimidines were measured in vitro on human breast adenocarcinoma cell line MCF7 using SulfoRhodamine-B stain (SRB) assay applying the method of 3-[4,5-dimethylthiazole- 2-yl]-2,5-dimethyltetrazolium bromide (MTT) technique [21] [22] . Exponentially grown cells from the selected cancer cell line were trypsinized, counted and seeded at the appropriate densities (2000 - 1000 cells/0.33 cm^2). Cells were then incubated in a humidified atmosphere at 37°C for 24 hours. Then, cells were exposed to different concentrations of the test compounds (0.1, 1, 10, 100, 1000 πM) for 72 hours. After that, the viability of treated cells was determined according to MTT technique. The viability of cells was expressed as percentage of control and the concentration that induces 50% inhibition of cell proliferation (IC_{50}). The relation between the surviving fraction and the compound concentration was plotted and the IC50 was calculated for each compound. Results are given in Table 1.

5. CONCLUSION

A series of novel 1-phenyl-3-methyl sulphonyl pyrazolo[3,4-d]pyrimidines 10 - 16 was synthesized. The antitumor activity of this new series was investigated against human breast adenocarcinoma cell line MCF7. Ten of the test compounds showed moderate activity relative to that of doxorubicin. The N-arylacetamide derivatives (13_{a-h}) exhibited better antitumor activity than all other series. Among this series, compound 13_a displayed the highest activity with IC_{50} equal to 23 μM. It is obvious from the results in Table 1 and Figure 2, increasing the linker length by one more CH_2 unit in compounds 13_{a-h} results in dramatic fall in the activity. Presence of hydrogen bond donor at the para position of the aromatic ring in the new derivatives $16_{a,b}$ is essential for the activity. This becomes clear comparing the high IC_{50} value of 16_a with that of 16_b which is only 47.8 μM. Further studies are required in order to determine the mechanism of the antitumor action and to identify the SAR of other positions of pyrazolo[3,4-d]pyrimidine nucleus.

ACKNOWLEDGEMENTS

The collaborative support of members of the Cancer Biology Department, National Institute of Cancer, Egypt in determining the antitumor activity of new compounds is highly appreciated.

CITE THIS PAPER

El-Morsy, A.M., El-Sayed, M.S. and Abulkhair, H.S. (2017) Synthesis, Characterization and In Vitro Antitumor Evaluation of New Pyrazolo [3,4-d]Pyrimidine Derivatives. Open Journal of Medicinal Chemistry, 7, 1-17.

REFERENCES

1. Gao, X., Lu, Y., Fang, L., Fang, X., Xing, Y., Gou, S. and Xi, T. (2013) Synthesis and Anticancer Activity of Some Novel 2-Phenazinamine Derivatives. European Journal of Medicinal Chemistry, 69, 1-9.
2. Seffrin, J.R., Hill, D., Burkart, W., Magrath, I., Badwe, R.A., Ngoma, T., Mohar, A. and Grey, N. (2009) It Is Time to Include Cancer and Other Noncommunicable Diseases in the Millennium Development Goals. CA: A Cancer Journal for Clinicians, 59, 282-284.
3. Narang, A.S. and Desai, D.S. (2009) In: Lu, Y. and Mahato, R.I., Eds., Pharmaceutical Perspectives of Cancer Therapeutics, Springer US, New York, 49-92.
4. Mishra, C.B., Mongre, R.K., Kumari, S., Jeong, D.K. and Tiwari, M. (2016) Synthesis, in Vitro and in Vivo Anticancer Activity of Novel 1-(4-Imino-1-Substituted-1H-Pyrazolo[3,4-d]Pyrimidin-5(4H)-Yl)Urea Derivatives. RSC Advances, 6, 24491-245x0.
5. Jorda, R., Havlicek, L., McNae, I.W., Walkinshaw, M.D., Voller, J., Sturc, A., Navratilova, J., Kuzma, M., Mistrik, M., Bartek, J., Strnad, M. and Krystof, V.J. (2011) Journal of Medicinal Chemistry, 54, 2980-2993.

6. Ismail, N.S.M., Ali, E.M.H., Ibrahim, D.A., Serya, R.A.T. and Abou El Ella, D.A. (2016) Pyrazolo[3,4-d]Pyrimidine Based Scaffold Derivatives Targeting Kinases as Anticancer Agents. Future Journal of Pharmaceutical Sciences, 2, 20-30.
7. Kandeel, M.M. (2012) Design, Synthesis, and Antitumor Evaluation of Novel Pyrazolo[3,4-d]Pyrimidine Derivatives. Scientia Pharmaceutica, 80, 531-545.
8. Abdellatif, K., Abdelall, E., Abdelgawad, M., Ahmed, R. and Bakr, R. (2014) Synthesis and Anticancer Activity of Some New Pyrazolo[3,4-d]Pyrimidin-4-One Derivatives. Molecules, 19, 3297-3309.
9. Ducray, R., Ballard, P., Barlaam, B.C., Hickinson, M.D., Kettle, J.G., Ogilvie, D.J. and Trigwell, C.B. (2008) Novel 3-Alkoxy-1H-Pyrazolo[3,4-d]Pyrimidines as EGFR and erbB2 Receptor Tyrosine Kinase Inhibitors. Bioorganic & Medicinal Chemistry Letters, 18, 959-962.
10. Kumar, A., Ahmad, I., Chhikara, B.S., Tiwari, R., Mandal, D. and Parang, K. (2011) Synthesis of 3-Phenylpyrazolopyrimidine-1,2,3-Triazole Conjugates and Evaluation of Their SRC Kinase Inhibitory and Anticancer Activities. Bioorganic & Medicinal Chemistry Letters, 21, 1342-1346.
11. Markwalder, J.A., Arnone, M.R., Benfield, P.A., Boisclair, M., Burton, C.R., Chang, C.-H., Cox, S.S., Czerniak, P.M., Dean, C.L., Doleniak, D., Grafstrom, R., Harrison, B.A., Kaltenbach, R.F., Nugiel, D.A., Rossi, K.A., Sherk, S.R., Sisk, L.M., Stouten, P., Trainor, G.L., Worland, P. and Seitz, S.P. (2004) Synthesis and Biological Evaluation of 1-Aryl-4,5-dihydro-1H-pyrazolo[3,4-d]pyrimidin-4-one Inhibitors of Cyclin-Dependent Kinases. Journal of Medicinal Chemistry, 47, 5894-5911.
12. Curran, K.J., Verheijen, J.C., Kaplan, J., Richard, D.J., Toral-Barza, L., Hollander, I., Lucas, J., Ayral-Kaloustian, S., Yu, K. and Zask, A. (2010) Pyrazolopyrimidines as Highly Potent and Selective, ATP-Competitive Inhibitors of the Mammalian Target of Rapamycin (mTOR): Optimization of the 1-Substituent. Bioorganic & Medicinal Chemistry Letters, 20, 1440-1444.
13. Meijer, L., Flajolet, M. and Greengard, P. (2004) Pharmacological Inhibitors of Glycogen Synthase Kinase 3. Trends in Pharmacological Sciences, 25, 471-480.
14. Peat, A.J., Boucheron, J.A., Dickerson, S.H., Garrido, D., Mills, W., Peckham, J., Preugschat, F., Smalley, T., Schweiker, S.L., Wilson, J.R., Wang, T.Y., Zhou, H.Q. and Thomson, S.A. (2004) Novel Pyrazolopyrimidine Derivatives as GSK-3 Inhibitors. Bioorganic & Medicinal Chemistry Letters, 14, 2121-2125.
15. El-Enany, M.M., Kamel, M.M., Khalil, O.M. and El-Nassan, H.B. (2010) Synthesis and Antitumor Activity of Novel 6-Aryl and 6-Alkylpyrazolo[3,4-d]pyrimidin-4-one Derivatives. European Journal of Medicinal Chemistry, 45, 5286-5291.
16. Abd El Hamid, M.K., Mihovilovic, M.D. and El-Nassan, H.B. (2012) Synthesis of Novel Pyrazolo[3,4-d]pyrimidine Derivatives as Potential Anti-Breast Cancer Agents. European Journal of Medicinal Chemistry, 57, 323-328.
17. Tominaga, Y., Honkawa, Y., Hara, M. and Hosomi, A.J. (1990) Synthesis of Pyrazolo[3,4-d]pyrimidine Derivatives Using Ketene Dithioacetals. Journal of Heterocyclic Chemistry, 27, 775-783.
18. Traxler, P., Bold, G., Frei, J., Lang, M., Lydon, N., Mett, H., Buchdunger, E., Meyer, T., Mueller, M. and Furet, P.J. (1997) Use of a Pharmacophore Model for the Design of EGF-R Tyrosine Kinase Inhibitors: 4-(Phenylamino)pyrazolo[3,4-d]pyrimidines. Journal of Medicinal Chemistry, 40, 3601-3616.
19. Davoodnia, A., Zhiani, R. and Tavakoli-Hoseini, N. (2008) Synthesis of Pyrazolo[4,3-e][1,2,4]triazolo[4,3-c]pyrimidines. Monatshefte für Chemie—Chemical Monthly, 139, 1405-1407.
20. Abdou, N.S., Serya, R.A.T., Esmat, A., Tolba, M.F., Ismail, N.S.M. and Abouzid, K.A.M. (2015) Synthesis and in Vitro Antiproliferative Activity of Novel Pyrazolo[3,4-d]pyrimidine Derivatives. MedChemComm, 6, 1518-1534.
21. Sharma, A., Chakravarti, B., Gupt, M.P., Siddiqui, J.A., Konwar, R. and Tripathi, R.P. (2010) Synthesis and Anti Breast Cancer Activity of Biphenyl Based Chalcones. Bioorganic & Medicinal Chemistry, 18, 4711-4720.
22. Freimoser, F.L., Jakob, C.A., Aebi, M. and Tuor, U. (1999) The MTT [3-(4,5-Dimethylthiazol-2-yl)-2,5-Diphenyltetrazolium Bromide] Assay Is a Fast and Reliable

Method for Colorimetric Determination of Fungal Cell Densities. Applied and Environmental Microbiology, 65, 3727-3729.

23. Ferrari, M. and Fornasiero, M.C. (1990) MTT Colorimetric Assay for Testing Macrophage Cytotoxic Activity in Vitro. Journal of Immunological Methods, 131, 165-172.
24. Sahu, N.P., Pal, C., Mandal, N.B., Banerjee, S., Raha, M., Kundu, A.P., Basu, A., Ghosh, M., Roy, K. and Bandyopadhyay, S. (2002) Synthesis of a Novel Quinoline Derivative, 2-(2-methylquinolin-4-ylamino)-N-phenylacetamide—A Potential Antileishmanial Agent. Bioorganic & Medicinal Chemistry, 10, 1687-1693.

CHAPTER 2

Tacrine, Trolox and Tryptoline as Lead Compounds for the Design and Synthesis of Multi-target Agents for Alzheimer's Disease Therapy

Gerard a. K. Teponnou, Jacques Joubert, Sarel F. Malan*

Pharmaceutical Chemistry, School of Pharmacy, University of the Western Cape, Private Bag X17, Bellville 7535, South Africa

ABSTRACT

The versatile biological activities of tacrine, trolox and β-carboline derivatives make them promising lead structures for the development of multifunctional Alzheimer's disease (AD) agents. Based on the topology of the active site of cholinesterases and other target proteins involved in the pathogenesis of AD, we have designed and synthesized tacrine-trolox and tacrine-tryptoline hybrids with various linker chain lengths. The hybrids containing the trolox moiety (**8a-8d**) showed moderate to high *Tc*AChE inhibition (IC_{50}: 17.37 - 2200 nM), eqBuChE inhibition (IC_{50}: 3.16 – 128.82 nM) and free radical scavenging activities (IC_{50}: 11.48 – 49.23 μM). The hybrids with longer linker chain lengths in general showed better ChE inhibitory activity. As expected, free radical scavenging activities were not significantly affected by varying linker chain lengths. The hybrid compound containing the tryptoline moiety linked with a 7 carbon spacer to tacrine (**14**) displayed the best AChE and BuChE inhibitory activity (IC_{50} = 17.37 and 3.16 nM). Docking experiments exhibited that compounds **8d** and **14** were able to bind to both the CAS and PAS of *Tc*AChE and eqBuChE, suggesting that they will be able to inhibit ChE induced Aβ aggregation. Novel multi-target agents that exhibit good ChE inhibition (**8d** and **14**) and anti-oxidant (**8d**) activity were identified as suitable candidates for further investigation.

Keywords: Alzheimer's disease, Cholinesterase, Molecular docking, Multi-target directed ligand, Tacrine, Trolox, Tryptoline.

1. INTRODUCTION

Alzheimer's disease (AD) is a chronic, multifactorial disease of the central nervous system [1]. Its characteristic symptoms include short-term memory impairment that worsens as the disease progresses and eventually leads to severe cognitive and physical disability [2]. AD represents the main cause of dementia and no curative drug is available. To date, it is estimated that 36.6 million people worldwide are suffering from the disease. This number is expected to double or triple by 2030 and 2050 [3]. The sequence of molecular events that underlies the occurrence of AD is still not well understood [4]. Current hypotheses suggest a decrease in the level of the neurotransmitter acetylcholine (ACh) in the brain regions involved in learning and memory, β-amyloid plaque formation (Aβ), oxidative stress, excitotoxicity, neuroinflammation and aggregation of tau proteins (τ-proteins) [5, 6]. The current treatment of AD is limited to acetylcholinesterase inhibitor drugs (donepezil, galantamine and rivastigmine) and a *N*-methyl-D-aspartate receptor (NMDAR) antagonist (memantine, Figure 1). These drugs act through the "one-molecule-one target" paradigm, offer only symptomatic treatment and do not stop the progression of AD [7].

A more appropriate strategy was proposed by developing multitarget directed ligands (MTDL) with the ability to modulate multiple targets involved in the pathogenesis of AD simultaneously [8, 9]. Numerous examples (Figure 2) of hybrid compounds with multifunctional activities have already been developed and published, including bis-7-tacrine dimers [10], galantamine-memantine hybrids [8], phenylthiazole-tacrine hybrids [9] and bivalent β-carboline hybrids [11] with pharmacological profiles offering promise of slowing or stopping neurodegenerative disease progression. These positive findings in the development of multifunctional hybrid compounds have prompted us to develop new hybrid compounds using tacrine, trolox and tryptoline as lead compounds (Figure 3).

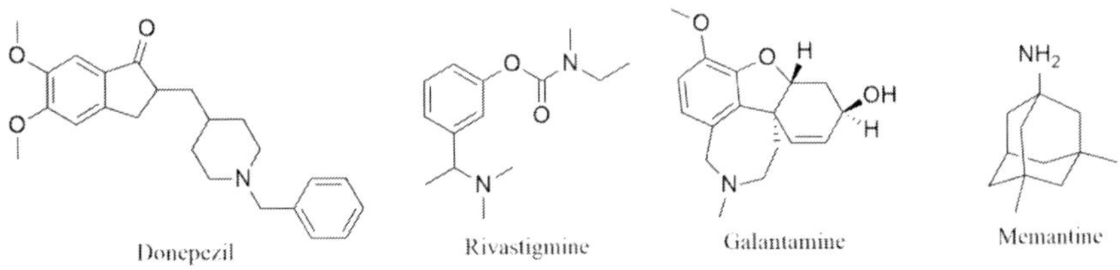

Figure 1. Chemical structures of current AD drugs.

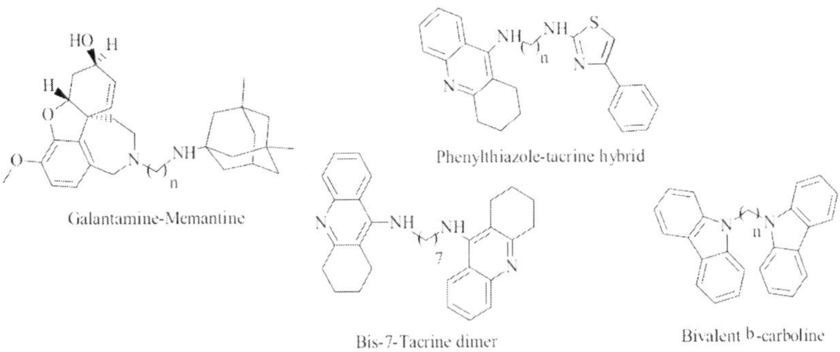

Figure 2. Chemical structures of multifunctional AD agents.

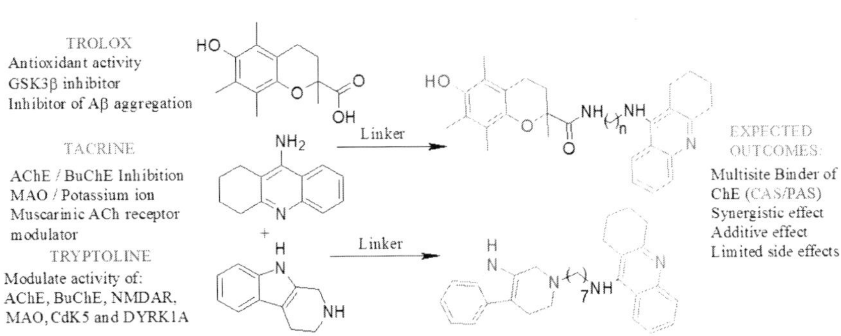

Figure 3. Design strategy towards tacrine-trolox and tacrine-tryptoline hybrids.

Tacrine is an attractive lead compound that has successfully been used in the multitarget design strategy [9, 10]. It has ChE inhibitory activity and additional biological activity such as monoamine oxidase (MAO) inhibition and ion channel modulating ability [12] that may be of value in the treatment of AD has been described for it. Recent studies have demonstrated that lead optimization of tacrine in the design of novel AD drugs can improve its biological profile and alleviate its hepatotoxicity [13]. Based on these studies, tacrine appears to be a suitable lead compound for the design of multitarget agents. Trolox is an analog of Vitamin E (α-tocopherol) with well documented antioxidant capacity which may neutralize free radical species believed to be linked to aging or diseases such as AD [14]. Trolox also has the ability to prevent neurotoxicity induced by Aβ and hydrogen peroxide [15]. Another source revealed that trolox could inhibit glycogen synthase kinase 3β (GSK 3β) whose hyperactivity causes neurofibrillary tangle formation [16]. These different biological activities of trolox consolidate its neuroprotective capacity and make it an excellent lead for the design of multifunctional drugs for treating AD. Tryptoline is a β-carboline derivative and derivatives thereof have been shown to inhibit ChE, MAO, NMDAR and other targets involved in the proposed pathogenesis of AD [17, 18].

Based on the topology of the active site of AChE we hypothesized that the combination of tacrine with trolox or tryptoline through linkers of varied chain lengths would lead to multifunctional hybrid compounds with the ability to modulate a number of important drug targets in the neurodegenerative cascade (Figure 4). It is expected that the tacrine moiety will interact with conserved aromatic residues in the catalytic anionic site (CAS) of AChE necessary for AChE inhibition while trolox or tryptoline will interact with conserved aromatic residues in the peripheral anionic site (PAS) necessary for the possible prevention of AChE-induced Aβ aggregation [19-22]. Varying the linker chain length explores the appropriate binding of the pharmacophores in the active site of AChE. The aim of these hybrids is to provide additive or synergistic therapeutic effects that might help overcome the limitation of current AD drugs and to aid in the search for novel AD lead compounds.

2. RESULT AND DISCUSSION

2.1. Synthesis

The synthesis of the tacrine-trolox hybrids **(8a – 8d)** required intermediate 9-chloro-1,2,3,4-tetrahydroacridine (**4**), which was obtained from two sequential reactions; condensation of anthranilic acid (**1**) and cyclohexanone (**2**), followed by a reaction of the Spiro intermediate (**3**) with phosphoryl chloride. The intermediate (**4**) was then aminated in the microwave with appropriate alkyl diamine linkers followed by conjugation to trolox through an amide bond using the activation agent *N,N*-carbonyldiimidazole to give compounds **8a - 8d** (Scheme 1). The tacrine-tryptoline hybrid (**14**) was synthesized by a two-step reaction including amination of 1,7-dibromoheptane with 9-amino-1,2,3,4-tetrahydroacridine to produce intermediate **12**, followed by conjugation to tryptoline through a further amination reaction. The tryptoline dimer **(16)** was obtained as a side-product from the reaction of tryptoline with an excess amount of 1,2-dibromoethane (Scheme 2) in a failed attempt to synthesize the tryptoline-bromoethane monomer. Compound **16** was added to the series of test compounds as a serendipitously synthesized compound that could show some biological activities because of possible fit inside the ChE active site cavity.

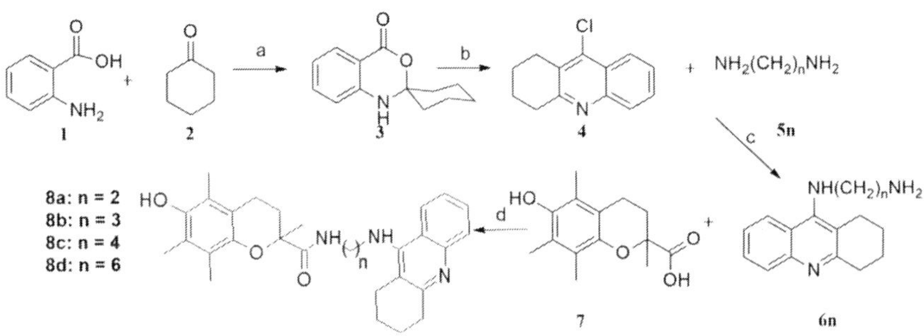

Scheme 1. Reagents and conditions: (a) anthranilic acid (1), cyclohexanone (2), toluene, reflux, 4 hrs; (b) spiro[2H-3,1-benzoxazine-2,1-cyclohexane]-4(1H)-one (3), POCl3, alkaline workup (KOH), 2 hrs; (c) 9-chloro-1,2,3,4-tetrahydroacridine (4), alkylenediamines in excess (5n), DCM, NaHCO3, microwave irradiation at 200 0C, 250 W and maximum pressure for 30 minutes; (d) N1-(1,2,3,4-tetrahydroacridin-9-yl)-alkane-1,n-diamine (6n), Trolox (7), N,N-carbonyldiimidazole, THF, rt, overnight, giving the tacrine-trolox hybrids (8a – 8d).

2.2. Cholinesterase Inhibition Studies

The selective loss of cholinergic neurons in AD results in a deficit of acetylcholine (ACh) in specific regions of the brain that mediate learning and memory functions [23]. Consequently, AD patients have been treated with AChE inhibitors [24-26] but unfortunately with limited therapeutic success, mainly because of the multifactorial nature of AD. Recent studies have also shown that compounds that are able to inhibit butyrylcholinesterase (BuChE) may be of value in the treatment of AD [27]. In AD, the ratio of BuChE/AChE gradually increases as the disease progresses, partially as a consequence of the progressive loss of the cholinergic synapses where AChE enzymes are located [28]. A compound able to inhibit both AChE and BuChE may thus be of more therapeutic value in AD.

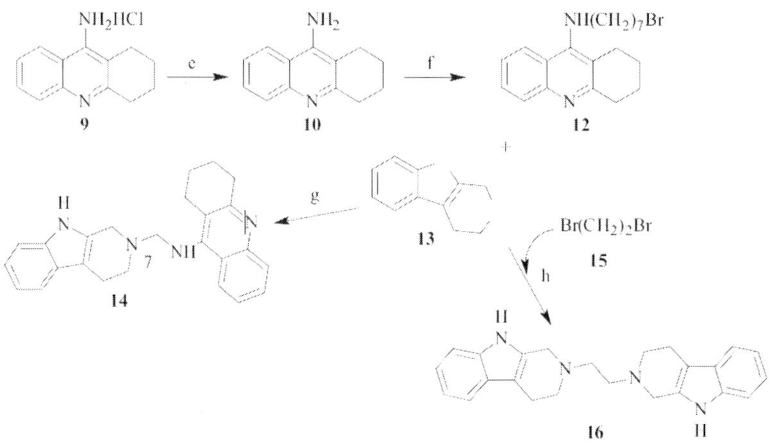

Scheme 2. Reagents and conditions: (e) tacrine hydrochloride (9), H2O, K2CO3, 1 hr; (f) tacrine (10), 1,7-dibromoheptane (11), CH3CN, KOH, rt, overnight; (g) N-(7-bromoheptyl)-1,2,3,4-tetrahydroacridin-9-amine (12), tryptoline (13), DMF, K2CO3, KI, microwave 160 0C, 250 PSI, 200 W, 1h, giving the tacrine-tryptoline hybrid (14); (h), tryptoline (13), 1,2-dibromoethane (15), KOH, CH3CN, microwave 100 0C, 250 PSI, 130 W, 1 hr, giving the bistryptoline dimer (16).

The inhibitory activity of the target compounds **8a – 8d**, **14** and **16** against *Tc*AChE (from electric eel) and eqBuChE (from equine serum) were measured according to the method of Ellman using tacrine and donepezil as reference compounds [29]. The IC_{50} values of the test compounds and reference compounds are given in Table 1. The tacrine-trolox hybrid compounds (**8a – 8d**) showed moderate to high AChE inhibitory activities (IC_{50}: 49.31 – 2200 nM) comparable to or better than the reference compounds donepezil, tacrine and previously reported β-carbolines [30]. Compound **16** displayed no inhibitory activity against AChE and though tryptoline is known to have AChE inhibitory activity, it is possible that the two carbon linker chain length limited the flexibility of the tryptoline moieties and prevented the dimer from adopting a stable conformation for effective activity. This scenario is slightly different for **8a** where the 2 carbon linker chain length is extended by the amide bond conferring moderate AChE inhibitory activity. Compound **8d** (IC_{50} = 49.31 nM) with a 6 carbon linker displayed higher activities than **8a, 8b** and **8c** with 2, 3 and 4 carbon linker chain lengths respectively. Compound **14** with the 7 carbon linker exhibited the highest activity (IC_{50} = 17.37 nM) and this correlates to the modelling study that clearly illustrate the interaction of the two pharmacophore moieties (tacrine and tryptoline) with Trp84 and Trp279, which are crucial for AChE activity (Figure 5). In this series, it is clear that hybrids with longer linker chain lengths show increased AChE inhibitory activities compared to the shorter ones. This correlates well with other studies described in the literature [8-11]. The idea of varying linker chain length seeks to assign to the two linked pharmacophore units, the ability to span the PAS and CAS of AChE, and obtain the appropriate orientation that would enable them to interact with crucial amino acids in the active site, generating better activities. As confirmed by these results, shorter linker chain lengths (less than 4 carbons) are less favorable for optimal activity.

Table 1. *IN VITRO* IC_{50} VALUES OF TEST COMPOUNDS 8A – 8D, 14 AND 16 FOR TCACHE, EQBUCHE AND DPPH.

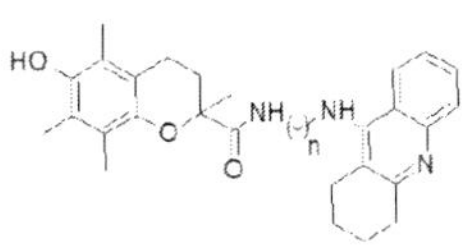

8a-d

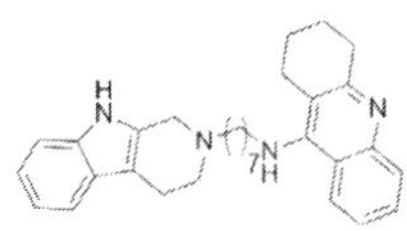

14

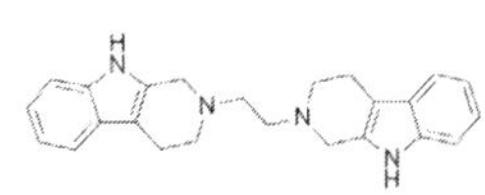

16

Compounds	AChE IC_{50} (nM)	BuChE IC_{50} (nM)	(SI)[a]	DPPH IC_{50} (μM)
8a (n = 2)	1530	128.82	11.87	46.23
8b (n = 3)	1530	12.58	121.62	11.48
8c (n = 4)	2200	11.22	196.07	14.38
8d (n = 6)	49.31	4.74	10.40	12.67
14	17.37	3.16	5.49	129.41
16	>100 μM	1698.24	>58.88	125.89
Tacrine	102	14.12	7.22	nd
Donepezil	220	2870	0.08	nd
Trolox	nd	nd	-	17.57

[a]BuChE selectivity index = IC_{50}(*Tc*AChE)/IC_{50}(eqBuChE). n.d. = not determined.

The test compounds showed moderate to high BuChE inhibitory activities comparable or higher than their reference compound tacrine (Table 1) and previously reported β-carbolines [30]. Compound **8d** (IC_{50} = 4.74 nM) and **14** (IC_{50} = 3.16 nM) with 6 and 7 carbon linker chain lengths respectively displayed higher activities than **8a**, **8b**, and **8c** with 2, 3 and 4 carbons linker chain lengths. It can be deduced from this observation that hybrids with longer linker chain lengths also seem to exhibit increased BuChE inhibitory activities compared to the shorter ones. This similarity of interaction of synthesized compounds and target proteins (AChE / BuChE) is consistent with the similarity of the topology of their active sites. Compounds **8d** and **14** exhibited BuChE inhibitor activity more than 3 fold higher than tacrine. All the synthesized compounds showed higher inhibitory activity for BuChE than for AChE. It has been demonstrated that the two proteins share 65% amino acid sequence similarity and the 35% difference in their amino sequence confer to BuChE the ability to accommodate bulkier substrates than AChE [31]. The similarity in amino acid sequence of AChE and BuChE may prove to be a challenge in efforts to design a 100% non-selective cholinesterase inhibitor. The goal of this study was however to discover new dual inhibitors of cholinesterase since both proteins are believed to contribute, to different degrees, to the depletion of ACh levels in AD brain [24-28]. Based on this observation, selective inhibitors could be less effective as therapeutic options.

2.3. Cholinesterase Molecular Modelling Studies

To provide insight into the binding mode of the synthesized compounds docking studies were carried out on *Tc*AChE and eqBuChE. The protein structure of *Tc*AChE co-crystallized with tacrine (PDB code: 2CMF) was retrieved from the Brookhaven Protein Data Bank (www.rcsb.org/pdb) and docking was carried out using the Dock application of Molecular Operating Environment (MOE 2015.10) software.

The BuChE molecular modeling experiments were conducted using a homology model because of the absence of a X-ray structure of eqBuChE. The automated homology-modeling program, SWISS-MODEL, was used to model the putative three-dimensional structure of eqBuChE based on the crystal structure of hBuChE (PDB: 2PM8) which shows an 89% sequence similarity [32]. The docking of the compounds into this homology model of eqBuChE was carried out with Autodock Vina software [33]. The docking results generated were directly loaded into and analyzed with MOE.

The docking revealed that the tacrine-trolox hybrids **8a** – **8d**, the tacrine-tryptoline hybrid (**14**) and the tryptoline dimer (**16**) show similar binding patterns. All the compounds spanned the CAS and PAS of *Tc*AChE but with different energy scores. Compounds with the short linker chain lengths tend to have higher energy scores of -8.10 kcal/mol **(16)** and -9.84 kcal/mol **(8a)** while **8b**, **8c**, **8d** and **14** have relatively lower energy scores of -10.20 kcal/mol, -10.30 kcal/mol, -11.43 kcal/mol and -10.40 kcal/mol, respectively. The difference in energy scores indicate a disparity in how strong / fast / stable the synthesized compounds bind to the active site. Based on this concept, compounds with longer linker chain lengths should display the highest activity. This is in line with the biological results where the same trend was observed. In all docking experiments, tacrine occupied the CAS and trolox / tryptoline occupied the PAS, which correlated well with our hypothesis and also demonstrated the high selectivity of the tacrine moiety for the CAS as supported in the literature [31]. Figure (4) shows the docking and binding interactions of compounds **8d** and **14** in *Tc*AChE. It is clear from the images that the quinoline ring of tacrine form π-π interactions with Trp84 in both compounds **8d** and **14** with the trolox and tryptoline moieties reaching the PAS region of the enzyme. An H-π interaction is also observed with compound (**14**) between 9-H of the tryptoline moiety and Trp279 in the PAS. These results confirm that the hybrid compounds are able to span both the CAS and PAS of AChE which may account for the promising AChE inhibitory activity observed. The molecular modeling experiments were also done with the test compounds in their protonated state. The only additional interaction observed was an ionic interaction between the protonated acridine nitrogen and HIS440. The binding energies and orientation of the molecules did not show any significant differences when compared to their protonated counterparts.

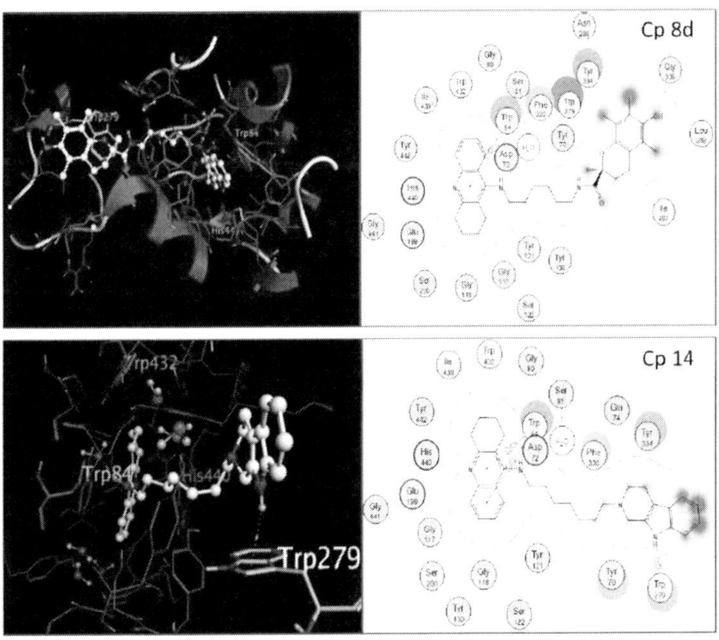

Figure 4. Compounds 8d (top) and 14 (bottom) docked into the active site of TcAChE (left) and possible ligands interactions with conserved aromatic residues (right). Compounds 8d and 14 below are shown in white. In pink and cyan are Trp84 (CAS) and Trp 279 (PAS), crucial amino acids regularly involved in inhibition of AChE.

To explore the binding mode in eqBuChE, compounds **8d** and **14** were used as representative compounds and were docked into the active site of the eqBuChE homology build model (Figure 5). The docking results of **8d** and **14** showed that the tacrine moiety binds in the CAS region, establishing π-π stacking interactions with Trp82. In both these compounds the trolox and tryptoline moieties spanned the active site gorge and were stabilized within the PAS region of eqBuChE. The binding energies of **8d** and **14** were -11.83 kcal/mol and -12.23 kcal/mol, respectively. The low binding energies and favourable positions of **8d** and **14** within the active site of eqBuChE may account for the high affinity and slight selectivity observed for BuChE over AChE. As observed for the AChE docking experiments, when the molecules were redocked in their protonated state, no significant difference was observed in their binding orientations or binding energies. An additional ionic interaction between the protonated acridine amine and HIS438 was the only additional binding interaction predicted.

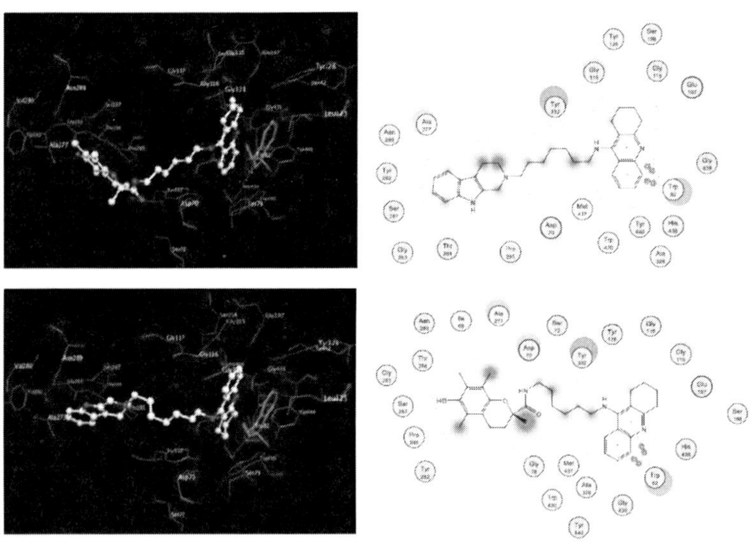

Figure 5. Complex of compound 8d (top) and 14 (bottom) with the eqBuChE homology model (left) and the interaction maps (right). Compounds 8d and 14 below are shown in white and Trp82 in the CAS site is shown in pink.

2.4. Anti-oxidant Studies

Reactive oxygen and nitrogen species contribute to the pathophysiology of neurodegenerative disorders [34]. Antioxidants are compounds capable of scavenging free radicals and antioxidant therapy is therefore considered as one of the options in neuroprotection [14]. The 2,2-diphenyl-1-picrylhydrazyl (DPPH) assay was used to assess the free radical scavenging or antioxidant effect of the test compounds. The synthetic nitrogen-centred $DPPH^{\cdot +}$ is not biologically relevant, but is used as indicator compound in testing hydrogen transfer capacity related to antioxidant activity [35]. The results of the ability of synthesized compounds to scavenge the DPPH free radical are included in Table 1. Trolox has well documented antioxidant properties and was used as reference compound in this assay. The conjugation of trolox to tacrine through varying linker chain lengths did not affect its ability to scavenge free radicals. The tacrine-trolox compounds (**8b** – **8d**, IC_{50}: 11.48 – 14.38 μM) exhibited activity with IC_{50} values in the same range as that of trolox (IC_{50} = 17.57 μM). The exception is compound **8a** (IC_{50}: 46.23 μM) which showed free radical scavenging activity of about half of that of trolox. Compounds **14** and **16** exhibited lower free radical scavenging activity with IC_{50} values of 129.41 μM and 125.24 μM. This was expected as the trolox moiety is replaced by the less potent free radical scavenging tryptoline moiety.

3. EXPERIMENTAL SECTION

Unless otherwise specified, materials were obtained from commercial suppliers and used without further purification. All reactions were monitored by thin-layer chromatography on 0.20 mm thick aluminium silica gel sheets (Alugram® SIL G/UV254, Kieselgel 60, Macherey-Nagel, Düren, Germany). Visualization was achieved using UV light (254 nm and 366 nm), an ethanol solution of ninhydrin or iodine vapours, with mobile phases prepared on a volume-to-volume basis. Chromatographic purifications were performed on silica gel (0.063–0.2 mm, Sigma Aldrich) except when otherwise stated. The MS spectra were recorded on a Perkin Elmer Flexar SQ 300 mass spectrometer by means of direct injection with a syringe pump. High resolution electron spray ionisation (HREI) mass spectra for all compounds were recorded on a Waters API Q-Tof Ultima mass spectrometer at 70 eV and 100 °C. The IR spectra were recorded on a Perkin Elmer Spectrum 400 spectrometer, fitted with a diamond attenuated total reflectance (ATR) attachment. Melting points were determined using a Stuart SMP-300 melting point apparatus and capillary tubes. All the melting points determined were recorded uncorrected. ^{1}H and ^{13}C NMR spectra were determined using a Bruker Avance III HD spectrometer at a frequency of 400 MHz and 100 MHz, respectively. Tetramethylsilane (TMS) was used as internal standard. All chemical shifts are reported in parts per million (ppm), relative to the internal standard. The following abbreviations are used to indicate the multiplicities of the respective signals: s - singlet; br s -

broad singlet; d - doublet; dd - doublet of doublets; t – triplet and m - multiplet.

3.1. Chemistry

3.1.1. Synthesis of spiro[2*H*-3,1-benzoxazine-2,1-cyclohexan]-4(1*H*)-one (3)

In 100 ml toluene in a 250 ml round bottom flask was dissolved 50 g (0.40 mol) anthranilic acid **(1)** and 45.3 ml (0.44 mol) cyclohexanone **(2)** was added. The mixture was refluxed for 4 hrs using a Dean-Stark trap until approximately 6.8 ml water was collected. The reaction mixture was cooled to room temperature and needle like-crystals were formed. The solid was collected by filtration and successively washed with 50 ml toluene and 50 ml ethanol. The solid was then dried *in vacuo* and 72.51 g white crystals were collected which afforded a yield of 90.91% of product **(3)** with physical characteristics similar to that described in the literature [37].

3.1.2. Synthesis of 9-chloro-1,2,3,4-tetrahydroacridine (4)

In a 250 ml round bottom flask, containing 120 ml (1.29 mol) of $POCl_3$ was added 67 g (0.309 mol) of **3** and the mixture was refluxed on an oil bath at 120 ^{0}C for 2 hrs. The reaction mixture was then cooled to room temperature and cautiously added to 540 g KOH in 1000 ml ice water while being stirred rapidly. CH_2Cl_2 (1500 ml) was added to dissolve the solid. The aqueous layer was extracted with 3×1000 ml CH_2Cl_2. The combined organic extracts were dried with $MgSO_4$. The mixture was filtered and the solvent was removed on a rotary evaporator to give a yellow solid, which was further recrystallized in acetone to give 64 g of pure yellow clear crystals (**4**) with similar physical characteristics as described in the literature [37].

3.1.3. General Synthesis of *N*^{1}(1,2,3,4-tetrahydroacridin-9-yl)alkane-1,n-diamines (n = 2, 3, 4 or 6)

To 0.43 g (1.98 mmol) of **4** contained in a microwave tube, was added 40.1 mmol of the appropriate alkylenediamine (**5n**). The mixture was reacted for 30 minutes at 200 ^{0}C with 250 W power and maximum pressure. After cooling, CH_2Cl_2 (30 ml) was added to the reaction mixture. Excess alkylenediamine was extracted from the mixture with 3×30 ml of distilled water acidified to pH 5-6 with diluted HCl. The organic layers were collected, dried with Na_2SO_4 and filtered. The organic solvent was evaporated on a rotary evaporator to yield the final products (**6n**). The physical characteristics of these compounds were similar to as described in the literature [38].

3.1.4. General Synthesis for the Tacrine-trolox Hybrids (8a – 8d)

To a solution of THF (20 ml) containing 0.75 g (3 mmol) of trolox (**7**) was added 0.54 g (3.3 mmol) of *N,N*-carbanoyldiimidazole. The mixture was stirred at room temperature for 30 minutes before adding 3 mmol of the appropriate N^1-(1,2,3,4-tetrahydroacridin-9-yl)alkane-1,n-diamine (**6n**) in 10 ml THF. The resulting solution was reacted overnight at room temperature. Thereafter, the solvent was removed *in vacuo* and the residue was added to DCM (25 ml) and extracted with acidified water (pH 3, 3×25 ml). The organic phase was further extracted with 5 N NaOH (aq) (3×25 ml). The combined organic phases were dried over $MgSO_4$ and the solvent was removed *in vacuo* rendering the crude products, which were first purified by flash column chromatography with acetone as mobile phase and then flushed with ethanol (250 ml) to render the impure products (**8n**). A final purification was carried out by column chromatography for each compound with ethyl acetate:ethanol (1:1) as mobile phase to yield the pure products (**8a** – **8d**).

3.1.4.1. 6-Hydroxy-2,5,7,8-tetramethyl-N-{2-[(1,2,3,4-tetrahydroacridin-9-yl)amino]ethyl}-3,4-dihydro-2H-1-benzopyran-2-carboxamide (8a)

$C_{29}H_{35}N_3O_3$; yield 16%; **Physical data**: **MP**: 33-35 ^{0}C; **Rf** (ethyl acetate:ethanol / 1:1) 0.3; **^{1}H NMR** (400 MHz, $CDCl_3$) δH: 7.95 (d, J = 8.97 Hz, 1H), 7.87 (d, J = 8.97 Hz, 1H), 7.55-7.53 (m, 1H), 7.32-7.26 (m, 2H), 6.86 (s, 1H), 3.61-3.56 (m, 4H), 3.08-3.05 (m, 2H), 2.67-2.63 (m, 2H), 2.35-2.31 (m, 1H), 2.12-2.03 (m, 9H), 1.89 (bs, 6H), 1.52 (s, 3H); **^{13}C-NMR** (100 MHz, $CDCl_3$), δC**:** 176.6, 148.9, 145.9, 143.9, 141.5, 133.3, 129.9, 129.7, 129.0, 128.7, 124.4, 122.9, 121.8, 117.7, 78.3, 50.3, 40.2, 31.9, 29.7, 29.4, 25.3, 24.7, 24.4, 22.6, 21.9, 20.4, 14.1, 12.3, 11.3. **HR-ESI** $[M+H]^+$: calcd. 474.2751, found. 474.2770; **IR** (ATR, cm^{-1}) V_{max}: 3326, 2924, 1447, 1091, 760.

3.1.4.2. 6-Hydroxy-2,5,7,8-tetramethyl-N-{3-[(1,2,3,4-tetrahydroacridin-9-yl)amino]propyl}-3,4-dihydro-2H-1-benzopyran-2-carboxamide (8b)

$C_{30}H_{37}N_3O_3$; Yield 25%; **Physical data**: **MP**: 118-122 ^{0}C; **Rf** (ethyl acetate:ethanol / 1:1) 0.33; **^{1}H NMR** (400 MHz, $CDCl_3$) δH: 7.93-7.87 (m, 2H), 7.51-7.49 (m, 1H), 7.34-7.31 (m, 2H), 6.71 (s, 1H), 3.47-3.35 (m, 4H), 3.08-3.04 (m, 2H), 2.68-2.60 (m, 5H), 2.07-2.03 (m, 9H), 1.87 (bs, 6H), 1.54 (s, 3H); **^{13}C-NMR** (100 MHz, $CDCl_3$), δC**:** 175.6, 157.8, 151.6, 146.0, 144.3, 129.1, 127.5, 124.3, 122.7, 122.1, 121.8, 120.0, 119.6, 118.0, 116.3, 78.6, 45.4, 36.6, 32.9, 31.4, 29.9, 25.1, 24.9, 23.1, 22.6, 20.8, 12.4, 12.2, 12.0, 11.6. **HR-ESI** $[M+H]^+$: calcd. 488.2908, found. 488.2914; **IR** (ATR, cm^{-1}) V_{max}: 3326, 2927, 1447, 760.

3.1.4.3. 6-Hydroxy-2,5,7,8-tetramethyl-N-{4-[(1,2,3,4-tetrahydroacridin-9-yl)amino]butyl}-3,4-dihydro-2H-1-benzopyran-2-carboxamide (8c)

$C_{31}H_{39}N_3O_3$; yield 13%; **Physical data**: **MP**: 154-159 ^{0}C; **Rf** (ethyl acetate:ethanol / 1:1) 0.36; **^{1}H NMR** (400 MHz, CD_3OD), δH: 8.15 (d, J = 8.45 Hz, 1H), 7.71-7.67 (m, 2H), 7.46-7.42 (m, 2H), 6.70 (s, 1H), 3.59-3.46 (m, 4H), 3.32-3.28 (m, 2H), 3.06-2.62 (m, 5H), 2.08-2.05-1.93 (m, 9H), 1.90-1.86 (m, 4H), 1.45-1.44 (m, 4H), 1.40 (s, 3H); **^{13}C-NMR** (100 MHz, CD_3OD), δC: 174.7, 153.9, 153.6, 145.9, 144.2, 142.0, 130.6, 124.6, 123.7, 123.6, 122.6, 121.6, 120.2, 117.8, 117.5, 113.0, 78.3, 48.1, 38.4, 30.5, 29.7, 28.0, 26.9, 24.5, 24.4, 21.5, 20.6, 12.6, 12.2, 12.0, 11.7. **HR-ESI** $[M+H]^+$: calcd. 502.3064, found. 502.3068; **IR** (ATR, cm^{-1}) V_{max}: 3370, 2928, 1520, 1088, 750, 530.

3.1.4.4. 6-Hydroxy-2,5,7,8-tetramethyl-N-{6-[(1,2,3,4-tetrahydroacridin-9-yl)amino]hexyl}-3,4-dihydro-2H-1-benzopyran-2-carboxamide (8d)

$C_{33}H_{43}N_3O_3$; yield 12%; **Physical data**: **MP**: 146-149 ^{0}C; **Rf** 0.46 (ethyl acetate-ethanol / 1:1); **^{1}H NMR** (400 MHz, CDCl3), δH: 8.29 (d, J = 8.55 Hz, 1H), 8.19 (d, J = 8.55 Hz, 1H), 7.61-7.57 (m, 2H), 7.40-7.336 (m, 1H), 6.40 (s, 1H), 3.77-3.40 (m, 5H), 2.68 (m, 2H), 2.52-2.43 (m, 4H), 2.17-2.08 (m, 9H), 1.85-1.77 (m, 6H), 1.51 (s, 3H), 1.23-1.19 (m, 6H); **^{13}C-NMR** (100 MHz, $CDCl_3$), δC: 174.5, 155.4, 151.4, 146.2, 144.4, 139.3, 131.9, 124.9, 124.5, 122.9, 121.4, 120.9, 118.1, 116.1, 11.4, 78.5, 48.2, 38.7, 38.5, 30.9, 29.7, 29.4, 28.8, 26.9, 26.3, 25.8, 25.4, 24.1, 22.1, 20.9, 13.1, 12.9, 11.9; **HR-ESI** $[M+H]^+$: calcd. 530.3377, found. 530.3387; **IR** (ATR, cm^{-1}) V_{max}: 3270, 2932, 1520, 1088, 576.

3.1.5. Synthesis of N*-(7-bromoheptyl)-1,2,3,4-tetrahydroacridin-9-amine (12)*

9-Amino-1,2,3,4-tetrahydroacridine (780 mg) (**10**) was obtained by precipitating 9-amino-1,2,3,4-tetrahydroacridine hydrochloride 1g (**9**) in a sodium hydrogen carbonate solution. To a solution of CH_3CN (10 ml) containing 400 mg (2.01 mmol) of **10** in a 100 ml round bottom flask was added 2.60 g (10.05 mmol) of 1,7-dibromoheptane (**11**) and 224 mg (4.02 mmol) of KOH. The mixture was stirred at room temperature overnight. The reaction was monitored by TLC (mobile phase: ethyl acetate:hexane:TEA (10:10:1)). After completion, the reaction mixture was poured into 10 ml of distilled water and extracted with DCM (3×20 ml). The combined organic layers were dried over anhydrous Na_2SO_4, filtered, and the solvent was evaporated under reduced pressure. The residue was purified by silica gel column chromatography using ethyl acetate:hexane:TEA (10:10:1) as eluent to produce 100 mg (13.20%) of a pure yellow oil (**12**) with similar characteristics as described in the literature [39].

3.1.6. Synthesis of N-(7-(3,4-dihydro-1H-pyrido[3,4-b]indol-2(9H)-yl)heptyl)-1,2,3,4-tetrahydroacridin-9-amine (14)

In a microwave tube containing 5 ml of DMF was dissolved (70 mg, 0.20 mmol) of **12**. Tryptoline (**13**, 26.28 mg, 0.15 mmol) and of K_2CO_3(38.80 mg, 0.28 mmol) were added with a catalytic amount of KI. The mixture was irradiated in a microwave reactor at 160 ^{0}C, 250 PSI and 200 W for 1 hr. The reaction was monitored by TLC (mobile phase: ethyl acetate-hexane-TEA in a ratio 10:10:1). After completion, the reaction mixture was poured into 5 ml of distilled water and extracted with ethyl acetate (3×10 ml). The combined organic phase was washed with water, and then with a saturated NaCl solution (3×10 ml). The organic fraction was dried with anhydrous Na_2SO_4 and evaporated under reduced pressure. The residue was purified by silica gel column chromatography using ethyl acetate-hexane-TEA (10:10:1) to produce 18 mg (25%) of a pure dark yellow solid (**14**).

$C_{31}H_{38}N_4$; yield 25%; **MP:** 98-101 ^{0}C; **Physical data**: **Rf** (ethyl acetate-hexane-TEA / 10:10:1) 0.43; **^{1}H NMR** (400 MHz, $CDCl_3$), δH: 8.37 (s, 1H), 7.98 (m, 2H), 7.59-7.55 (m, 1H), 7.45 (d, J = 7.29 Hz, 1H), 7.37-7.30 (m, 2H), 7.13-7.044 (m, 2H), 3.58-3.54 (m, 4H), 3.08 (s, 2H), 2.83-2.80 (m, 4H), 2.68 (t, J = 6.02 Hz, 2H), 2.54 (t, J = 7.67 Hz, 2H), 1.90 (m, 4H), 1.67-1.65 (m, 2H), 1.53 (m, 2H), 1.35 (m, 6H); **^{13}C-NMR** (100 MHz, $CDCl_3$), δC 157.7, 151.6, 136.2, 132.0, 129.0, 127.8, 123.9, 123.1, 121.3, 119.7, 117.9, 115.2, 110.9, 108.3, 57.9, 51.4., 50.4, 49.4, 33.3, 31.7, 29.2, 27.4, 27.3, 26.8, 24.8, 23.0, 22.6, 21.4,, 18.6; **HR-ESI** $[M+H]^+$: calcd. 467.3169, found. 467.3166; **IR** (ATR, cm^{-1}) V_{max}: 3214, 2926, 1563, 740.

3.1.7. Synthesis of 1,2-Bis(3,4-dihydro-1H-pyrido[3,4-b]indol-2(9H)-yl)ethane (16)

To a solution of CH_3CN (20 ml) containing 0.4 g (2.32 mmol) of **13** in a 100 ml round bottom flask was added 2.20 g (11.61 mmol) of 1,2-dibromoethane (**15**) and KOH (0.30 mg, 4.63 mmol). The mixture was irradiated in microwave reaction at 130 W, 250 PSI and 100 ^{0}C for 1 hr). The reaction was monitored with TLC (mobile phase: methanol). After completion of the reaction, the solvent was evaporated under reduced pressure and the resulting solid residue was dissolved in methanol and purified by silica gel column chromatography (mobile phase: methanol) to yield an impure powder-like product, which was further subjected to 2 cycles of recrystallization in ethanol to afford 150 mg of a pure pale-white solid **(16).**

$C_{24}H_{26}N_4$; yield 17%; **Physical data**: **MP**: 255-260 ^{0}C; **Rf** (methanol) 0.57; **^{1}H NMR** (400 MHz, $(CD_3)_2SO$): δH: 10.70 (s, 2H), 7.35 (m, 4H), 6.97(m, 4H), 3.69 (s, 4H), 3.35 (s, 4H), 2.80 (bs, 4H), 2.68 (s, 4H); **^{13}C-NMR** (100 MHz, $(CD_3)_2SO$), δC: 135.8, 132.9, 126.7, 120.2, 118.2, 117.3, 110.9, 106.4, 55.3,

51.2, 50.5, 21.3; **HR-ESI** $[M+H]^+$: calcd. 371.2230, found. 371.2233; IR (ATR, cm^{-1}) V_{max}: 3374, 2815, 1450, 741.

3.2. Cholinesterase Assay

AChE from electric eel, BuChE from equine serum, S-butylthiocholine iodide (BTCI), acetylthiocholine iodide (ATCI), 5,5′-dithiobis-(2-nitrobenzoic acid) (Ellman's reagent, DTNB), donepezil and tacrine hydrochloride were purchased from Sigma-Aldrich. The inhibitory activities of the test compounds were evaluated by Ellman's method [29]. The compounds were dissolved in DMSO and diluted with the buffer solution (50 mM Tris–HCl, pH = 8.0) to yield the corresponding test concentrations (DMSO less than 0.01%). In each well of the plate, 160 μL of 1.5 mM DTNB, 50 μL of AChE (0.22 U/mL eeAChE) or 50 μL of BuChE (0.12 U/mL eqBuChE) were incubated with 10 μL of different concentrations of test compounds (0.01–100 μM) at 37°C for 10 min. After this period, acetylthiocholine iodide (15 mM) or S-butyrylthiocholine iodide (15 mM) as the substrate (30 μL) was added, incubated for a further 10 minutes, and thereafter the absorbance was measured at a wavelength of 405 nm at different time intervals (0, 60, 120, and 180 s). Five different concentrations of test compounds were used and the assays were run in triplicate. IC_{50} values were calculated as concentration of the compound that produces 50% enzyme activity inhibition, using the Graph Pad Prism 6 software (San Diego, CA, USA).

3.3. Anti-oxidant Assay

Antioxidant scavenging activity was studied using 1,1-diphenyl-2-picrylhydrazyl free radical ($DPPH^{\cdot+}$) [40, 41]. Various concentrations of test solutions in 20 μl DMSO were added to 180 μL of 0.1 mM solution of $DPPH^{\cdot+}$ in methanol. Neat methanol (20 μl) was used as experimental control. After 30 min of incubation at room temperature, the reduction of free radical was measured by reading the absorbance at 517 nm. Trolox® was used as reference standard. The scavenging activity of the samples corresponded to the intensity of the quenching of $DPPH^{\cdot+}$. For statistical viability, the assay was performed three times with 3 readings for five different concentrations. IC_{50} values were calculated as concentration of the compound that produces 50% enzyme activity inhibition, using the Graph Pad Prism 6 software (San Diego, CA, USA).

3.4. Cholinesterase Docking Studies

3.4.1. Acetylcholinesterase

Modeling was performed using the Molecular Operating Environment (MOE 2015.10) software package. The X-ray crystal structure of the AChE in complex with bistacrine (code: 2CMF) was downloaded from the protein data bank (PDB) and loaded in the working environment. The bistacrine ligand was removed and the enzyme structure was checked for missing atoms, bonds and contacts. Hydrogens and partial charges were added using the protonate 3D application in MOE. The hybrid compounds were accurately drawn in ACD Chemsketch and saved as MDL mole files (V3000) and imported into the MOE database. The structures were energy minimized in implicit solvated environment under AMBER99 force field and docked into the active site of the protein. The binding site was defined with the conserved aromatic residues Trp279 and Trp84. All other parameters were left as default values. Finally, the conformation with the lowest docking score was selected for analysing the interactions between the AChE and the hybrid compounds.

3.4.2. Butyrylcholinesterase

Because no X-ray structure exists for eqBuChE, a homology model was used to rationalize the experimental data. The eqBuChE model was retrieved from the SWISS-MODEL Repository. This is a database of annotated three-dimensional comparative protein structure models generated by the fully automated homology-modelling pipeline SWISS-MODEL. A putative three-dimensional structure of eqBuChE was created based on the crystal structure of hBuChE (PDB ID: 2PM8), these two enzyme exhibit 89% sequence identity [32]. Proper bonds, bond orders, hybridization and charges were assigned and CHARMm force field was applied using MOE. The prepared protein was directly loaded into AutoDockTools (ADT; version 1.5.4), hydrogens were added and partial charges for proteins and ligands were calculated using Gasteiger charges. Flexible torsions in the ligands were assigned with the AutoTors module, and the acyclic dihedral angles were allowed to rotate freely. Docking calculations were performed with the program Autodock Vina [33]. Because VINA uses rectangular boxes for the binding site, the box center was defined and the docking box was displayed using ADT. All dockings were performed where a cube of 75 Å with grid points separated by 1 Å, was positioned over the active site of the protein (x = 29.885; y = -54.992; z = 58.141). Default parameters were used except num_modes, which was set to 40. The lowest docking-energy conformation was considered as the most stable orientation. Finally, the docking results generated were analyzed using MOE.

4. CONCLUSION

In this study, novel multi-target tacrine-trolox and tacrine-tryptoline hybrids were designed, synthesized and evaluated for ChE inhibition and free radical scavenging activity. All the hybrid compounds showed nanomolar to micromolar activity in AChE, BuChE and free radical scavenging assays, some even better than their references compounds. The most potent are compounds **8d** and **14.** Compound **8d** demonstrated good activity in all three assays with low SI for AChE / BuChE. Compound **14** showed the best dual cholinesterase inhibitor activity but with lower free radical scavenging activity. As evident in the bar graph (Figure 6), **8d** is the best lead molecule from this study as it demonstrated the best ability to inhibit AChE, BuChE and act as a potent antioxidant. This result is supported by recently published work wherein the non-hepatoxicity of tacrine-trolox hybrids has been demonstrated [36]. Novel compound **14** showed promising activity for ChE inhibition and hybrid molecules in this series could also be further explored. Compounds **8d** and **14** further showed interactions with both the CAS and PAS of AChE in docking studies and this indicate that **8a-d** and **14** may be able to prevent cholinesterase induced Aβ aggregation [19-22]. Overall, these results indicate that **8d** and **14** as potential multifunctional agents for the treatment of AD, should be explored further.

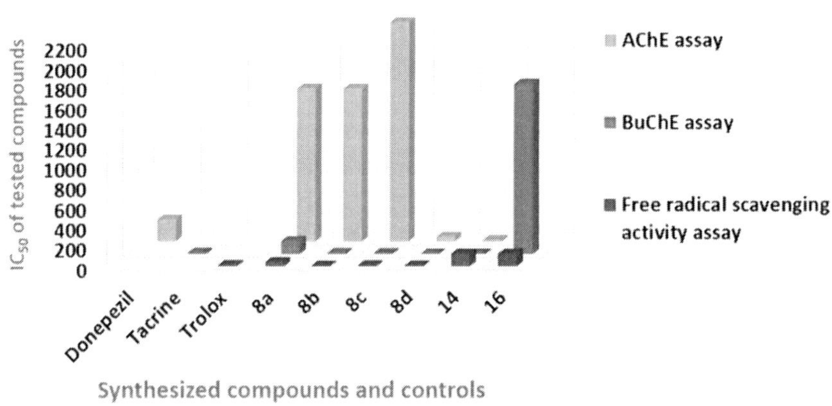

Figure 6. Bar graph showing comparative study of IC50 values depicting the multifunctional ability of the synthesized compounds. Cholinesterase IC50 (nM); free radical scavenging activity IC50 (μM).

CONFLICT OF INTEREST

The authors confirm that this article content has no conflict of interest.

ACKNOWLEDGEMENTS

We are grateful for the financial support from the University of the Western Cape and The Medical Research Council of South Africa.

REFERENCES

1. Goedert, M.; Spillantini, M.G. A century of Alzheimers disease. Science, 2006, 314(5800), 777-781.
2. Silvestrelli, G.; Lanari, A.; Parnetti, L.; Tomassoni, D.; Amenta, F. Treatment of Alzheimers disease: from pharmacology to a better understanding of disease pathophysiology. Mech. Ageing Dev., 2006, 127(2), 148-157.
3. Duthey, B. Background paper 6.11: Alzheimer disease and other dementias. A Public Health Approach to Innovation, 2013, pp: 1-74.
4. Grimm, O.W.; Hartmann, T. Recent understanding of the molecular mechanisms of alzheimer's disease. J. Addict. Res. Ther., 2012.
5. [Skovronsky, D.M.; Lee, V.M.; Trojanowski, J.Q. Neurodegenerative diseases: new concepts of pathogenesis and their therapeutic implications. Annu. Rev. Pathol., 2006, 1, 151-170.
6. Crews, L.; Masliah, E. Molecular mechanisms of neurodegeneration in Alzheimers disease. Hum. Mol. Genet., 2010, 19(R1), R12-R20.
7. Capurro, V.; Busquet, P.; Lopes, J.P.; Bertorelli, R.; Tarozzo, G.; Bolognesi, M.L.; Piomelli, D.; Reggiani, A.; Cavalli, A. Pharmacological characterization of memoquin, a multi-target compound for the treatment of Alzheimers disease. PLoS One, 2013, 8(2), e56870.
8. Simoni, E.; Daniele, S.; Bottegoni, G.; Pizzirani, D.; Trincavelli, M.L.; Goldoni, L.; Tarozzo, G.; Reggiani, A.; Martini, C.; Piomelli, D.; Melchiorre, C.; Rosini, M.; Cavalli, A. Combining galantamine and memantine in multitargeted, new chemical entities potentially useful in Alzheimers disease. J. Med. Chem., 2012, 55(22), 9708-9721.
9. Dias, K.S.; Viegas, C., Jr Multi-target directed drugs: a modern approach for design of new drugs for the treatment of Alzheimers disease. Curr. Neuropharmacol., 2014, 12(3), 239-255.
10. Wang, H.; Carlier, P.R.; Ho, W.L.; Wu, D.C.; Lee, N.T.; Li, C.P.; Pang, Y.P.; Han, Y.F. Effects of bis(7)-tacrine, a novel anti-Alzheimers agent, on rat brain AChE. Neuroreport, 1999, 10(4), 789-793.
11. Rook, Y.; Schmidtke, K.U.; Gaube, F.; Schepmann, D.; Wünsch, B.; Heilmann, J.; Lehmann, J.; Winckler, T. Bivalent β-carbolines as potential multitarget anti-Alzheimer agents. J. Med. Chem., 2010, 53(9), 3611-3617.
12. Kozurkova, M.; Hamulakova, S.; Gazova, Z.; Paulikova, H.; Kristian, P. Neuroactive multifunctional tacrine congeners with cholinesterase, anti-amyloid aggregation and neuroprotective properties. Pharmaceuticals, 2011, 4(2), 382-418.
13. Szymański, P.; Skibiński, R.; Inglot, T.; Bajda, M.; Jończyk, J.; Malawska, B.; Mikiciuk-Olasik, E. New tacrine analogs as acetylcholinesterase inhibitors - theoretical study with chemometric analysis. Molecules, 2013, 18(3), 2878-2894.
14. Halliwell, B. Free radicals and other reactive species in disease. eLS, 2001, 1-7.

15. Radesäter, A.C.; Johansson, S.; Öberg, C.; Luthman, J. The vitamin-E analog trolox and the NMDA antagonist MK-801 protect pyramidal neurons in hippocampal slice cultures from IL-1β-induced neurodegeneration. Neurotox. Res., 2003, 5(6), 433-442.
16. Muñoz, F.J.; Opazo, C.; Gil-Gómez, G.; Tapia, G.; Fernández, V.; Valverde, M.A.; Inestrosa, N.C. Vitamin E but not 17β-estradiol protects against vascular toxicity induced by β-amyloid wild type and the Dutch amyloid variant. J. Neurosci., 2002, 22(8), 3081-3089.
17. Herraiz, T.; González, D.; Ancín-Azpilicueta, C.; Arán, V.J.; Guillén, H. β-Carboline alkaloids in Peganum harmala and inhibition of human monoamine oxidase (MAO). Food Chem. Toxicol., 2010, 48(3), 839-845.
18. Frost, D.; Meechoovet, B.; Wang, T.; Gately, S.; Giorgetti, M.; Shcherbakova, I.; Dunckley, T. β-carboline compounds, including harmine, inhibit DYRK1A and tau phosphorylation at multiple Alzheimers disease-related sites. PLoS One, 2011, 6(5), e19264.
19. Inestrosa, N.C.; Alvarez, A.; Pérez, C.A.; Moreno, R.D.; Vicente, M.; Linker, C.; Casanueva, O.I.; Soto, C.; Garrido, J. Acetylcholinesterase accelerates assembly of amyloid-beta-peptides into Alzheimers fibrils: possible role of the peripheral site of the enzyme. Neuron, 1996, 16(4), 881-891.
20. De Ferrari, G. V.; Canales, M. A.; Shin, I.; Weiner, L. M. A structural motif of acetylcholinesterase that promotes amyloid beta-peptide fibril formation. Biochemistry, 2001, 40, 10447-10457.
21. Rees, T.M.; Berson, A.; Sklan, E.H.; Younkin, L.; Younkin, S. nBrimijoin, S.; Soreq, H. Memory deficits correlating with nAcetylcholinesterase splice shift and amyloid burden in doubly nTransgenic mice. Curr. Alzheimer Res., 2005, 2, 291-300.
22. Dinamarca, M.C.; Sagal, J.P.; Quintanilla, R.A.; Godoy, J.A. n Arrázola, M. S.; Inestrosa, N. C. Amyloid-beta-Acetylcholinesterasen complexes potentiate neurodegenerative changes induced by the nAbeta peptide. Implications for the Pathogenesis of Alzheimersn Disease. Mol. Neurodegener., 2010, 5, 4.
23. Talesa, V.N. Acetylcholinesterase in Alzheimers disease. Mech. Ageing Dev., 2001, 122(16), 1961-1969.
24. Birks, J.; Harvey, R.J. Donepezil for dementia due to Alzheimers disease. Cochrane Database Syst. Rev., 2006, (1), CD001190.
25. Loy, C.; Schneider, L. Galantamine for Alzheimers disease. Cochrane Database Syst. Rev., 2004, (4), CD001747.
26. Birks, J.; Iakovidou, V.; Tsolaki, M. Rivastigmine for Alzheimers disease. Cochrane Database Syst. Rev., 2000, (4), CD001191.
27. Greig, N.H.; Utsuki, T.; Ingram, D.K.; Wang, Y.; Pepeu, G.; Scali, C.; Yu, Q.S.; Mamczarz, J.; Holloway, H.W.; Giordano, T.; Chen, D.; Furukawa, K.; Sambamurti, K.; Brossi, A.; Lahiri, D.K. Selective butyrylcholinesterase inhibition elevates brain acetylcholine, augments learning and lowers Alzheimer beta-amyloid peptide in rodent. Proc. Natl. Acad. Sci. USA, 2005, 102(47), 17213-17218.
28. Giacobini, E. Cholinergic function and Alzheimers disease. Int. J. Geriatr. Psychiatry, 2003, 18(Suppl. 1), S1-S5.
29. Ellman, G.L.; Courtney, K.D.; Andres, V., Jr; Feather-Stone, R.M. A new and rapid colorimetric determination of acetylcholinesterase activity. Biochem. Pharmacol., 1961, 7, 88-95.
30. Schott, Y.; Decker, M.; Rommelspacher, H.; Lehmann, J. 6-Hydroxy- and 6-methoxy-β-carbolines as acetyl- and butyrylcholinesterase inhibitors. Bioorg. Med. Chem. Lett., 2006, 16(22), 5840-5843.
31. Bajda, M.; Więckowska, A.; Hebda, M.; Guzior, N.; Sotriffer, C.A.; Malawska, B. Structure-based search for new inhibitors of cholinesterases. Int. J. Mol. Sci., 2013, 14(3), 5608-5632.
32. Kiefer, F.; Arnold, K.; Künzli, M.; Bordoli, L.; Schwede, T. The SWISS-MODEL Repository and associated resources. Nucleic Acids Res., 2009, 37(Database issue), D387-D392.
33. Trott, O.; Olson, A.J. AutoDock Vina: improving the speed and accuracy of docking with a new scoring function, efficient optimization, and multithreading. J. Comput. Chem., 2010, 31(2), 455-461.

34. Nishida, J.; Kawabata, J. DPPH radical scavenging reaction of hydroxy- and methoxychalcones. Biosci. Biotechnol. Biochem., 2006, 70(1), 193-202.
35. Karali, N.; Güzel, O.; Ozsoy, N.; Ozbey, S.; Salman, A. Synthesis of new spiroindolinones incorporating a benzothiazole moiety as antioxidant agents. Eur. J. Med. Chem., 2010, 45(3), 1068-1077.
36. Nepovimova, E.; Korabecny, J.; Dolezal, R.; Babkova, K.; Ondrejicek, A.; Jun, D.; Sepsova, V.; Horova, A.; Hrabinova, M.; Soukup, O.; Bukum, N.; Jost, P.; Muckova, L.; Kassa, J.; Malinak, D.; Andrs, M.; Kuca, K. Tacrine-trolox hybrids: a novel class of centrally active, nonhepatotoxic multi-target-directed ligands exerting anticholinesterase and antioxidant activities with low in vivo toxicity. J. Med. Chem., 2015, 58(22), 8985-9003.
37. Chao, X.; He, X.; Yang, Y.; Zhou, X.; Jin, M.; Liu, S.; Cheng, Z.; Liu, P.; Wang, Y.; Yu, J.; Tan, Y.; Huang, Y.; Qin, J.; Rapposelli, S.; Pi, R. Design, synthesis and pharmacological evaluation of novel tacrine-caffeic acid hybrids as multi-targeted compounds against Alzheimers disease. Bioorg. Med. Chem. Lett., 2012, 22(20), 6498-6502.
38. Lan, J.S.; Xie, S.S.; Li, S.Y.; Pan, L.F.; Wang, X.B.; Kong, L.Y. Design, synthesis and evaluation of novel tacrine-(β-carboline) hybrids as multifunctional agents for the treatment of Alzheimers disease. Bioorg. Med. Chem., 2014, 22(21), 6089-6104.
39. Savini, L.; Gaeta, A.; Fattorusso, C.; Catalanotti, B.; Campiani, G.; Chiasserini, L.; Pellerano, C.; Novellino, E.; McKissic, D.; Saxena, A. Specific targeting of acetylcholinesterase and butyrylcholinesterase recognition sites. Rational design of novel, selective, and highly potent cholinesterase inhibitors. J. Med. Chem., 2003, 46(1), 1-4.
40. Cotelle, N.; Bernier, J.L.; Catteau, J.P.; Pommery, J.; Wallet, J.C.; Gaydou, E.M. Antioxidant properties of hydroxy-flavones. Free Radic. Biol. Med., 1996, 20(1), 35-43.
41. Brand-Williams, W.; Cuvelier, M.E.; Berset, C. Leben.-Wiss, W. Use of a free radical method to evaluate antioxidant activity. Technol Food Sci Technol, 1995, 28, 25-30.

CHAPTER 3

Computational Methods Applied to Rational Drug Design

David Ramírez*

Centro de Bioinformática y Simulación Molecular, Universidad de Talca, 2 Norte 685, Casilla, Talca, Chile

ABSTRACT

Due to the synergic relationship between medical chemistry, bioinformatics and molecular simulation, the development of new accurate computational tools for small molecules drug design has been rising over the last years. The main result is the increased number of publications where computational techniques such as molecular docking, *de novo* design as well as virtual screening have been used to estimate the binding mode, site and energy of novel small molecules. In this work I review some tools, which enable the study of biological systems at the atomistic level, providing relevant information and thereby, enhancing the process of rational drug design.

Keywords: Bioinformatics, *de novo* design, medical chemistry, molecular docking, molecular modeling, rational drug design, virtual screening.

1. INTRODUCTION

Synergic relationship between experimental biochemistry sciences and computational theoretical methods has increased in recent years, providing significant benefits in different biological sciences areas, specially biochemistry and molecular biology. It has also brought advantages in medicinal chemistry for the rational design of drugs [1]. Regarding theoretical study of biochemical systems, one of its greatest profits is the molecular knowledge of the currently studied structures; these studies allow an exhaustive system analysis, and find meaningful answers to research questions that scientists face in their experimental laboratories.

Before exploring some of the computational methodologies currently used in drug design, it is important to consider that researchers have to draw from relevant information such as the three-dimensional structure of one or several substrates and molecular targets in order to start the theoretical study of certain biological systems. This information is essential, and in many cases, it determines the success or failure of the ongoing theoretical study. For instance, if a researcher wants to study the system formed by an endogenous or an exogenous inhibitor with its respective molecular target, the three-dimensional structure of the molecules involved in the interaction to be simulated has to be known.

The number of molecular targets with a well known structure is exponentially increasing due to the dramatic progress of spectroscopic techniques such as X-ray crystallography, Nuclear Magnetic Resonance (NMR) [2], and the development of the super-resolved fluorescence microscopy that shows a 3D image of a single molecule [3]. In addition, structures which have been produced from structural genomics have also become valuable tools for the study of systems whose molecular targets have not been solved yet [4, 5]. This growth has allowed a massive and constant use of computational tools in research centers worldwide. Furthermore, the developed methodologies for the production and optimization of small molecules (as most of the substrates studied in biochemistry sciences), are already registered in special databases and have provided vital information to research and find inherent characteristics which have never been described before [6, 7].

Nowadays, there are multiple computational methodologies used as bioinformatics tools for the study of biological systems and drug discovery. The use of one or another tool is linked to the researcher interests, the calculation level intended, possible technical limitations, how and also which kind of information can be extracted from data processing and analysis. Some of the main methodologies for these studies are: Molecular Docking, *de novo* design, Virtual Screening (VS), Quantitative structure-activity relationships (QSAR), Molecular Dynamics simulation (MDs) and Molecular Modeling (MM) [8]. These methodologies are wieldy used for rational drug design and discovery processes, where the ligand receptor binding mode is imperative to understand the molecular interaction mechanism and the structural factors related with the bioactivity of each inhibitor in detail [9].

In this review, basic and significant elements of molecular docking, *de novo* design and virtual screening are shown. These are some of the greatest computational methodologies used for the analyses of biochemical sciences, which also improve the common techniques used to study proteins structures, the design of new molecules biologically relevant for pharmacological application, the structure activity relationship, among others.

2. MOLECULAR DOCKING

Molecular Docking dates from the middle of the 19th century. Chemists Archibald Scott Couper, Friedrich August Kekulé and Aleksandr Mikhailovich introduced the valence in organic chemistry, and submitted the first structures with graphical representations of carbon atoms [10]. In 1861 Johann Josef Loschmidt produced the largest molecules collection at that time (368), including the first accurate benzene structure [11]. Due to the technological development in computational sciences, the concept of 'force field' was introduced from the vibrational spectroscopy, which consider the forces acting all over the atoms in a molecule. The scientific community did not adopt the concept of 'force field' until 1946, when the first premise about the combination of steric interactions and the Newtonian mechanics model of vibrational modes, bonds, and angles emerged.

Many other scientific advances were developed at that time, but it was in 1953 when Metropolis *et al.* published the study: "Equation of State Calculations by Fast Computing Machines" [12]. For the last century, the molecular modeling foundations were set, and due to the technological advances of our time, it now allows us to perform better theoretical studies and computational calculations and validates the need of using computational methods for science progress.

Molecular docking main goal is to predict as accurately as possible the best conformations (poses) of a ligand - by using a score function - in a conformational area, which is delimited by the molecular target binding site [13]. This is applied in different stages of the drug design process in order to predict the binding mode of the ligands already known [14]; it is also used to identify novel and potent ligands [15] and as a binding affinity predictive tool [16]. The first molecular docking algorithms considered translation and orientation only (degrees of freedom), and used the ligand and protein as rigid bodies. Improvements in computer science allowed the creation of new and accurate algorithms in order to use the ligand as flexible with a rigid (or nearly rigid) receptor [17].

This methodology is described as a multi-step process, where each step has one or several complexity levels. The process starts with the implementation of the molecular docking algorithms responsible for placing the ligands in the binding sites. This process is challenging, since it depends directly on the freedom levels granted to the ligands. The freedom levels sampling must be done accurately in order to identify the best conformation for the receptor-binding site. In addition, it has to be done fast enough to assess several compounds in a relatively short period of time [14].

These algorithms are complemented with scoring functions to sort different poses with the receptor regarding the union force. The scoring functions have some disadvantages because they do not take into account all the interfering forces between the ligand and the receptor; in addition, they also fail in accounting the ligand solvation, the entropic changes when bonding the receptor, the receptor flexibility and the receptor's conformational changes induced by the same ligand interaction. Based on this, molecular docking is a technique quite fast and effective in terms of time and computational cost [18].

The main software and algorithms developers performing molecular docking are focused on improving aspects such as: receptor's flexibility, solvation, coupling fragment, post-processing, molecular docking in homology models and comparisons between couplings. Main progress regarding this technique has taken place in the scoring functions area. The use of this methodology has been enhanced over the years; for instance, by doing a single search in PubMed using the word 'docking' as a search criterion, it shows the evolution of the publications where molecular docking has been used (Figure 1).

Many authors have demonstrated that molecular docking software can produce "accurate" binding modes, providing acceptable solutions to the sampling problem; however, the main challenge between the programs and algorithms developers is the poses scoring problem [19]. In this sense, the molecular docking tools are powerful in the prediction of correct binding poses but have critical issues to estimate accurately the corresponding binding affinities.

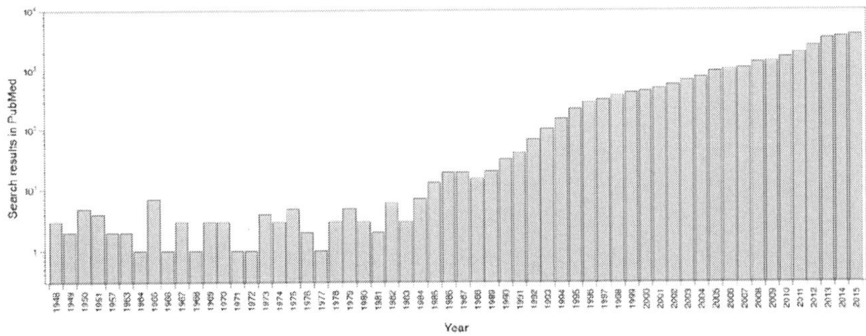

Figure 1. Reported publications number where molecular docking was used. Results from PubMed using the word 'docking' as search criteria.

An effort to improve affinity prediction, a rescoring process with other simple functions or solvated-based scoring functions is typically performed. The poses generated by docking program are taken, and methods such as MM/PBSA (Molecular Mechanics/ Poisson-Boltzmann Surface Area) and MM/GBSA (Molecular Mechanics/ Generalized-Born Surface Area) [19-21] include implicit solvent and can be used in order to correct scoring function values and improve docking accuracy. Other strategy is the use of molecular dynamics simulation (MDs) to get conformational sampling of the complex obtained using docking, performing then a subsequent calculation of the binding energy by averaging the score values for different poses extracted from the trajectory [22-24]. Under this approach, the receptor flexibility and the presence of explicit water molecules contribute to a more realistic description of the system, which could have an influence in binding energy calculations. However, the biggest challenge of molecular docking (and also the principal problem) is the accurate prediction of the binding energies, which has major implications for the prediction of novel effective drugs. This process is performed by using scoring functions that score the predicted poses [25].

Several classes of scoring functions have been developed [25]. These consider the protein as a rigid body [26, 27], or as a soft body [28-30]. They also may consider flexible side chains [31] or certain flexible domains in the target [32-34]. Other empirical scoring functions have shown be useful in the virtual screening and filtering of databases of drug-like compounds at the early stage of drug development processes [34]. Frequently, scoring functions try to reproduce experimental binding affinities, but the ones with popular docking programs do not always yield the best predictions. New programs have arisen in recent years advertising the importance of Molecular Docking in the study of biochemistry sciences and drug design, the most widely used are AutoDock [33, 34] with 29.5% followed by GOLD [37] with 17.5% and Glide [27, 36] with 13.2% (Figure 2).

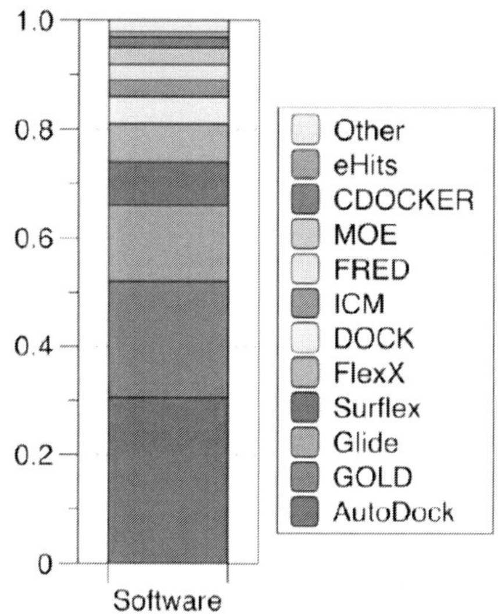

Figure 2. Use of molecular docking programms between 2010 and 2011. Based on a PubMed search. (Taken and adapted from [18]).

Generally speaking, theoretical molecular docking aspects are within the thermodynamics principles, where the main goal is to accurately predict the complex Receptor-Ligand structure [*R-L*] in equilibrium conditions (Equation 1). As an example in Figure (3) it was set the Receptor [*R*] (acetylcholinesterase), the ligand [*L*] (donepezil), and the complex [*R-L*] for this particular system.

$$R - L_{\ aq} \rightleftharpoons R_{\ aq} + [L]_{aq} \tag{1}$$

Figure (3) illustrates the ligand Donepezil [39] union (drug used in pharmacological therapy against Alzheimer's disease [40]) to the receptor Acetylcholinesterase described through the Gibbs free energy of binding (ΔG_{Bind}); this energy is related to the binding affinity by the ratio of Equation 2:

$$\Delta G_{Bind} = -RTlnK_A \qquad K_A = \frac{1}{K_i} = \frac{R - L}{R \ \ L} \tag{2}$$

K_A is the affinity constant and K_i is the inhibition constant. The prediction for the accurate structure of the [*R*-*L*] complex does not require the experimental information of the constants; however, the prediction of the ligand biological activity does. The terms influencing the scoring function can be classified as follows: Consideration of the [*R*-*L*] complex, where aspects such as the ligand and receptor deformation, hydrogen bonds, and electrostatic and steric factors are important. When the equilibrium is considered (Equation 1), the most important factors are: desolvation, rotational and translational entropy.

Molecular docking has been widely used in rational drug design in the past decades with extensive applications in several issues such as:

- The theoretical study of antitubulin agents with anti-cancer activity [41].
- The study of the estrogen receptor binding domains [42].
- Shityakov *et al.* have studied the blood-barrier permeation by *In silico* predictive models [43, 44], the dimer complexes of dopamine and levodopa derivatives to assess drug delivery to the central nervous system [45], and the interaction of nanoparticles with multidrug resistance protein [46].
- The prediction of anti-cancer drugs for the treatment of hypopharyngeal cancer [47].
- Sabogal *et al.* established a protocol to identify an interaction pattern between the sea anemone neurotoxins and the potassium channel Kv1.3 [48].
- Yi *et al.* studied the metabolic behavior of anisole by using Molecular Docking [49].

Academically, the amount of research groups who use this technique to study biological systems have increased. In biology, computational techniques have taken great strength, especially in the structural biochemistry field. To prove it, a large number of articles such as the Structural Biochemistry Group at the University of Edinburgh for instance (website: http://www.bch.ed.ac.uk/index.php) have been published. In addition, at multidisciplinary centers of study such as the Department of Molecular and structural Biochemistry of the North Carolina State University - USA (website: http://biochem.ncsu.edu), study programs like biochemistry, molecular biology and biophysics in undergraduate, masters and doctorate levels have been created. This allows students to familiarize with the technique, including it in their academic activities. For these reasons, molecular docking is not only used in the pharmaceutical industry for drug discovery, but also for anyone interested in the study of biological systems and receptor-ligand interactions.

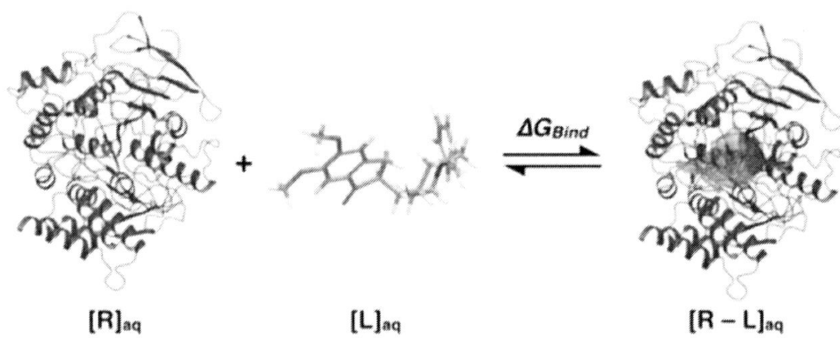

Figure 3. Ligand donepezil binding to acetylcholinesterase (based on the crystalline structure PDB: 1EVE [39]).

3. *DE NOVO* DESIGN

Another computer-based drug design method known as '*de novo* design' have been implemented to find new potent and selective ligands. *de novo* design serves as a tool for the discovery of new ligands for biological targets as well as optimization of new ligands [50]. These tools produce (from scratch) novel molecular structures with desired pharmacological properties [51]; subsequently the new structures are docked to find the binding affinities and interacting modes. This new tools and algorithms identify potent ligand-protein interaction characteristics in the binding site, and construct novel molecular structures by assembling atoms and molecular fragments either combined or sequentially. *de novo* design is a powerful tool, and some authors have reported most potent ligands designed by this tool that their precursors [52], revealing the importance of this tool in the drug design process. Some of these new tools are shown in Table. 1.

Automated *de novo* design is a very useful tool for hit and lead-ligand identification; design molecules, provide ideas for medical chemist and pharmaceutical industries, and to develop novel leads with desired chemical characteristics.

Table 1.*de novo* design tools (taken and adapted from [53]).

Tool	Concept	Reference
Builder	Combinatory search by recombination of docked molecules	[54]
Caveat	Database search for fragment fitting	[55]
Concerts	Fragment-based, stochastic search	[56]
Dynamic ligand design	Atom-based, structure sampling by simulated annealing	[57]
GenStar	Atom-Based, molecules growth based on an enzyme contact model	[58]
GroupBuild	Fragment-based, combinatorial search	[59]
Grow	Peptide design, sequential growth	[60]
Growmol	Fragment-based, sequential growth, stochastic search	[61]
Hook	Linker search for fragments placed by MCSS	[62]
Legend	Atom-based, stochastic search	[63]
LUDI	Fragment-based, combinatorial search	[64]
MCDNLG	Atom-based, stochastic search	[65]
MCSS	Fragment-based, stochastic sampling	[66]
MolMaker	Graph-theoretical 3D design	[67]
NewLead	Fragment-based, builds on 3D pharmacophore-models	[68]
Pro-Ligand	Fragment-based search	[69]
Pro-Select	Fragment-based, scaffold-linker approach	[70]
Skelgen	Small-fragment based, Monte-Carlo search	[71]
SME	Peptide design, whole-molecule optimization	[72]
SMoG	Fragment-based, sequential growth, stochastic search	[73]
Splice	Recombination of ligands retrieved by a 3D database search	[74]
Sprout	Fragment-based, sequential growth, combinatorial search	[75]
Topas	Fragment-based, evolutionary search	[76]

4. VIRTUAL SCREENING

A computational technique widely used in drug design is Virtual Screening (VS). Its main objective is to search for specific information in compounds or molecules libraries with similar structural properties that can acceptably interact with a therapeutic target and is an important tool to access novel drug-like compounds [77]. Prior to the VS, massive molecular docking was used but it had a big computational cost and spent a lot of computational time; due to this, a technique - which enables the efficiency improvement to search for compounds in large molecular libraries - was needed. The use of VS (as well as molecular docking) has increased in recent years, rising from an average of 10 publications reported in the 70s - 80s, to an average of more than 1000 publications reported in previous years (Figure 4).

Virtual screening can be performed through two approaches, Ligand-Based Virtual Screening (LBVS) and Structure-Based Virtual Screening (SBVS). LBVS is used when there is not knowledge about the three-dimensional structure of the therapeutic target (receptor), and yet you have a group of molecules with some biological activity against that target; in that sense, structure relationships can be determined. The molecules group with reported activity can also establish the molecular characteristics. These characteristics allow the interaction between the molecules and the molecular target. Based on those (such as acceptor and donors hydrogen bonds, metals, hydrophobic moieties, aromatic rings among others), researchers can screen databases to find other molecules to fit in the profile of the initial group – with the same molecular descriptors – which can possibly present biological activity against the studied therapeutic target [78]. In the LBSV the lead-ligands generated are ranked based on their similarity score obtained by different methods or algorithms [77].

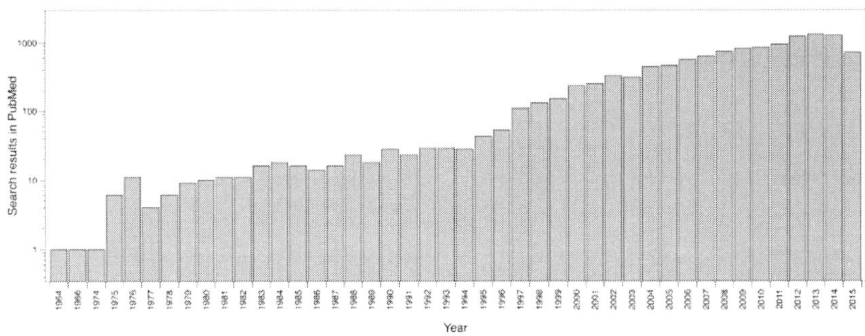

Figure 4. Reported publications number where virtual screening was used. Results from PubMed using the words 'virtual screening' as search criteria.

On the other hand, Structure-Based Virtual Screening (SBVS) establishes its action mechanism in the studied molecular target structure and begins with the identification of the potential ligand-binding site on the target. Then, one of its greatest challenges is to determine the target 3D structure by NMR, X-ray crystallography or molecular modeling. Later, an automated and rapid docking of a large number of chemical compounds against the 3D structure of the molecular target is done. This technique has helped with the identification of potential therapeutic molecules for specific pathologies [77].

To use this technique in an effective way, a protocol with several stages should be established:

1) Ligands Database Preparation

   It is necessary to prepare all the molecules to be studied with the SBVS. Tautomers, isomers, protonation states, ionization states, enantiomers, *etc.* must be considered. This database must be prepared in accordance with the software to use.

2) Receptor Preparation

   Next, the molecular target (Receptor) must be provided. It may uses the same ligands database preparation software.

3) Molecular Docking

   It is necessary to determine the common pharmacophore in the ligand database; after this - and through the appropriate program - massive molecular docking is carried out.

4) Post-Processing

   The post-processing is performed in accordance with the user requirements through the analysis of the score function, geometric analysis, shape complementarity, solvation corrections, entropic changes, pose clustering, *etc.*

5) Compounds Selection

   The final step is the compounds selection for biological tests, or to be further analyzed by other computational techniques [79].

Nowadays, there are several comprehensive programs used to perform VS such as: AutoDock Vina [35], Glide [28, 38, 80], DOCK [81], FlexX [82], GOLD [37], ICM [83], among others. Additionally - and due to the high computational cost for the SBVS implementation - several research groups around the world have developed free software and algorithms to be accessed through virtual platforms. These developers also have the advantage of having extensive updated databases, which allows the virtual screening of millions of molecules. For instance, among the major online servers to perform VS there is the ZINCPharmer, sponsored by the University of Pittsburgh (website: http://zincpharmer.csb.pitt.edu) [84]. This software allows researchers to perform online SBVS for free, mainly based on the established ligand pharmacophore used as input. The screening is basically performed in the ZINC database [7], a non-commercial database with more than 22 million compounds.

In recent years, researchers who use this technique in their computational research have increased, and the establishment of data sets for potential therapeutic agents from VS can be observed, for example: SENP2 inhibitors [85], kinases proteins agonists [86], inhibitors with vasodilator activity [87], inhibitors of HIV-1 reverse transcriptase [88], among others.

The development of these dataset has allowed the access and use of information more efficiently, by evaluating compounds against new molecular targets and finding many potential drug candidates (better than those before the implementation of this technique).

Table 2.Some active compounds identified by computational methods.

Compound structure	Use	Computational method	References
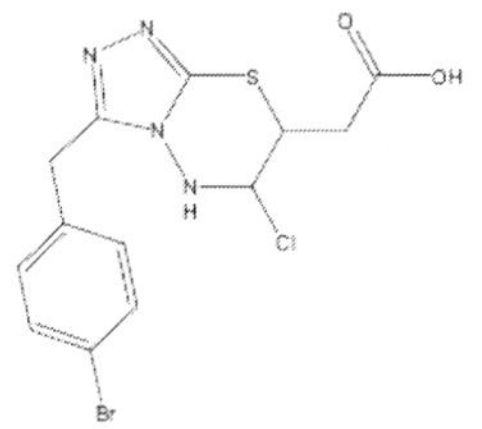	Antimycobacterial agents	Virtual Screening	[89]
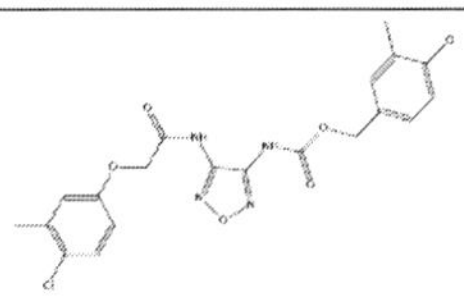	SENP2 inhibitors	Virtual screening, Molecular Docking	[85]
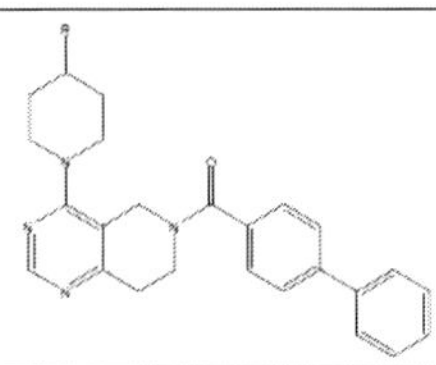	TASK-3 channel antagonist	Pharmacophore based virtual screening	[90]
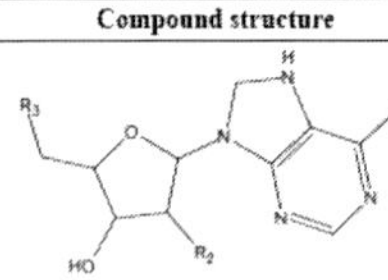	GAPGH inhibitor	Combinatorial docking	[91]
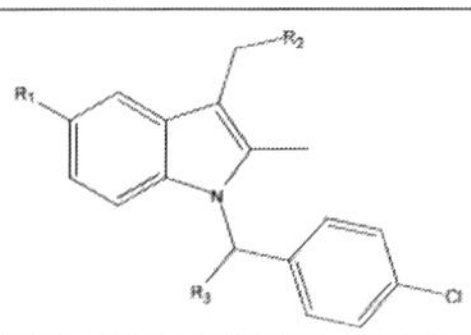	Aldose reductase inhibitor	LBVS	[92]
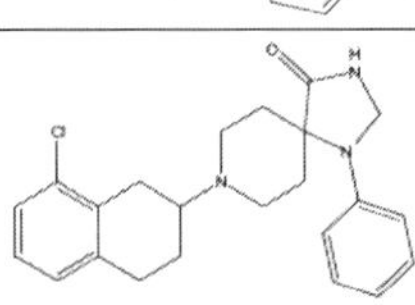	Ca2+ antagonist	Pharmacophore searching	[93]
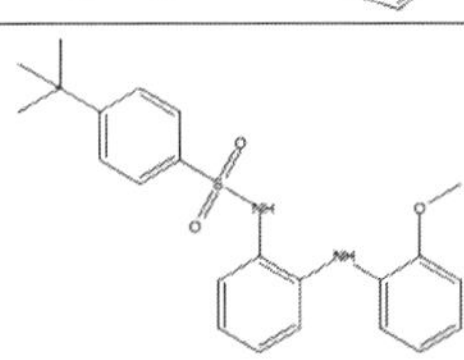	Kv1.5 channel blocker	Fragment based, *de novo* design	[76]

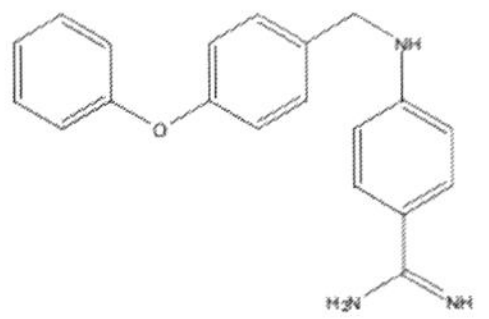	Thrombin inhibitor	Combinatorial docking, *de novo* design	[94]
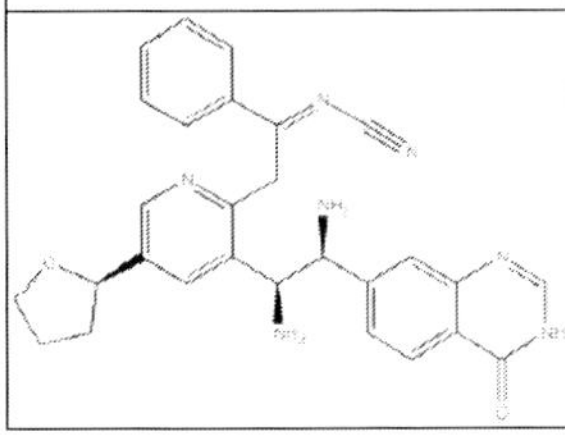	Antiretroviral agent	Virtual Screening	[95]

On the other hand, VS is used as a mechanism for the evaluation of new potential drugs against certain molecular targets in rational drug design. In biochemical terms, this enables the computational evaluation of large molecules quantities in a short time with a relatively low computational cost. It means that after a screening, score, post-processing and visual inspection, it can continue to the laboratory to perform *in vitro* testing with the best candidates obtained *via* VS allowing the properties evaluation such as the biological activity, the enzyme kinetics, among others. This contributes with the biological system study in the drug discovery process.

Several bioactive molecules identified, optimized or designed by using molecular docking, *de novo* design or VS methods (Table. 2) let us understand the importance of the application and use of them as computational tools for drug discovery to explore the chemical space of a binding site and present novel, potent and selective pharmacological alternatives in medical chemistry and drug discovery.

5. CONCLUSION

Although in appearance, a direct relationship between the computational methodologies - molecular docking and virtual screening - with research in biological and medical sciences does not exist, in recent years most of the authors whom develop biological systems research in areas such as biochemistry, medicinal chemistry, pharmacology, organic chemistry, theoretical chemistry, *etc.*, use computational methods to support, improve and reinforce their research. The structural studies of these systems at theoretical level assist with the construction of a wide and accurate perspective of the issue, and enable the knowledge consolidation.

In addition, there is still a long way to develop methodologies with "thinner" approaches (accurate and precise). Although methodologies such as the molecular docking and virtual screening are effective (in terms of time and computational cost), they have also a poor predictive capability. There are other methods involving a higher calculation level for researchers who study dynamic biological systems (Molecular Dynamics), or bonds and interactions in an electronic level (Quantum Mechanics). These like the other methods are more time-consuming in terms of computational cost, but may provide better results; it all depends on the researcher interests and funding. However, to improve computational techniques such as Molecular Docking and Virtual Screening, a change in the structural paradigm is needed, and this will be achieved when the understanding of the interactions between a drug and a receptor improves.

CONFLICT OF INTEREST

The author confirms that this article content has no conflict of interest.

ACKNOWLEDGEMENTS

I thank to Universidad de Talca for doctoral scholarship.

REFERENCES

1. Zheng, M.; Liu, X.; Xu, Y.; Li, H.; Luo, C.; Jiang, H. Computational methods for drug design and discovery: focus on China. Trends Pharmacol. Sci., 2013, 34(10), 549-559.
2. Blundell, T.L.; Jhoti, H.; Abell, C. High-throughput crystallography for lead discovery in drug design. Nat. Rev. Drug Discov., 2002, 1(1), 45-54.
3. Hsiao-lu, D.L.; Sahl, S.J.; Lew, M.D.; Moerner, W.E. The double-helix microscope microscope super-resolves extended biological structures by localizing single blinking molecules in three dimensions with nanoscale precision. Appl. Phys. Lett., 2012, 100(15), 153701.
4. Berman, H.M.; Bhat, T.N.; Bourne, P.E.; Feng, Z.; Gilliland, G.; Weissig, H.; Westbrook, J. The protein data bank and the challenge of structural genomics. Nat. Struct. Mol. Biol., 2000, 7, 957-959.Westbrook, J.; Feng, Z.; Chen, L.; Yang, H.; Berman, H.M. The protein data bank and structural genomics. Nucleic Acids Res., 2003, 31(1), 489-491.
5. Seiler, K.P.; George, G.A.; Happ, M.P.; Bodycombe, N.E.; Carrinski, H.A.; Norton, S.; Brudz, S.; Sullivan, J.P.; Muhlich, J.; Serrano, M. others. Chembank: a small-molecule screening and cheminformatics resource database. Nucleic Acids Res., 2008, 36(Suppl. 1), D351-D359.
6. Irwin, J.J.; Shoichet, B.K. ZINC-a free database of commercially available compounds for virtual screening. J. Chem. Inf. Model., 2005, 45(1), 177-182.

7. Mandal, S.; Moudgil, M.; Mandal, S.K. Rational drug design. Eur. J. Pharmacol., 2009, 625(1), 90-100.
8. De-la-Torre, P.; Treuer, A. V; Gutierrez, M.; Poblete, H.; Alzate-Morales, J. H.; Trilleras, J.; Astudillo-Saavedra, L.; Caballero, J. Synthesis and in silico analysis of the quantitative structure-activity relationship of heteroaryl-acrylonitriles as ache inhibitors J. Taiwan Inst. Chem. Eng. , 2015.
9. Brown, H.C. Foundations of the structural theory. J. Chem. Educ., 1959, 36(3), 104.
10. Wiswesser, W.J. Johann Josef Loschmidt (1821-1895): A forgotten genius. Aldrichim Acta, 1989, 22, 17-19.
11. Metropolis, N.; Rosenbluth, A.W.; Rosenbluth, M.N.; Teller, A.H.; Teller, E. Equation of state calculations by fast computing machines. J. Chem. Phys., 1953, 21(6), 1087-1092.
12. Takahashi, O.; Masuda, Y.; Muroya, A.; Furuya, T. Theory of docking scores and its application to a customizable scoring function. SAR QSAR Environ. Res., 2010, 21(5-6), 547-558.
13. Kitchen, D.B.; Decornez, H.; Furr, J.R.; Bajorath, J. Docking and scoring in virtual screening for drug discovery: Methods and applications. Nat. Rev. Drug Discov., 2004, 3(11), 935-949.
14. Mohan, V.; Gibbs, A.C.; Cummings, M.D.; Jaeger, E.P.; DesJarlais, R.L. Docking: successes and challenges. Curr. Pharm. Des., 2005, 11(3), 323-333.
15. Krovat, E.M.; Steindl, T.; Langer, T. Recent advances in docking and scoring. Curr. Comput. Aided. Drug Des., 2005, 1(1), 93-102.
16. Leach, A.R.; Shoichet, B.K.; Peishoff, C.E. Prediction of protein-ligand interactions. Docking and scoring: successes and gaps.docking and scoring in virtual screening for drug discovery: Methods and applications. J. Med. Chem., 2006, 49(20), 5851-5855.
17. Yuriev, E.; Ramsland, P.a. Latest developments in molecular docking: 2010-2011 in review. J. Mol. Recognit., 2013, 26(5), 215-239.
18. Yuriev, E.; Agostino, M.; Ramsland, P.A. Challenges and advances in computational docking: 2009 in review. J. Mol. Recognit., 2011, 24(2), 149-164.
19. Lee, M.R.; Sun, Y. Improving docking accuracy through molecular mechanics generalized born optimization and scoring. J. Chem. Theory Comput., 2007, 3(3), 1106-1119.
20. Adasme-Carreño, F.; Muñoz-Gutierrez, C.; Caballero, J.; Alzate-Morales, J.H. Performance of the mm/gbsa scoring using a binding site hydrogen bond network-based frame selection: the protein kinase case. Phys. Chem. Chem. Phys., 2014, 16(27), 14047-14058.
21. Mena-Ulecia, K.; Vergara-Jaque, A.; Poblete, H.; Tiznado, W.; Caballero, J. Study of the affinity between the protein kinase pka and peptide substrates derived from kemptide using molecular dynamics simulations and MM/GBSA. PLoS One, 2014, 9(10), e109639.
22. Pak, Y.; Wang, S. Application of a molecular dynamics simulation method with a generalized effective potential to the flexible molecular docking problems. J. Phys. Chem. B, 2000, 104(2), 354-359.
23. Caballero, J.; Alzate-Morales, J.H. Molecular dynamics of protein kinase-inhibitor complexes: a valid structural information. Curr. Pharm. Des., 2012, 18(20), 2946-2963.
24. Elokely, K.M.; Doerksen, R.J. Docking challenge: protein sampling and molecular docking performance. J. Chem. Inf. Model., 2013, 53(8), 1934-1945.
25. McGann, M.R.; Almond, H.R.; Nicholls, A.; Grant, J.A.; Brown, F.K. Gaussian docking functions. Biopolymers, 2003, 68(1), 76-90.
26. McGaughey, G.B.; Sheridan, R.P.; Bayly, C.I.; Culberson, J.C.; Kreatsoulas, C.; Lindsley, S.; Maiorov, V.; Truchon, J-F.; Cornell, W.D. Comparison of topological, shape, and docking methods in virtual screening. J. Chem. Inf. Model., 2007, 47(4), 1504-1519.
27. Friesner, R.A.; Banks, J.L.; Murphy, R.B.; Halgren, T.A.; Klicic, J.J.; Mainz, D.T.; Repasky, M.P.; Knoll, E.H.; Shelley, M.; Perry, J.K. others. Glide: a new approach for rapid, accurate docking and scoring. 1. method and assessment of docking accuracy. J. Med. Chem., 2004, 47(7), 1739-1749.
28. Jones, G.; Willett, P.; Glen, R.C. Molecular recognition of receptor sites using a genetic algorithm with a description of desolvation. J. Mol. Biol., 1995, 245(1), 43-53.
29. Jones, G.; Willett, P.; Glen, R.C.; Leach, A.R.; Taylor, R. Development and validation of a genetic algorithm for flexible docking. J. Mol. Biol., 1997, 267(3), 727-748.

30. Barreca, M.L.; Iraci, N.; De Luca, L.; Chimirri, A. Induced-fit docking spproach provides insight into the binding mode and mechanism of action of HIV-1 integrase inhibitors. Chem. Med. Chem, 2009, 4(9), 1446-1456.
31. Cavasotto, C.N.; Abagyan, R.A. Protein flexibility in ligand docking and virtual screening to protein kinases. J. Mol. Biol., 2004, 337(1), 209-225.
32. Davis, I.W.; Baker, D. Rosettaligand docking with full ligand and receptor flexibility. J. Mol. Biol., 2009, 385(2), 381-392.
33. Shityakov, S.; Förster, C. In silico structure-based screening of versatile p-glycoprotein inhibitors using polynomial empirical scoring functions. Adv. Appl. Bioinform. Chem., 2014, 7, 1.
34. Trott, O.; Olson, A.J. Autodock vina: improving the speed and accuracy of docking with a new scoring function, efficient optimization, and multithreading. J. Comput. Chem., 2010, 31(2), 455-461.
35. Goodsell, D.S.; Morris, G.M.; Olson, A.J. Automated docking of flexible ligands: applications of auto dock. J. Mol. Recognit., 1996, 9(1), 1-5.
36. Verdonk, M.L.; Cole, J.C.; Hartshorn, M.J.; Murray, C.W.; Taylor, R.D. Improved protein-ligand docking using gold. Proteins Struct. Funct. Bioinforma., 2003, 52(4), 609-623.
37. Halgren, T.A.; Murphy, R.B.; Friesner, R.A.; Beard, H.S.; Frye, L.L.; Pollard, W.T.; Banks, J.L. Glide: a new approach for rapid, accurate docking and scoring. 2. Enrichment factors in database screening. J. Med. Chem., 2004, 47(7), 1750-1759.
38. Kryger, G.; Silman, I.; Sussman, J.L. Structure of acetylcholinesterase complexed with E2020 (Aricept®): implications for the design of new anti-alzheimer drugs. Structure, 1999, 7(3), 297-307.
39. Krishnan, K.R.; Charles, H.C.; Doraiswamy, P.M.; Mintzer, J.; Weisler, R.; Yu, X.; Perdomo, C.; Ieni, J.R.; Rogers, S. Randomized; Placebo-Controlled Trial of the Effects of Donepezil on Neuronal Markers and Hippocampal Volumes in Alzheimer's Disease, 2014.
40. Penthala, N.R.; Zong, H.; Ketkar, A.; Madadi, N.R.; Janganati, V.; Eoff, R.L.; Guzman, M.L.; Crooks, P.A. Synthesis, anticancer activity and molecular docking studies on a series of heterocyclic trans-cyanocombretastatin analogues as antitubulin agents. Eur. J. Med. Chem., 2015, 92, 212-220.
41. Subramanian, S.R.; Zuckerman, D.M. Docking to the highly flexible estrogen receptor ligand binding domain via mixed-resolution Monte Carlo. Biophys. J., 2015, 108(2), 319a.
42. Shityakov, S.; Neuhaus, W.; Dandekar, T.; Förster, C. Analysing molecular polar surface descriptors to predict blood-brain barrier permeation. Int. J. Comput. Biol. Drug Des., 2013, 6(1-2), 146-156.
43. Shityakov, S.; Förster, C. In silico predictive model to determine vector-mediated transport properties for the blood-brain barrier choline transporter. Adv. Appl. Bioinform. Chem., 2014, 7, 23.
44. Shityakov, S.; Broscheit, J.; Foerster, C. α-cyclodextrin dimer complexes of dopamine and levodopa derivatives to assess drug delivery to the central nervous system: adme and molecular docking studies.in silico predictive model to determine vector-mediated transport properties for the blood--brain barrier choline transporter Int. J. Nanomedicine, 2012, 7, 3211.
45. Shityakov, S.; Förster, C. Multidrug resistance protein P-Gp interaction with nanoparticles (fullerenes and carbon nanotube) to assess their drug delivery potential: A theoretical molecular docking study. Int. J. Comput. Biol. Drug Des., 2013, 6(4), 343-357.
46. Gao, J.; Chao, L.V. Molecular docking to predict hypopharyngeal cancer treatment drugs: tiliroside-noscapine interaction. Lat. Am. J. Pharm., 2014, 33(8), 1397-1399.
47. Sabogal-arango, A.; Barreto, G.E.; Ramírez-sánchez, D.; González-mendoza, J.; Barreto, V.; Morales, L.; González, J. Computational insights of the interaction among sea anemones neurotoxins and Kv1.3 channel. Bioinform. Biol. Insights, 2014, 8, 73-81.
48. Yi, Z.; Lu, M. Deep understanding of metabolic behaviour of alisol a using molecular docking. Lat. Am. J. Pharm, 2014, 33(3), 522-524.
49. Kutchukian, P.S.; Shakhnovich, E.I. de novo design: balancing novelty and confined chemical space. Expert Opin. Drug Discov., 2010, 5(8), 789-812.

50. Schneider, G.; Fechner, U. Computer-based de novo design of drug-like molecules. Nat. Rev. Drug Discov., 2005, 4(8), 649-663.
51. Chirapu, S.R.; Pachaiyappan, B.; Nural, H.F.; Cheng, X.; Yuan, H.; Lankin, D.C.; Abdul-Hay, S.O.; Thatcher, G.R.; Shen, Y.; Kozikowski, A.P. others. Molecular modeling, synthesis, and activity studies of novel biaryl and fused-ring bace1 inhibitors. Bioorg. Med. Chem. Lett., 2009, 19(1), 264-274.
52. Schneider, G.; Böhm, H.J. Virtual screening and fast automated docking methods. Drug Discov. Today, 2002, 7(1), 64-70.
53. Roe, D.C.; Kuntz, I.D. BUILDER v. 2: Improving the chemistry of a de novo design strategy. J. Comput. Aided Mol. Des., 1995, 9(3), 269-282.
54. Lauri, G.; Bartlett, P.A. CAVEAT: A program to facilitate the design of organic moleculesimproving the chemistry of a de novo design strategy.. J. Comput. Aided Mol. Des., 1994, 8(1), 51-66.
55. Pearlman, D.A.; Murcko, M.A. CONCERTS: Dynamic connection of fragments as an approach to de novo ligand design. J. Med. Chem., 1996, 39(8), 1651-1663.
56. Miranker, A.; Karplus, M. An automated method for dynamic ligand design. Proteins Struct. Funct. Bioinforma., 1995, 23(4), 472-490.
57. Rotstein, S.H.; Murcko, M.A. GenStar: a method for de novo drug design. J. Comput. Aided Mol. Des., 1993, 7(1), 23-43.
58. Rotstein, S.H.; Murcko, M.A. GroupBuild: a fragment-based method for de novo drug design. J. Med. Chem., 1993, 36(12), 1700-1710.
59. Moon, J.B.; Howe, W.J. Computer design of bioactive molecules: a method for receptor-based de novo ligand designan automated method for dynamic ligand design.. Proteins Struct. Funct. Bioinforma., 1991, 11(4), 314-328.
60. Bohacek, R.S.; McMartin, C. Multiple highly diverse structures complementary to enzyme binding sites: results of extensive application of a de novo design method incorporating combinatorial growth. J. Am. Chem. Soc., 1994, 116(13), 5560-5571.
61. Eisen, M.B.; Wiley, D.C.; Karplus, M.; Hubbard, R.E. HOOK: a program for finding novel molecular architectures that satisfy the chemical and steric requirements of a macromolecule binding site. Proteins Struct. Funct. Bioinforma., 1994, 19(3), 199-221.
62. Nishibata, Y.; Itai, A. Automatic creation of drug candidate structures based on receptor structure. Starting point for artificial lead generation. Tetrahedron, 1991, 47(43), 8985-8990.
63. Böhm, H-J. The computer program LUDI: s new method for the de novo design of enzyme inhibitors. J. Comput. Aided Mol. Des., 1992, 6(1), 61-78.
64. Gehlhaar, D.K.; Moerder, K.E.; Zichi, D.; Sherman, C.J.; Ogden, R.C.; Freer, S.T. de novo design of enzyme inhibitors by monte carlo ligand generation. J. Med. Chem., 1995, 38(3), 466-472.
65. Miranker, A.; Karplus, M. Functionality maps of binding sites: A multiple copy simultaneous search method. Proteins Struct. Funct. Bioinforma., 1991, 11(1), 29-34.
66. Clark, D.E.; Firth, M.A.; Murray, C.W. MOLMAKER: de novo generation of 3D databases for use in drug design. J. Chem. Inf. Comput. Sci., 1996, 36(1), 137-145.
67. Tschinke, V.; Cohen, N.C. The NEWLEAD program: A new method for the design of candidate structures from pharmacophoric hypotheses. J. Med. Chem., 1993, 36(24), 3863-3870.
68. Clark, D.E.; Frenkel, D.; Levy, S.A.; Li, J.; Murray, C.W.; Robson, B.; Waszkowycz, B.; Westhead, D.R. PRO_LIGAND: An approach to de novo molecular design. 1. Application to the design of organic molecules. J. Comput. Aided Mol. Des., 1995, 9(1), 13-32.
69. Murray, C.W.; Clark, D.E.; Auton, T.R.; Firth, M.A.; Li, J.; Sykes, R.A.; Waszkowycz, B.; Westhead, D.R.; Young, S.C. PRO_SELECT: Combining structure-based drug design and combinatorial chemistry for rapid lead discovery. 1. Technology. J. Comput. Aided Mol. Des., 1997, 11(2), 193-207.
70. Todorov, N.P.; Dean, P.M. Evaluation of a method for controlling molecular scaffold diversity in de novo ligand design. J. Comput. Aided Mol. Des., 1997, 11(2), 175-192.

71. Schneider, G.; Schrödl, W.; Wallukat, G.; Müller, J.; Nissen, E.; Rönspeck, W.; Wrede, P.; Kunze, R. Peptide design by artificial neural networks and computer-based evolutionary search. Proc. Natl. Acad. Sci. USA, 1998, 95(21), 12179-12184.
72. DeWitte, R.S.; Ishchenko, A.V.; Shakhnovich, E.I. SMoG: de novo design method based on simple, fast, and accurate free energy estimates. 2. Case studies in molecular design. J. Am. Chem. Soc., 1997, 119(20), 4608-4617.
73. Ho, C.M.; Marshall, G.R. FOUNDATION: A program to retrieve all possible structures containing a user-defined minimum number of matching query elements from three-dimensional databases. J. Comput. Aided Mol. Des., 1993, 7(1), 3-22.
74. Mata, P.; Gillet, V.J.; Johnson, A.P.; Lampreia, J.; Myatt, G.J.; Sike, S.; Stebbings, A.L. SPROUT: 3D structure generation using templates. J. Chem. Inf. Comput. Sci., 1995, 35(3), 479-493.
75. Schneider, G.; Lee, M-L.; Stahl, M.; Schneider, P. de novo design of molecular architectures by evolutionary assembly of drug-derived building blocks. J. Comput. Aided Mol. Des., 2000, 14(5), 487-494.
76. Reddy, a, S.; Pati, S. P.; Kumar, P. P.; Pradeep, H. N.; Sastry, G. N. Virtual screening in drug discovery - a computational perspective. Curr. Protein Pept. Sci., 2007, 8(4), 329-351.
77. Koeppen, H.; Kriegl, J.; Lessel, U.; Tautermann, C. S.; Wellenzohn, B. Structure-based virtual screening: an overview. Drug Discov., 2011, 7(20), 1047-1055.
78. Lyne, P.D. Structure-based virtual screening: an overview. Drug Discov. Today, 2002, 7(20), 1047-1055.
79. Friesner, R.A.; Murphy, R.B.; Repasky, M.P.; Frye, L.L.; Greenwood, J.R.; Halgren, T.A.; Sanschagrin, P.C.; Mainz, D.T. Extra precision glide: docking and scoring incorporating a model of hydrophobic enclosure for protein-ligand complexes. J. Med. Chem., 2006, 49(21), 6177-6196.
80. Ewing, T.J.; Makino, S.; Skillman, A.G.; Kuntz, I.D. DOCK 4.0: search strategies for automated molecular docking of flexible molecule databases. J. Comput. Aided Mol. Des., 2001, 15(5), 411-428.
81. Schellhammer, I.; Rarey, M. FlexX-Scan: fast, structure-based virtual screening. PROTEINS Struct. Funct. Bioinforma., 2004, 57(3), 504-517.
82. Abagyan, R.; Totrov, M.; Kuznetsov, D. ICM-a new method for protein modeling and design: applications to docking and structure prediction from the distorted native conformation. J. Comput. Chem., 1994, 15(5), 488-506.
83. Koes, D. R.; Camacho, C. J. ZINC Pharmer: Pharmacophore search of the ZINC database. Nucleic Acids Res., 2012, 40(Web Server issue), W409-W414.
84. Kumar, A.; Ito, A.; Takemoto, M.; Yoshida, M.; Zhang, K.Y. Identification of 1, 2, 5-oxadiazoles as a new class of SENP2 inhibitors using structure based virtual screening. J. Chem. Inf. Model., 2014, 54(3), 870-880.
85. Jain, A.; Trivedi, V. Docking and virtual screening to identify pkc agonists: potentials in anticancer therapeutics. Curr. Comput. Aided. Drug Des., 2014, 10(1), 50-58.
86. Mittal, A.; Paliwal, S.; Sharma, M.; Singh, A.; Sharma, S.; Yadav, D. Pharmacophore based virtual screening, molecular docking and biological evaluation to identify novel PDE5 inhibitors with vasodilatory activity. Bioorg. Med. Chem. Lett., 2014, 24(14), 3137-3141.
87. Hecht, F.M.; Grant, R.M.; Petropoulos, C.J.; Dillon, B.; Chesney, M.A.; Tian, H.; Hellmann, N.S.; Bandrapalli, N.I.; Digilio, L.; Branson, B. others. Sexual transmission of an HIV-1 variant resistant to multiple reverse-transcriptase and protease inhibitors. N. Engl. J. Med., 1998, 339(5), 307-311.
88. Kincaid, V.A.; London, N.; Wangkanont, K.; Wesener, D.A.; Marcus, S.A.; Heroux, A.; Nedyalkova, L.; Talaat, A.M.; Forest, K.T.; Shoichet, B.K. others. Virtual screening for UDP-Galactopyranose mutase ligands identifies a new class of antimycobacterial agents. ACS Chem. Biol., 2015.
89. Coburn, C.; Luo, Y.; Cui, M.; Wang, J.; Soll, R.; Dong, J.; Hu, B.; Lyon, M.; Santarelli, V. P.; Kraus, R. L.; Gregan, Y.; Wang, Y.; Fox, S. V; Binns, J.; Doran, S. M.; Reiss, D. R.; Tannenbaum, P. L.; Gotter, A. L.; Meinke, P. T.; Renger, J. J. Discovery of a pharmacologically active antagonist of the two-pore-domain potassium channel K2P9.1 (TASK-3). Chem. Med. Chem, 2012, 7(1), 123-133.

90. Bressi, J.C.; Verlinde, C.L.; Aronov, A.M.; Le Shaw, M.; Shin, S.S.; Nguyen, L.N.; Suresh, S.; Buckner, F.S.; Van Voorhis, W.C.; Kuntz, I.D. others. Adenosine analogues as selective inhibitors of glyceraldehyde-3-phosphate dehydrogenase of trypanosomatidae via structure-based drug design. J. Med. Chem., 2001, 44(13), 2080-2093.
91. Iwata, Y.; Arisawa, M.; Hamada, R.; Kita, Y.; Mizutani, M.Y.; Tomioka, N.; Itai, A.; Miyamoto, S. Discovery of novel aldose reductase inhibitors using a protein structure-based approach: 3D-database search followed by design and synthesis. J. Med. Chem., 2001, 44(11), 1718-1728.
92. Schneider, G.; Clément-Chomienne, O.; Hilfiger, L.; Schneider, P.; Kirsch, S.; Böhm, H-J.; Neidhart, W. Virtual screening for bioactive molecules by evolutionary de novo design. Angew. Chem. Int. Ed., 2000, 39(22), 4130-4133.
93. Böhm, H-J.; Banner, D.W.; Weber, L. Combinatorial docking and combinatorial chemistry: design of potent non-peptide thrombin inhibitors. J. Comput. Aided Mol. Des., 1999, 13(1), 51-56.
94. Shityakov, S.; Dandekar, T. Lead Expansion and virtual screening of indinavir derivate HIV-1 protease inhibitors using pharmacophoric-shape similarity scoring function. Bioinformation, 2010, 4(7), 295.

CHAPTER 4

Monitoring the Level of ^{14}C-Labelled Selegiline Following Oral Administration

Huba Kalász[1,*], Kornélia Tekes[2], Erzsébet B. Faigl[3], Zita Pöstényi[1], Eszter Berekméri[1], Gellért Karvaly[4], Ernest Adeghate[5]

[1] *Department of Pharmacology and Pharmacotherapy, Semmelweis University, Budapest, Hungary*

[2] *Department of Pharmacodynamics, Semmelweis University, Budapest, Hungary*

[3] *Laboratory of Synthesis, Institute of Isotopes, Budapest, Hungary*

[4] *Department of Laboratory Medicine, Semmelweis University, Budapest, Hungary*

[5] *Department of Anatomy, College of Medicine and Health Sciences, United Arab Emirates University, Al Ain, United Arab Emirates*

ABSTRACT

Background

Selegiline [(-)-deprenyl] is widely used for the treatment of Parkinson's disease in humans.

Objective

Time-dependence of tissue distribution of selegiline following per os administration to rats.

Method

Oral administration of radiolabeled selegiline to rats resulted in a pattern of tissue distribution similar to that following intraperitoneal injection. Analyses were done using both reversed-phase HPLC and also by counting radioactivity in various body compartments of rats.

Results

As a consequence of oral administration of 30 mg/kg of selegiline, its level in the stomach was extremely high (179.57 μg/g tissue through 54.67 μg/g at 15 min to 120 min), that is one magnitude higher than that in the serum level. High selegiline concentrations were also detected in the lacrimal glands (7.45 μg/g), kidneys (6.87 μg/g), livers (6.01 μg/g) and lungs (3.47 μg/g) after 30 minutes of application, which were higher than after intraperitoneal injections.

CONCLUSION

The relatively high tissue levels remained for 120 min monitoring. Selegiline levels in the brain (1.69 μg/g) and in the testes (1.88 μg/g) were also considerably higher than following intraperitoneal administration during the entire period of observation (15 to 120 min).

Keywords: Selegiline, (-)-Deprenyl, Rats, Distribution, HPLC, Oral administration.

1. INTRODUCTION

Selegiline [(-)-deprenyl hydrochloride] (Figure 1) is widely used for the treatment of Parkinson's disease in humans [1]. It is a specific and irreversible inhibitor of the monoamine oxidase B enzyme when used in therapeutic doses [2-4].

Ehringer and Hornykiewicz [5, 6] detected that the dopaminergic tone of the human brain progressively declined with age and also in certain diseases. This phenomenon is characteristic of Parkinson's disease. Selegiline exerts its effect in the dopaminergic system of the brain counterbalancing this decline by multiple mechanisms. Selectively and irreversibly inhibiting the MAO enzyme, selegiline is responsible for the inhibition of the degradation of dopamine [3], moreover, selegiline inhibits the high-affinity reuptake of biogenic amines (such as dopamine) [7, 8] and also facilitates dopamine release from the *corpus striatum* [9]. Acute selegiline treatment produces either no or only moderate effect [10]. Following chronic administration, a definite increase in both the dopamine content and the potassium-induced dopamine release in the brain tissue could be observed [9].

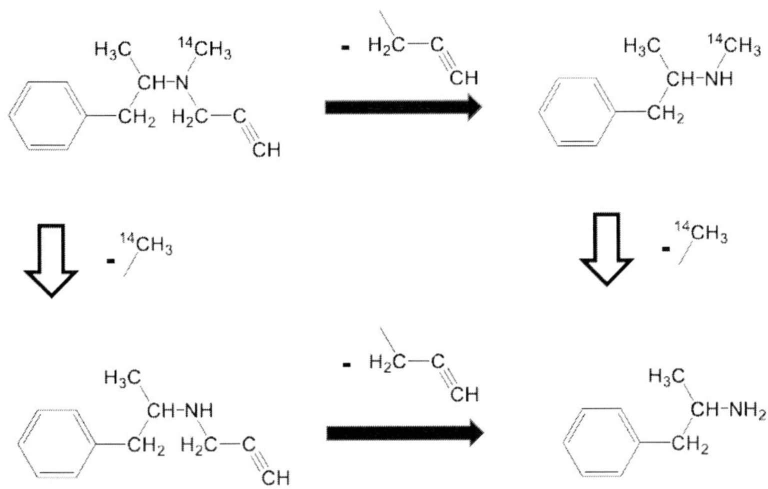

Figure 1. The chemical structure of selegiline (top left) and its metabolic pathway.

In addition to the therapeutic use of selegiline in Parkinsonism, its further beneficial effects have also been published. Knoll, *et al.* [11-13] found an increased life span of rats as a result of chronic treatments with selegiline and p-fluoroselegiline [14]. Independent research groups [15-17] confirmed Knoll's results, and claimed that longevity of rats was originated from the inhibition of oxidative enzymes by selegiline [17, 18]. Knoll and Dalló published the positive correcting effect of selegiline and p-fluoroselegiline on sluggish sexual behavior of male rats [14, 19].

Definite incorporation during drug distribution takes place in silent binding sites (such as serum proteins) [20, 21] and active binding sites, such as receptors and various receptor-like binding sites, such as special sites including enzymes [22]. Active binding sites can be the target of drug for drug discovery [23]. The effect of drugs serves to locate their binding site(s). Our recent publications definitely suggested incorporation of selegiline in the brain and testes following i.p. administration [24, 25]. Similarly, deuterated fluoroselegiline showed incorporation and served for PET studies of monoamino oxidase B in the cynomolgus monkey brain [26]. In pharmacological experiments on the distribution and metabolism of selegiline, intravenous and intraperitoneal injections are administered in order to avoid the first-pass effect of the liver. However, the majority of human treatments prefer oral administration to injection if it is not contraindicated. This paper is devoted to determine time-dependence of tissue distribution of selegiline following per os administration to rats.

2. MATERIALS AND METHODS

2.1. Materials

2.1.1. Chemicals and Solvents

Selegiline hydrochloride (Figure 1) was the kind gift of Chinoin Pharmaceutical and Chemical Company, Budapest, Hungary (its recent name is: Chinoin Sanofi-Aventis, Budapest, Hungary). Radiolabeled selegiline (^{14}C-methyl) was synthetized in the Isotope Institute (Budapest, Hungary) with a specific activity of 1.851 MBq/mg. Sodium citrate, disodium phosphate, octane sulfonic acid sodium, Na_2-EDTA, hydrogen peroxide and HPLC grade acetonitrile were purchased from Sigma-Aldrich Kft. (Budapest, Hungary).

Soluene-350 and Optiphase scintillation cocktails were purchased from Perkin Elmer (PER-FORM Hungária Kft, Budapest, Hungary).

2.2. Animals

Male Wistar rats were supplied by Toxicoop (Budapest, Hungary) weighing 195 – 205 gram. The experimental protocol was approved by the ethical committee of ANTSz (Budapest, Hungary) with a permission number 1810/003/2004. All experimental conditions conformed to 86/509/EEC regulation.

Rats (in each group n = 3) were orally (per os) treated with 30 mg/kg ^{14}C-labeled selegiline dissolved in saline. The rats were sacrificed following 15, 30, 60 and 180 minutes of treatment. Blood, CSFs, urine, faeces, the brains, lacrimal glands, salivary glands, testes, kidneys, livers, lungs, hearts, stomachs, eyes and cochleas were isolated or dissected. The samples were kept at -80 ºC until analysis.

2.2.1. Sample Preparation

The samples were homogenized in 0.3 M of perchloric acid using Ultra Turrax T25 homogenizer (IKA Labortechnik, Staufen, Germany) at 20,000 rpm for 30 sec at room temperature. Homogenates were centrifuged at 14,000 rpm for 20 min at 4 ºC in an Eppendorf centrifuge, (A. Hettich, Tuttlingen, Germany) and supernatants were used for selegiline determination by RP-HPLC.

Calibration curve was constructed with triple injections of 6 spiked standard samples from 0.1 μg/ml through 5 μg/ml selegiline (Figure 2). Selegiline levels in the samples were calculated by the mean of the results of triple injections.

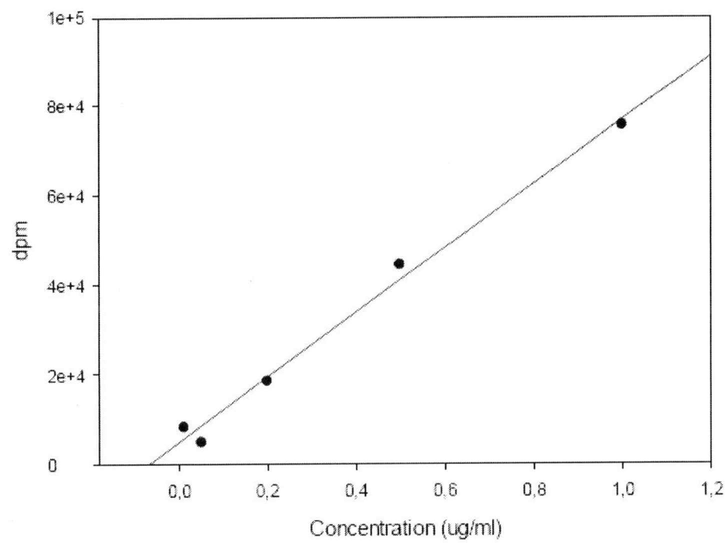

Figure 2. Calibration curve for determination of radioactivity.

2.2.2. HPLC Determination of Selegiline Content

The set-up for HPLC was purchased from ABL&E Jasco (Budapest, Hungary). It consisted of a JASCO DG-2080-54 Degasser, JASCO PU-1580 Pump, JASCO AS-2057 Plus Automatic Sample Injector, an Intro EC detector (Antec, Leyden, the Netherlands), a JMBS Hercule 2000 Chromatography Interface and a Borwin 1.50 Controller and Evaluation Software.

The stationary phase was a Zorbax Rx-C18 (4.6 mm x 250 mm x 5 μm) column purchased from Kromat Kft (Budapest, Hungary).

The mobile phase was a 2:8 (v/v) mixture of acetonitril: aqueous buffer at pH 3.7, while the buffer was the solution of 10 g citric acid, 10 g trisodium phosphate, 100 mg OSA, dissolved in 600 ml HPLC quality of water, which was completed to 800 ml. Temperature of the column was kept at 35 ºC and analyses were done using 1 ml/min flow rate. An injection volume of 75 µl was applied.

An Intro EC detector using amperometric detection mode at 0.95 V was used and the total range was 2 nA.

2.3. Determination of Radioactivity in Various Samples

2.3.1. Treatment of Tissue Samples

About 100 mg of each liver, brain, lacrimal glands, salivary gland and testes were transferred into 20 ml scintillation vials. 1.5 ml Soluene-350 was added to each sample, kept at a 60 ºC water bath for 90 minutes and was then cooled down to room temperature. Afterwards, 0.2 ml of 30% hydrogen peroxide was given in two portions (that is twice 0.1 ml) and was shaken well after each portion. The samples were kept at room temperature for 30 minutes, followed by keeping at 60 ºC for 30 minutes. After cooling down to room temperature, 10 ml scintillation cocktail was given to each sample and radioactivity counting was performed after 12 hours in a Beckman LS-5000TA counter. The treatment of serum and urine samples was similar except that 50 µl of the samples was taken. Moreover, 0.5 ml isopropanol was also given to the samples following treatment with 30% hydrogen peroxide and provision of the scintillation cocktail.

3. RESULTS

3.1. Calibration for ^{14}C-Deprenyl Content

Calibration curve for the determination of radioactivity is given in Figure (2). Figure (3) gives an HPLC determined calibration curve of non-metabolized selegiline. Figure (4) triplet gives a representative chromatogram of selegiline in testis. Figure (4a) shows the chromatogram of blank (untreated) testis. Figure (4b) shows the chromatogram of testis spiked with selegiline and Figure (4c) gives the chromatogram of testis from a selegiline-treated rat.

Table 1 provides the results of radioactivity determinations in the tissues and body fluids. Radioactivity is expressed in µg/g or µg/ml, based on dpm/g wet tissue or dpm/ml body fluid. Radioactivity indicates the presence of the metabolically intact selegiline plus its major metabolite and the level of (-)-methamphetamine (Figure 1).

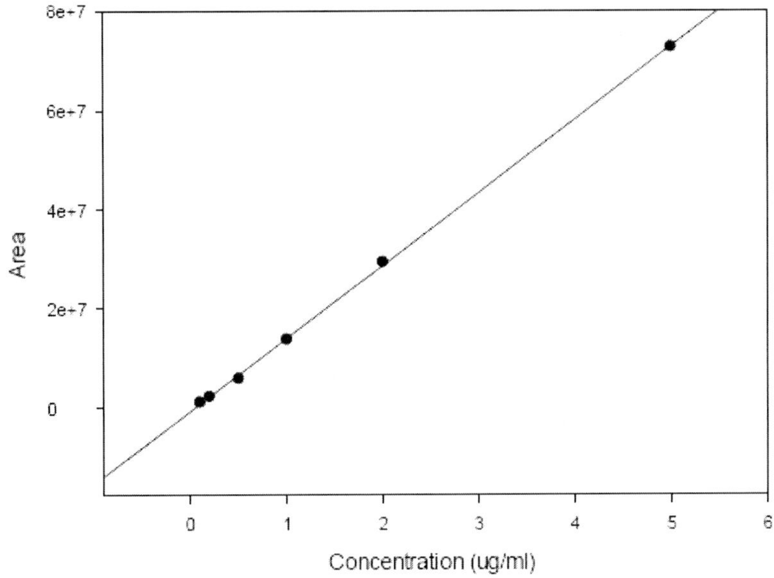

Figure 3. Calibration curve for HPLC determination of selegiline.

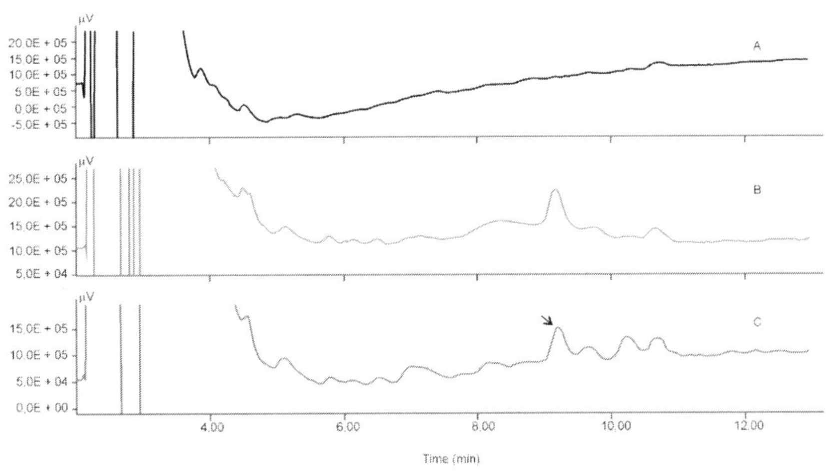

Figure 4. Representative chromatograms for HPLC determination of selegiline in testes.

Table 1. Selegiline (and methamphetamine) content of various body compartments of rats following p.o. Treatment with ^{14}c-labeled selegiline. All data are calculated using the calibration curve and from determination of radioactivity.

Tissue or Body Fluid	Selegiline (or methamphetamine) level, μg/ml or μg/g following 15 min	Selegiline (or methamphetamine) level, μg/ml or μg/g following 30 min	Selegiline (or methamphetamine) level, μg/ml or μg/g following 60 min	Selegiline (or methamphetamine) level, μg/ml or μg/g following 120 min
Serum	7.348	3.844	5.123	8.541
Brain	5.842	2.258	2.234	4.266
Cochlea	3.040	3.793	4.825	6.571
Eyes	2.123	1.364	2.850	2.245
Lacrimal gland	15.172	7.617	9.636	19.284
Salivary gland	5.764	3.332	5.080	6.414
Testis	3.264	2.354	2.259	3.923
Kidney	14.461	7.277	6.489	12.392
Liver	49.323	26.842	32.208	81.143
Lung	11.051	4.495	5.282	8.444
Heart	5.678	2.879	2.694	4.978
Stomach	263.229	98.048	106.372	77.322
Urine	---	474.172	90.512	769.814
Faeces	1.240	0.809	0.782	3.727

Table 2. On-metabolized selegiline content of various body compartments of rats following p.o. Treatment with [14]c-labeled selegiline. All data are calculated using the hplc.

Tissue or Body Fluid	Selegiline level, µg/ml or µg/g following 15 min	Selegiline level, µg/ml or µg/g following 30 min	Selegiline level, µg/ml or µg/g following 60 min	Selegiline level, µg/ml or µg/g following 120 min
Serum	2.06	1.51	1.66	2.20
Brain	1.69	1.45	1.65	1.54
CSF	1.50	1.96	1.64	1.53
Cochlea	1.25	1.18	1.37	1.83
Eyes	1.92	1.85	3.82	5.68
Lacrimal glands	9.88	7.45	3.75	5.22
Salivary glands	2.27	2.49	1.90	2.24
Testis	2.01	1.88	1.86	2.03
Kidney	9.81	6.87	2.48	4.60
Liver	6.61	6.01	2.51	2.33
Lung	4.56	3.47	3.62	3.99
Heart	2.26	1.85	1.89	2.22
Stomach	179.57	76.16	89.69	54.67

Table 2 provides the non-metabolized selegiline content of body fluids and tissues of rats on the basis of HPLC determinations. Triplicate determinations were carried out. The tissue level of selegiline in the stomachs was extremely high at each sample (15, 30, 60 and 120 mins) as selegiline was administered orally and the decline during the experimental period indicates its transfer into the ileum. High contents of selegiline were detected in the lacrimal glands, the kidneys, the livers, and in the lungs at all of the time points detected.

Pharmacokinetics shows the upmost increase of selegiline in the livers, as well as a definite (but moderate) increase in the lacrimal glands, cochleas and kidneys.

4. DISCUSSION

The chemical structure of radiolabeled selegiline (Figure 1) shows that the N-methyl group contains the source of radioactivity in the form of ^{14}C. However, the major metabolite of selegiline is methamphetamine which maintains the radioactivity keeping the ^{14}C-methyl group as well [27, 28].

In Ehrlich's beloved Latin terminology: *Corpora non agunt nisi fixata* (agents only work when they are bound) [29]. The key–lock principle led to the side-chain theory of antibody formation, which became a cornerstone of the newly-emerging discipline of immunology recognized by the Nobel Prize. Ehrlich formulated his ideas in chemical terms, speaking of "chemical affinities" in biological processes (*e.g.* toxin/antitoxin or antigen/antiserum complexes, "magic bullets" seeking specific targets). Some three decades of his experience were epitomized in his maxim, "*Corpora non agunt nisi fixata*, *i.e.* a substance is not (biologically) active unless it is "fixed" (bound by a receptor)". Penetration of selegiline into the testis is thereby of basic importance from the point of view of MAO enzyme inhibition.

Roberge, *et al.* [30] reported the dependence of activity of human seminal plasma on its level of monoamine oxidase. As early as thirty years ago, an observation was made according to which the MAO enzyme was present in the seminal plasma and a higher MAO activity was detected in infertile men than in fertile ones. Monoamine oxidase (MAO) content and activity in the testes show changes during sexual differentiation [31, 32]. MAO inhibitors affect the copulatory behavior of male rats [33]. At the same time, there is a blood-testis barrier and its function is to select penetration of both xenobiotics and endogenous compounds [34-37].

A recent publication by Mihalik, *et al.* [38] claims the favorable effect of selegiline on the increase in the number and motility of sperms of male rats following chronic intraperitoneal treatment with selegiline.

Pharmacokinetics of orally given selegiline are subjected to the alterating influence of its first-pass effect. Oral administration of various drugs used in humans is generally preferred to any other (i.p., i.v., s.c.) except in the special cases. This is the reason why oral treatments are carried out. However, pharmacology can use either i.p. or i.v. administration to avoid the first-pass effect of the drug. This method has been used and reported in our preliminary publications [9, 24, 25, 27]. However, to monitor the pharmacokinetics related to the use of selegiline in humans, oral administration should be applied. Certain fluctuations in the selegiline levels in the serum, brain, etc. must be the formation and completion/saturation of the first-pass effect of the liver.

5. CONCLUSION

A high deprenyl level was found in the stomachs, kidneys, lacrimal glands, livers and lungs after 15 minutes of administration, indicating the site of absorption as well as the major excretion routes. The relatively constant selegiline levels in the brain and testes show the sites where selegiline is bound by the monoamine oxidase enzyme. This interaction can cause an inhibition of this enzyme resulting in an elevated number and motility of sperms.

CONFLICT OF INTEREST

The authors confirm that this article content has no conflict of interest.

ACKNOWLEDGEMENTS

Kalász Teaching and Research Co (Budapest, Hungary) and the Hungarian National Granting Agency have financially sponsored this project with OTKA No. 100155 grant. Cooperation with and advice by Dr. Tibor Zelles and Mr. János Horváth are highly appreciated.

REFERENCES

1. Standaert, D.G.; Roberson, E.D. Goodman & Gilman's The Pharmacological Basis of Therapeutics, 12th ed; Brunton, L.; Chabner, B.; Knollman, B., Eds.; McGraw Hill Medical, 2011, pp. In: 609-628.

2. Ecseri, Z.; Kósa, I.; Knoll, J.; Somfai, É. Process to develop a new drug containing optically active derivative of phenyl isopropylamine. Hangarian Patent, 154.655., 1965.
3. Magyar, K.; Knoll, J. Selective inhibition of the B form of monoamine oxidase. Pol. J. Pharmacol. Pharm., 1977, 29(3), 233-246.
4. Knoll, J.; Ecsery, Z.; Magyar, K.; Sátory, E. Novel (-)deprenyl-derived selective inhibitors of B-type monoamine oxidase. The relation of structure to their action. Biochem. Pharmacol., 1978, 27(13), 1739-1747.
5. Ehringer, H.; Hornykiewicz, O. Verteilung von Noradrenalin und Dopamin (3-Hydroxytyramin) im Gehirn des Menschen und ihr Verhalten bei Erkrankungen des extrapyramidalen Systems. Klin. Wochenschr., 1960, 38, 1236-1239.
6. Hornykiewicz, O. The discovery of dopamine deficiency in the parkinsonian brain. J. Neural Transm. Suppl., 2006, (70)(Suppl.), 9-15.
7. Tekes, K.; Magyar, K. Effect of MAO inhibitors on the high-affinity reuptake of biogenic amines in rat subcortical regions. Neurobiology (Bp.), 2000, 8(3-4), 257-264.
8. Magyar, K.; Szende, B.; Lengyel, J.; Tekes, K. The pharmacology of B-type selective monoamine oxidase inhibitors; milestones in (-)-deprenyl research. J. Neural Transm. Suppl., 1996, 48(48)(Suppl.), 29-43.
9. Kalász, H.; Kerecsen, L.; Knoll, J.; Pucsok, J. Chromatographic studies on the binding, action and metabolism of (-)-deprenyl. J. Chromatogr. A, 1990, 499, 589-599.
10. Hársing, L.G., Jr; Vizi, E.S. Release of endogenous dopamine from rat isolated striatum: effect of clorgyline and (-)-deprenyl. Br. J. Pharmacol., 1984, 83(3), 741-749.
11. Knoll, J. Sexual performance and longevity. Exp. Gerontol., 1997, 32(4-5), 539-552.
12. Knoll, J. The striatal dopamine dependency of life span in male rats. Longevity study with (-)deprenyl. Mech. Ageing Dev., 1988, 46(1-3), 237-262.
13. Knoll, J. The pharmacology of selegiline ((-)deprenyl). New aspects. Acta Neurol. Scand. Suppl., 1989, 126, 83-91.
14. Dalló, J.; Knoll, J. Effect of (-)-para-fluoro-deprenyl on survival and copulation in male rats. Acta Physiol. Hung., 1992, 79(2), 125-129.
15. Kitani, K.; Kanai, S.; Sato, Y.; Ohta, M.; Ivy, G.O.; Carrillo, M.C. Chronic treatment of (-)deprenyl prolongs the life span of male Fischer 344 rats. Further evidence. Life Sci., 1993, 52(3), 281-288.
16. Kitani, K.; Kanai, S.; Ivy, G.O.; Carrillo, M.C. Assessing the effects of deprenyl on longevity and antioxidant defenses in different animal models. Ann. N. Y. Acad. Sci., 1998, 854, 291-306.
17. Kitani, K.; Kanai, S.; Ivy, G.O.; Carrillo, M.C. Pharmacological modifications of endogenous antioxidant enzymes with special reference to the effects of deprenyl: a possible antioxidant strategy. Mech. Ageing Dev., 1999, 111(2-3), 211-221.
18. Budni, P.; de Lima, M.N.; Polydoro, M.; Moreira, J.C.; Schroder, N.; Dal-Pizzol, F. Antioxidant effects of selegiline in oxidative stress induced by iron neonatal treatment in rats. Neurochem. Res., 2007, 32(6), 965-972.
19. Yen, T.T.; Dalló, J.; Knoll, J. The aphrodisiac effect of low doses of (-) deprenyl in male rats. Pol. J. Pharmacol. Pharm., 1982, 34(5-6), 303-308.
20. Müller, W.E.; Wollert, U. Human serum albumin as a silent receptor for drugs and endogenous substances. Pharmacology, 1979, 19(2), 59-67.
21. Szökö, E.; Kalász, H.; Kerecsen, L.; Magyar, K. Binding of (-) deprenyl to serum proteins. Pol. J. Pharmacol. Pharm., 1984, 36(4), 413-421.
22. Rang, H.P.; Ritter, J.M.; Flower, R.J.; Henderson, G. Rang and Dale's Pharmacology; Rang, H.P.; Dale, M.M.; Ritter, J.M.; Flower, R.J., Eds.; Elsevier Science B.V: Amsterdam, 2016, pp. 482-497.
23. Waldman, S.A.; Terzic, A. Pharmacology and Therapeutics - Principles to Practice; Saunders: Philadelphia, 2009, p. 7.
24. Tekes, K.; Pöstényi, Z.; Faigl, E.B.; Magyar, K.; Polyák, A.; Trencsényi, G.; Balogh, L.; Kalász, H. Distribution of N-methyl-(14)C-labeled selegiline in the rat. J. Pharm. Biomed. Anal., 2015, 111, 147-152.

25. Kalász, H.; Tekes, K.; Pöstényi, Z.; Proszenikov, A.; Vizvári, E.; Tóth-Molnár, E. Pharmacokinetics of selegiline in a rabbit model. Lett. Drug Des. Discov., 2016, 13, 752-756.
26. Nag, S.; Varrone, A.; Tóth, M.; Thiele, A.; Kettschau, G.; Heinrich, T.; Lehmann, L.; Halldin, C. In vivo evaluation in cynomolgus monkey brain and metabolism of [18F]fluorodeprenyl: a new MAO-B pet radioligand. Synapse, 2012, 66(4), 323-330.
27. Kalász, H.; Magyar, K.; Szőke, É.; Adeghate, E.; Adem, A.; Hasan, M.Y.; Nurulain, S.M.; Tekes, K. Metabolism of selegiline [(-)-deprenyl)]. Curr. Med. Chem., 2014, 21(13), 1522-1530. [(-)-deprenyl)].
28. Szathmari, I. Milestones in Monoamine Oxidase Research: Discovery of (-)-Deprenyl; Magyar, K.; Vizi, E.S., Eds.; Medicina Publishing House Co.: Budapest, 2000, pp. 61-80.
29. Ehrlich, P. Chemotherapics, Scientific principles, methods, and results. Lancet, 1913, 2, 445-451.
30. Roberge, A.G.; Moufarege, A.; Lavoie, J.; Roberge, C.; Tremblay, R.R. Biochemical properties and kinetic parameters of monoamine oxydase in human seminal plasma. Int. J. Fertil., 1984, 29(3), 180-185.
31. Gaziri, L.C.; Ladosky, W. Monoamine oxidase variation during sexual differentiation. Neuroendocrinology, 1973, 12(4), 249-256.
32. Urry, R.L.; Dougherty, K.A.; Cockett, A.T. Age-related changes in male rat reproductive organ weights and plasma testosterone concentrations after administration of a monoamine oxidase inhibitor. Fertil. Steril., 1976, 27(11), 1326-1334.
33. Dewsbury, D.A.; Davis, H.N., Jr; Jansen, P.E. Effects of monoamine oxidase inhibitors on the copulatory behavior of male rats. Bull. Menninger Clin., 1972, 36(1), 209-217.
34. Mruk, D.D.; Cheng, C.Y. The mammalian blood-testis barrier: its biology and regulation. Endocr. Rev., 2015, 36(5), 564-591.
35. Gerber, J.; Heinrich, J.; Brehm, R. Blood-testis barrier and Sertoli cell function: Lessons from SCCx43KO mice. Reproduction, 2015. pii: REP-15-0366. [Epub ahead of print]
36. Alves, M.G.; Dias, T.R.; Silva, B.M.; Oliveira, P.F. Metabolic cooperation in testis as a pharmacological target: from disease to contraception. Curr. Mol. Pharmacol., 2014, 7(2), 83-95.
37. Jiang, X.H.; Bukhari, I.; Zheng, W.; Yin, S.; Wang, Z.; Cooke, H.J.; Shi, Q.H. Blood-testis barrier and spermatogenesis: lessons from genetically-modified mice. Asian J. Androl., 2014, 16(4), 572-580.
38. Mihalik, J.; Mašlanková, J.; Solár, P.; Horváthová, F.; Hubková, B.; Almášiová, V.; Šoltés, J.; Švaňa, M.; Rybárová, S.; Hodorová, I. The effect of R-(-)-deprenyl administration on reproductive parameters of rat males. Eur. J. Pharmacol., 2015, 754, 148-152.

CHAPTER 5

Synthesis and Evaluation of Antituberculosis Activity of Substituted 2,7-Dimethylimidazo [1,2-a]Pyridine-3-Carboxamide Derivatives

Bhagwat Jadhav[1], R. Kenny[2], Y. Nivid[3], Mustapha Mandewale[1], Ramesh Yamgar[2*]

[1]*P.G. and Research Centre, Department of Chemistry, Government of Maharashtra, Ismail Yusuf College of Arts, Science and Commerce, Mumbai, India*

[2]*Department of Chemistry, Chikitsak Samuha's Patkar-Varde College of Arts, Science and Commerce, Mumbai, India*

[3]*Terna Medical College, Mumbai, India*

ABSTRACT

A series of substituted 2,7-dimethylimidazo[1,2-a]pyridine-3-carboxamides derivatives 5a-5m were synthesized through multi-step reactions. To achieve the synthesis of the desired compounds monobromo and dibromo substituted 2-amino-γ-picoline was reacted with ethyl 2-chloroacetoacetate. The crude ethyl ester subjected to hydrolysis in presence of lithium hydroxide to get 2a and 2b, with imidazo[1,2-a]pyridine-3-carboxylic acid to get 3a-3b, on treatment with substituted amines 4a-4g to get desired product 5a-5m in presence of EDCI and HOBt. The substituted imidazo[1,2-a]pyridine-3-carboxamides are characterized by

FTIR, ^{1}H-NMR, ^{13}C-NMR and mass spectra. These newly synthesized compounds were tested in vitro for their antimycobacterial activity. The preliminary results of antituberculosis study showed that most of the synthesized compounds 5a-5m demonstrated moderate to good antituberculosis activity. Among the tested compounds 5b, 5d and 5e were found to be the most active with minimum inhibitory concentration (MIC) of 12.5 μg/mL against Mycobacterium tuberculosis (H37 RV strain) ATCC No-27294.

Keywords: Carboxamides, Imidazo[1,2-a]Pyridine, Tuberculosis

1. INTRODUCTION

The imidazo[1,2-a]pyridines are found to possess various potential biological activities. A large number of substituted imidazo[1,2-a]pyridines have been synthesized and tested for various biological activities such as antiulcer [1] , antibacterial [2] , antimicrobial [3] , antifungal [4] , and antiviral [5] [6] agents. Various imidazo fused heterocycles with aryloxy alkylamines side chain including imidazo[1,2-a]pyridines act as calcium channel blockers or as local anaesthetics [7] . Byth and co-workers [8] have synthesized various pyrimidine sulfonamide substituted imidazo[1,2-a]pyridine as cyclin- dependent kinase (CDK) inhibitors which led to the identification of potent and selective inhibitors of CDK2 and CDK1. Several 8-arylimidazo[1,2-a]pyridines have been synthesized and evaluated as cardiotonic agents [9] . Some highly substituted imidazo[1,2-a]pyridines have been substituted as gastric antisecretory agents useful as gastric antiulcer agents [10] . Zhonghui Lu, Gregory R. Ott. and co-worker [11] [12] have synthesized a series of 2,3-substituted imidazo[1,2-a]pyridine derivatives which resemble (3S, 4S)pyrrolidine-N-hydroxy-3-carboxamide and (3R, 4R)pyran-N-hydroxy-3-car- boxamide has been reported to have potent selectivity against tumor necrosis factor converting enzyme (TACE). The synthesis and tuberculosis activity of various substituted imidazo[1,2-a]pyridine-3-carboxamides have been reported [13] . Katherine A. Abrahams et al. [14] described the synthesis and biological activity of imidazo[1,2-a] pyridine-3-carboxamides against Mycobacterium tuberculosis. Garrett C. Moraski et al. synthesized a series of nine 2,7-dimethylimidazo[1,2-a]pyridine-3-car- boxamides and one 2,6-dimethylimidazo[1,2-a]pyrimidine-3-carboxamide and evaluated for their in vitro antituberculosis activity against replicating, nonreplicating, multi- and extensive drug resistant Mtb strains [15] . Linhu Li et al. [16] have developed the synthesis and antituberculosis activity of various imidazo[1,2-a]pyridine-3-car- boxamide hybrids with extended amine functionality. Recently Samala G. et al. [17] synthesized 2-methy-limidazo[1,2-a]-pyridine-3-carboxamides as Mycobacterium tuberculosis Pantothenate synthetase inhibitors.

In continuation of our research for new antimicrobial agents [18] [19] [20] [21] [22] , we have undertaken research studies on synthesis and biological screening of some new imidazo[1,2-a]pyridine derivatives.

2. MATERIALS AND METHODS

Melting points of compounds were determined in open capillary tubes in silicon oil bath using a Veego melting point apparatus and are uncorrected. ^{1}H-NMR and ^{13}C-NMR spectra were acquired on Bruker (400 MHz), chemical shifts were given on the delta scale as parts per million (ppm) with tetramethylsilane (TMS) as the internal standard. Electron impact ionization mass spectra were recorded on Agilent Technologies 5975C MSD detector at 70 eV. Infrared spectra were measured with KBr pellet on a FTIR-7600 Lambda Scientific Pvt. Ltd. in the range 4000 - 400 cm^{-1}. Thin layer chromatography (TLC) was performed on silica F254 coated aluminum plates (Merck) as adsorbent and the spots were visualized using ultraviolet light or iodine chamber.

3. CHEMISTRY

We synthesized imidazo[1,2-a]pyridine-3-carboxamide derivatives and evaluated them for Antituberculosis Activity. Figure 1 envisages the schematic representation for the synthesis of (un)substituted 2,7-dimethylimidazo[1,2-a]pyridine-3-carboxamides (5a- 5m). For the synthesis of the desired molecules, mono bromo and dibromo substituted 2-amino-γ-picoline (1a-1b) were reacted with ethyl 2-chloroacetoacetate in presence of base to give ethyl imidazo[1,2-a]pyridine-3-carboxylates (2a-2b). These esters were then hydrolized to give imidazo[1,2-a]pyridine-3-carboxylic acids (3a-3b). These acid derivatives were treated with (un)substituted anilines (4a-4f) using hydroxybenzotriazole (HOBt) and EDCI as coupling agent to afford (un)substituted 2,7-di-methylimi-dazo[1,2-a]pyridine-3-carboxamides (5a-5m).

The FT-IR spectrum of hydrazone 5a, as a representative example, showed strong absorption bands at 1654 and 1595 cm^{-1} due to amide, C=O stretching and N-H bending respectively. The broad peak at 3138 cm^{-1} is attributed to amide N-H stretching. Its ^{1}H-NMR spectrum revealed, in addition to expected aromatic signals, three singlets at δ 2.69, 2.87, 9.67

ppm are assignable to the two methyl groups and amide proton (-NH-C=O), respectively. In addition, the ^{13}C-NMR spectrum of 5a displayed characteristic peak at δ 158.5 ppm assignable to carbonyl carbon of amide. Moreover the EIMS spectrum of 5a revealed molecular ion peak at m/z 423.9 (M+) corresponding to molecular formula [$C_{16}H_{13}Br_2N_3O$]. In a similar manner, compounds 5b-5m were prepared and characterized as shown in Table 1 and Table 2.

4. EXPERIMENT

4.1. Preparation of Ethyl 6-Bromo-2,7-Dimethylimidazo[1,2-a] Pyridine-3-Carboxylate (2a)

A mixture of 5-bromo-4-methylpyridin-2-amine (10.0 g, 53.5 mmol) and ethyl 2-chlo- roacetoacetate (8.13 ml, 58.8 mmol) were taken in 1,2-dimethoxyethane (107 mL) and heated to reflux for 16 hours. The reaction mixture was concentrated under reduced pressure to get crude compound.

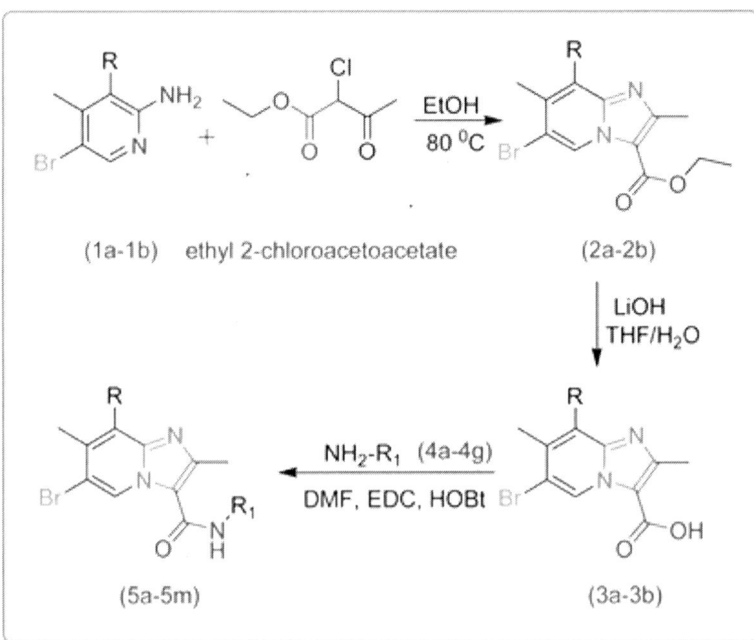

Figure 1. Reaction scheme for preparation of compounds 5a-5m.

Table 1. Structures of compounds 5a-5g.

Entry	Amine [4]	Product [5]
a	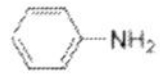	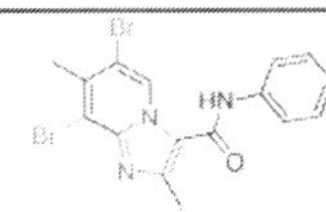
b		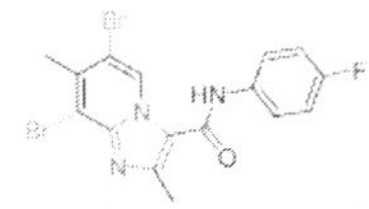
c	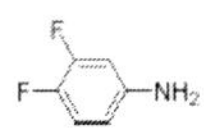	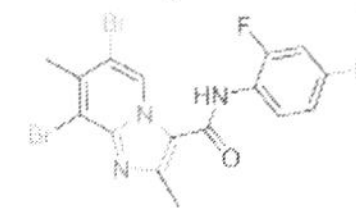
d	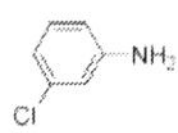	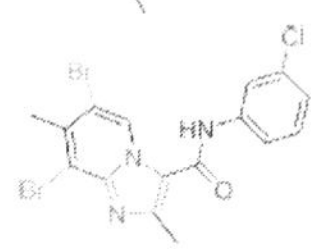
e	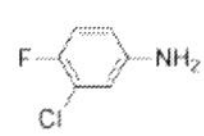	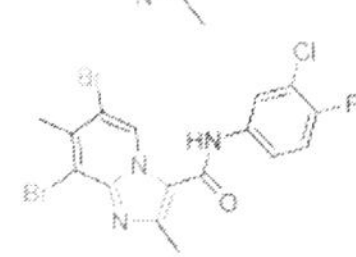
f	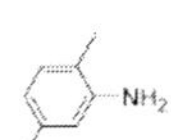	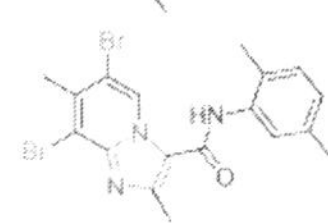
g	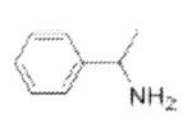	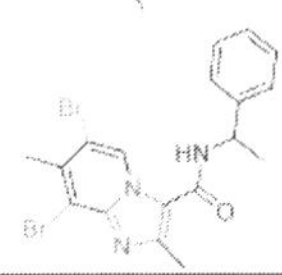

Table 2. Structures of compounds 5h-5m.

Entry	Product [5]	Entry	Product [5]
h	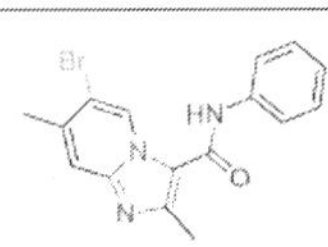	k	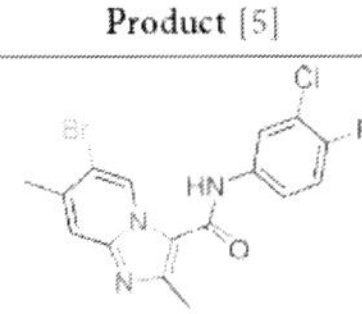
i	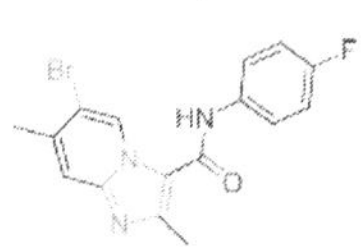	l	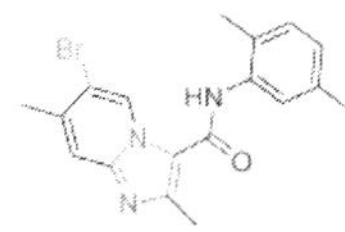
j	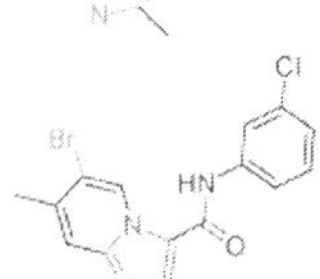	m	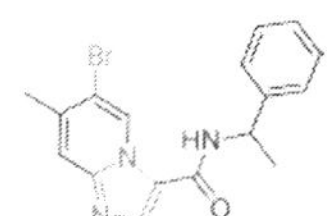

The crude compound was then purified by column chromatography using 15% ethylacetate (EtOAc) in hexane as eluent to get ethyl 6-bromo-2,7-dimethylimidazo- [1,2-a]pyridine-3-carboxylate (2a) (9.50 g, 66%) as an Off-white solid; ESIMS m/z: 299 [M + 2].

4.2. Preparation of Ethyl 6,8-Dibromo-2,7-Dimethylimidazo[1,2-a] Pyridine-3-Carboxylate (2b)

Compound (2b) prepared by similar process 4.1. (11.47 g, 57%) as an off-white solid.

4.3. Preparation of 6-Bromo-2,7-Dimethylimidazo[1,2-a] Pyridine-3-Carboxylic Acid (3a)

A mixture of 6-bromo-2,7-dimethylimidazo[1,2-a]pyridine-3-carboxylate (9.000 g, 30.3 mmol) and Lithium hydroxide (1.088 g, 45.4 mmol) were taken in THF/Water (1:1) (60 mL) and stirred at room temperature for 14 h. The reaction mixture extracted three times with 25 ml of ethyl acetate, further organic layers were discarded. Aqueous layer acidified with 2 NHCl at 5°C - 10°C till the reaction mixture turned to pH ~5, the precipitate formed was filtered and dried to get 6-bromo-2,7-dimethylimidazo [1,2-a]pyri- dine-3-carboxylic acid (3a) (7.34 g, 92%) as an off-white solid. EIMS m/z: 271 [M + 2].

4.4. Preparation of 6,8-Dibromo-2,7-Dimethylimidazo[1,2-a] Pyridine-3-Carboxylic Acid (3b)

Compound (3b) is prepared by process 4.3. (8.35 g, 82%) as an off-white solid. EIMS m/z: 348 [M + 1].

4.5. General Procedure for the Preparation of 6,8-Dibromo-2, 7-Dimethyl-N-Phenylimidazo[1,2-a]Pyridine-3-Carboxamides (5a-5m)

To the stirred solution of carboxylic acid (3a-3b) (1.0 equiv), EDCI (1.2 equiv), HOBt (1.2 equiv) and Et_3N (2.5 equiv) in DMF at 0°C, was added compound (4a-4g) (1.05 equiv) and allowed to stir at room temperature for 22 h. The reaction mixture was diluted with dichloromethane, washed with water and the separated organic layer was concentrated under reduced pressure. The crude compound was further purified by column chromatography using 30% EtOAc/Hexanes as the eluent.

4.5.1. Preparation of 6,8-Dibromo-2,7-Dimethyl-N-Phenylimidazo[1,2-a] Pyridine-3-Carboxamide (5a)

The stirred solution of 6,8-dibromo-2,7-dimethylimidazo[1,2-a]pyridine-3-carboxylic acid (0.5 g, 1.437 mmol) in DMF (14.37 ml) at 0°C was mixed with EDCI (0.331 g, 1.724 mmol), HOBt (0.233 g, 1.724 mmol) and Triethylamine (0.505 ml, 3.59 mmol). The reaction mixture was stirred for 30 min and then aniline 4a (0.138 ml, 1.509 mmol) was added. The reaction mixture was allowed to stir further at room temperature for 22 h. The reaction mixture was then diluted with dichloromethane and washed with water and the separated organic layer was concentrated under reduced pressure, purified by column chromatography using 30% EtOAc/Hexanes as eluent to get title compound in 70% yield (0.42 g) White solid; m.p. = 134°C - 137°C; ^{1}H-NMR (400 MHz, $CDCl_3$): δ 2.69 (s, 3H), 2.87 (s, 3H), 7.22 - 7.18 (m, 1H), 7.43 - 7.39 (t, 2H), 7.61 - 7.59 (d, J = 8.0 Hz, 1H), 7.69 - 7.78 (m, 2H), 9.67 (s, 1H); ^{13}C-NMR (400 MHz,$CDCl_3$): δ 16.1, 22.7, 110.3, 110.9, 116.5, 120.0, 122.4, 123.1, 124.2, 124.5, 126.5, 128.8, 136.8, 137.3, 145.5, 158.5; IR (cm^{-1}): 3448 (Aromatic C-H Stretch), 3138 (Amide N-H Stretch), 1654 (Amide C=O Stretch), 1595 (Amide N-H bending); EIMS m/z: 423.9 [M^+].

4.5.2. Preparation of 6,8-Dibromo-2,7-Dimethyl-N-(4-Fluorophenyl) Imidazo[1,2-a]Pyridine-3-Carboxamide (5b)

73% yield (0.46 g), White solid; m.p. = 142°C - 143°C; ^{1}H-NMR (400 MHz, $CDCl_3$): δ 2.69 (s, 3H), 2.86 (s, 3H), 7.13 - 7.09 (m, 2H), 7.48 - 7.50 (m, 1H), 7.57 - 7.54 (dd, 2H), 9.63 (s, 1H); ^{13}C-NMR (400 MHz, $CDCl_3$): δ 16.3, 22.7, 110.7, 111.1, 115.4, 115.7, 116.2, 122.0, 122.1, 123.4, 124.1, 126.2, 126.6, 145.6, 158.2, 158.6; IR (cm^{-1}): 3443 (Aromatic C-H Stretch), 3141 (Amide N-H Stretch), 1653 (Amide C=O Stretch), 1607 (Amide N-H bending); EIMS m/z: 441.8 [M^+].

4.5.3. Preparation of 6,8-Dibromo-N-(2,4-Difluorophenyl)-2, 7-Dimethylimidazo[1,2-a]Pyridine-3-Carboxamide (5c)

68% yield (0.44 g), White solid; m.p. = 151°C - 154°C; ^{1}H-NMR (400 MHz, $CDCl_3$): δ 2.70 (s, 3H), 2.88 (s, 3H), 6.99 - 6.93 (m, 2H), 8.40 - 8.34 (m, 2H), 9.69 (s, 1H); IR (cm^{-1}): 3292 (Aromatic C-H Stretch), 3122 (Amide N-H Stretch), 1664 (Amide C=O Stretch), 1611 (Amide N-H bending); EIMS m/z: 459.8 [M^+].

4.5.4. Preparation of 6,8-Dibromo-N-(3-Chlorophenyl)-2, 7-Dimethylimidazo[1,2-a]Pyridine-3-Carboxamide (5d)

69% yield (0.45 g), White solid; M.P = 161°C - 163°C; ^{1}H-NMR (400 MHz, $CDCl_3$): δ 2.70 (s, 3H), 2.87 (s, 3H), 7.19 - 7.16 (m, 2H), 7.44 - 7.42 (m, 2H), 7.76 - 7.75 (t, 1H), 9.66 (s, 1H); IR (cm^{-1}): 3413 (Aromatic C-H Stretch), 3129 (Amide N-H Stretch), 1651 (Amide C=O Stretch), 1587 (Amide N-H bending); EIMS m/z: 457.8 [M^+].

4.5.5. Preparation of 6,8-Dibromo-N-(3-Chloro-4-Fluorophenyl)-2, 7-Dimethylimidazo[1,2-a]Pyridine-3-Carboxamide (5e)

61% yield (0.41 g) White solid; m.p. = 146°C - 148°C; ^{1}H-NMR (400 MHz, $CDCl_3$): δ 2.70 (s, 3H), 2.86 (s, 3H), 7.20 - 7.15 (t, 1H), 7.42 - 7.38 (m, 2H), 7.82 - 7.80 (dd, 1H), 9.63 (s, 1H); IR (cm^{-1}): 3426 (Aromatic C-H Stretch), 3129 (Amide N-H Stretch), 1650 (Amide C=O Stretch), 1500 (Amide N-H bending); EIMS m/z: 475.8 [M^+].

4.5.6. Preparation of 6,8-Dibromo-N-(2,5-Dimethylphenyl)-2, 7-Dimethylimidazo[1,2-a]Pyridine-3-Carboxamide (5f)

70% yield (0.45 g) White solid; m.p. = 159°C - 163°C; ^{1}H-NMR (400 MHz, $CDCl_3$): δ 2.32 (s, 3H), 2.38 (s, 3H), 2.69 (s, 3H), 2.88 (s, 3H), 6.97 - 6.95 (d, 1H), 7.15 - 7.13 (m, 2H), 7.83 (s, 1H), 9.72 (s, 1H); IR (cm^{-1}): 3452 (Aromatic C-H Stretch), 3127 (Amide N-H Stretch), 1651 (Amide C=O Stretch), 1580 (Amide N-H bending); EIMS m/z: 451.9 [M^+].

4.5.7. Preparation of 6,8-Dibromo-2,7-Dimethyl-N-(1-Phenylethyl)Imidazo [1,2-a]Pyridine-3-Carboxamide (5g)

68% yield (0.44 g) White solid; m.p. = 132°C - 134°C; ^{1}H-NMR (400 MHz, $CDCl_3$): δ 1.65 - 1.64 (t, 3H), 2.66 (s, 3H), 2.73 (s, 3H), 5.36 - 5.29 (q, 2H), 7.41 - 7.28 (m, 5H), 9.64 (s, 1H); IR (cm^{-1}): 3451 (Aromatic C-H Stretch), 3125 (Amide N-H Stretch), 1655 (Amide C=O Stretch); EIMS m/z: 451.9 [M^+].

4.5.8. Preparation of 6-Bromo-2,7-Dimethyl-N-Phenylimidazo[1,2-a] Pyridine-3-Carboxamide (5h)

54% yield (0.34 g) as white solid; m.p. = 121°C - 123°C; ^{1}H-NMR (400 MHz, $CDCl_3$): δ 2.48 (s, 3H), 2.81 (s, 3H), 7.20 - 7.16 (t, 1H), 7.44 - 7.37 (m, 3H), 7.48 (brs, 1H), 7.61 - 7.58 (d, 2H), 9.64 (s, 1H); IR (cm^{-1}): 3305 (Aromatic C-H Stretch), 3044 (Amide N-H Stretch), 1625 (Amide C=O Stretch), 1597 (Amide N-H bending); EIMS m/z: 346.20 [M + 2].

4.5.9. Preparation of 6-Bromo-N-(4-Fluorophenyl)-2,7-Dimethyl Imidazo[1,2-a]Pyridine-3-Carboxamide (5i)

64% yield (0.43 g), White solid; m.p. = 144°C - 146°C; ^{1}H-NMR (400 MHz, $CDCl_3$): δ 2.50 (s, 3H), 2.82 (s, 3H), 7.12 - 7.08 (m, 2H), 7.51 (s, 1H), 7.60 - 7.56 (m, 2H), 7.70 (brs, 1H), 9.60 (s, 1H); ^{13}C-NMR (400 MHz, $CDCl_3$): δ 16.3, 22.5, 112.3, 115.1, 115.4, 121.6, 122.0, 127.8, 138.5, 158.1, 158.8, 160.5; IR (cm^{-1}): 3292 (Aromatic C-H Stretch), 3043 (Amide N-H Stretch), 1625 (Amide C=O Stretch); EIMS m/z: 364.0 [M + 2].

4.5.10. Preparation of 6-Bromo-N-(3-Chlorophenyl)-2,7-Dimethyl Imidazo[1,2-a]Pyridine-3-Carboxamide (5j)

62% yield (0.45 g), as white solid; m.p. = 144°C - 146°C; ^{1}H-NMR (400 MHz, $CDCl_3$): δ 2.50 (s, 3H), 2.82 (s, 3H), 7.17 - 7.15 (d, 1H), 7.34 - 7.30 (t, 1H), 7.49 - 7.43 (m, 3H), 7.76 - 7.76 (m, 1H), 9.65 (s, 1H); IR (cm^{-1}): 3326 (Aromatic C-H Stretch), 3025 (Amide N-H Stretch), 1632 (Amide C=O Stretch), 1589 (Amide N-H bending); EIMS m/z: 379.9 [M + 2].

4.5.11. Preparation of 6-Bromo-N-(3-Chloro-4-Fluorophenyl)-2, 7-Dimethylimidazo[1,2-a]Pyridine-3-Carboxamide (5k)

62% yield (0.45 g), White solid; m.p. = 136°C - 138°C; ^{1}H-NMR (400 MHz, $CDCl_3$): δ 2.50 (s, 3H), 2.82 (s, 3H), 7.19 - 7.15 (t, 1H), 7.42 - 7.38 (m, 2H), 7.47 (t, 1H), 7.82 - 7.79 (dd, 1H), 9.64 (s, 1H); IR (cm^{-1}): 3336 (Aromatic C-H Stretch), 3027 (Amide N-H Stretch), 1630 (Amide C=O Stretch), 1599 (Amide N-H bending); EIMS m/z: 397.9 [M + 2].

4.5.12. Preparation of 6-Bromo-N-(2,5-Dimethylphenyl)-2, 7-Dimethylimidazo[1,2-a]Pyridine-3-Carboxamide (5l)

66% yield (0.45 g) White solid; m.p. = 151°C - 154°C; ^{1}H-NMR (400 MHz, $CDCl_3$): δ 2.32 (s, 3H), 2.41 (s, 3H), 2.50 (s, 3H), 2.82 (s, 3H), 7.19 - 7.15 (t, 1H), 7.42 - 7.38 (m, 2H), 7.47 (t, 1H), 7.82 - 7.79 (dd, 1H), 9.64 (s, 1H); IR (cm^{-1}): 3453 (Aromatic C-H Stretch), 3135 (Amide N-H Stretch), 1644 (Amide C=O Stretch), 1582 (Amide N-H bending); EIMS m/z: 374.0 [M + 2].

4.5.13. Preparation of 6-Bromo-2,7-Dimethyl-N-(1-Phenylethyl)Imidazo [1,2-a]Pyridine-3-Carboxamide (5m)

64% yield (0.44 g), White solid; m.p. = 181°C - 186°C; ^{1}H-NMR (400 MHz, $CDCl_3$): δ 1.65 - 1.63 (d, 3H), 2.48 (s, 3H), 2.69 (s, 3H), 5.37 - 5.30 (q, 1H), 6.01 - 5.99 (d, 1H), 7.42 - 7.28 (m, 6H), 9.64 (s, 1H); IR (cm^{-1}): 3317 (Aromatic C-H Stretch), 3027 (Amide N-H Stretch), 1615 (Amide C=O Stretch); EIMS m/z: 374.0 [M + 2].

5. RESULTS AND DISCUSSION

5.1 Antimycobacterial activity

The anti-tubercular activity of the synthesized compounds (5a-5m) against Mycobacterium tuberculosis (H37 RV strain) ATCC No-27294, were assessed at the Department of Microbiology, Maratha Mandal's NGH Institute of Dental Sciences and Research Centre, Belgaum-590010, India. The method applied is similar to that reported by Maria and Lourenco [23] .

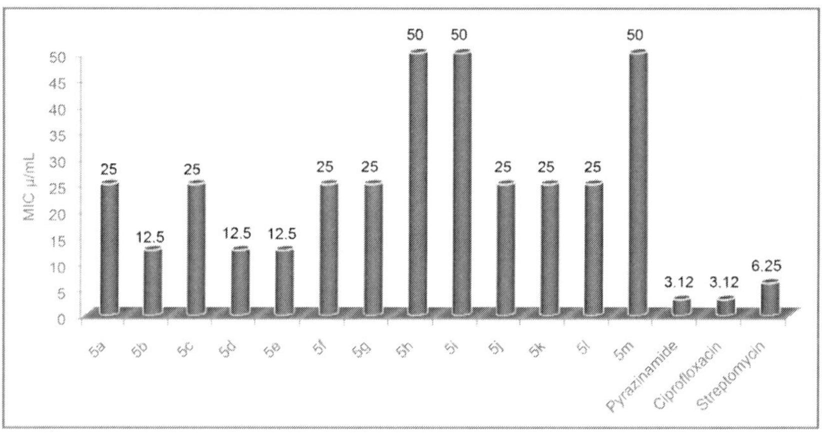

Figure 2. MIC in mg/mL for test samples 5a-5m.

The M. Tuberculosis stain H37Rv (ATCC 27294) was cultured at 37°C in Lowestein- Jensenn medium until log phase growth. Then a cell suspension was prepared at a concentration of about 2×10^6 UFC Ml-CM and further diluted 1:20 in Middlebrook 7H9 medium. The later was supplemented with 10% OADC (oleic acid-albumin-dex- trose-catalase) and 0.001% Tween 80. One mL bactarial suspention was added to each tube (Capped, glass) to gather with the sample solutions of various concentrations. The final concentrations of the compounds under test ranged from 0.8 to 100 µg per mL and adjusted to a final 2 mL volume. After a 7 day incubation 100 µl of 3-(4,5-dime- thylthiazol-2-yl)-2,5-diphenyltetrazolium bromide (MTT) (5 mg/mL) with 20% Tween 80 was added to the glass tubes. A violet colour indicated bacterial growth. The tubes were evaluated for colour change on day 8. For standard tests MIC value of Pyrazinamide-3.125 µg/mL, Streptomycin-6.25 µg/mL and Ciprofloxacin-3.125 µg/mL, were determined each time. The MIC of each sample corresponded to the lowest concentration at which the bacteria tested did not show growth. Susceptibility testing was performed three times. The results were expressed as the mean of three tests.

However, compounds 5a-5m showed moderate to good anti-tubercular activity as represented in Figure 2. This is due to the presence of active structural moieties like imidazo[1,2-a]pyridine and carboxamide group, which might interfere in the mechanism of cell mitosis and hence stop further growth of Mycobacterium tuberculosis. All the studied samples are showing dissimilar potency due to the effective barrier of cell wall membrane of Mycobacterium tuberculosis for entrance of external substances like test compounds under this study. In addition, these compounds disturb the respiration process of the cell and thereby restrict the biosynthesis of proteins. If the biosynthesis of proteins is blocked then formation of bacterial cell wall is not possible which ultimately results in cell death and therefore restricts further growth and infection due to bacteria [24] .

6. CONCLUSION

In conclusion, new derivatives of imidazo[1,2-a]pyridine have been synthesized successfully with good yield. The experimental data for the evaluation of anti-tuberculosis activity revealed that most of the synthesized compounds 5a-5m demonstrated moderate to good antituberculosis activity. Among the tested compounds 5b, 5d and 5e were found to be the most active with minimum inhibitory concentration (MIC) of 12.5 μg/mL against Mycobacterium tuberculosis (H37 RV strain) ATCC No-27294 with minimum inhibitory concentration of 12.5 μg/ml. This study would pave the way for future development of more effective imidazo[1,2-a]pyridine analogs for applications in biological science.

ACKNOWLEDGEMENTS

The authors thank Principal and Head Department of Chemistry, Government of Maharashtra, Ismail Yusuf Arts, Science and Commerce College for providing research and library facilities. The authors also thank Management and Principal of C. S.'s Patkar-Varde College, Goregaon (W), Mumbai for their constant encouragement and support. The authors also acknowledge the help of Dr. Kishore Bhat of Governmental Dental College, Belgaum, for facilitating anti-TB assays and providing the procedure for the same.

CITE THIS PAPER

Jadhav, B., Kenny, R., Nivid, Y., Mandewale, M. and Yamgar, R. (2016) Synthesis and Evaluation of Antituberculosis Activity of Substituted 2,7-Dime- thylimidazo[1,2-a]Pyridine-3-Carboxamide Derivatives. Open Journal of Medicinal Che- mistry, 6, 59-69.

REFERENCES

1. Katritzky, A., Xu, Y. and Tu, H. (2003) Regiospecific Synthesis of 3-Substituted Imidazo[1,2-a]Pyridines, Imidazo[1,2-a]Pyrimidines, and Imidazo[1,2-c]Pyrimidine. The Journal of Organic Chemistry, 68, 4935-4937.
2. Rival, Y., Grassy, G. and Michel, G. (1992) Synthesis and Antibacterial Activity of Some Imidazo[1,2-a]Pyrimidine Derivatives. Chemical and Pharmaceutical Bulletin, 40, 1170-1176.
3. Moraski, G., Garrett, C., Lowell, M., Jeffrey, C., Philip, H., Helena, B., Mai, B., Torey, A., Juliane, O., Tanya, P. and Marvin, M. (2013) Advancement of Imidazo[1,2-a]Pyridines with Improved Pharmacokinetics and nM Activity vs. Mycobacterium tuberculosis. ACS Medicinal Chemistry Letters, 4, 675-679.
4. Rival, Y., Grassy, G., Taudou, A. and Ecalle, R. (1991) Antifungal Activity in Vitro of Some Imidazo[1,2-a]Pyrimidine Derivatives. European Journal of Medicinal Chemistry, 26, 13-18.
5. Chaouni-Bendallah, A., Galtier, C., Allouchi, H. and Kherbeche, A. (2001) 3-Benzamido, Ureido and Thioureidoimidazo[1,2-a]Pyridine Derivatives as Potential Antiviral Agents. Chemical and Pharmaceutical Bulletin, 49, 1631-1635.
6. Chezal, J., Paeshuyse, J., Gaumet, V., Canitrota, D., Maisonial, A., Lartigue, C., Gueiffier, A., Moreaua, E., Teulade, J., Chavignon, O. and Neyts, J. (2010) Synthesis and Antiviral Activity of an Imidazo[1,2-a]Pyrrolo[2,3-c]Pyridine Series against the Bovine Viral Diarrhea Virus. European Journal of Medicinal Chemistry, 45, 2044-2047.
7. Sanfilippo, P.J., Urbanski, M., Press, J.B. and Moore, J.B.J. (1998) Synthesis of (Aryloxy)Alkylamines. 2. Novel Imidazo-Fused Heterocycles with Calcium Channel Blocking and Local Anesthetic Activity. Journal of Medicinal Chemistry, 31, 2221-2227.
8. Byth, K.F., Culshaw, J.D., Green, S., Oakes, S. and Thomaset, A.P. (2004) Imidazo[1,2-a]-Pyridines. Part 2: SAR and Optimisation of a Potent and Selective Class of Cyclin-Dependent Kinase Inhibitors. Bioorganic & Medicinal Chemistry Letters, 14, 2245-2248.
9. Dvey, D., Erhardt, P.W., Lumma, W.C. and Cantor, E. (1987) Cardiotonic Agents. 1. Novel 8-aryl Substituted Imidazo[1,2-a]- and -[1,5-a]Pyridines and Imidazo[1,5-a]Pyridinones as Potential Positive Inotropic Agents. Journal of Medicinal Chemistry, 30, 1337-1342.
10. Kaminnski, J.J. and Doweyko, A.M. (1997) Antiulcer Agents. 6. Analysis of the in Vitro Biochemical and in Vivo Gastric Antisecretory Activity of Substituted Imidazo[1,2-a]-pyridines and Related Analogues Using Comparative Molecular Field Analysis and Hypothetical Active Site Lattice Methodologies. Journal of Medicinal Chemistry, 40, 427-436.
11. Zhonghui, L., Gregory, R., Anand, R. and Rui-Qin, L. (2008) Potent, Selective, Orally Bioavailable Inhibitors of Tumor Necrosis Factor-α Converting Enzyme (TACE): Discovery of Indole, Benzofuran, Imidazopyridine and Pyrazolopyridine P1′ Substituents. Bioorganic & Medicinal Chemistry Letters, 18, 1958-1962.
12. Gregory, R., Anand R., Zhonghui, L. and Rui-Qin, L. (2008) Potent, Exceptionally Selective, Orally Bioavailable Inhibitors of TNF-α Converting Enzyme (TACE): Novel 2-Substituted-

1H-Benzo[d]Imidazol-1-yl)Methyl)Benzamide P1' Substituents. Bioorganic & Medicinal Chemistry Letters, 18, 1577-1582.

13. Moraski, G.C., Miller, P.A., Bailey M.A., Ollinger, J., Parish, T., Boshoff, H.I., Cho, S., Anderson, J.R., Mulugeta, S., Franzblau, S.G. and Miller, M.J. (2015) Putting Tuberculosis (TB) to Rest: Transformation of the Sleep Aid, Ambien, and "Anagrams" Generated Potent Antituberculosis Agents. ACS Infectious Diseases, 13, 85-90.
14. Abrahams, K.A., Jonathan A., Cox, G., Vickey, L., Loman, N.J., Pallen, M.J., Constantinidou, C., Fernández, R., Alemparte, C., Remuinán, M.J., Barros, D., Ballell, L. and Besra, G.S. (2012) Identification of Novel Imidazo[1,2-α]Pyridine Inhibitors Targeting M. tuberculosis QcrB. PLoS ONE, 7, e52951.
15. Moraski, G.C., Markley, L.D., Hipskind, P.A., Boshoff, H., Cho, S., Franzblau, S.G. and Miller, M.J. (2011) Advent of Imidazo[1,2-α]Pyridine-3-Carboxamides with Potent Multi- and Extended Drug Resistant Antituberculosis Activity. ACS Medicinal Chemistry Letters, 2, 466-470.
16. Li, L., Li, Z., Liu, M., Shen, W., Wang, B., Guo, H. and Lu, Y. (2016) Design, Synthesis and Antimycobacterial Activity of Novel Imidazo[1,2-α]Pyridine Amide-Cinnamamide Hybrids. Molecules, 21, 49.
17. Samala, G., Nallangi, R., Devi, P.B., Saxena, S., Yadav, R., Sridevi, J.P., Yogeeswari, P. and Sriram, D. (2014) Identification and Development of 2-Methylimidazo[1,2-α]Pyridine -3-Carboxamides as Mycobacterium tuberculosis Pantothenate Synthetase Inhibitors. Bioorganic & Medicinal Chemistry, 22, 4223-4232.
18. Mandewale, M.C., Thorat, B.R., Shelke, D. and Yamgar, R.S. (2015) Synthesis and Biological Evaluation of New Hydrazone Derivatives of Quinoline and Their Cu(II) and Zn(II) Complexes against Mycobacterium tuberculosis. Bioinorganic Chemistry and Applications, 2015, Article ID: 153015.
19. Mandewale, M.C., Thorat, B.R., Nivid, Y., Jadhav, R., Nagarsekar, A.S. and Yamgar, R.S. (2016) Synthesis, Structural Studies and Antituberculosis Evaluation of New Hydrazone Derivatives of Quinoline and Their Zn(II) Complexes. Journal of Saudi Chemical Society, In Press.
20. Mandewale, M.C., Kokate, S., Thorat, B.R., Sawant, S.S. and Yamgar, R.S. (2016) Zinc Complexes of Hydrazone Derivatives Bearing 3,4-Dihydroquinolin-2(1h)-One Nucleus as New Anti-Tubercular Agents. Arabian Journal of Chemistry, In Press.
21. Mandewale, M.C., Thorat B.R. and Yamgar, R.S. (2015) Synthesis and Anti-Mycobacterium Study of Some Fluorine Containing Schiff Bases of Quinoline and Their Metal Complexes. Der Pharma Chemica, 7, 207-215.
22. Nivid, Y., Jadhav, B.S., Kenny, R.S., Nazirkar, B.P., Thorat, B.R., Mulgoankar, B.S. and Yamgar, R.S. (2015) Synthesis and Biological Activity Studies of Some Novel Substituted Imidazo[1,2-A]Pyridine. Heterocyclic Letters, 5, 177-184.
23. Lourenco, M.S., DeSouza, M.N., Pinheiro, A.C., Ferreira, M.L., Goncalves, R.B., Nogneira T.M. and Peralta, M.A. (2007) Evaluation of Anti-Tubercular Activity of Nicotinic and Isoniazid Analogues. Arkivoc. 15, 181-191.
24. Raman, N. (2005) Antibacterial Study of the Mannich Base, N-(1-Morpholino-Benzyl) Semicarbazide and Its Transition Metal(II) Complexes. Research Journal of Chemistry and Environment, 4, 9.

CHAPTER 6

Synthesis, Characterization, Antimicrobial and Antioxidant Activities of The Homocyclotrimer Of 4-Oxo-4h-Thieno[3,4-C]Chromene-3-Diazonium Sulfate

Emmanuel Sopbue Fondjo[*, a], Djeukoua Dimo Kamal Sorel[a], Tamokou Jean-de-dieu[b], Tsemeugne Joseph[a], Kouamo Sylvian[a], Ngouanet Doriane[a], Chouna Jean Rodolphe[a], Nkeng-Efouet-Alango Pepin[a], Kuiate Jules-Roger[b], Ngongang Ndjintchui Arnaud[c], Sondengam Beibam Lucas[d]

[a] *Laboratory of Applied Synthetic Organic Chemistry, Department of Chemistry, Faculty of Science, University of Dschang, P.O. Box 67 Dschang, Republic of Cameroon*

[b] *Laboratory of Microbiology and antimicrobial Substances, Department of Biochemistry, Faculty of Science, University of Dschang, PO Box 067 Dschang, Republic of Cameroon*

[c] *Technical University of Munich, Department of Chemistry Lichtenbergstrasse 4, D-85747 Garching, Germany*

[d] *Department of Organic Chemistry, University of Yaounde I, P.O. Box 812 Yaounde, Republic of Cameroon*

ABSTRACT

The *in situ* formed 4-oxo-4H-thieno[3,4-c]chromene-3-diazonium sulfate (**5**) in the coupling reactions involving the parent 2-aminothiophene (**4**) and various phenolic and arylamines' couplers, readily undergoes homocyclotrimerization at low temperature to afford in fairly good yield the first ever reported eighteen member ring heteroaromatic holigomer **6**. Compound **6** was fully characterized by its elemental analysis, IR, UV-Vis, ^{1}H-NMR, ^{13}C-NMR and HRMS spectral data. The HMBC and HSQC techniques were used to ascertain the structural assignments. A comparative study on the antimicrobial and antioxidant activities of compounds **3**, **4** and **6** was carried out to assess the SAR due to the transformations (from **3** to **6** *via* **4**) on the tested compounds. It was found that compounds **6** and **4** were respectively the most active compounds against bacteria (MIC = 32-64 μg/ml) and yeasts (MIC = 16–64 μg/ml). Compound **6** also showed high radical-scavenging activities and ferric reducing power when compared with vitamin C and BHT used as reference antioxidants.

Keywords: 2-aminothiophene, Azo compounds, Antimicrobial, Antioxidant, Coumarin, Homocyclotrimerization, Holigomer, SAR.

1. INTRODUCTION

As a result of the dramatic increase in microbial infections, serious attention has been directed in recent years towards the discovery and development of new antimicrobial drugs.

In medicinal chemistry, thiophene derivatives have been well known for their therapeutic applications. A review of the recent literature showed that many effective antimicrobial agents contain an heterocyclic moiety within their structure [1, 2] with special emphasis on those agents incorporating a thiophene moiety [3-7]. The thiophene ring system is notably a structural component of the commercial imidazole antifungal agent sertaconazole [8]. In addition, the antimicrobial profile of condensed thiophenes, particularly their thieno[2,3-d]pyrimidin-4-one analogues, is also well categorized in the literature [9-11].

Coumarin and its derivatives represent one of the most active classes of compounds possessing a wide spectrum of biological activity [12-18]. The interesting biological activity of these coumarins made these compounds attractive targets in organic synthesis.

Azo compounds constitute one of the largest classes of industrially synthesized organic compounds, potent in drug and cosmetics [19]. Azo dyes are used widely in dying textile fibers, biomedical studies, advanced applications in organic synthesis and high technological areas like lasers, liquid crystalline displays, electro-optical devices and ink jet printer [20-22] as well as shows variety of interesting biological activities including antibacterial [23-26] and pesticidal activities [27]. The azo dyes possess antiseptic and antiprotozoal properties and also promote wound healing.

Based on the above considerations, we investigated the synthesis of hybrid molecular architectures containing thiophene, coumarin, and azo moieties likely to combine the biological features of all the components.

2. RESULTS AND DISCUSSION

2. CHEMISTRY

The starting 2-aminothiophene reagent **4** was prepared by applying the third version of the Gewald methodology [28, 29], whereby the first step of the procedure consisted of a Knoevenagel [30] condensation of 2-hydroxy-1-acetonaphthon (**1**) with ethyl cyanoacetate (**2**) to give the fused benzocoumarin **3** (Scheme 1). In the second step of the reaction sequence, compound **3** was treated with elemental sulphur and ammonia as catalytic amine base to afford compound **4** in excellent yield as previously reported [31]. The melting point, as well as all the elemental and all the other spectroscopic data were in all respects comparable to those originally reported for this compound.

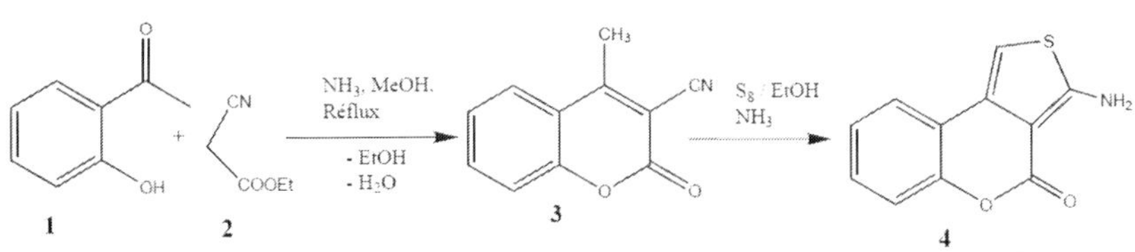

Scheme 1. Reaction sequences to compound 4.

The *in situ* formed intermediate thienyl diazonium sulphate **5** (Scheme 2) was generated from the diazotization of compound **4** using nitrosyl sulphuric acid at very low temperature (0-5 °C). The freshly prepared diazonium solution was then coupled with 2-aminothiophene **4**. The resulted mixtures were worked-up as usual to yield the azo compound **6** (Scheme 2), which was fully characterized by its physical, elemental and spectroscopic data.

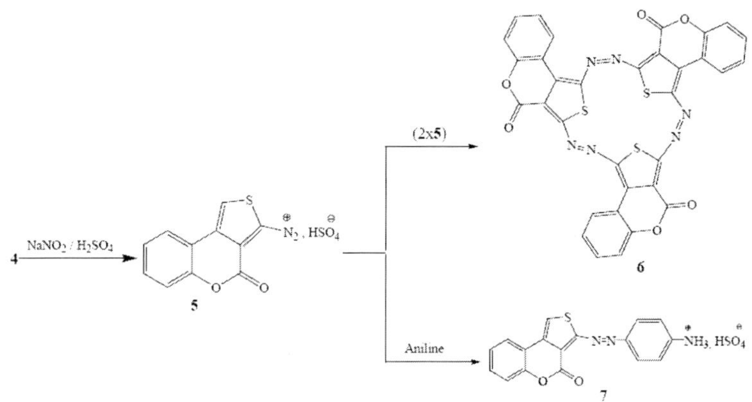

Scheme 2. Reaction sequences to compounds 6.

Compound **6** was obtained as green yellowish powder with a sharp melting point at 213.5 ^{0}C. Its UV-visible spectrum recorded in DMSO, showed absorptions in the range 212-500 nm, in agreement with the highly conjugated (18π)-electrons of the heteroaromatic chain, partly extended to the three annelated benzene rings. This also justified the strong bathochromic effect of the main absorption bands from the starting 2-aminothiophene compound **4** into the final product **6**.

On the IR-spectrum, the Csp^2-H stretching frequencies due to the aromatic rings were observed around 2950 cm^{-1}, whereas the combined strong absorption frequencies of the C=O groups was exhibited around 1724 cm^{-1}. Strong to medium size stretching frequencies of the benzenic and thiophenic C=C bonds were exhibited at 1677 and 1588 cm^{-1}, and in the range 1544-1500 cm^{-1} respectively. The *ortho* disubstitution pattern of the benzene rings was characterized by the strong absorption band appearing at 767cm^{-1}. The presence of the –N=N– bonds was confirmed by a medium size band at 1490 cm^{-1}.

The structure of compound **6** is further strongly supported by its HRMS ESI$^+$-mode, which shows the pseudo molecular ion peak at m/z 685 (29%) corresponding to [M+H]; this is in full agreement with the combustion analysis data. The mass spectrum of compound **6** contained fragments ions at m/z 640 (3%), 596 (1%), 552 (3%), 659 (8%), 627 (1%), 620 (1%), 588 (1%), 532 (1%), 628 (1%), 600 (1%), 572 (2%), 544 (1%), 516 (40%), 484 (1%), 452 (1%) and 248 (94%) which could be assigned to [M^+ -CO_2], [M^+ -2CO_2], [M^+ -3CO_2], [M^+ -N_2], [M^+ -2N_2], [M^+-2S], [M^+ -3S], [M^+ -3S -2N_2], [M^+ -2CO], [M^+ -3CO], [M^+ -3CO -N_2], [M^+ -3CO -2N_2], [M^+ -3CO -3N_2], [M^+ -3CO -3N_2-S], [M^+ -3CO -3N_2 -2S] and [M^+ -3CO -3N_2 -3S -$C_{10}H_4$]. The Scheme 3 below rationalizes the formation of the fragments at *m/z* 516 (40%), 437 (65%), 393 (19%) and 339 (42%) respectively.

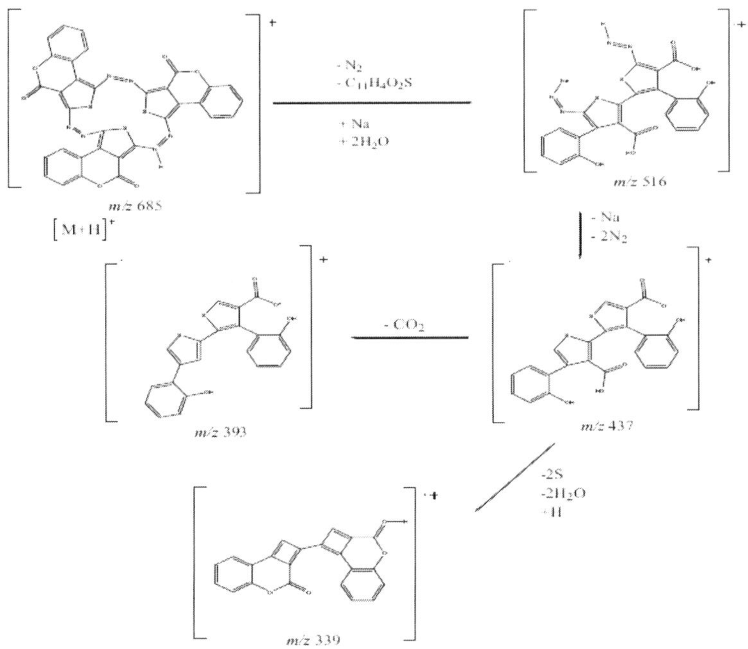

Scheme 3. Some important ESI+-mode fragments of compound 6.

The absence of thiophenic protons in the ^{1}H-NMR spectrum of compound **6** which normally appear in the range 6.5-7.2 ppm, indicated that these protons were involved in the cyclisation process. The ^{1}H-NMR spectrum showed four sets of aromatic Csp^{2}-H protons at 8.73 (dd; *J* = 8; 1.6 Hz; H-7, H-21 et H-35), 7.85 (ddd; *J* = 8.4; 6.4; 1.6 Hz; H-9, H-23 et H-37), 7.53 (dd; *J* = 5.2; 1.2 Hz; H-10, H-24 et H-38) and 7.51 ppm (ddd; *J* = 8.1; 7.2; 1.2 Hz; H-8, H-22 et H-36). The presence of the signal of four protons instead of twelve could be explained by the symmetric nature of compound **6**.

The ^{13}C-NMR spectrum of compound **6** exhibited eleven signals instead of thirty three as deduced from the molecular formula. The symmetrical nature of compound **6** could also explain this observation. The ^{13}C-NMR spectrum consisted of 11 signals among which four [135.5 (C-9, C-23 and C-37); 128.9 (C-7, C-21 and C-35); 125.4 (C-8, C-22 and C-36) and 117.1 (C-10, C-24 and C-38)] could be assigned to tertiary Csp^{2}-H aromatic carbons. The remaining other seven signals [184.9 (C-13, C-27 and C-41); 154.9 (C-11, C-

25 and C-39); 153.7 (C-5, C-19 and C-33); 153.5 (C-14, C-28 and C-42); 147.8 (C-1 ; C-15 and C-29); 119.9 (C-6, C-20 and C-34) and 114.9 (C-4, C-18 and C-32)] were attributed to the quaternary carbon atoms by comparing the experimental values with the simulated ones.

A remarkable feature in the ^{13}C-NMR spectrum of this compound is the intense downfield shifts ($\Delta\delta$ = 19.2 ppm) of the $\delta_{C=O}$ of the carbonyl of the coumarin rings respectively from 165.70 ppm in the parent thiophene substrate **4** [31] to 184.9 ppm in the coupling product. This abnormally high value could be explained by the strong diamagnetic anisotropy [32, 33] of the 18 π-electrons of the heteroaromatic ring system comprising the alternating three fused thiophene rings and the alternating three –N=N– bonds.

It should be mentioned that, in cyclic conjugated systems such as annulenes, the effect of ring currents is particularly illustrative. In fact, with (4n+2) π-electrons the annulenes exhibit aromatic character. For example (18)-annulene [32, 33] exhibits a similar ring current as found in six membered aromatic or heteroaromatic compounds. In these cases too, the magnetic field induced by the circulation of π-electrons opposes the applied magnetic field in the center of the ring and reinforces it around the periphery. As a result, the protons and carbons lying inside the ring of (18)-annulenes are strongly shielded while those on the outer side are strongly deshielded. In the case of compound **6**, the 18-membered heteroaromatic skeleton is planar and is assumed to have a similar behavior in the applied magnetic field as with normal (18)-annulenes.

The HMBC spectra [34, 35] were very helpful for locating the tertiary and quaternary carbons by identifying the various protons interacting with them through two-bond ($^{2}J_{CH}$), three-bond ($^{3}J_{CH}$) and occasionally four-bond ($^{4}J_{CH}$) couplings as displayed in Table 1 and Figure (1). Additionally, the HSQC experiments were performed in order to assign accurately the direct connectivities of the aromatic sp2 carbons to the corresponding protons.

3. BIOLOGY

3.1 Antimicrobial Activity

The results of the antimicrobial activity are summarized in Table 2. Compounds **4** and **6** displayed weak to moderate antibacterial and antifungal activities (MIC = 16–64 μg/ml). Compounds **4** and **6** were respectively the most active compounds against bacteria (MIC = 32–64 μg/ml) and yeasts (MIC = 16–64 μg/ml). No activity was noticed for compound **3** against all the tested microorganisms (results not shown).

Previous studies have shown antibacterial [36] and antifungal [37] activities of azo compounds. The lowest MIC value of 16 µg/mL was recorded on *Candida albicans* with compound **4** while the lowest MBC value of 32 µg/mL was obtained on *Shigella flexneri* and *Escherichia coli* with compound **6** and on *Candida albicans* and *Candida parapsilosis* with compound **4**. However, the highest MIC and MMC values of 128 µg/mL were recorded on *Candida glabrata* with compound **6**. A lower MMC/MIC (≤4) value signifies that a minimum amount of compound is used to kill the microbial species, whereas, a higher MMC/MIC (>4) value signifies the use of comparatively more amount of sample for the control of any microorganism [38]. The findings of the present study suggest that compounds **4** and **6** possess microbicidal activity against the sensitive microorganisms.

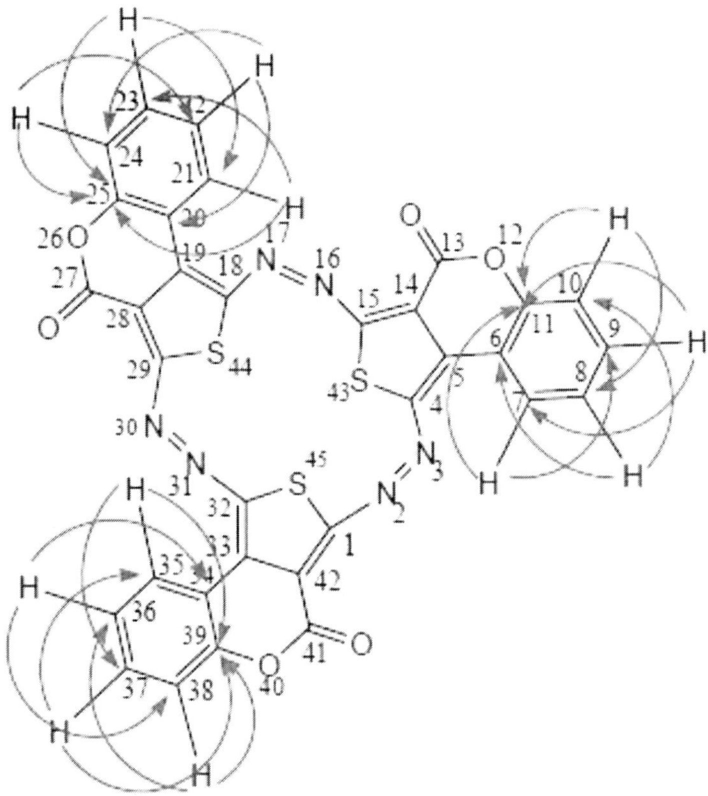

Figure 1. HMBC interactions in compound 6.

Table 1.HSQC and HMBC interactions and ^{1}H and ^{13}C chemical shifts δ in compound **6** in DMSO-d_6 as the solvent (25 °C).

C-atom	δ ^{13}C	HSQC (H→C)	HMBC (H→C)
6, 20 and 34	119.9		7.51 (H-8, H-22, H-36)
14, 28 and 42	153.5		
1, 15 and 29	147.8		
13, 27 and 41	184.9		
11, 25 and 39	154.9		8.73 (H-7, H-21, H-35), 7.85 (H-9, H-23, H-37), 7.53 (H-10, H-24, H-38)
10, 24 and 38	117.1	7.53 (H-10, H-24, H-38)	7.51 (H-8, H-22, H-36)
9, 23 and 37	135.5	7.85 (H-9, H-23, H-37)	8.73 (H-7, H-21, H-35)
8, 22 and 36	125.4	7.51 (H-8, H-22, H-36)	7.53 (H-10, H-24, H-38)
7, 21 and 35	128.9	8.73 (H-7, H-21, H-35)	7.85 (H-9, H-23, H-37)
4, 18 and 32	114.9		
5, 19 and 33	153.7		

Table 2. Antimicrobial activity (MIC and MMC in µg/mL) of compounds **4** and **6** against bacterial and yeast species.

Compounds	Inhibition Parameters	ST	SF	EF	KP	EC	STB	STA	CPn	CP	CA	CG
6	MIC	64	32	64	32	32	64	32	64	64	32	128
	MMC	64	32	64	64	32	64	64	64	64	64	128
	MMC/MIC	1	1	1	2	1	1	2	1	1	2	1
4	MIC	>256	32	>256	>256	64	>256	64	32	32	16	64
	MMC	nd	64	nd	nd	64	nd	64	64	32	32	64
	MMC/MIC	nd	2	nd	nd	1	nd	1	2	1	2	1
*Standard drugs	MIC	8	2	2	2	8	8	1	4	2	4	2
	MMC	8	4	4	4	8	8	2	4	8	4	4
	MMC/MIC	1	2	2	2	1	1	2	1	4	1	2

Compound **3** was not active at concentrations up to 256 µg/ml; nd: not determined; * ciprofloxacin and nystatin were used as standard drugs for bacteria and yeasts respectively; MIC: Minimum inhibitory concentration; MMC: minimum microbicidal concentration; ST: *Salmonella typhi* ATCC6539; SF: *Shigella flexneri* ; EF: *Enterococcus faecalis* ATCC10541; KP: *Klebsiella pneumoniae* ATCC13883; EC: *Escherichia coli* ATCC11775; STB: *Salmonella paratyphi* B; STA: *Salmonella paratyphi* A ; CPn: *Candida parapsilosis*; CP: *Candida parapsilosis ATCC 22019*; CA: *Candida albicans* ATCC 9002; CG: *Candida glabrata* IP35.

3.2 Ferric Reducing Antioxidant Power (FRAP)

The FRAP assay measures the ability of an antioxidant to reduce ferric (III) to ferrous (II) in a redox-linked colourimetric reaction that involves single electron transfer [39]. The reducing power of a compound serves as a significant indicator of its potential antioxidant activity. The results of the reducing power assay of synthesized compounds are summarized in Figure (2). All the investigated compounds showed dose-dependent reducing power. Interestingly, compound **4** showed the lowest reducing power, while compound **6** exhibited the highest reducing power at the different concentrations tested. The antioxidant power of compound **6** is almost equal to that of butylated hydroxytoluene (BHT) used as standard antioxidant.

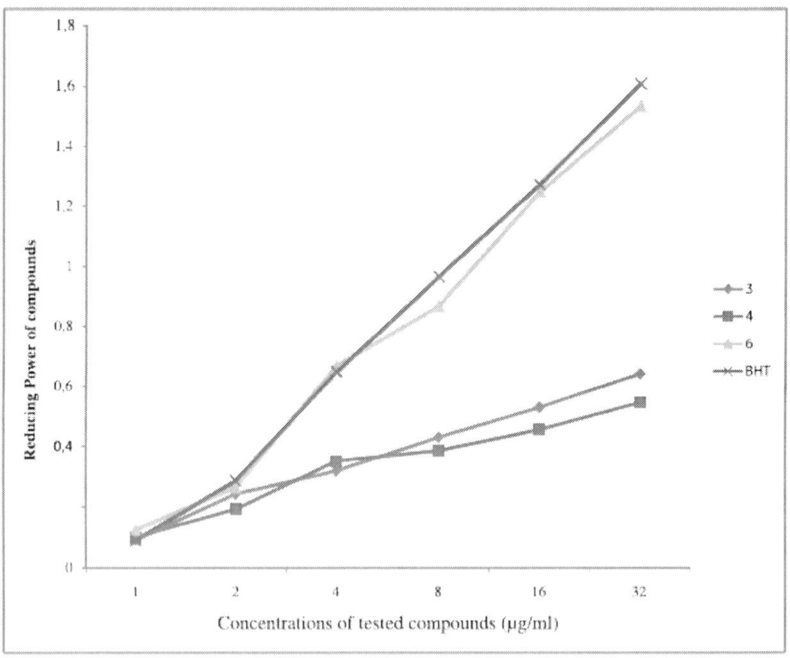

Figure 2. Reducing power activities of the tested compounds as well as BHT.

3.3 DPPH Free Radical Scavenging Activity

The concentrations which decreased by 50% the absorbance of DPPH (EC_{50}) are presented in Figure (3). These results showed that the compound **4** had the highest EC_{50} (*i.e.*, the lowest activity) while compound **6** had the lowest EC_{50} (*i.e.* had the highest activity). The DPPH free radical scavenging activity of compound **6** was comparable to that of the standard antioxidant vitamin C. These results corroborate the FRAP assay, where this compound exhibited the best antioxidant activity. Previous study has shown strong antioxidant activity of azo compounds performed by using DPPH free radical scavenging assay and metal chelating method [40]. The results of the antioxidant study show that these azo compounds may have great relevance in the prevention and therapies of diseases in which oxidants or free radicals are implicated [40].

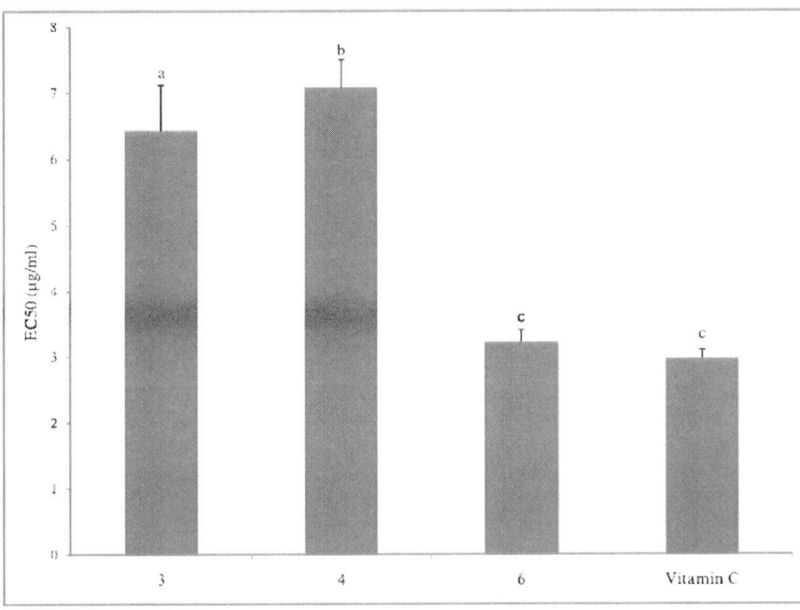

Figure 3. Equivalent concentrations of test compounds scavenging 50% of DPPH radical (EC50)

From the structure activity relationship (SAR) point of view, one can logically say that, the thiolation of the carbonitrile **3** at its methyl group, followed by the intramolecular heterocyclisation and aromatization of the fused five member ring to afford compound **4** – resulted into the significant improvement of the antibacterial and antifungal activities of the latter as compared to its precursor. However, these chemical transformations have slightly reduced the antioxidant activity of **4** compare to that of compound **3**.

Furthermore, the diazotization of compound **4**, which led to the homocyclotrimerization of the *in situ* formed thienyl diazonium salts **5**, resulted in enhancing significantly and globally the evaluated biological activities of the final product **6**. The results are in accordance with other reports in the literature [41, 42], which showed the importance of thiolation and diazotization in antimicrobial and antioxidant activities.

4. MATERIALS AND METHODS

4.1 General Information

All Melting points are corrected and were determined with a STUART SCIENTIFIC Melting Point Apparatus Model SMP3. The TLCs were carried out on Eastman Chromatogram Silica Gel Sheets (13181; 6060) with fluorescent indicator. A mixture of ethyl acetate and methylene chloride (1:1) was used as eluent and iodine was used as revelator for the chromatograms. The IR spectra were measured with a Fourier Transform Infrared spectrometer Brucker Alpha. The UV spectra were recorded with a JENWAY 6715 UV-Vis Spectrophotometer. Combustion analyses were carried out with a C, H, N, S Euro EA from Hekatech company, their results were found to be in good agreement (±0.3%) with the calculated values. EIMS spectra were recorded on a double focusing mass spectrometer (Varian MAT 311A). ^{1}H-NMR spectra were recorded in DMSO-d_6 on a Bruker DRX spectrometer operating at 500 MHz. ^{13}C-NMR spectra were recorded in DMSO-d_6 on a Bruker DRX spectrometer operating at 125 MHz. TMS was used as internal reference.

4.2 Preparation of the Reagents and Starting Materials

All the reagents mentioned in this work were purchased from Aldrich and Fluka and were used without further purification. Starting material **4** has been prepared according to literature procedures as published earlier [31a].

4.3 3-Amino-4H-thieno [3,4-c] [1] Benzopyran-4-one (4)

A mixture of 2-ceto-4-methyl-2H- [1] benzopyran-3-carbonitrile (4.65 g) and sulphur (1.22 g, in excess) in ethanol (30 ml) was stirred using a magnetic plate shaker thermostated at 45 °C. Ammonia (6 ml) was added drop wise during the first 10 min of the reaction. After 7 h of reaction, the resulting precipitate (5.08 g, 93%) was collected by filtration, washed with distilled water and recrystallised in benzene to yield a yellow powder, mp 195–196 °C (Lit.18 198–199 °C, from benzene); IR (potassium bromide): 3449 (-NH2); 1687 (C=O); 1603 and 1547 (C=C); 1224 (C-O); 1H NMR: 7.87 (1H; dd; J = 8.0; 1.5 Hz; H-8); 7.36 (1H; ddd; J = 8.0; 7.4; 1.8 Hz; H-6); 7.22-7.17 (2H; m; H-7 and H-5); 7.78 (2H; br s; NH2); 6.89 (1H; s; thiophenic H); 13C(1H) NMR: 158.98 (C-2); 118.05 (C-2a); 166.70 (C-3); 151 (C-4a); 97.54 (C-5); 124.39 (C-6); 117.09 (C-7); 123.91 (C-8); 130.90 (C-8a); 98.07 (C-8b); 129.26 (C-9); ms: (EI) m / z (%) 217 (M+; 100%). Anal. Calcd for $C_{11}H_7NO_2S$ C: 60.83; H, 3.23; N, 6.45; S, 14.75. Found C, 61.17; H, 3.36; N, 6.49; S, 14.45.

4.4 Preparation of Diazonium Salt Solution

Dry sodium nitrite (2.07 g, 3 mmol) was slowly added over a period of 30 minutes to concentrated sulphuric acid (10 mL) with occasional stirring. The solution was cooled to 0-5 °C. Compound **4** (0.22 g, 1 mmol) was dissolved in DMSO (10 mL) and cooled to 0-5 °C. The nitrosyl sulphuric acid solution was added to the solution of **4** and the temperature was maintained to 0-5 °C. The clear diazonium salt solution thus obtained consisting of the *in situ* formed intermediates **5**, was used immediately in the coupling reactions.

4.5 Procedure for the Preparation of the Coupling Product 6

Aniline (0.279 g; 3 mmol) was dissolved in DMSO (10 mL) and then cooled in an ice-bath at 0-5 °C. The diazonium solution of **4** previously prepared was added drop wise over 1 hour, and then 15 mL of sodium acetate solution (10%) was added in the mixture. The solid precipitate consisting of the previously reported [31b] (4H)-2-(p-Aminophenylazo)thieno[3,4-c]chromen-4-one hydrogen sulfate (**7**) was collected on a filter and crystallised from methanol to give the corresponding thienyl azo aryl dye. Work up of the resulted filtrate gave compound **6** in average yield of 11%.

12,26,40-trioxa-43,44,45-trithia-2,3,16,17,30,31-hexaazadecacyclo [30.10.1. $1^{4,15}.1^{18,29}.0^{5,14}.0^{6,11}.0^{19,28}.0^{20,25}.0^{33,42}.0^{34,39}$] pentatetraconta-1(42), 2 ,4,

6,8, 10,14, 16,18,20,22,24,28,30,32, 34, 36,38-octadecaene-13,27,41-trione (6).

Compound **6** (124 mg, 11%) was obtained as green yellowish powder. m.p. 213.5 °C; Rf = 0.43 in CH_2Cl_2/AcOEt 50% v/v; IR (potassium bromide): 2588 (=C-H), 1724 (C=O), 1677-1588 (C_{Ar}=C_{Ar}), 1544-1500 (thiophenic C=C), 1490 (-N=N-), 767 (=C_{Ar}-H) cm^{-1}; λmax (DMSO) (Log ε): 400 (4.70), 282 (4.49), 229 (4.71); ^{1}H-NMR (DMSO-d_6, 500 MHz) δ 8.73 (dd, 3H, 7- H, 21-H and 35-H, *J* = 8 and 1.6 Hz), 7.85 (ddd, 3H, 9-H, 23-H and 37-H, *J* = 8.4, 6.4 and 1.6 Hz), 7.53 (dd, 3H, 10-H, 24-H and 38-H, *J* = 5.2 and 1.2 Hz), 7.51 (ddd, 3H, 8-H, 22-H and 36-H, *J* = 8.1, 7.2 and 1.2 Hz). ^{13}C (^{1}H)- NMR (DMSO-d_6, 125 MHz) δ 135.5 (C-9, C-23 and C-37), 128.9 (C-7, C-21 and C-35), 125.4 (C-8, C-22 and C-36), 117.1 (C-10, C-24 and C-38), 184.9 (C-13, C-27 and C-41), 153.7 (C-5, C-19 and C-33), 153.5 (C-14, C-28 and C-42), 147.8 (C-1, C-15 and C-29), 119.9 (C-6, C-20 and C-34), 114.9 (C-4, C-18 and C-32), 154.9 (C-11, C-25 and C-39); ms: (ESI$^+$) *m/z* (%) 685 (29), 516 (40), 437 (65), 393 (19), 339 (42), 248 (94). Anal. Calcd. for $C_{33}H_{12}N_6O_6S_3$: C, 57.89; H, 1.77; N, 12.27; S, 14.05. Found C, 57.82; H, 1.75; N, 12.30; S, 14.11.

5. BIOLOGICAL ASSAY

5.1 Microorganisms

The studied microorganisms were reference strains (from the ATCC) of *Salmonella typhi* ATCC6539, *Enterococcus faecalis* ATCC10541, *Escherichia coli* ATCC11775, *Klebsiella pneumoniae* ATCC13883, *Candida parapsilosis ATCC 22019*, *C. albicans* ATCC 9002. Also, included were four clinical isolates of *Shigella flexneri*, *Salmonella paratyphi* B, *Salmonella paratyphi* A, and *Candida parapsilosis collected* from Pasteur Centre (Yaounde-Cameroon) and one clinical strain of *Candida glabrata* IP35 obtained from Pasteur Institute (Paris-France). The bacterial and fungal species were maintained on agar slant at +4 °C and subcultured at 37 °C on nutrient agar (NA, Conda, Madrid, Spain) and Sabouraud Dextrose Agar (SDA, Conda) slants respectively, prior to any antimicrobial test.

5.2 Determination of Minimum Inhibitory Concentration (MIC) and Minimum Microbicidal Concentration (MMC)

MIC values were determined by a broth micro-dilution method as described earlier [43] with slight modifications. Each test sample was dissolved in dimethylsulfoxide (DMSO) and the solution was then added to Mueller Hinton Broth (MHB) for bacteria or Sabouraud Dextrose Broth (SDB) for yeasts to give a final concentration of 1024 μg/mL. This was serially diluted twofold to obtain a concentration range of 0.25 - 1024 μg/mL. Then, 100 μL of each concentration were added in each well (96-well microplate) containing 95 μL of MHB or SDB and 5 μL of inoculum for final concentrations varying from 0.25–512 μg/mL. The inoculum was standardized at 1.5×10^6 CFU/mL by adjusting the optical density to 0.1 at 600 nm using a JENWAY 6105 UV/Vis spectrophotometer. The final concentration of DMSO in each well was <1% [preliminary analyses with 1% (v/v) DMSO did not inhibit the growth of the test organisms]. The negative control well consisted of 195 μL of MHB or SDB and 5 μL of the standard inoculum. The plates were covered with sterile lids, then agitated to mix the contents of the wells using a plate shaker and incubated at 37 °C for 24 h (for bacteria) or for 48 h (for yeasts). The assay was repeated three times. The MIC values of samples were determined by adding 50 μL of a 0.2 mg/mL *p*-iodonitrotetrazolium violet solution followed by incubation at 37 °C for 30 min. Viable microorganisms reduced the yellow dye to a pink color. MIC values were defined as the lowest sample concentrations that prevented this change in color indicating a complete inhibition of microbial growth. For the determination of MMC values, a portion of liquid (5 μL) from each well that showed no growth of microorganism was plated on Mueller Hinton Agar or SDA and incubated at 37 °C for 24 h (for bacteria) or 37 °C for 48 h (for yeasts). The lowest concentrations that yielded no growth after this subculturing were taken as the MMC values [44]. Ciprofloxacin (Sigma-Aldrich, Steinheim, Germany) and nystatin (Merck, Darmstadt, Germany) were used as positive controls for bacteria and yeasts, respectively.

5.3 Ferric Reducing Antioxidant Power (FRAP) Assay

The ferric reducing power was determined by the Fe^{3+}-Fe^{2+} transformation in the presence of of compounds **3**, **4** and **6**. The Fe^{2+} was monitored by measuring the formation of Perl's Prussian blue at 700 nm. The method reported by Padmaja *et al.* [45] was used, with slight modification. Briefly, volumes of test solutions were mixed with 500 μL of phosphate buffer (pH 6.6) and 500 μL of 1% potassium ferricyanide and incubated at 50 °C for 20 min. Then 500 μL of 10% trichloroacetic acid was added to the mixture and centrifuged at 3000 rpm for 10 min. Supernatant (500 μL) was diluted with 500 μL of water and shaken with 100 μL of freshly prepared 0.1% ferric chloride. The absorbance was measured at 700 nm. Butylated hydroxytoluene (BHT) was used as a positive control. Increased absorbance of the reaction mixture indicates higher reduction capacity of the compounds.

6. ANTIOXIDANT ASSAY

6.1 DPPH Free radical Scavenging Assay

The free radical scavenging activity of compounds **3**, **4** and **6** was evaluated according to described methods [38]. Briefly, the test samples, prior dissolved in DMSO (SIGMA) beforehand, were mixed with a 20 mg/l 2,2-diphenyl-1-picryl-hydrazyl (DPPH) methanol solution, to give final concentrations of 1, 10, 25, 50, 75, 100, and 200 μg/mL. After 30 min at room temperature, the absorbance values were measured at 517 nm and converted into percentage of antioxidant activity. L-ascorbic acid was used as a standard control. The percentage of decolouration of DPPH (%) was calculated as follows:

$$\% \text{ decolouration of DPPH} = \frac{(\text{Absorbance of control} - \text{Absorbance of test sample}) \times 100}{\text{Absorbance of control}}$$

The radical scavenging percentages were plotted against the logarithmic values of the concentration of test samples and a linear regression curve was established in order to calculate the EC_{50} (μg/ml), which is the amount of sample necessary to inhibit by 50% the absorbance of free radical DPPH. All the analyses were carried out in triplicate.

6.2 Statistical Analysis

Data were analyzed by one-way analysis of variance followed by Waller-Duncan Post Hoc test. The experimental results were expressed as the mean ± Standard Deviation (SD). Differences between groups were considered significant when $p < 0.05$. All analyses were performed using the Statistical Package for Social Sciences (SPSS, version 12.0) software.

7. CONCLUSION

The findings described in the present work have confirmed the synthetic potential of 2-aminothiophenes and the very high reactivity of the title diazonium salt **5.** In addition to the normally anticipated coupling products described the previous investigations, this reactions' intermediate further reacts with itself to give the homocoupling product **6.** Compounds **6** and **4** were respectively the most active compounds against bacteria (MIC = 32-64 μg/ml) and yeasts (MIC = 16–64 μg/ml). Compound **6** also showed high radical-scavenging activity and ferric reducing power when compared with vitamin C and BHT used as reference antioxidants suggesting that this compound could be used as a tool for the design of novel antioxidant derivatives.

CONFLICT OF INTEREST

The authors confirm that this article content has no conflict of interest.

ACKNOWLEDGEMENTS

E.S.F. gratefully acknowledges financial support from DAAD (grant N° A/09/07421) for a scholarship. A.D.N. is grateful to his supervisors Prof. Dr. J. S. Glaser and Dr. R. Marx for helpful suggestions in performing the NMR experiments. Additional financial supports for the work were obtained from the University of Dschang research grant committee and the Cameroonian Ministry of Higher Education special research allocation.

REFERENCES

1. Daidone, G.; Maggio, B.; Schillaci, D. Salicylanilide and its heterocyclic analogues. A comparative study of their antimicrobial activity. Pharmazie, 1990, 45(6), 441-442.
2. Franchini, C.; Muraglia, M.; Corbo, F.; Florio, M.A.; Di Mola, A.; Rosato, A.; Matucci, R.; Nesi, M.; van Bambeke, F.; Vitali, C. Synthesis and biological evaluation of 2-mercapto-1,3-benzothiazole derivatives with potential antimicrobial activity. Arch. Pharm. (Weinheim), 2009, 342(10), 605-613.
3. Pinto, E.; Queiroz, M.J.; Vale-Silva, L.A.; Oliveira, J.F.; Begouin, A.; Begouin, J.M.; Kirsch, G. Antifungal activity of synthetic di(hetero)arylamines based on the benzo[b]thiophene moiety. Bioorg. Med. Chem., 2008, 16(17), 8172-8177.
4. Daniel, V.P.; Murukan, B.; Kumari, B.S.; Mohanan, K. Synthesis, spectroscopic characterization, electrochemical behaviour, reactivity and antibacterial activity of some transition metal complexes with 2-(N-salicylideneamino)-3-carboxyethyl-4,5-dimethylthiophene. Spectrochim. Acta A Mol. Biomol. Spectrosc., 2008, 70(2), 403-410.
5. Ghorab, M.M.; Amin, N.E.; El-Gaby, M.S.; Taha, N.M.; Shehab, M.A.; Faker, I.M. Synthesis, antimicrobial, and antitumor activities of some New bisheterocyclic compounds containing biologically active diphenylsulfone moiety. Phosphorus Sulfur Silicon Relat. Elem., 2008, 183, 2918-2928.
6. Darwish, E.S. Facile synthesis of heterocycles via 2-picolinium bromide and antimicrobial activities of the products. Molecules, 2008, 13(5), 1066-1078.
7. Queiroz, M.J.; Ferreira, I.C.; De Gaetano, Y.; Kirsch, G.; Calhelha, R.C.; Estevinho, L.M. Synthesis and antimicrobial activity studies of ortho-chlorodiarylamines and heteroaromatic tetracyclic systems in the benzo[b]thiophene series. Bioorg. Med. Chem., 2006, 14(20), 6827-6831.

8. Carrillo-Muñoz, A.J.; Giusiano, G.; Ezkurra, P.A.; Quindós, G. Sertaconazole: updated review of a topical antifungal agent. Expert Rev. Anti Infect. Ther., 2005, 3(3), 333-342.
9. Chambhare, R.V.; Khadse, B.G.; Bobde, A.S.; Bahekar, R.H. Synthesis and preliminary evaluation of some N-[5-(2-furanyl)-2-methyl-4-oxo-4H-thieno[2,3-d]pyrimidin-3-yl]-carboxamide and 3-substituted-5-(2-furanyl)-2-methyl-3H-thieno[2,3-d]pyrimidin-4-ones as anti-microbial agents. Eur. J. Med. Chem., 2003, 38(1), 89-100.
10. El-Sharief, S.A.; Micky, J.A.; Shmeiss, N.A.; El-Gharieb, G. Synthesis and reactions of some tetrahydrobenzothieno[2,3-d]pyrimidine derivatives with biological interest. Phosphorus Sulfur Silicon Relat. Elem., 2003, 178, 439-451.
11. El-Sherbeny, M.A.; El-Ashmawy, M.B.; El-Subbagh, H.I.; El-Emam, A.A.; Badria, F.A. Synthesis, antimicrobial and antiviral evaluation of certain thienopyrimidine derivatives. Eur. J. Med. Chem., 1995, 30, 445-449.
12. El-Agrody, A.; El-Latif, A.M.; El-Hady, N.; Fakery, A.; Bedair, A. Heteroaromatization with 4-hydroxycoumarin part II: synthesis of some new pyrano[2,3-d]pyrimidines, [1,2,4]triazolo[1,5-c]pyrimidines and pyrimido[1,6-b][1,2,4]triazine derivatives. Molecules, 2001, 6, 519-527.
13. Rositca, D.N.; Vayssilov, G.N.; Rodios, N.; Bojilova, A. Regio- and stereoselective [2 + 2] photodimerization of 3-substituted 2-alkoxy-2-oxo-2H-1,2-benzoxaphosphorines. Molecules, 2002, 7, 420-432.
14. Al-Amiery, A.A.; Al-Bayati, R.; Saour, K.; Radi, M. Cytotoxicity, antioxidant and antimicrobial activities of novel 2-quinolone derivatives derived from coumarins. Res. Chem. Intermed., 2011, 38, 559-569.
15. Al-Amiery, A.A.; Musa, A.Y.; Kadhum, A.A.; Mohamad, A.B. The use of umbelliferone in the synthesis of new heterocyclic compounds. Molecules, 2011, 16(8), 6833-6843.
16. Kadhum, A.A.; Al-Amiery, A.A.; Musa, A.Y.; Mohamad, A.B. The antioxidant activity of new coumarin derivatives. Int. J. Mol. Sci., 2011, 12(9), 5747-5761.
17. Kadhum, A.H.; Al-Amiery, A.A.; Sikara, M.; Mohamad, A. Synthesis, structure elucidation and DFT studies of new thiadiazoles. Int. J. Phys. Sci., 2011, 6, 6692-6697.
18. Kadhum, A.H.; Al-Amiery, A.A.; Aday, H.; Al-Majedy, Y.K.; Al-Temimi, A.A.; Al-Bayati, R.; Mohamad, A. Co-crystal structure of mixed molecules. Int. J. Phys. Sci., 2012, 7, 1564-1570.
19. Combes, R.D.; Haveland-Smith, R.B. A review of the genotoxicity of food, drug and cosmetic colours and other azo, triphenylmethane and xanthene dyes. Mutat. Res., 1982, 98(2), 101-248.
20. Chhowala, T.N.; Desai, K.R. Synthesis and studies of eco-friendly acid dye metal complexes and its application on woolen fabrics. IOSR J. Appl. Chem., 2015, 8, 5-10.
(a) Cioanca , E-R.; Carlescu, I; Lisa, G.; Scutaru, D. Synthesis, liquid crystalline properties and thermal stability of 4-(4-alkyloxyphenylazo) benzoic acids. Anal. Univ. Buc. Chim., 2010, 19, 39-46. (serie nouă) (b) Savchenko, I. Synthesis and electrooptical properties of side-chain polymethacrylates and polycomplexes containing azobenzene moieties with different length spacers. Chemistry & Chemical Technology, 2014, 8, 383-388.
21. Fuji, Y.; Hanaki, N.; Fujiwara, T.; Tanaka, S.; Noro, M.; Tateishy, K.; Usami, K.; Hibino, A.; Wachi, N.; Taguchi, T.; Yabuk, Y. Development of high durability cyan and magenta dyes for ink jet printing. Fujifilm Research & Development, 2009, 54, 35-42.
22. Awad, I.M.; Aly, A.A.; Abdel-Alim, A.M.; Abdel-Aal, R.A.; Ahmed, S.H. Synthesis of some 5-azo(4′-substituted benzene-sulphamoyl)-8-hydroxyquinolines with antidotal and antibacterial activities. J. Inorg. Biochem., 1988, 33(2), 77-89.
23. Makhsumov, A.G.; Normatov, F.A.; Ergashev, M.S.; Garib, F.Y.; Adylov, S.K. Synthesis of certain hydroxyazobenzenes and investigation of their antimicrobial activity. Pharm. Chem. J., 1991, 25, 542-555.
24. Ibrahim, S.A.; el-Gahami, M.A.; Khafagi, Z.A.; el-Gyar, S.A. Structure and antimicrobial activity of some new azopyrazolone chelates of Ni(II) and Cu(II) acetates, sulfates, and nitrates. J. Inorg. Biochem., 1991, 43(1), 1-7.
25. Jarrahpour, A.A.; Motamedifar, M.; Pakshir, K.; Hadi, N.; Zarei, M. Synthesis of novel azo Schiff bases and their antibacterial and antifungal activities. Molecules, 2004, 9(10), 815-824.

26. Samadhiya, S.; Halve, H. Synthetic utility of Schiff bases as potential herbicidal agents. Orient. J. Chem., 2001, 17, 119-122.
27. Gewald, K. Zur reaktion von α-oxo-mercaptanen mit nitrilen. Angew. Chem., 1961, 73, 114-116.
28. Gewald, K. Heterocyclen aus CH-aciden nitrilen, VII. 2-amino-thiophene aus α-oxo-mercaptanen und methylenaktiven nitrilen. Chem. Ber., 1965, 98, 3571-3577.
29. Ried, W.; Nyiondi, B.E. Über die gemeinsame einwirkung von schwefel und methylenaktiven nitrilen oder ammoniak auf 2-hydroxyacetophenon. Liebigs Ann. Chem., 1973, 1, 134-140.
(a) Fondjo, S.E.; Döpp, D.; Henkel, G. Reactions of some annelated 2-aminothiophenes with electron poor acetylenes. Tetrahedron, 2006, 62, 7121-7131.
30. [http://dx.doi.org/10.1016/j.tet.2006.04.037] (b) Fondjo, S.E.; Tsemeugne, J.; Sondengam, B.L.; Oppenlaender, T.; Kamdem, W.H.; Tane, P.; Connolly, J.D.; Wim, D.; Taoufik, R.; Haruhisa, K.; Yoshiteru, O. Coupling of Two Diazotized 3-Aminothieno[3,4-c]coumarins with Aromatic Amines. J. Heterocycl. Chem., 2011, 48, 1295-1301.
31. Atta-Ur, -Rahman Chapter 1: Chemical Shifts in 1H-NMR Spectroscopy. In Nuclear Magnetic Resonance, Basic Principles;; Springer-Verlag: New York Inc, 1986, pp. 24-5.
32. Atta-Ur, -Rahman Chapter 4: Chemical Shifts and Spin-Spin Couplings in 13C-NMR Spectroscopy. In Nuclear Magnetic Resonance, Basic Principles;; Springer-Verlag: New York Inc, 1986, pp. 140-6.
33. Atta-Ur, -Rahman; Choudhary, I.M. Chapter 7: Recent Developments in NMR Spectroscopy. In Solving Problems with NMR Spectroscopy; Academic Press, Inc.: San Diego: California, 1996, pp. 273-5.
34. Atta-Ur, -Rahman; Choudhary, I.M. Chapter 7: Recent Developments in NMR Spectroscopy. In Solving Problems with NMR Spectroscopy; Academic Press, Inc.: San Diego: California, 1996, pp. 376-7.
35. Amir, M.; Alamkhan, S.; Drabo, S. Synthesis and antibacterial activity of 4-arylazo-1-substituted-3,5-diphenylpyrazoles. J. Indian Chem. Soc., 2002, 79(3), 280-281.
36. Dhingra, V.; Bhatwadekar, R.; Pendse, S. Synthesis and biological activity of some 1-isonicotionoyl-3,5- diphenyl-4-4 substituted phenyl azo pyrazoles. Asian J. Chem., 1993, 5, 515-518.
37. Djouossi, M.G.; Tamokou, J.D.; Ngnokam, D.; Kuiate, J.R.; Tapondjou, A.L.; Harakat, D. Laurence- Nazabadioko, L.V.: Antimicrobial and antioxidant flavonoids from the leaves of Oncoba spinosa Forssk (Salicaceae). BMC Complement. Altern. Med., 2015, 15(134), 1-8.
38. Li, Y.; Guo, C.; Yang, J.; Wei, J.; Xu, J.; Cheng, S. Evaluation of antioxidant properties of pomegranate peel extract in comparison with pomegranate pulp extract. Food Chem., 2006, 96(2), 254-260.
39. Kumar, K.C.; Keshavayya, J.; Rajesh, T.; Peethambar, S.K. Synthesis, characterization and biological activity of heterocyclic azo dyes derived from 2-aminobenzothiozole. Int. J. Pharm. Pharm. Sci., 2013, 5(1), 296-301.
40. Fogue, P.S.; Lunga, P.K.; Fondjo, E.S.; De Dieu Tamokou, J.; Thaddée, B.; Tsemeugne, J.; Tchapi, A.T.; Kuiate, J-R. Substituted 2-aminothiophenes: antifungal activities and effect on Microsporum gypseum protein profile. Mycoses, 2012, 55(4), 310-317.
41. Metwally, M.A.; Suleiman, Y.A.; Gouda, M.A.; Harmal, A.N.; Khalil, A.M. Synthesis, antitumor and antioxidant evaluation of some new antipyrine based azo dyes incorporating pyrazolone moiety. Int. J. Mod. Org. Chem., 2012, 1(3), 213-225.
42. Nyaa, T.B.; Tapondjou, A.L.; Barboni, L.; Tamokou, J.D.; Kuiate, J.R.; Tane, P.; Park, H. NMR assignment and antimicrobial/antioxidant activities of 1β-hydroxyeuscaphic acid from the seeds of Butyrospermum parkii. J. Nat. Prod. Sci., 2009, 15, 76-82.
43. Tamokou, Jde.D.; Kuiate, J.R.; Tene, M.; Kenla Nwemeguela, T.J.; Tane, P. The antimicrobial activities of extract and compounds isolated from Brillantaisia lamium. Iran. J. Med. Sci., 2011, 36(1), 24-31.
44. Padmaja, M.; Sravanthi, M.; Hemalatha, K.P.J. Evaluation of antioxidant activity of two Indian medicinal plants. J. phytol., 2011, 3(3), 86-91.

CHAPTER 7

Synthesis of Pyrroles and Condensed Pyrroles as Anti-Inflammatory Agents with Multiple Activities and Their Molecular Docking Study

M. T. Sarg[1*], M. M. Koraa[2], A. H. Bayoumi[3], S. M. Abd El Gilil[1]

[1]*Department of Organic Chemistry, Faculty of Pharmacy-Girls, Al Azhar University, Cairo, Egypt*

[2]*Department of Organic Chemistry, Faculty of Pharmacy, Cairo University, Giza, Egypt*

[3]*Department of Organic Chemistry, Faculty of Pharmacy-Boys, Al Azhar University, Cairo, Egypt*

ABSTRACT

We herein disclose a series of novel pyrrole derivatives as well as fused pyrrolopyridines 6a,b and 7a,b, pyrrolopyrazoles 8a, b, pyrrolo[2,3-d]pyrimidine derivatives 10a-d, 12a,b, 14a,b, 18a,b, 20a,b, 21a,b, 22a,b, 23a,b, 24a,b, 31a,b, 36a,b, 40a,b, pyrrolo[1,2,6]thiadiazine derivatives 19a,b, pyrrolotriazolopyrimidines 25a,b, 26a,b, 27a,b and 28a,b, pyrrolo[2,3-d][1,2,3]triazine derivatives 32a,b and pyrrolo[2,3-d][1,3]oxazine derivatives 39a,b as novel compounds. All compounds were evaluated for their anti-inflammatory, analgesic (compared to the reference drug Indomethacin) and antimicrobial activities (compared to the reference drug Ampicillin and Fluconazole). Compounds 4d, 5b-d, 6a,b, 9c,d, 10d, 12ab,

13b, 19a,b, 21b, 23b, 31a,b, 38b and 40a were found to be the most active anti-inflammatory drugs exhibiting potency ranging from 1 - 1.01 compared to the reference drug indomethacin. In addition to docking study of these highly active twenty compounds against the active site of cyclooxygenase-2 enzyme (COX-2), among the tested compounds, compounds 5d, 9d, 11b, 12a, 13b and 32a showed multiple activities; anti-inflammatory, analgesic and anti-bacterial activities.

Keywords: Anti-Inflammatory Activity, Pyrrole, Pyrrolo[2,3-d]Pyrimidine, Molecular Modeling

1. INTRODUCTION

Nitrogen heterocycles are of special interest as they constitute an important class of natural and non-natural products as they occupy a key position in the area of drugs and pharmaceuticals [1] [2] . Pyrroles have drawn considerable attention due to their synthetic importance and useful biological activities that are extensively used in drug discovery [3] . Pyrrole derivatives exhibited a vital role in many pharmacological activities including anti-inflammatory [4] - [13] , anti-microbial [14] - [19] , anti-fungal [20] - [22] , antiviral [23] - [25] and anti-cancer [26] - [28] activities.

It is well known that the anti-inflammatory activity is due to the ability to inhibit the cyclooxygenase (COX) activity of prostaglandin H synthase, an enzyme which mediates the production of prostanoids (including prostaglandins, prostacyclins and thromboxanes) from arachidonic acid. Prostaglandins act as mediators in the process of inflammation, thus the discovery of COX-2 specific inhibitors (Coxibs), which their pharmacological properties are correlated to their ability to decrease the COX-2 dependent prostanoid biosynthesis, providing a rational for the development of drug devoid of GIT disorders while retaining clinical efficacy as anti-inflammatory agent [29] . The recent market withdrawal of some coxibs such as rofecoxibs (Vioxx®) and valdecoxib (Bextra®) due to their adverse cardiovascular side effects [30] clearly delineates the need to explore and evaluate alternative templates with COX-2 inhibitory activity.

Therefore, our aim was to design derivatives of existing clinically used NSAIDs, such as Tolmetin and Ketorolac [31] [32] which are well known pyrrole derivatives acting as anti-inflammatory drugs. In the light of these facts, this paper deals with the synthesis of novel pyrrole and condensed pyrrole derivatives and evaluates them for their anti-inflammatory activity. Furthermore, the extent of the pharmacological effects of pyrrole derivatives is reported to depend on the active groups which are attached to it, as several scientists have elucidated that in pyrrole system positions 2 and 3, it can be suitably modified by the introduction of groups [28] , aromatic [19] [25] or heterocyclic moieties to show excellent pharmacological results [12] .

2. RESULTS AND DISCUSSION

2.1. Chemistry

Reaction of 4-chlorophenacyl bromide 1 either with 3-trifluoromethyl aniline 2a in refluxing ethanol or by stirring with 3-amino pyridine 2b in diethyl ether afforded 2-(substituted amino)-1-(4-chlorophenyl)ethanones 3a,b; respectively which upon stirring with the active methylene bearing carbonitriles namely; malononitrile and ethyl cyanoacetate in sodium ethoxide afforded 3-substituted-2-amino-4-(4-chlorophenyl)-1-(substitutedaryl)-1H-py- rroles 4a-d which were utilized as building units for novel substituted pyrrole compounds (Scheme 1).

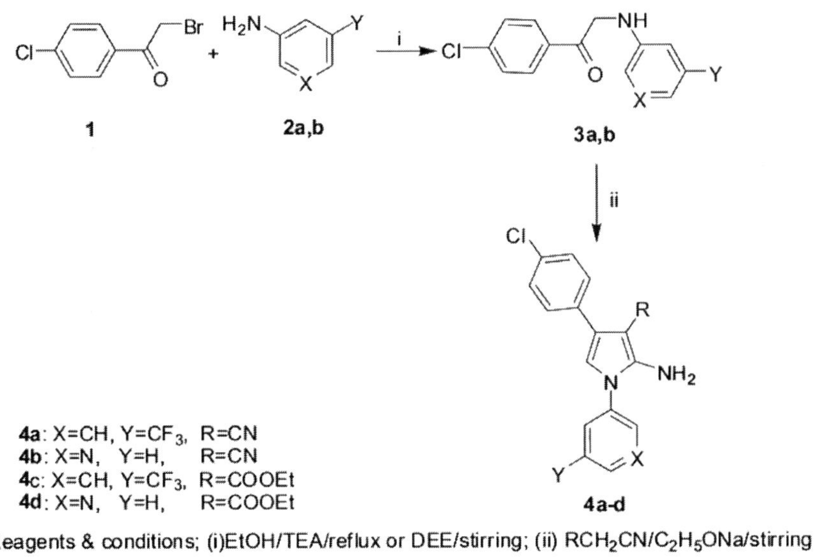

Scheme 1. Synthetic pathways for compounds 3a,b and 4a-d.

Furthermore, the o-aminonitrile derivatives 4a,b were condensed with the appropriate aromatic aldehyde namely; benzaldehyde and 2,4-

dichlorobenzaldehyde in absolute ethanolto afford the corresponding benzylidenimino derivatives 5a-d (Scheme 2). Moreover, reaction of compounds 4a,b either with ethyl cyanoacetate or malononitrile in refluxing ethanol yielded 6-oxopyrrolo[2,3-b]pyridines 6a,b and 4,6-diamino-pyrrolo[2,3- b]pyridine derivatives 7a,b; respectively. Additionally, cyclization of compounds 4a,b with hydroxylamine hydrochloride in boiling sodium ethoxide afforded the aminopyrazole derivatives 8a,b.

Furthermore, compounds 4a,b were reacted with urea/thiourea in sodium ethoxide to give the target compounds 9a-d (Scheme 3), which were cyclized upon refluxing in pyridine to afford the pyrrolopyrimidine derivatives 10a-d which can also be obtained through refluxing of the o-aminonitrile derivatives 4a,b with urea/ thiourea in a mixture of glacial acetic acid and concentrated hydrochloric acid (3:1). Furthermore, the target phenylthioureaderivatives 11a,b was obtained via refluxing of compounds 4a,b with phenyl isothiocyanate in absolute ethanol. Also, refluxing of compounds 11a,b in pyridine afforded the cyclized compounds 4-imino- pyrrolopyrimidine-2-thiones 12a,b which were also prepared directly via refluxing of compounds 4a,b with phenyl isothiocyanate in dry pyridine.

Moreover, our goal was directed to synthesize various substituted pyrrole-3-carbonitrile derivatives bearing different substituted imino side chains in position two of pyrrole (Scheme 4). Therefore, refluxing of compounds 4a,b in a mixture of triethyl orthoformate and acetic anhydride furnished compounds 13a,b with a replaceable ethoxymethyleneimino function which were cyclized upon reaction with ammonia in refluxing methanol via elimination of an ethanol moiety to yield 4-aminopyrrolopyrimidine derivatives 14a,b which were also obtained directly through refluxing of compounds 4a,b with formamide. On the other hand, stirring of compounds 13a,b with dimethylamine in absolute ethanol afforded N,N-dimethyl formimidamide derivatives 15a,b which was also synthesized directly via refluxing compounds 4a,b with dimethylformamide/dimethylacetal in xylene. Moreover, acylation of compounds 4a,b was accomplished through refluxing of compounds 4a,b with p-toluene sulphonyl chloride in toluene to afford 4-methylbenzenesulfonamides 16a,b via elimination of a hydrochloride molecule.

Also, partial hydrolysis of compounds 4a,b was accomplishedby stirring at room temperature with concentrated sulfuric acid to afford substituted-2-aminopyrrol-3-carboxamides 17a,b (Scheme 5), which undergo cyclization into pyrrolo[2,3-d]pyrimidin-4-ones 18a,b, pyrrolo[3,2-d]1,2,6-thiadiazine-2,4-diones 19a,b and 2-me- thyl-3H-pyrrolo[2,3-d]pyrimidin-4(7H)-ones 20a,b upon their refluxing with formamide, thionyl chloride and excess acetic anhydride; respectively. However, the target compounds 18a,b was also obtained directly by refluxing the o-aminonitrile derivatives 4a,b with formic acid.

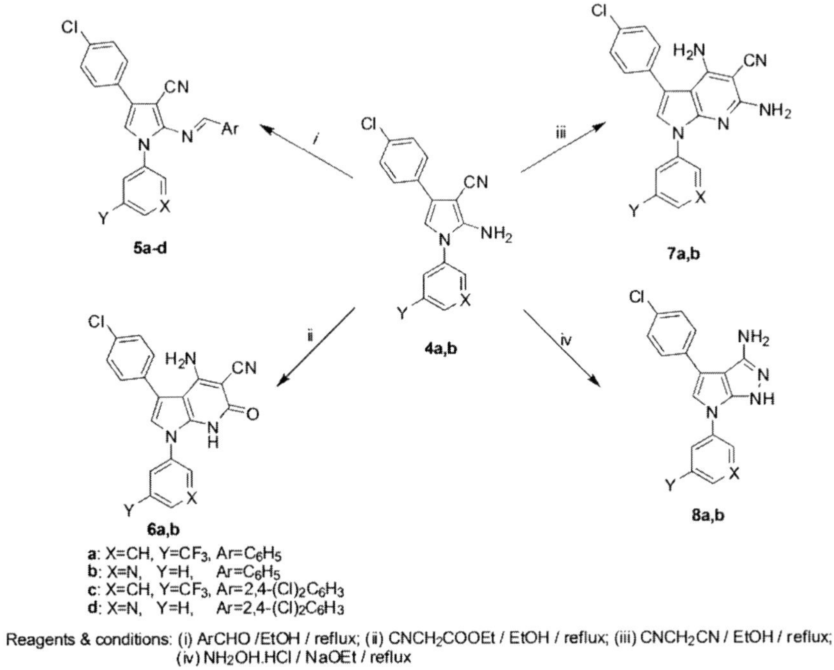

Scheme 2. Synthetic pathways for compounds 5a-d, 6a,b, 7a,b and 8a,b.

Moreover, chlorination of pyrrolopyrimidin-4-one derivatives 18a,b with excess phosphorus oxychloride followed by alkalinization to pH 10 using sodium bicarbonate furnished 4-chloropyrrolo[2,3-d]pyrimidines 21a,b (Scheme 6). However, the chlorinated compounds 21a,b were reacted with morpholine in ethanol under reflux in presence of a catalytic amount of triethylamine to furnish the 4-morpholino-pyrrolo[2,3-d]pyrimidines 22a,b. Furthermore, hydrazinolysis of chloro derivatives 21a,b was carried out by refluxing with hydrazine hydrate 98% in ethanol in presence of triethylamine as a catalyst to afford 4-hydrazinyl-7H-pyrrolo[2,3-d]pyrimidines 23a,b.

Furthermore, stirring of compounds 13a,b with hydrazine hydrate 98% in absolute ethanol afforded 4-imino- 4,7-dihydro-3H-pyrrolo[2,3-d]pyrimidines 24a,b (Scheme 7) which were used as good starting materials for preparation of several pyrrolo[3,2-e][1,2,4]triazolo[1,5-c]pyrimidine derivatives. Thus, compounds 24a,b were reacted with ethyl cyanoacetate in absolute ethanol containing few drops of glacial acetic acid to yield the cyanomethyl triazolopyrimidine derivatives 25a,b. While, refluxing of compounds 24a,b with carbon disulphide in absolute ethanol containing potassium hydroxide yielded the corresponding triazolopyrimidine-2-thione derivatives 26a,b. Moreover, the reaction of the precursors 24a,b with triethyl orthoformate in absolute ethanol gave the unsubstituted triazolopyrimidine compounds 27a,b. While, upon refluxing of 24a,b with phenyl isothiocyanate in absolute ethanol in presence of a catalytic amount of triethylamine afforded the phenyl amine substituted triazolopyrimidine derivatives 28a,b.

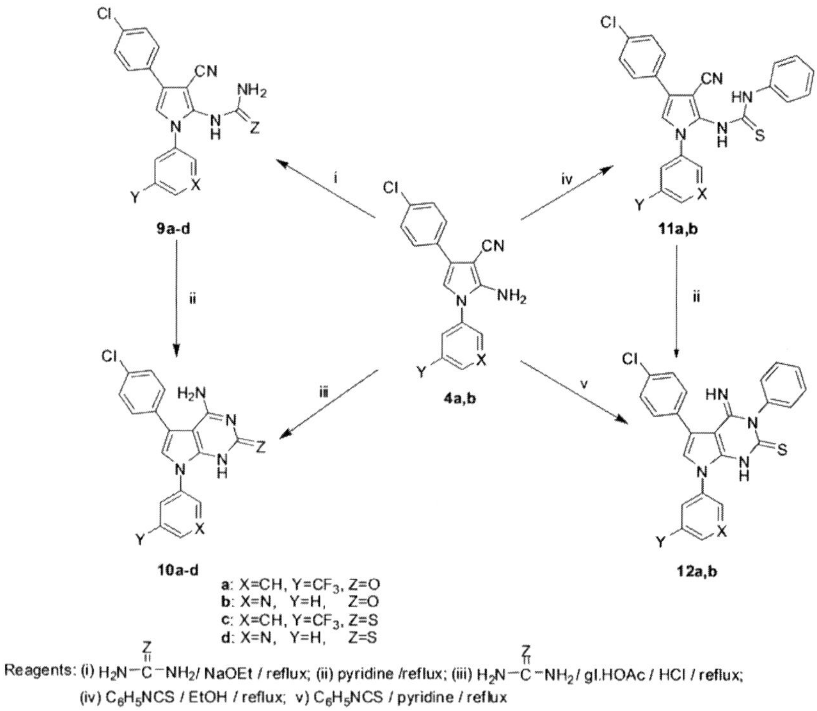

Scheme 3. Synthetic pathways for compounds 9a-d, 10a-d, 11a,b and 12a,b.

Heating of the o-aminonitrile derivatives 4a,b with sodium azide in presence of ammonium chloride in dimethylformamide as a solvent to furnish 2-aminopyrrol-3-tetrazole derivatives 29a,b (Scheme 8). However, heating compounds 4a,b with ethylene diamine in the presence of carbon disulphide afforded the target compounds 2-aminopyrrol-3-imidazole derivatives 30a,b. Also, the o-aminonitrile derivatives 4a,b were refluxed with carbon disulphide in dry pyridine to yield the pyrrolopyrimidine-2,4-dithiones 31a,b. Besides, pyrrolo[2,3- d][1,2,3]triazin-4-ones 32a,b were obtained via stirring of compounds 4a,b with sodium nitrite solution in a mixture of glacial acetic acid and concentrated hydrochloric acid at 0°C - 5°C. The reaction mechanism is suggested to proceed first through partial hydrolysis of the cyano function to afford the corresponding o-aminocarboxamide derivatives that undergo subsequent diazotization then coupling with amino function of the carboxamide to yield the target 4-oxotriazine derivatives 32a & 32b.

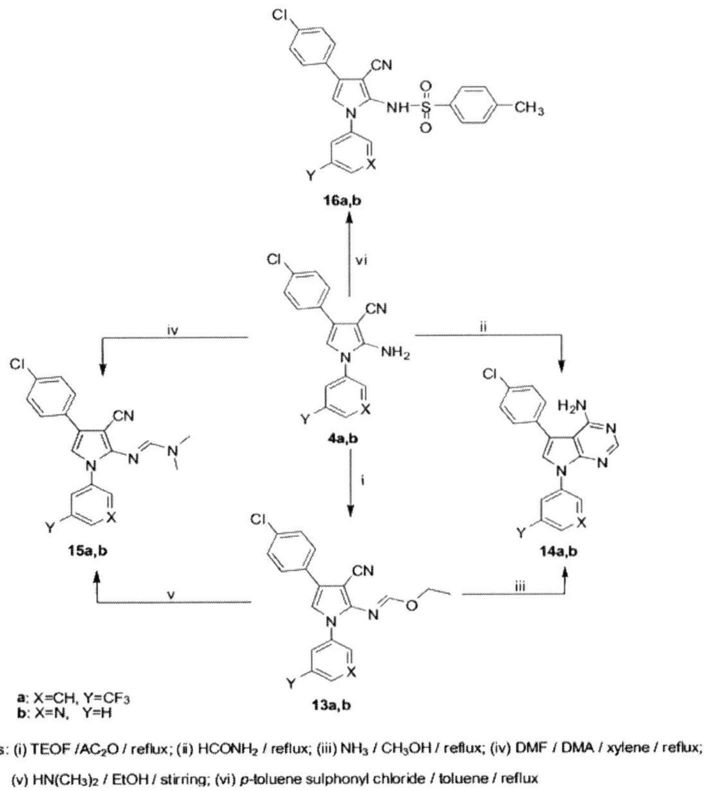

Scheme 4. Synthetic pathways for compounds 13a,b, 14a,b, 15a,b and 16a,b.

This work was extended to shed more light on the activity and synthetic potential of the amino and carboxylate groups in compounds 4c,d (Scheme 9). Thus, compounds 4c,d reacted with hydrazine hydrate in absolute ethanol to afford the corresponding acid hydrazide derivatives 33a,b.

However, acetylation of amino group in compounds 4c,d with acetic anhydride afforded the target acetamido derivatives 34a,b. However, the synthesis of different hydrazone derivatives through the condensation of hydrazine compounds with monosaccharide is well documented in the literature [33] - [35] . Therefore, this information prompted us to explore the effect of glucose hydrazone derivatives 35a,b whichwere synthesized through the condensation of the hydrazide derivatives 33a,b with glucose in absolute ethanol. Additionally, the reaction of o-acetamidocarboxylate derivatives 34a,b were refluxed with excess hydrazine hydrate in ethanol to afford the target substituted pyrrolopyrimidinone derivatives 36a,b. The reaction mechanism is suggested to proceed first through formation of the hydrazone derivatives followed by tautomerism then intramolecular cyclization via elimination of an ethanol molecule.

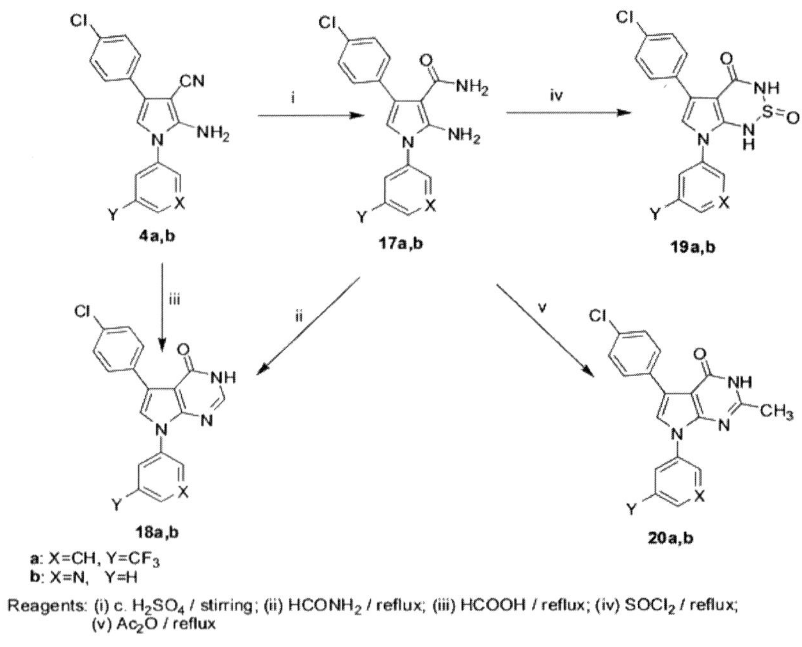

Scheme 5. Synthetic pathways for compounds 17a,b, 18a,b, 19a,b and 20a,b.

Finally, the title carboxylic acid derivatives 37a,b was obtained via refluxing of the o-aminoester compounds 4c,d with ethanolic sodium hydroxide (Scheme 10). Also, our aim was extended to develop novel ethyl 1-(substitutedaryl)-4-(4-chlorophenyl)-2-(3-phenylthioureido)-1H-pyrrol-3-carboxylates 38a,b which were obtained via refluxing of 2-aminopyrrole derivatives 4c,d with phenylisothiocyanate in absolute ethanol. Furthermore, fused 2-methyloxazine-4-one rings are reported to be prepared from the reaction of different o-amino- carboxylic acid derivatives with excess acetic anhydride. Thus, the target 2-methyloxazine-4-one derivatives 39a,b were synthesized via refluxing of the o-aminocarboxylic acid compounds 37a,b in excess acetic anhydride. Finally, the thiourea derivatives 38a,b were heated with sodium ethoxide under reflux to furnish the target 2-thioxopyrimidin-4-one derivatives 40a,b.

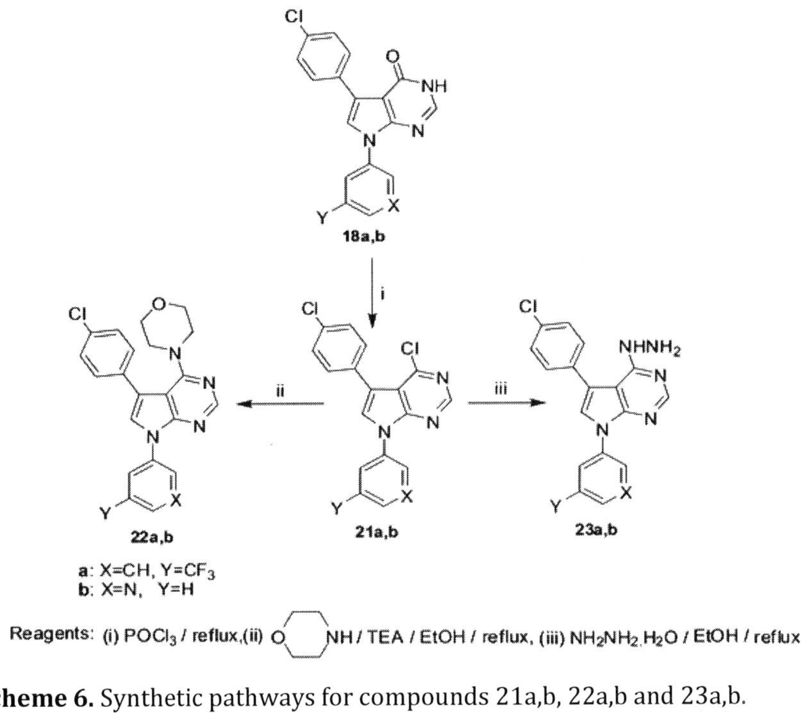

Scheme 6. Synthetic pathways for compounds 21a,b, 22a,b and 23a,b.

2.2. Biology

All the newly synthesized compounds 4a-40b were preliminarily evaluated for their anti-inflammatory and analgesic activities (using rat paw edema method and writhing test; respectively) as well as their gastric ulcerative effect (ulcerogenicity) an in-vitro antibacterial activity against Staphylococcus aureus (ATCC 25923) as a representative of Gram-positive bacteria; Pseudomonas aeruginosa (ATCC 27853) and Escherichia coli (ATCC 8739) as representatives of Gram-negative bacteria. The compounds were also evaluated for their in-vitro antifungal activity against Candida albicans (ATCC 10231) (using the cup diffusion technique).

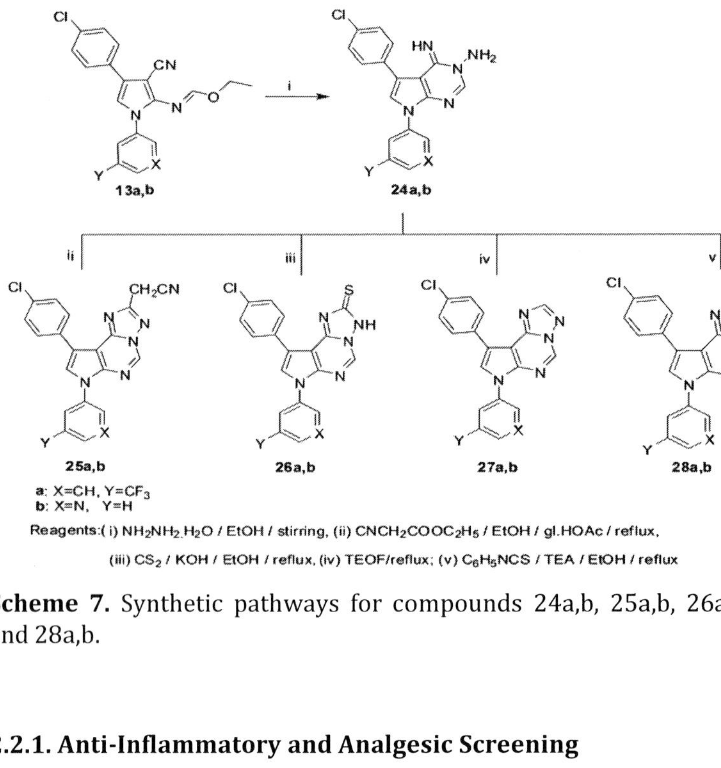

Scheme 7. Synthetic pathways for compounds 24a,b, 25a,b, 26a,b, 27a,b and 28a,b.

2.2.1. Anti-Inflammatory and Analgesic Screening

For the tested compounds 4a-40b, the percent of edema inhibition after 1-6 h and the percent inhibition of the writhing movements are presented in Tables 1-6.

It was revealed from the results that, compounds 4d, 5b, 5c, 5d, 6a, 6b, 9c, 9d, 10d, 12a, 12b, 13b, 19a, 19b, 21b, 23a, 23b, 31a, 31b, 38b and 40a exerted highly potent anti-inflammatory effect, comparable to that of indomethacin (Indocin®) at 6 h interval post carrageenan showing inhibition potency ranging from 1.01% - 1.05%. While, compounds 4b, 4c, 5a, 7a, 7b, 8b, 9a, 9b, 9c, 10b, 10c, 11a, 11b, 13a, 13b, 14b, 15b, 16a, 16b, 18a, 18b, 20b, 21a, 24a, 25b, 26a, 26b, 28b, 32a, 32b, 34a, 34b, 38a and 40b exerted moderate anti-inflammatory activity at 6 h interval post carrageenan, comparable with that of indomethacin (Indocin®) showing inhibition potency ranging from 0.68% - 1%.

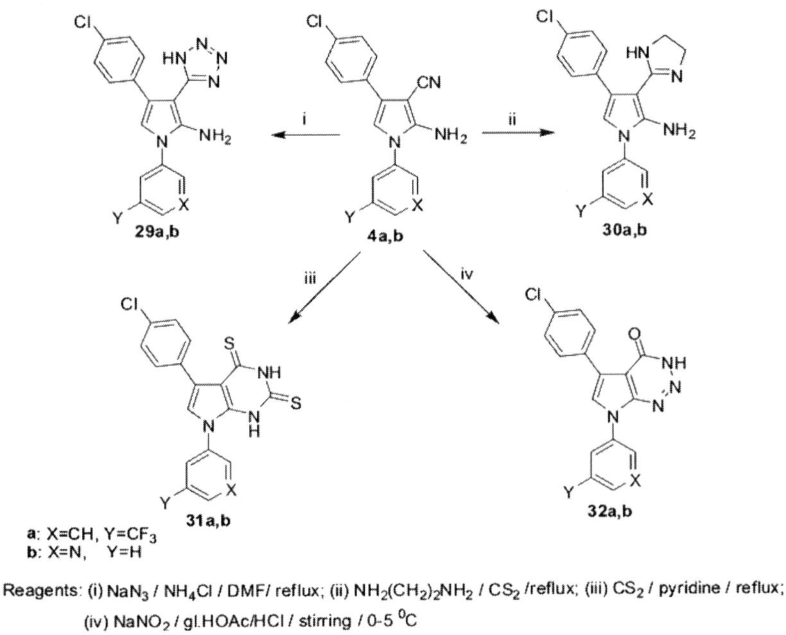

Scheme 8. Synthetic pathways for compounds 29a,b, 30a,b, 31a,b and 32a,b.

In addition to, compounds 4a, 8a, 10a, 14a, 15a, 22a, 22b, 25a, 27a, 27b, 28a, 29a, 29b, 30a, 30b, 33a, 33b, 35a, 35b, 36a, 36b, 37a, 37b, 39a and 39b which showed weak anti-inflammatory activity at 6 h interval less than indomethacin showing inhibition potency ranging from 0.36% - 0.67%. The activity profiles of all the previous compounds were the same as indomethacin (response increasing by time).

It is worth mentioning that, the highly potent compounds were those comprising 3-cyanopyrrole rings attached to different side chains in the 2 position, among these chains are the aryl imino function as in compounds 5b-d, thiourea group as in compounds 9c,d and ethoxymethyleneimino chain in compound 13b. Also, compounds containing pyrrole-3-carboxylate with the 2 position either unsubstituted as in compound 4d or substituted with thiourea side chain as compound 38b were highly potent.

Furthermore, among the highly potent compounds were thefused pyrrolopyrimidine compounds bearing 2-thioxo function with different substituents in the 4 position. Among these groups in the 4 position was the amino group as compound 10d, imino function in compounds 12a,b, 4-thioxo function as in compounds 31a,b and 4-oxo group as in compound 40a. Also, the 4-chloropyrrolopyrimidine derivative 21b as well as the 4-

hydrazinopyrrolopyrimidine analogue 23b exhibited potent activity comparable to the reference drug Indomethacin (Indocin®).

Moreover, other fused pyrrole compounds such as the pyrrolopyridine-2-one derivatives 6a,b and the pyrrolo[1,2,6]thiadiazine-2,4-dione analogues 19a,b exerted excellent activity.

As revealed from the results presented in Tables 1-3 that, compounds 4d, 5c, 5d, 6b, 10c, 10d, 17b, 31a and 31b exhibited the most potent analgesic activity with potency ranging from 1 - 1.10 to the reference drug Indomethacin. It is to be noted that some functions are assumed to be responsible for the highly potent analgesic activity of these compounds. Among these functions are the ester function in 3-position as in compound 4d, the 2,4-dichlorobenzylidine imino function in pyrrole-2-position as in compounds 5c and 5d and pyrrolopyridine moiety in compound 6b. Also, the pyrrolopyrimidine thione function as in compounds 10c and 10d.

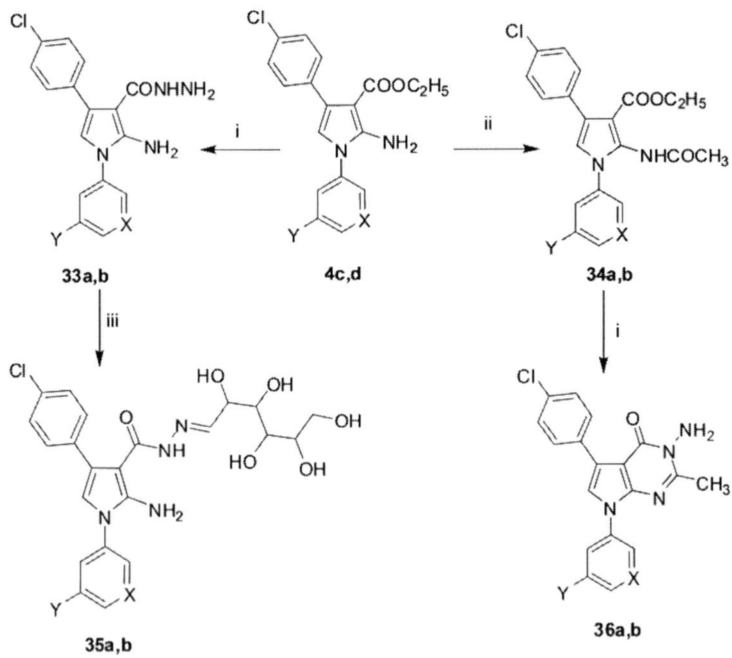

a: X=CH, Y=CF_3
b: X=N, Y=H

Reagents: (i) $NH_2NH_2.H_2O$ / EtOH /reflux; (ii) Ac_2O / reflux; (iii) glucose/ EtOH/ reflux

Scheme 9. Synthetic pathways for compounds 33a,b, 34a,b, 35a,b and 36a,b.

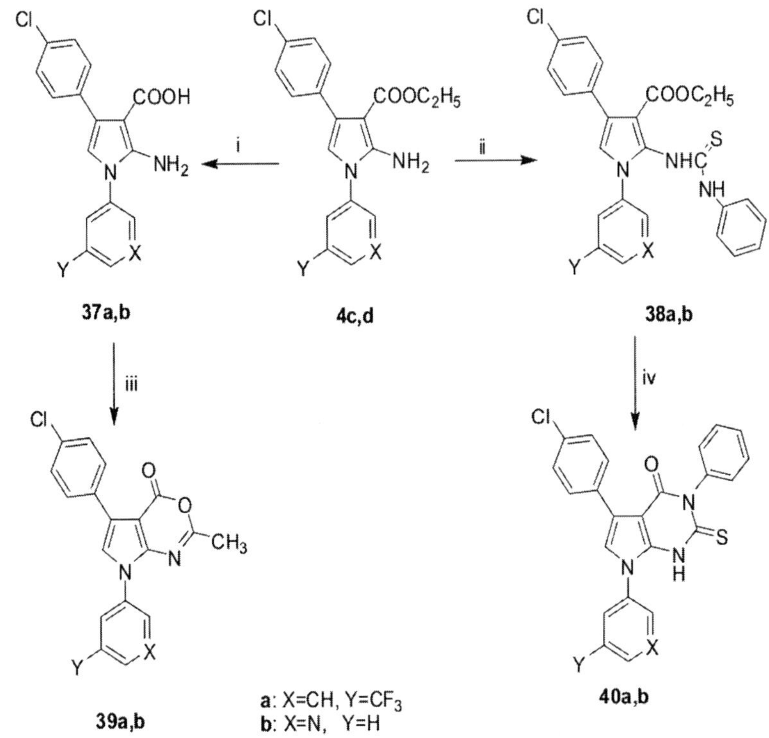

Reagents: (i) NaOH /EtOH / HOAc/ reflux; (ii) C_6H_5NCS/EtOH/reflux; (iii) Ac_2O/ reflux; (iv) NaOEt / reflux

Scheme 10. Synthetic pathways for compounds 37a,b, 38a,b, 39a,b and 40a,b.

Furthermore, the carboxamide function as in compound 17b led to high analgesic activity. Additionally, pyrrolopyrimidine dithione function as in compounds 31a and 31b exerted the most potent analgesic activity.

2.2.2. Ulcerogenicity

Five compounds that exhibited the most potent anti-inflammatory activity; 12b, 23b, 31a, 31b and 38b were evaluated for their ulcerative effect on rats as revealed in Table 7.

In general, all the tested compounds showed better results than the reference drug Indomethacin. Especially, compounds 12b and 31b which were devoid of any ulcerative effect compared to 94% of that of Indomethacin as illustrated in the previous table.

Table 1. Anti-inflammatory and analgesic results for compounds of Scheme 1 and Scheme 2.

Comp. No.	Mean ± S.D. (Percent edema inhibition)							Analgesic activity		
	1 h	2 h	3 h	4 h	5 h	6 h	Potency	No. of writhing movements	Percent inhibition	Potency
Control	0.23 ± 0.03 (----)	0.26 ± 0.05 (-----)	0.45 ± 0.01 (-----)	0.54 ± 0.08 (-----)	0.63 ± 0.04 (-----)	0.78 ± 0.12 (-----)	----	55	---	---
Indocin	0.22 ± 0.03 (11.73)	0.14 ± 0.03 (20.50)	0.21 ± 0.02 (44.12)	0.23 ± 0.04 (52.71)	0.22 ± 0.08 (69.5)	0.05 ± 0.02 (88.70)	1	9	83.63	**1**
4a	0.21 ± 0.03 (0.71)	0.25 ± 0.04 (4.23)	0.37 ± 0.03 (32.84)	0.5 ± 0.002 (21.06)	0.35 ± 0.06 (37)	0.32 ± 0.03 (50)	0.56	23	58.18	0.69
4b	0.34 ± 0.02 (6.31)	0.35 ± 0.03 (22.12)	0.36 ± 0.07 (37.20)	0.24 ± 0.02[a,b] (55.50)	0.16 ± 0.08[a,b] (73.5)	0.13 ± 0.03[a] (61.21)	0.69	15	72.72	0.86
4c	0.22 ± 0.03 (5.60)	0.26 ± 0.02 (13.21)	0.18 ± 0.02 (38.21)	0.33 ± 0.03 (43.10)	0.17 ± 0.04[a] (70.4)	0.13 ± 0.004[a] (74.16)	0.83	10	81.81	0.97
4d	0.21 ± 0.02 (8.81)	0.10 ± 0.02 (43.16)	0.22 ± 0.09 (56.32)	0.14 ± 0.01[a] (66)	0.22 ± 0.02 (72.4)	0.07 ± 0.03[a] **(90.97)**	**1.02**	6	89.09	**1.06**
5a	0.22 ± 0.02 (5.65)	0.21 ± 0.04 (23.08)	0.35 ± 0.04 (36.21)	0.25 ± 0.01[a,b] (54)	0.36 ± 0.05 (69.84)	0.13 ± 0.04[a,b] (79.42)	0.89	27	50.90	0.60
5b	0.23 ± 0.05 (8.70)	0.19 ± 0.04 (15.94)	0.38 ± 0.02 (29.50)	0.26 ± 0.05[a,b] (52.20)	0.24 ± 0.04[a,b] (61.90)	0.19 ± 0.01 **(89.07)**	**1.010**	18	67.27	0.80

5c	0.19 ± 0.06 (17.39)	0.44 ± 0.01 (36.50)	0.36 ± 0.01 (42.85)	0.37 ± 0.05 (55.36)	0.29 ± 0.02[a,b] (63.50)	0.22 ± 0.08[a] **(89.54)**	**1.011**	7	87.27	**1.04**
5d	0.21 ± 0.02 (6.19)	0.20 ± 0.02 (32.41)	0.26 ± 0.02 (23.47)	0.20 ± 0.03 (52.38)	0.40 ± 0.04 (59.61)	0.13 ± 0.004 **(89.67)**	**1.012**	8	85.45	**1.02**
6a	0.23 ± 0.02 (13.04)	0.21 ± 0.04 (19.23)	0.41 ± 0.01 (39.20)	0.37 ± 0.05 (43.56)	0.20 ± 0.07 (67.32)	0.36 ± 0.05 **(91.08)**	**1.02**	14	74.54	0.89
6b	0.37 ± 0.03 (11.15)	0.48 ± 0.08 (25.47)	0.32 ± 0.04 (48.89)	0.21 ± 0.02 (61.05)	0.21 ± 0.06 (71.08)	0.41 ± 0.01 **(91.96)**	**1.03**	8	85.45	**1.02**
7a	0.21 ± 0.02 (13.41)	0.44 ± 0.02 (17.31)	0.19 ± 0.04 (37.21)	0.22 ± 0.02 (51.31)	0.23 ± 0.02 (59.10)	0.36 ± 0.01 (61.24)	0.69	19	65.45	0.78
7b	0.22 ± 0.06 (11.35)	0.23 ± 0.07 (21.82)	0.41 ± 0.03 (19.07)	0.38 ± 0.02 (49.14)	0.19 ± 0.06 (61.24)	0.33 ± 0.02 (88.13)	0.99	11	80	0.95
8a	0.21 ± 0.03 (6.19)	0.20 ± 0.03 (27.81)	0.18 ± 0.06 (38.21)	0.21 ± 0.03 (40.11)	0.2 ± 0.08 (47.12)	0.34 ± 0.06 (56.02)	0.63	37	32.72	0.39
8b	0.14 ± 0.06 (7.41)	0.12 ± 0.06 (18.47)	0.32 ± 0.03[a] (22.46)	0.19 ± 0.04 (37.28)	0.33 ± 0.03 (43.62)	0.12 ± 0.06 (61.35)	0.69	33	40	0.47

Table 2. Anti-inflammatory and analgesic results for compounds of Scheme 3.

Comp. No.	Mean ± S.D. (Percent edema inhibition)							Analgesic activity		
	1 h	2 h	3 h	4 h	5 h	6 h	Potency	No. of writhing movements	Percent inhibition	Potency
9a	0.21 ± 0.03 (9.50)	0.25 ± 0.04 (5.17)	0.18 ± 0.02[a] (13.03)	0.43 ± 0.02 (28.67)	0.28 ± 0.08 (43.39)	0.21 ± 0.02[a] (60.66)	0.68	37	32.72	0.39
9b	0.22 ± 0.02 (2.17)	0.14 ± 0.05 (3.07)	0.48 ± 0.07 (11.4)	0.45 ± 0.09 (27.85)	0.12 ± 0.04 (31.59)	0.54 ± 0.08 (69.52)	0.78	27	50.90	0.60
9c	0.15 ± 0.06 (10.3)	0.44 ± 0.01 (16.5)	0.18 ± 0.04 (36.5)	0.34 ± 0.02 (43.41)	0.17 ± 0.03[a,b] (73.52)	0.46 ± 0.09 (89.10)	1	25	54.54	0.65
9d	0.19 ± 0.04 (15.9)	0.31 ± 0.02 (19.7)	0.42 ± 0.09 (32.8)	0.36 ± 0.02 (52.36)	0.25 ± 0.01[a,b] (78.65)	0.21 ± 0.04 **(89.72)**	**1.01**	23	58.18	0.69
10a	0.27 ± 0.02 (6.08)	0.43 ± 0.06 (22.5)	0.45 ± 0.05 (31.52)	0.23 ± 0.08 (41.37)	0.36 ± 0.05 (41.49)	0.22 ± 0.02 (59.43)	0.67	14	74.54	0.89
10b	0.19 ± 0.06 (17.3)	0.14 ± 0.03 (20.9)	0.37 ± 0.01 (31.73)	0.41 ± 0.07 (39.24)	0.23 ± 0.06 (43.26)	0.23 ± 0.08 (61.42)	0.69	12	78.81	0.94
10c	0.44 ± 0.08 (0.00)	0.23 ± 0.05 (16.4)	0.36 ± 0.05 (41.28)	0.21 ± 0.02 (48.31)	0.34 ± 0.02 (66.82)	0.22 ± 0.04 (80.42)	0.90	9	83.63	**1**
10d	0.41 ± 0.03 (4.18)	0.32 ± 0.01 (27.2)	0.12 ± 0.02 (39.31)	0.21 ± 0.06 (52.45)	0.45 ± 0.05 (71.34)	0.45 ± 0.05 **(89.83)**	**1.01**	7	87.27	**1.04**
11a	0.42 ± 0.01 (3.48)	0.24 ± 0.02 (8.31)	0.41 ± 0.01 (23.21)	0.33 ± 0.03 (45.32)	0.24 ± 0.02 (69.45)	0.21 ± 0.04 (88.40)	0.97	24	56.36	0.67
11b	0.16 ± 0.04 (7.32)	0.42 ± 0.04 (11.4)	0.48 ± 0.08 (22.91)	0.21 ± 0.05 (35.37)	0.22 ± 0.06 (63.49)	0.36 ± 0.05 (88.67)	0.99	23	58.18	0.69
12a	0.49 ± 0.003 (21.0)	0.32 ± 0.04 (33.6)	0.25 ± 0.06 (49.56)	0.34 ± 0.03 (61.52)	0.25 ± 0.11 (81.63)	0.12 ± 0.09[a] **(91.89)**	**1.03**	16	70.90	0.84
12b	0.19 ± 0.03 (17.4)	0.34 ± 0.03 (26.8)	0.44 ± 0.08 (58.32)	0.25 ± 0.04 (71.43)	0.09 ± 0.06 (83.51)	0.15 ± 0.12[a] **(92.04)**	**1.04**	13	76.36	0.91

Table 3. Anti-inflammatory and analgesic results for compounds of Scheme 4.

Comp. No.	Mean ± S.D. (Percent edema inhibition)							Analgesic activity		
	1 h	2 h	3 h	4 h	5 h	6 h	Potency	No. of writhing movements	Percent inhibition	Potency
13a	0.22 ± 0.06 (17.9)	0.23 ± 0.01 (0)	0.53 ± 0.08 (51.37)	0.21 ± 0.05 (41.32)	0.54 ± 0.04 (53.67)	0.37 ± 0.03 (61.40)	0.69	27	50.90	0.60
13b	0.19 ± 0.02 (12.4)	0.21 ± 0.05 (8.61)	0.47 ± 0.08 (39.55)	0.46 ± 0.05 (52.47)	0.18 ± 0.03 (62.43)	0.24 ± 0.02 (89)	1	25	54.54	0.65
14a	0.27 ± 0.02 (0)	0.18 ± 0.04 (0)	0.34 ± 0.01 (19.32)	0.36 ± 0.01 (27.74)	0.24 ± 0.06 (33.27)	0.21 ± 0.02 (42.85)	0.48	57	---	---
14b	0.36 ± 0.07 (3.69)	0.22 ± 0.03 (10.2)	0.35 ± 0.05 (17.86)	0.41 ± 0.07 (38.31)	0.21 ± 0.03 (41.84)	0.61 ± 0.04 (61.90)	0.69	43	21.81	0.26
15a	0.22 ± 0.07 (3.40)	0.17 ± 0.03 (7.39)	0.46 ± 0.03 (21.74)	0.34 ± 0.04 (11.15)	0.23 ± 0.02 (42.58)	0.22 ± 0.04 (55.80)	0.63	33	40	0.47
15b	0.22 ± 0.02 (7.39)	0.48 ± 0.08 (18.2)	0.35 ± 0.02 (34.58)	0.24 ± 0.02[a,b] (45.63)	0.35 ± 0.05 (51.43)	0.43 ± 0.01 (69.20)	0.78	29	47.27	0.56
16a	0.37 ± 0.03 (0)	0.31 ± 0.01 (13.4)	0.38 ± 0.02 (31.23)	0.24 ± 0.05 (43.51)	0.29 ± 0.02 (52.10)	0.36 ± 0.05 (66)	0.74	27	50.90	0.60
16b	0.46 ± 0.08 (11.5)	0.24 ± 0.03[a,b] (17.3)	0.35 ± 0.08 (41.30)	0.33 ± 0.04 (48.37)	0.24 ± 0.01[a,b] (61)	0.28 ± 0.03[a] (84)	0.94	23	58.18	0.69

[a,b]: Significantly different from control value and reference value at $P < 0.05$. •S.D. = Standard deviation.

Table 4. Anti-inflammatory and analgesic results for compounds of Scheme 5 and Scheme 6.

	Mean ± S.D. (Percent edema inhibition)							Analgesic activity		
Comp. No.	1 h	2 h	3 h	4 h	5 h	6 h	Potency	No. of writhing movements	Percent inhibition	Potency
17a	0.05 ± 0.04 (76.40)	0.08 ± 0.01 (68.42)	0.18 ± 0.003 (66.41)	0.22 ± 0.02 (52)	0.14 ± 0.06 (49.42)	0.12 ± 0.06 (40.71)	0.46	10	81.81	0.97
17b	0.21 ± 0.05 (88.87)	0.21 ± 0.02 (71.31)	0.12 ± 0.04[a] (62.74)	0.13 ± 0.04 (58.04)	0.19 ± 0.04 (50.10)	0.18 ± 0.04 (55.37)	0.62	8	85.45	**1.02**
18a	0.21 ± 0.03 (45.32)	0.09 ± 0.003[a] (47.48)	0.25 ± 0.03 (51.21)	0.17 ± 0.05 (53.42)	0.48 ± 0.07 (57.39)	0.12 ± 0.06[a] (85)	0.96	19	65.45	0.78
18b	0.25 ± 0.04 (55)	0.32 ± 0.01 (56)	0.32 ± 0.01 (61.36)	0.10 ± 0.01 (50.21)	0.45 ± 0.09 (72.35)	0.43 ± 0.02 (87.21)	0.98	14	74.54	0.89
19a	0.27 ± 0.05 (17.40)	0.33 ± 0.04 (23.64)	0.45 ± 0.08 (59.41)	0.45 ± 0.05 (64)	0.45 ± 0.07 (70.26)	0.25 ± 0.04 **(90.63)**	**1.02**	12	78.81	0.94
19b	0.15 ± 0.06 (20.51)	0.23 ± 0.07 (31.36)	0.21 ± 0.02 (46.07)	0.39 ± 0.05 (57.42)	0.23 ± 0.07 (73)	0.24 ± 0.04[a,b] **(90.88)**	**1.02**	10	81.81	0.97
20a	0.24 ± 0.02 (4.21)	0.23 ± 0.03 (11.32)	0.25 ± 0.03 (30.22)	0.25 ± 0.04 (47)	0.26 ± 0.02 (59.41)	0.47 ± 0.07 (55.61)	0.63	27	50.90	0.60

20b	0.23 ± 0.04 (6.26)	0.23 ± 0.01 (14.21)	0.47 ± 0.06 (34.52)	0.22 ± 0.022 (44.39)	0.23 ± 0.03 (57.23)	0.47 ± 0.1 (63)	0.71	21	61.81	0.73
21a	0.23 ± 0.02 (11.65)	0.31 ± 0.04 (21.64)	0.33 ± 0.03 (33.61)	0.22 ± 0.03[a,b] (59)	0.36 ± 0.05 (68.32)	0.14 ± 0.03[a,b] (89)	1	25	54.54	0.65
21b	0.23 ± 0.05 (8.70)	0.29 ± 0.05 (26.87)	0.25 ± 0.05 (34.38)	0.25 ± 0.03[a,b] (62.43)	0.24 ± 0.05[a,b] (61.90)	0.18 ± 0.01 **(91.70)**	**1.03**	23	58.18	0.69
22a	0.17 ± 0.04 (0)	0.45 ± 0.02 (11.50)	0.24 ± 0.02 (22.84)	0.37 ± 0.05 (33.44)	0.19 ± 0.03[a,b] (42.31)	0.23 ± 0.08[a] (48.21)	0.54	55	---	---
22b	0.24 ± 0.02 (2.40)	0.25 ± 0.02 (8.43)	0.13 ± 0.02 (23.36)	0.21 ± 0.03 (38.36)	0.50 ± 0.04 (46.53)	0.12 ± 0.05 (49.57)	0.56	57	---	---
23a	0.23 ± 0.03 (12.04)	0.23 ± 0.06 (21.23)	0.37 ± 0.02 (44.20)	0.37 ± 0.05 (52.27)	0.23 ± 0.03 (72.21)	0.37 ± 0.03 **(89.93)**	**1.01**	12	78.81	0.94
23b	0.36 ± 0.04 (14.15)	0.43 ± 0.04 (27.31)	0.35 ± 0.03 (50)	0.21 ± 0.02 (63.25)	0.22 ± 0.06 (74.30)	0.42 ± 0.02 **(92.56)**	**1.04**	10	81.81	0.97

28b	0.23 ± 0.03 (0)	0.21 ± 0.08 (6.41)	0.15 ± 0.03 (29.30)	0.21 ± 0.02 (37.05)	0.40 ± 0.01 (51.56)	0.13 ± 0.07 (65.47)	0.74	18	67.27	0.80
29a	0.12 ± 0.02 (0)	0.24 ± 0.05 (2.51)	0.34 ± 0.02 (11.53)	0.39 ± 0.02 (26)	0.65 ± 0.03 (33.96)	0.35 ± 0.03 (41.32)	0.47	56	---	---
29b	0.35 ± 0.02 (3.21)	0.27 ± 0.02 (7.31)	0.25 ± 0.06 (0)	0.16 ± 0.02[a,b] (22.41)	0.14 ± 0.07[a,b] (40)	0.21 ± 0.04[a,b] (52.67)	0.59	54	---	---
30a	0.27 ± 0.01 (5.42)	0.07 ± 0.01[a] (10.42)	0.32 ± 0.03 (21.36)	0.14 ± 0.06 (33.12)	0.35 ± 0.08 (41.44)	0.14 ± 0.05[a] (44)	0.49	25	54.54	0.65
30b	0.24 ± 0.03 (0)	0.33 ± 0.03 (12.53)	0.45 ± 0.02 (23.64)	0.08 ± 0.02[a] (41.26)	0.32 ± 0.08 (32.36)	0.76 ± 0.03 (56.31)	0.63	20	63.63	0.76
31a	0.26 ± 0.03 (16.41)	0.32 ± 0.04 (22.64)	0.44 ± 0.07 (57.46)	0.45 ± 0.05 (62)	0.55 ± 0.07 (75.26)	0.23 ± 0.05 **(93.04)**	**1.04**	6	89.09	**1.06**
31b	0.14 ± 0.05 (20.89)	0.22 ± 0.07 (32.38)	0.31 ± 0.02 (47.09)	0.28 ± 0.04 (67.42)	0.23 ± 0.06 (77)	0.23 ± 0.05[a,b] **(93.86)**	**1.05**	4	92.72	**1.10**
32a	0.23 ± 0.03 (7.31)	0.24 ± 0.037 (16.42)	0.24 ± 0.03 (0)	0.24 ± 0.04 (43.39)	0.25 ± 0.07 (61.47)	0.46 ± 0.06 (80)	0.90	18	67.27	0.80
32b	0.27 ± 0.03 (6.27)	0.23 ± 0.02 (24.23)	0.37 ± 0.05 (37.42)	0.27 ± 0.03 (48.58)	0.22 ± 0.04 (73.63)	0.36 ± 0.09 (82.41)	0.93	15	72.72	0.86

Table 5. Anti-inflammatory and analgesic results for compounds of Scheme 7 & Scheme 8.

	Mean ± S.D. (Percent edema inhibition)							Analgesic activity		
Comp. No.	1 h	2 h	3 h	4 h	5 h	6 h	Potency	No. of writhing movements	Percent inhibition	Potency
24a	0.11 ± 0.04 (4.46)	0.23 ± 0.03 (27.35)	0.27 ± 0.06 (33.19)	0.31 ± 0.02 (44.12)	0.24 ± 0.07 (67.13)	0.42 ± 0.06 (79.47)	0.89	36	34.54	0.41
24b	0.25 ± 0.04 (16.32)	0.45 ± 0.05 (31.40)	0.58 ± 0.07 (22.38)	0.21 ± 0.05 (45.77)	0.21 ± 0.06 (65.09)	0.37 ± 0.05 (88.02)	0.99	29	47.27	0.56
25a	0.20 ± 0.01 (9.03)	0.42 ± 0.05 (12.57)	0.35 ± 0.05 (23.52)	0.22 ± 0.07 (0)	0.25 ± 0.04 (31.47)	0.21 ± 0.03 (51.38)	0.58	23	58.81	0.69
25b	0.18 ± 0.04 (13.40)	0.11 ± 0.02 (10.9)	0.36 ± 0.02 (19.83)	0.42 ± 0.10 (35.30)	0.22 ± 0.05 (43.06)	0.22 ± 0.07 (60.26)	0.68	20	63.63	0.76
26a	0.26 ± 0.03 (7.41)	0.14 ± 0.05 (13.45)	0.47 ± 0.06 (27.30)	0.46 ± 0.02 (47.44)	0.12 ± 0.02 (51.45)	0.53 ± 0.06 (66.42)	0.75	12	78.81	0.94
26b	0.36 ± 0.03 (11.15)	0.46 ± 0.02 (21.74)	0.31 ± 0.04 (38.61)	0.23 ± 0.02 (41.03)	0.22 ± 0.04 (53.02)	0.12 ± 0.04 (71.68)	0.81	10	81.81	0.97
27a	0.24 ± 0.02 (1.37)	0.46 ± 0.07 (5.49)	0.31 ± 0.02 (27.63)	0.22 ± 0.02 (32.10)	0.22 ± 0.05 (21.08)	0.43 ± 0.01 (37.35)	0.42	43	21.81	0.26
27b	0.15 ± 0.05 (7.31)	0.45 ± 0.01 (14.52)	0.34 ± 0.02 (31.82)	0.35 ± 0.04 (26.41)	0.28 ± 0.03[a,b] (33.50)	0.22 ± 0.02[a] (41.95)	0.47	36	34.54	0.41
28a	0.15 ± 0.03 (0)	0.46 ± 0.01 (15.30)	0.22 ± 0.02 (26.84)	0.36 ± 0.02 (32.45)	0.12 ± 0.05[a,b] (43.05)	0.22 ± 0.04[a] (59.31)	0.67	20	63.63	0.76

Table 6. Anti-inflammatory and analgesic results for compounds of Scheme 9 & Scheme 10.

Comp. No.	Mean ± S.D. (Percent edema inhibition)							Analgesic activity		
	1 h	2 h	3 h	4 h	5 h	6 h	Potency	No. of writhing movements	Percent inhibition	Potency
33a	0.03 ± 0.02 (3.24)	0.07 ± 0.04 (11)	0.17 ± 0.03[a] (23.51)	0.23 ± 0.02 (43)	0.16 ± 0.06 (44.61)	0.17 ± 0.06 (45.71)	0.51	22	60	0.71
33b	0.23 ± 0.06 (7.42)	0.22 ± 0.03 (15.41)	0.12 ± 0.04[a] (41)	0.12 ± 0.05 (0)	0.18 ± 0.03 (48.21)	0.19 ± 0.06 (59.42)	0.67	19	65.45	0.78
34a	0.34 ± 0.03 (7.31)	0.37 ± 0.01 (17.42)	0.17 ± 0.03 (33.76)	0.33 ± 0.06 (0)	0.16 ± 0.08 (68.41)	0.17 ± 0.05 (75.30)	0.85	15	72.72	0.86
34b	0.24 ± 0.05 (11.42)	0.14 ± 0.03 (0)	0.28 ± 0.05 (42.87)	0.15 ± 0.09 (59.42)	0.21 ± 0.03 (0)	0.03 ± 0.03 (77)	0.87	18	67.27	0.80
35a	0.36 ± 0.03 (10)	0.35 ± 0.03 (1.75)	0.46 ± 0.08 (18)	0.38 ± 0.05 (23.51)	0.44 ± 0.05 (25.61)	0.26 ± 0.07 (31.78)	0.36	60	---	---
35b	0.14 ± 0.05 (7.31)	0.34 ± 0.06 (4.67)	0.27 ± 0.03 (21.45)	0.34 ± 0.04 (0)	0.25 ± 0.08 (33.21)	0.26 ± 0.05[a,b] (36)	0.41	55	---	---
36a	0.24 ± 0.01 (0)	0.22 ± 0.02 (8.53)	0.23 ± 0.06 (22.31)	0.26 ± 0.04 (31.54)	0.23 ± 0.03 (39.52)	0.45 ± 0.06 (43.46)	0.49	37	32.72	0.39
36b	0.27 ± 0.04 (6.80)	0.24 ± 0.03 (13.45)	0.43 ± 0.08 (29.45)	0.25 ± 0.03 (35)	0.22 ± 0.06 (43.52)	0.46 ± 0.07 (59.42)	0.67	32	41.81	0.49

37a	0.21 ± 0.05 (0)	0.38 ± 0.04 (10.54)	0.17 ± 0.03 (21.56)	0.30 ± 0.02 (33.76)	0.29 ± 0.03 (44)	0.25 ± 0.03 (53.67)	0.60	44	20	0.23
37b	0.25 ± 0.04 (6.31)	0.25 ± 0.05 (18.65)	0.26 ± 0.08 (27)	0.16 ± 0.08 (31.43)	0.22 ± 0.09 (39.54)	0.24 ± 0.06 (55)	0.62	37	32.72	0.39
38a	0.23 ± 0.04 (7.65)	0.23 ± 0.06 (21.56)	0.54 ± 0.05 (37.78)	0.26 ± 0.02 (57.62)	0.12 ± 0.09 (73.24)	0.22 ± 0.09 (88)	0.99	17	69.09	0.82
38b	0.43 ± 0.02 (10.43)	0.25 ± 0.05 (25.64)	0.25 ± 0.03 (0)	0.28 ± 0.04 (50.42)	0.25 ± 0.09 (77.34)	0.49 ± 0.10[a,b] **(92.17)**	**1.04**	12	78.81	0.94
39a	0.22 ± 0.04 (0)	0.28 ± 0.03 (17.35)	0.23 ± 0.01 (23)	0.25 ± 0.03 (34.90)	0.28 ± 0.01 (0)	0.47 ± 0.07 (50)	0.56	39	29.09	0.34
39b	0.24 ± 0.02 (5.90)	0.21 ± 0.06 (11)	0.46 ± 0.08 (19.31)	0.21 ± 0.07 (39)	0.29 ± 0.05 (43.86)	0.47 ± 0.08 (58.31)	0.66	33	40	0.47
40a	0.28 ± 0.03 (0)	0.14 ± 0.04 (21.43)	0.43 ± 0.07 (37.30)	0.50 ± 0.05 (49)	0.17 ± 0.08 (72)	0.60 ± 0.04 **(90)**	**1.01**	18	67.27	0.80
40b	0.32 ± 0.04 (10.32)	0.44 ± 0.10 (37.32)	0.33 ± 0.03 (0)	0.27 ± 0.02 (41.03)	0.24 ± 0.04 (53.02)	0.12 ± 0.03 (86.68)	0.98	13	76.36	0.91

Table 7. Ulcerogenic effects of the tested compounds.

Compounds	Control	Indomethacin	12b	23b	31a	31b	38b
% Ulceration	0.0	94	0.0	51.6	33.2	0.0	27.8

In general, all the tested compounds showed better results than the reference drug Indomethacin. Especially, compounds 12b and 31b which were devoid of any ulcerative effect compared to 94% of that of Indomethacin as illustrated in the previous table.

2.2.3. Anti-Microbial Screening

For the tested compounds 4a-40b, the resulting inhibition zones were measured in mm diameter, Tables 8-10.

Among the tested compounds, compounds 5d, 9d, 11b, 12a, 13b, 15a, 15b, 17a, 22a, 27a, 32a, 33a, 39a and 39b were found to be the most active.

Table 8. Inhibition zones (IZ) in mm diameter for compounds of Schemes 1-4.

Comp. No.	*S. aureus*	*E. Coli*	*Ps. aeruginosa*	*C. albicans*
4a	9	10	12	12
4b	11	10	15	17
4c	11	10	13	11
4d	12	10	17	9
5a	10	10	15	16
5b	11	15	13	14
5c	11	20	18	16
5d	11	**25**	**26**	10
6a	10	20	15	16
6b	11	11	18	10
7a	11	11	10	15
7b	12	11	10	10
8a	10	20	20	11
8b	10	18	**22**	11
9a	10	12	14	20
9b	9	20	**20**	10
9c	9	18	**20**	9
9d	10	**25**	11	10
10a	10	20	14	13

10b	9	20	18	9
10c	11	14	**20**	10
10d	11	11	**20**	21
11a	12	9	**20**	9
11b	9	13	**30**	9
12a	10	**30**	14	20
12b	9	23	13	9
13a	10	24	16	18
13b	11	12	**30**	17
14a	9	14	11	9
14b	10	9	10	17
15a	9	**28**	17	9
15b	10	**40**	**20**	15
16a	10	13	17	14
16b	10	10	**20**	22
Ampicillin	30	22	27	-
Fluconazole	-	-	-	32

Table 9. Inhibition zones (IZ) in mm diameter for compounds of Schemes 5-8.

Comp. No.	*S. aureus*	*E. Coli*	*Ps. aeruginosa*	*C. albicans*
17a	15	18	**29**	20
17b	14	20	17	18
18a	10	23	13	**25**
18b	15	20	13	20
19a	13	20	18	**25**
19b	10	18	19	21
20a	10	13	11	16
20b	10	16	14	15
21a	9	13	19	9
21b	11	13	21	17
22a	11	12	**32**	14
22b	11	17	23	16
23a	10	12	18	12
23b	11	16	20	10
24a	13	20	14	10
24b	12	11	11	10
25a	12	15	19	10
25b	17	14	11	12
26a	9	23	19	9

26b	10	17	20	15
27a	10	24	**30**	15
27b	10	23	24	11
28a	9	10	12	14
28b	10	15	10	15
29a	10	13	12	13
29b	15	13	20	17
30a	19	12	12	15
30b	**26**	13	16	18
31a	10	13	13	10
31b	10	20	17	15
32a	9	12	**30**	17
32b	20	12	13	9
Ampicillin	30	22	27	-
Fluconazole	-	-	-	32

Table 10. Inhibition zones (IZ) in mm diameter for compounds of Scheme 9 & Scheme 10.

Comp. No.	***S. aureus***	***E. Coli***	***Ps. aeruginosa***	***C. albicans***
33a	10	**30**	10	9
33b	9	13	14	8
34a	9	11	11	10
34b	9	10	15	9
35a	9	10	10	17
35b	9	10	10	13
36a	10	14	10	**23**
36b	10	12	24	14
37a	10	11	14	13
37b	9	11	12	10
38a	10	18	18	14
38b	10	20	12	9
39a	10	16	**40**	9
39b	10	19	**30**	10
40a	12	12	13	12
40b	12	10	10	9
Ampicillin	30	22	27	-
Fluconazole	-	-	-	32

Among the different functions attached to the free pyrrole ring which exerted potent antimicrobial activity against Gram negative bacteria are the 2,4-dichlorobenzylidineimino, thiourea, phenylurea, ethoxymethyleneimino as well as dimethylaminomethyleneimino groups as in compounds 5d, 9d, 11b, 13b and 15a, 15b; respectively, in addition to, the carboxamide and the acid hydrazide side chains attached to the pyrrole ring in compounds 17a and 33a; respectively.

Furthermore, functions surmounted on the pyrrolopyrimidine skeleton as 4-imino-3-phenyl-2-thione and 4- morpholino groups as in compounds 12a and 22a; respectively, also displayed significant antimicrobial activity. Besides to, the triazolopyrrolopyrimidine derivative 27a, pyrrolotriazine analogue 32a as well as the pyrroloxazine compounds 39a, 39b which exerted high antimicrobial activities against Gram negative bacteria. It is worth mentioning that, all compounds exerted weak activity against Gram positive bacteria except for the 3- imidazolidine substituted pyrrole derivative 30b which exerted moderate antimicrobial activity. However, only the pyrrolopyrimidinone derivative 18a and the 1,2,6-thiadiazine-2,4-dione analogue 19a exerted moderate antifungal activity against C. albicans.

2.3. Computer Aided Docking

The most active twenty compounds as anti-inflammatory agents 4d, 5b, 5c, 5d, 6a, 6b, 9c, 9d, 10d, 12a, 12b, 13b, 19a, 19b, 21b, 23b, 31a, 31b, 38b and 40a were subjected to docking using Molecular Operating Environment (MOE) program [36] on the 3D structure of the cyclooxygenase-2 enzyme (COX-2) in a trial to predict their mode of action as anti-inflammatory drugs.

2.3.1. Docking on the Active Site of Cyclooxygenase-2 Enzyme (COX-2)

1) Diclofenac interactions with the active site of COX-2:

Diclofenac interacted as hydrogen bond acceptor via four hydrogen bonds via both the oxygen atoms of carboxyl group with the amino acid residues Tyr 385 (2.73 A°) and Ser 530 (2.65 A°, 2.91 A° and 3.04 A°) as shown in Figure 1.

2.3.2. Docking of Compound 4d into COX-2

Active site revealed that several molecular interactions were considered to be responsible for the observed affinity, as the N of pyridine moiety acted as a hydrogen bond acceptor with the side chain residue; His 90 (2.25 A°) with a strength of 81.3%. In addition to a hydrogen bond interaction between the hydrogens of the amino group which acted as a hydrogen bond donor with the side chain residue Tyr 355 (2.61 A°) with a strength of 5.3%.

Besides to, hydrophobic interactions involving the pyridine C_6 and p-chlorophenyl C_3 carbon as well the CH_3 group of ester function and the following amino acid residues: His 90, Met 113, Val 116, Leu 117, Arg 120, Val 349, Leu 352, Ser 353, Tyr 355, Leu 359, Leu 384, Tyr 385, Trp 387, Phe 518, Met 522, Val 523, Gly 526, Ala 527, Ser 530 and Leu 531 as shown in Figure 2.

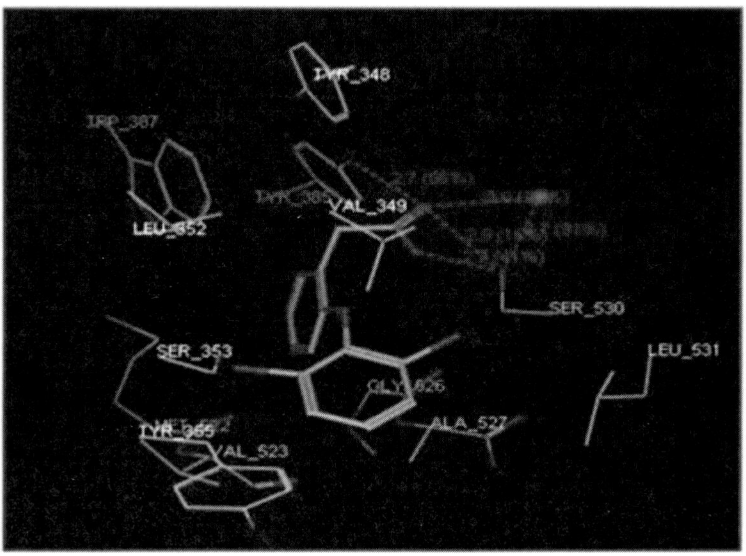

Figure 1. Docking of Diclofenac into the active site of COX-2.

2.3.3. Docking of Compound 5b into COX-2

Active site revealed the presence of arene cation interaction between the pyrrole ring and the amino acid residue Arg 120. In addition to, hydrophobic interactions involving pyridine C_4 and C_6carbons and the following amino acid residues: His 90, Met 113, Val 116, Leu 117, Arg 120, Val 349, Leu 352, Ser 353, Tyr 355, Leu 359, Trp 387, Arg 513, Phe 518, Met 522, Val 523, Ala 527, Ser 530 and Leu 531 as shown in Figure 3.

2.3.4. Docking of Compound 5c into COX-2

Active site illustrated the presence of several interactions of the cyano group with different amino acid residues as it acted as a hydrogen bond acceptor with the side chain residues; His 90, Tyr 355 and Arg 513 (3.35 A°, 2.43 A° and 3.16 A°; respectively) at a strength of 2.1%, 90.6% and 13.4%;

respectively. This beside hydrophobic interactions among the cyano function and the following amino acid residues: His 90, Val 116, Leu 117, Arg 120, Gln 192, Val 349, Leu 352, Ser 353, Tyr 355, Leu 359, Tyr 385, Trp 387, Arg 513, Ala 516, Ile 517, Phe 518, Val 523, Gly 526, Ala 527, Ser 530 and Leu 531 as shown in Figure 4.

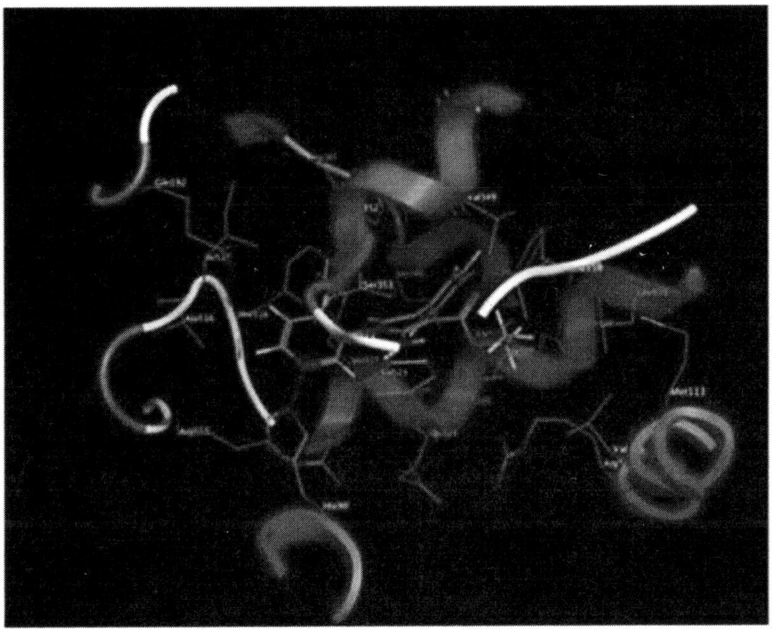

Figure 2. Docking of compound 4d into the active site of COX-2.

2.3.5. Docking of Compound 5d into COX-2

Active site revealed the presence of hydrogen bond interaction between the cyano group, as it acted as a hydrogen bond acceptor with the amino acid residue Ser 530 residue (3.32 A°) with a strength of 1.8%. In addition to, arene cation interactions involving the p-chlorophenyl ring with the amino acid residue Phe 518 and the 2,4-dichlorophenyl ring with the amino acid residue Arg 120. There are also hydrophobic interactions involving the pyridine C_2, C_5, C_6 as well as the pyridine nitrogen atom with the following amino acid residues: His 90, Val 116, Arg 120, Val 349, Leu 352, Ser 353, Tyr 355, Leu 359, Leu 384, Phe 518, Met 522, Gly 526, Ala 527, Ser 530 and Leu 531 as shown in Figure 5.

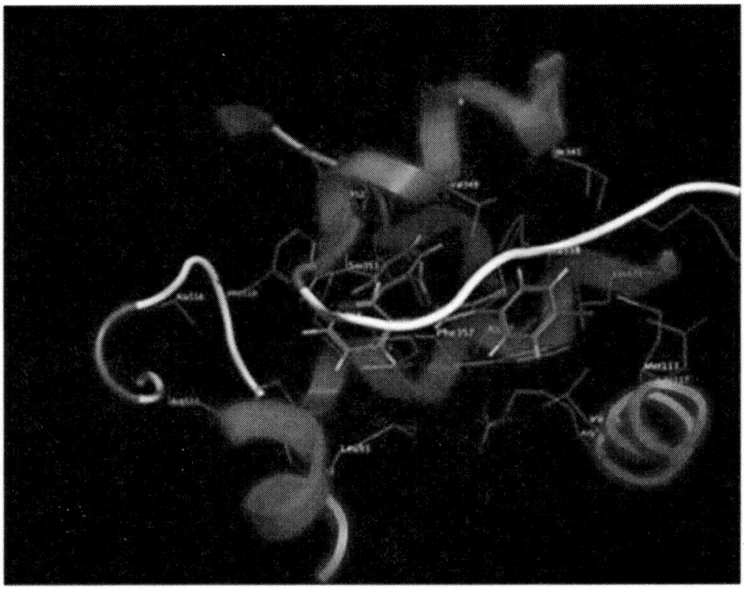

Figure 3. Docking of compound 5b in the active site of COX-2.

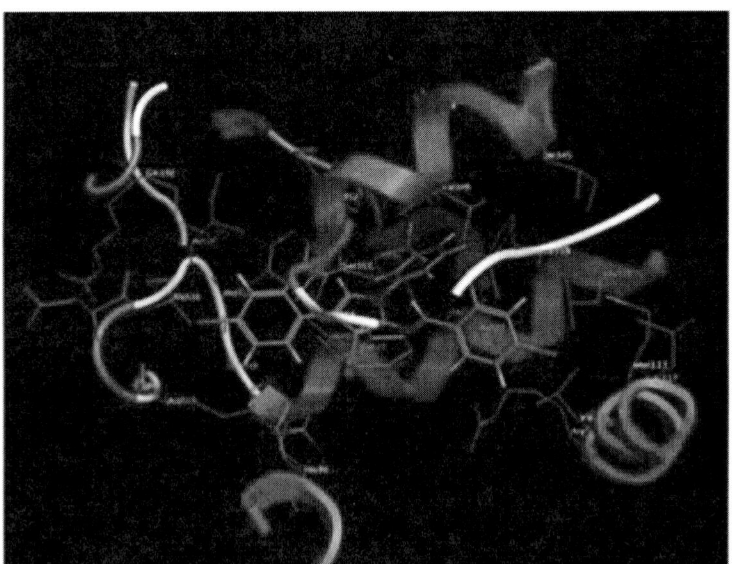

Figure 4. Docking of compound 5c in the active site of COX-2.

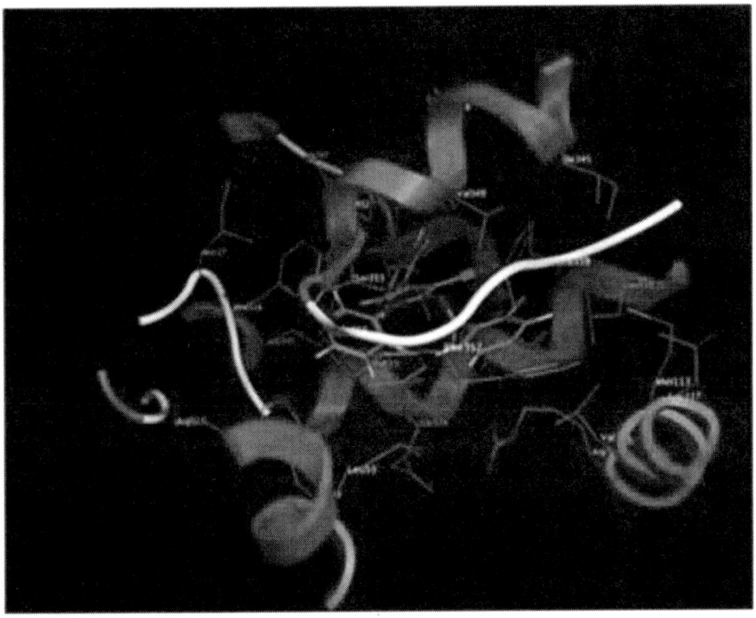

Figure 5. Docking of compound 5d in the active site of COX-2.

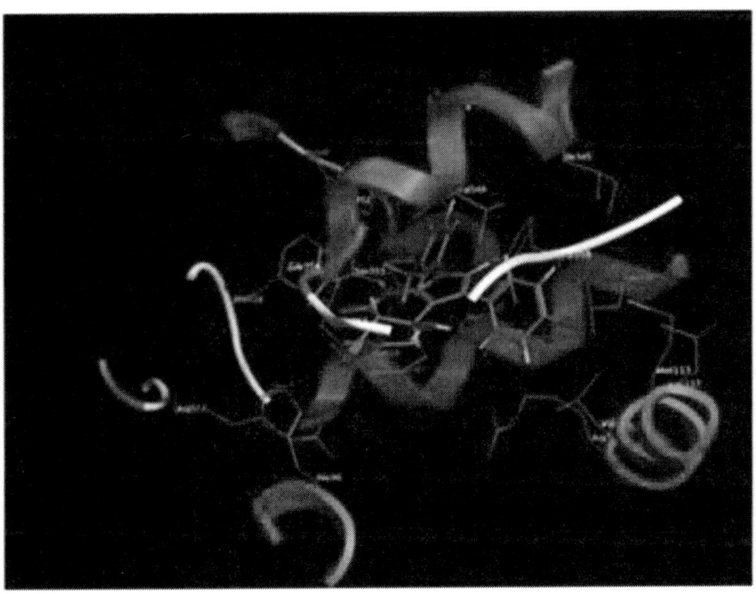

Figure 6. Docking of compound 6a in the active site of COX-2.

2.3.6. Docking of Compound 6a into COX-2

Active site revealed the presence of four hydrogen bonds in which the cyano nitrogen acted as a hydrogen bond acceptor for three hydrogen bonds with the amino acid residues His 90, Arg 513 and Tyr 355 (2.82 A°, 3.44 A° and 2.91 A°; respectively) with a strength of 20.4%, 7% and 7.3%; respectively. While, the amino group acted as a hydrogen bond acceptor with the amino acid residue Tyr 355 residue (3.24 A°) with a strength 5.5%. In addition to, hydrophobic interactions involving the carbonyl oxygen with the following amino acid residues: His 90, Met 113, Val 116, Leu 117, Arg 120, Ser 530, Val 349, Leu 352, Ser 353, Tyr 355, Leu 384, Try 385, Ala 527 and Leu 531 as shown in Figure 6.

2.3.7. Docking of Compound 6b into COX-2

Active site revealed the presence of four hydrogen bonds and two arenes cation interactions. In which the amino group acted as a hydrogen acceptor via three hydrogen bonds with the amino acid residues His 90, Tyr 355 and Arg 513 (2.25 A°, 3.32 A° and 3.43 A°; respectively) with a strength of 3.5%, 9.1% and 43.2%; respectively. While the cyano nitrogen atom acted as a hydrogen bond acceptor with the amino acid residue His 90 (3.41 A°) with strength of 2.2%. In addition to, two arene cation interactions among the p-chlorophenyl moiety and the amino acid residue Arg 120 and Arg 513. Besides to hydrophobic interactions involving the pyridine ring, pyrrole C_2 carbons as well as the chlorine atom and the p-chlorophenyl C_3 carbon with the following amino acid residues: Pro 86, Val 89, His 90, Arg 120, Val 349, Leu 352, Tyr 355, Arg 513, Ala 516, Phe 518, Val 523, Glu 524, Gly 526, Ala 527 and Ser 530 as shown in Figure 7.

2.3.8. Docking of Compound 9c into COX-2

Active site revealed the presence of two hydrogen bond interactions between the cyano nitrogen, as it acted as a hydrogen bond acceptor with the amino acid residue side Arg 120 and Tyr 355 (3.21 A° and 1.60 A°; respectively) with a strength of 12.3% and 95.9%; respectively. In addition to, hydrophobic interactions involving the cyano nitrogen and chlorine atom with many amino acid residues: His 90, Val 349, Leu 352, Ser 353, Tyr 355, Leu 359, Phe 381, Tyr 385, Trp 387, Arg 513, Ala 516, Val 523, Gly 526, Ala 527, Ser 530 and Leu 531 as shown in Figure 8.

2.3.9. Docking of Compound 9d into COX-2

Active site revealed the presence of one hydrogen bond between the pyridyl nitrogen atom as it acted as a hydrogen bond acceptor with the amino acid residue His 90 (2.82 A°) with a strength of 4.8%. In addition to, hydrophobic interactions concerning 4-chlorophenyl C_2 carbon, pyridine $C_2$and C_4

carbons, the thiourea amino group and sulphur atom with the following amino acid residues: His 90, Arg 120, Val 349, Leu 352, Ser 353, Tyr 355, Phe 381, Leu 384, Tyr 385, Trp 387, Met 522, Val 523, Gly 526 and Ala 527 as shown in Figure 9.

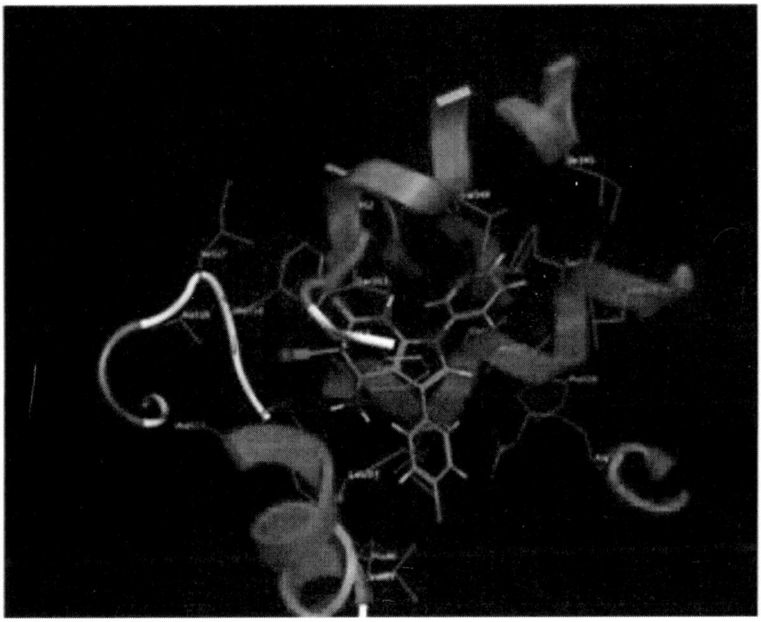

Figure 7. Docking of compound 6b in the active site of COX-2.

2.3.10. Docking of Compound 10d into COX-2

Active site revealed hydrogen bond interaction between the N atom of pyridine moiety as it acted as a hydrogen bond acceptor with the side chain residues His 90 and Arg 513 (2.74 A° and 3.54 A°; respectively) with a strength of 21.7% and 1.2%; respectively. Besides to, arene cation interaction between the p-chlorophenyl ring and the amino acid residue Arg 120. In addition to, hydrophobic interactions among the p-chloro phenylC_2 carbon, chlorine atom as well as the pyrimidine thioxo function, N_3 atom and C_4 carbon with the following amino acid residues: His 90, Val 116, Arg 120, Val 349, Leu 352, Ser 353, Leu 359, Trp 387, Arg 513, Ala 516, Ile 517, Phe 518, Met 522, Val 523 and Ala 527 as shown in Figure 10.

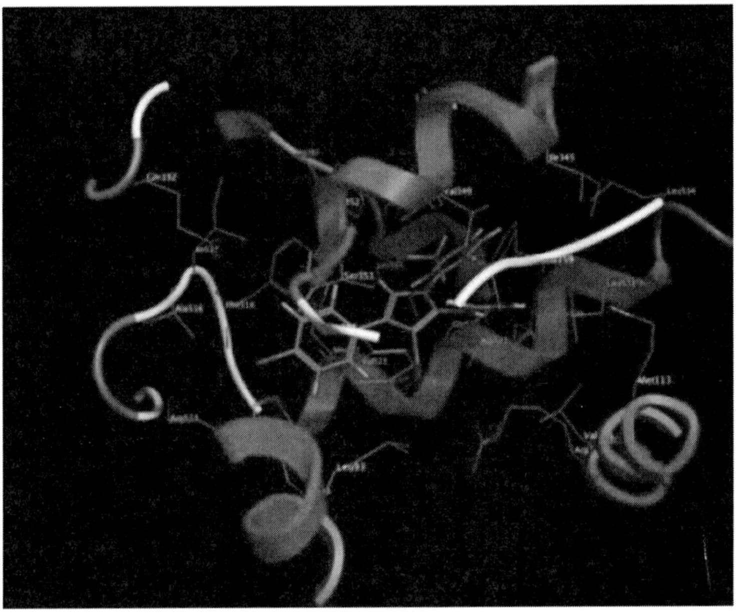

Figure 8. Docking of compound 9c in the active site of COX-2.

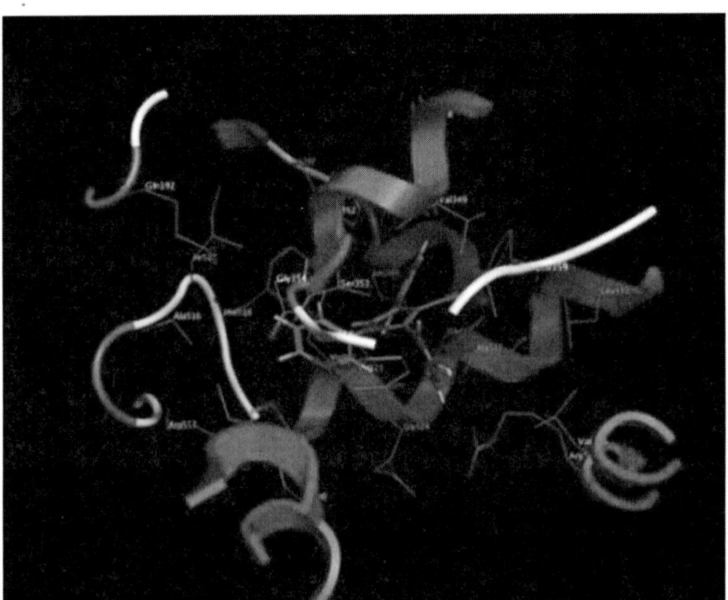

Figure 9. Docking of compound 9d in the active site of COX-2.

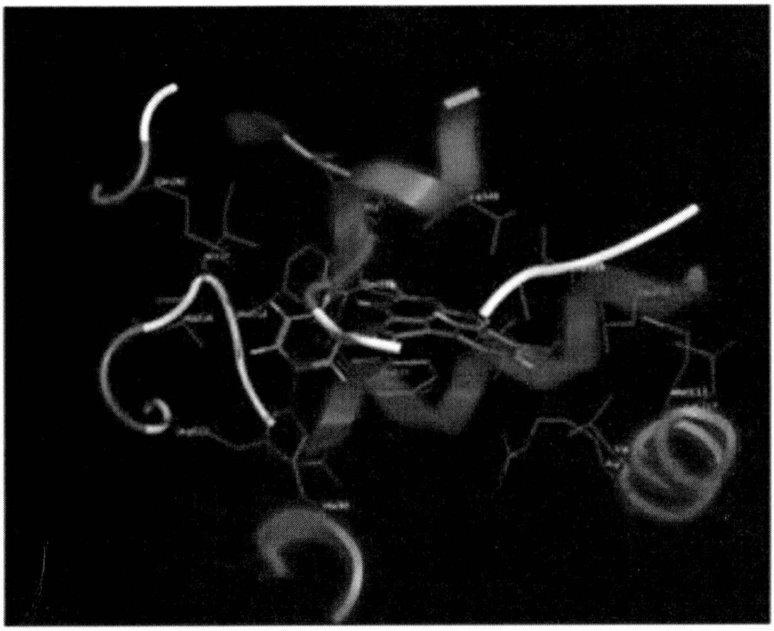

Figure 10. Docking of compound 10d in the active site of COX-2.

2.3.11. Docking of Compound 12a into COX-2

Active site revealed the presence of only hydrophobic interactions involving the phenyl C_2 carbon and the p-chlorophenyl C_3 and C_6 carbons as well as the chlorine atom with the following amino acid residues: His 90, Met 113, Val 116, Arg 120, Ile 345, Val 349, Leu 352, Ser 353, Tyr 355, Leu 359, Tyr 385, Arg 513, Ala 516, Phe 518, Met 522, Val 523, Gly 526, Ala 527, Ser 530 and Leu 531 as shown in Figure 11.

2.3.12. Docking of Compound 12b into COX-2

Active site showed the presence of a hydrogen bond interaction between imine function, as it acted as a hydrogen bond acceptor with the side chain Tyr 355 residue (3.15 A°) with a strength of 6.3%. In addition to hydrophobic interactions involving the chlorine atom and the pyridyl C_6 carbon with many amino acid residues: His 90, Val 116, Arg 120, Val 349, Leu 352, Ser 353, Leu 384, Tyr 385, Trp 387, Phe 518, Met 522, Val 523, Gly 526, Ala 527 and Leu 531 as shown in Figure 12.

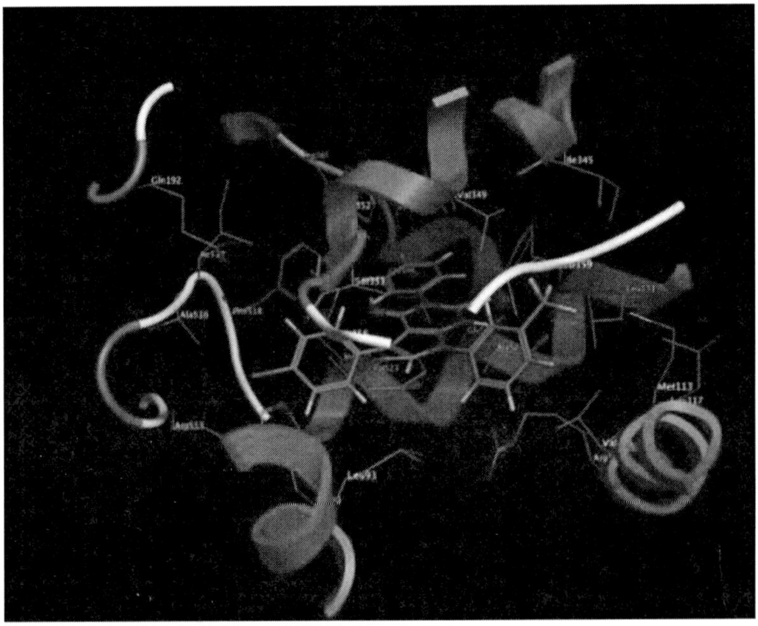

Figure 11. Docking of compound 12a in the active site of COX-2.

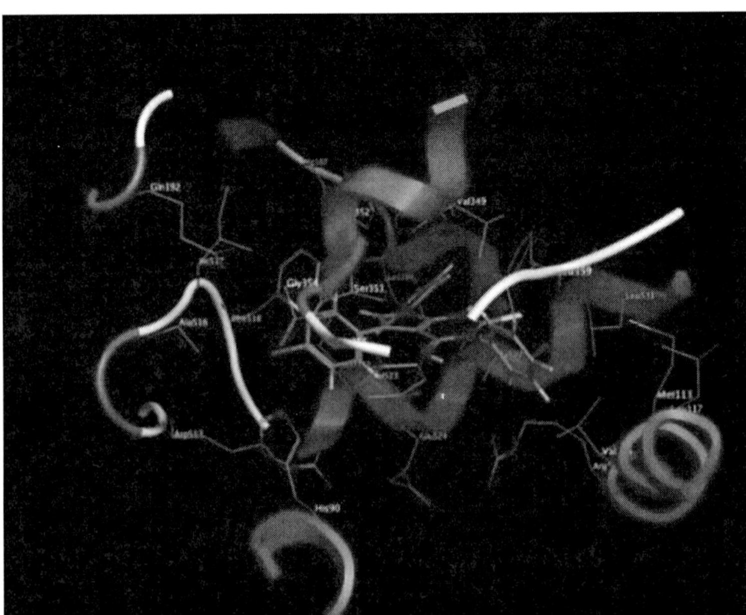

Figure 12. Docking of compound 12b in the active site of COX-2.

2.3.13. Docking of Compound 13b into COX-2

Active site showed hydrophobic interactions concerning the chlorine atom, cyano nitrogen and the methylene imino side chain with many amino acid residues: His 90, Met 113, Val 116, Arg 120, Leu 331, Ile 345, Val 349, Leu 352, Ser 353, Tyr 355, Leu 359, Val 523, Ala 527, Ser 530 and Leu 534 as shown in Figure 13.

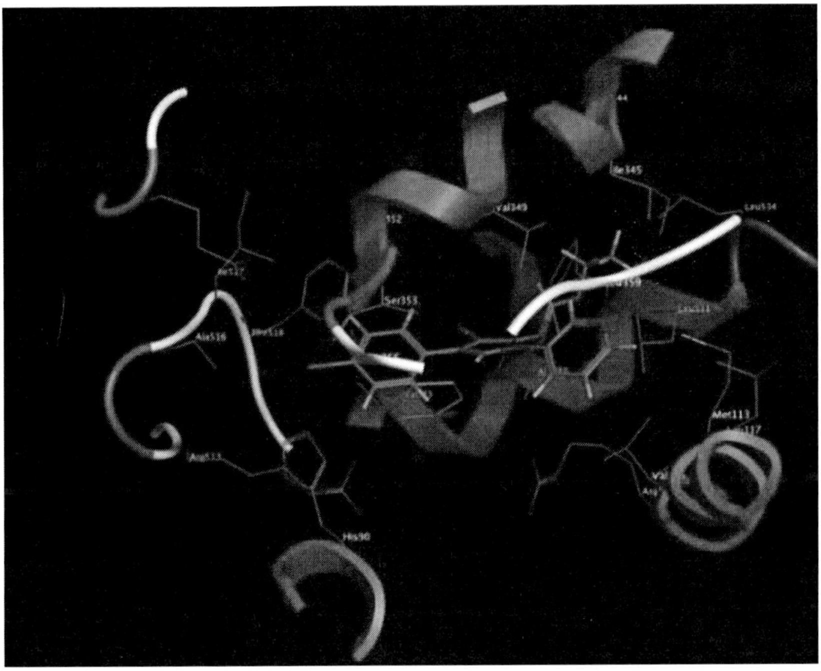

Figure 13. Docking of compound 13b in the active site of COX-2.

2.3.14. Docking of Compound 19a into COX-2

Active site revealed the presence of two hydrogen bond interactions between the oxygen atom of pyrimidine C_4 oxo function as it acted as a hydrogen bond acceptor with the side chain residues Arg 120 and Tyr 355 (2.63 A° and 3.04 A°; respectively) with a strength of 19.6% and 12.5%; respectively. In addition to hydrophobic interactions involving the pyrimidine N_3 and 2-oxo functions as well as pyrrole C_5 and p-chlorophenyl C_2 and C_6 carbons with many amino acid residues: Pro 86, Val 89, His 90, Arg 120, Tyr 348, Val 349, Leu 352, Tyr 355, Tyr 385, Trp 387, Arg 513, Met 522, Val 523, Glu 524, Gly 526, Ala 527 and Leu 531 as shown in Figure 14.

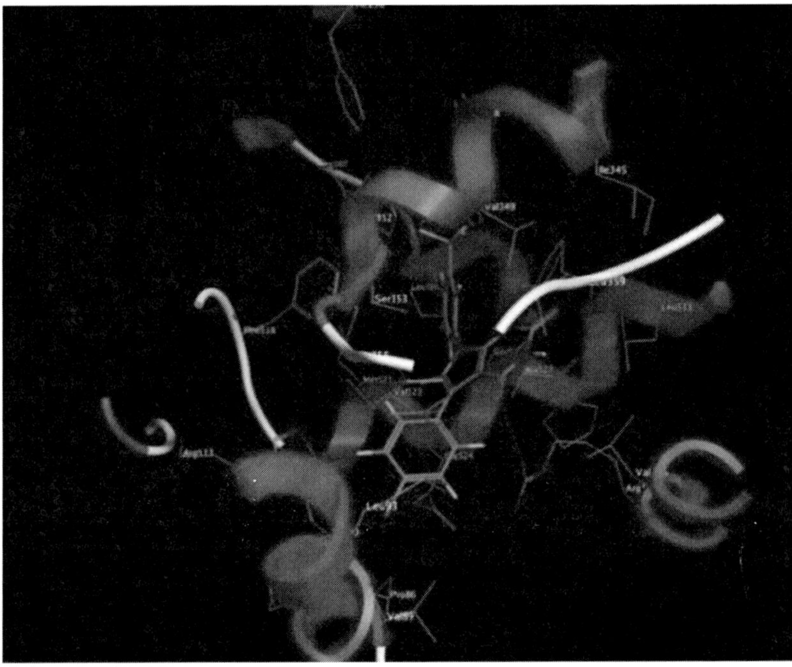

Figure 14. Docking of compound 19a in the active site of COX-2.

2.3.15. Docking of Compound 19b into COX-2

Active site revealed the presence of hydrogen bond interactions between the oxygen atoms of pyrimidine C_2 and C_4 oxo functions as they acted as hydrogen bond acceptor with the side chain residue Arg 513 (3.27 A°, 2.3%) and Arg 120 (2.28 A°, 9.5%); respectively. In addition to, another hydrogen bond between the pyridyl nitrogen atom as it acted as hydrogen bond acceptor with the amino acid residue Arg 513 (3.45 A°, 1.7%). In addition to, hydrophobic interactions concerning the carbon atoms of p-chlorophenyl moiety and the following amino acid residues: Pro 86, His 90, Arg 120, Gln 192, Val 349, Leu 352, Ser 353, Tyr 355, Arg 513, Ala 516, Val 523, Glu 524, Ala 527, Ser 530 and Leu 531 as shown in Figure 15.

2.3.16. Docking of Compound 21b into COX-2

Active site revealed the presence of hydrogen bond interaction between the N_1 of the pyrimidine moiety, as it acted as a hydrogen bond acceptor with the side chain residue Arg 120 (2.35 A°) with strength of 64.2%. In addition to, arene cation interaction between the pyridine moiety and the amino acid residue Arg 120. Besides to, hydrophobic interactions involving the pyrimidine C_2carbon, pyrrole C_5 carbon and pyridine C_6 carbon with many amino acid residues: Arg 120, Val 349, Leu 352, Tyr 355, Tyr 385, Trp 387, Arg 513, Val 523, Glu 524, Gly 526, Ala 527 and Leu 531 as shown in Figure 16.

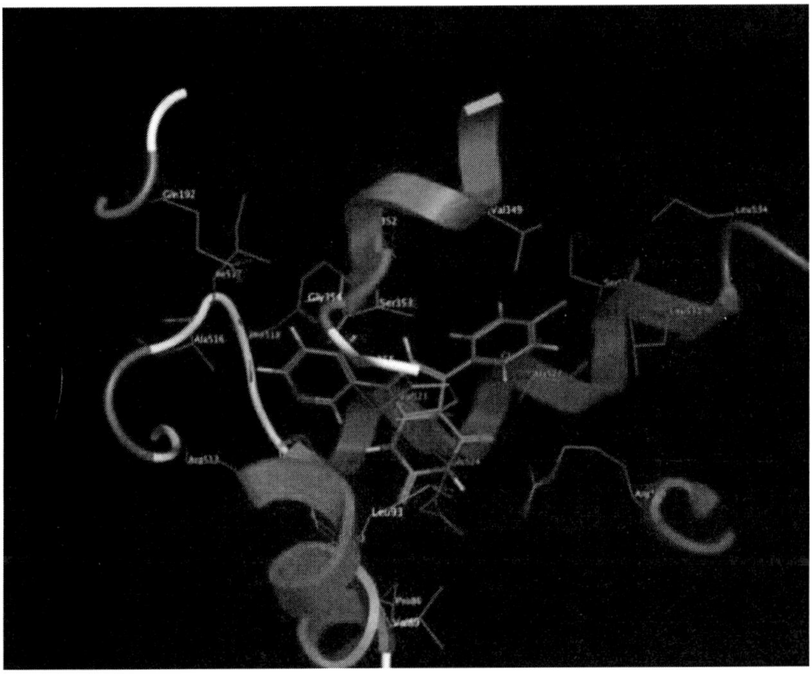

Figure 15. Docking of compound 19b in the active site of COX-2.

2.3.17. Docking of Compound 23b into COX-2

Active site revealed the presence of three hydrogen bond interactions in which the hydrazine NH_2proton acted as a hydrogen bond donor with the amino acid residue Val 349 (3.21 A°, 1.5%), while the nitrogen atom acted as a hydrogen bond acceptor with the amino acid residue Ser 353 (2.62 A°, 2.5%). Also, the hydrazine NH proton acted as a hydrogen bond donor to the amino acid residue Val 349 (1.38 A°, 93.7%). Besides to, hydrophobic interactions among the p-chlorophenyl moiety and the pyrimidine C_4 carbon with the following amino acid residues: His 90, Tyr 348, Val 349, Leu 352, Ser 353, Tyr 355, Phe 381, Leu 384, Tyr 385, Phe 518, Met 522, Val 523, Gly 526, Ala 527 and Ser 530 as shown in Figure 17.

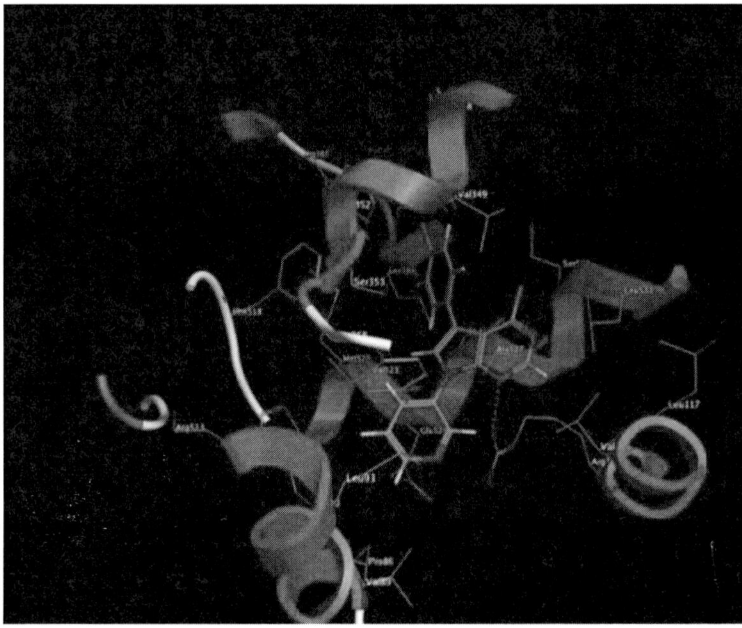

Figure 16. Docking of compound 21b in the active site of COX-2.

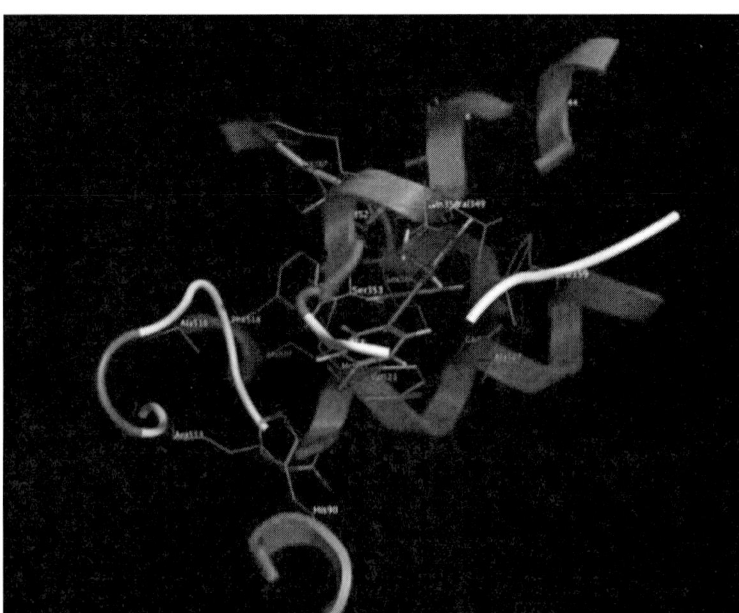

Figure 17. Docking of compound 23b in the active site of COX-2.

2.3.18. Docking of Compound 31a into COX-2

Active site showed arene cation interaction between the 3-triflouromethylphenyl ring and the amino acid residue Arg 513. In addition to, hydrophobic interactions involving the 3-trifluoromethylphenyl C_6 and fluorine atom as well as the two sulphur atoms of the two thioxo functions of pyrimidine ring with the following amino acid residues: His 90, Gln 192, Val 349, Leu 352, Ser 353, Tyr 355, Leu 384, Tyr 385, Trp 387, Arg 513, Ala 516, Ile 517, Phe 518, Met 522, Val 523, Gly 526, Ala 527 and Ser 530 as shown in Figure 18.

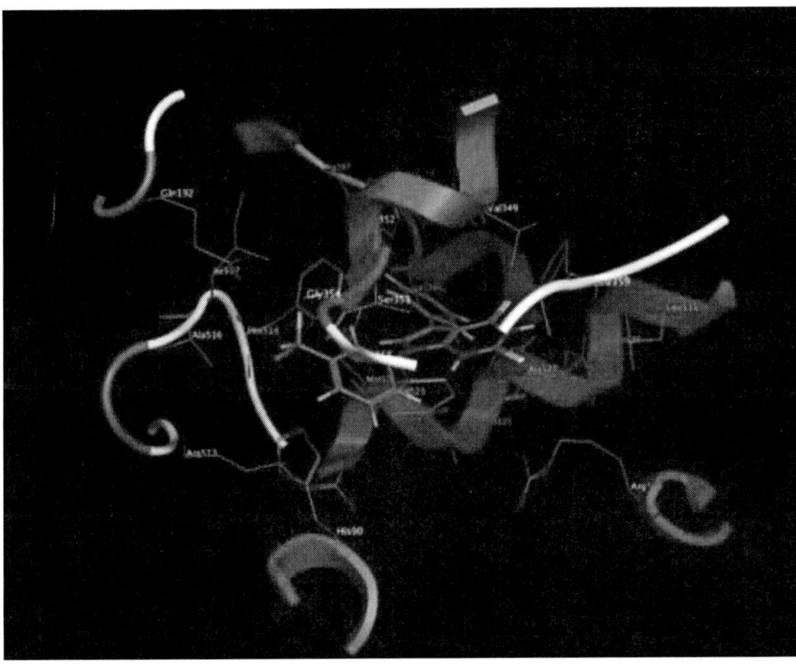

Figure 18. Docking of compound 31a in the active site of COX-2.

2.3.19. Docking of Compound 31b into COX-2

Active site revealed the presence of hydrophobic interactions involving the pyridine C_6 carbon, chlorine atom, the sulphur atom of the 2-thioxo function of pyrimidine ring as well as the pyrimidine N_3 NH moiety with the following amino acid residues: His 90, Val 349, Leu 352, Ser 353, Tyr 355, Leu 384, Tyr 385, Trp 387, Arg 513, Phe 518, Met 522, Val 523, Gly 526, Ala 527 and Leu 531 as shown in Figure 19.

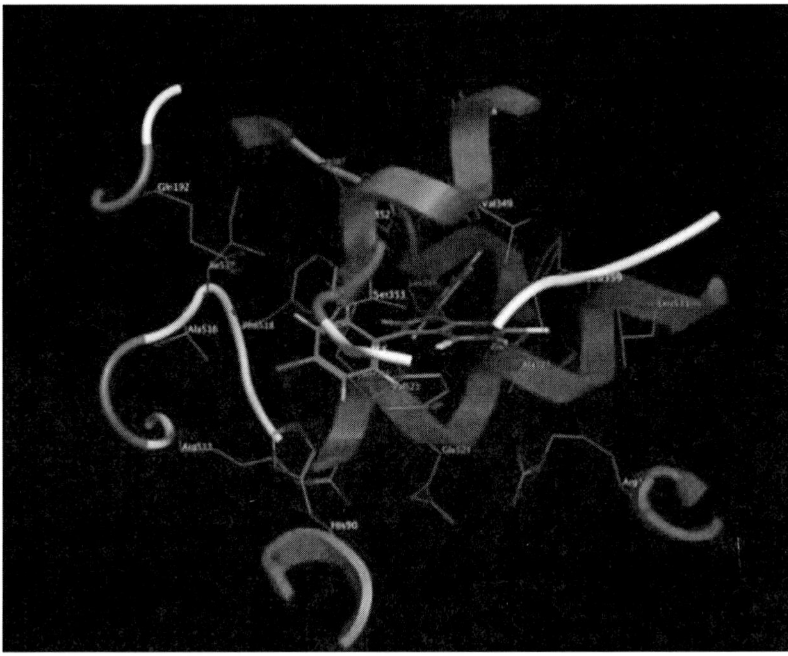

Figure 19. Docking of compound 31b in the active site of COX-2.

2.3.20. Docking of Compound 38b into COX-2

Active site showed one hydrogen bond between NH-phenyl proton which acted as a hydrogen bond donor with the side chain residue Tyr 355 (2.03 A°, 0.8%). In addition to, two arene cation interactions between both pyridine and phenyl rings and the amino acid residue Arg 120. Besides to, hydrophobic interactions involving the phenyl C_4 carbon and the sulphur atom with many amino acid residues: Pro 86, Val 116, Arg 120, Tyr 348, Val 349, Leu 352, Ser 353, Tyr 355, Leu 359, Tyr 385, Trp 387, Arg 513, Val 523, Glu 524, Gly 526, Ala 527, Ser 530 and Leu 531 as shown in Figure 20.

2.3.21. Docking of Compound 40b into COX-2

Active site revealed the presence of hydrophobic interactions involving the p-chlorophenyl C_2carbon and pyrrole C_2 carbon with many amino acid residues: His 90, Met 113, Val 116, Leu 117, Arg 120, Ile 345, Val 349, Leu 352, Ser 353, Tyr 355, Leu 359, Tyr 385, Trp 387, Arg 513, Ala 516, Val 523, Gly 526, Ala 527, Ser 530 and Leu 531 as shown in Figure 21.

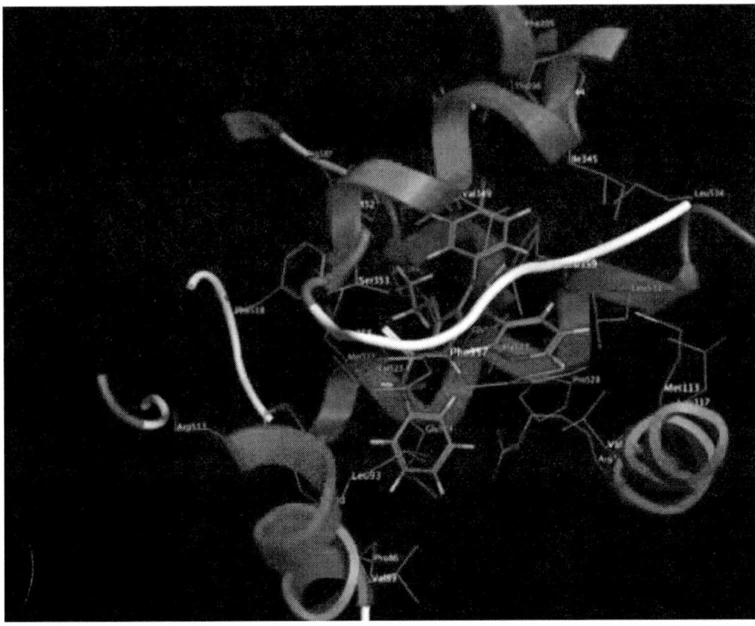

Figure 20. Docking of compound 38b in the active site of COX-2.

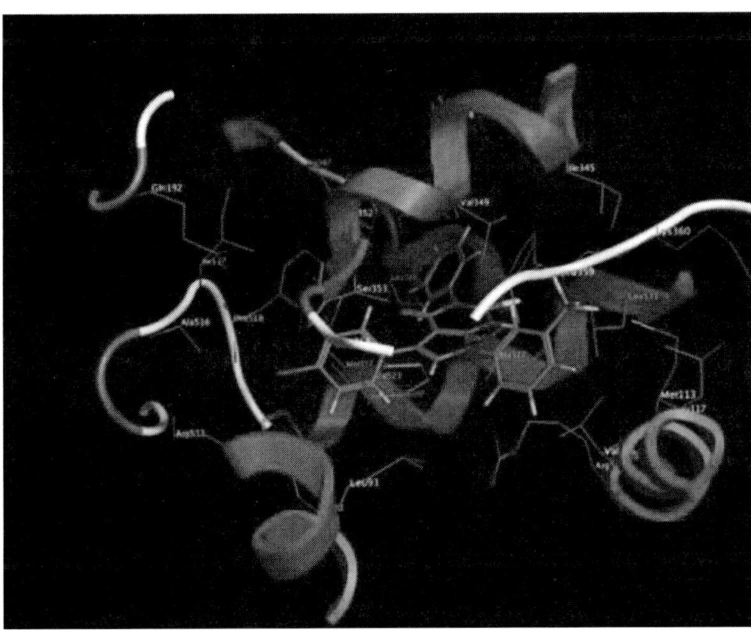

Figure 21. Docking of compound 40b in the active site of COX-2 enzyme.

3. EXPERIMENTAL

3.1. Chemistry

All melting points were measured on Electro thermal LA 9000 SERIS, Digital Melting point Apparatus and are uncorrected. IR spectra (KBr) were recorded on FT-IR 200 spectrophotometer ($\acute{\upsilon}\cdot cm^{-1}$), pharmaceutical analytical unit, Faculty of Pharmacy, Al-Azhar University. ^{1}H-NMR spectra were recorded in (DMSO-d_6) at 300 MHz on a Varian Gemini NMR spectrometer (δ ppm) using TMS as an internal standard, ResearchService Unit, Faculty of Science, Cairo University. Mass spectra were recorded on GC MS-QP 5050A mass spectrometer at 70 eV andmicroanalytical data were performed on Elementar Vario EI III CHN analyzer at the microanalytical unit, in Regional center for Mycology and Biotechnology, Al-Azhar University. Thin layer chromatography was performed on precoated (0.25 mm) silica gel GF254 plates (E. Merck, Germany). Compounds were detected with UV lamp at λ 254 nm.

3.1.1. 1-(4-Chlorophenyl)-2-(3-Trifluoromethylphenylamino)Ethanone; 3a

Equimolar amounts of 4-chlorophenacyl bromide (0.38 g, 2 mmol.) and 3-trifloromethylaniline (0.32 g, 0.25 mL, 2 mmol.) were refluxed for 4 h in absolute ethanol (25 mL) in presence of drops of TEA, then left to cool. The solid product was collected by filtration after cooling and recrystallized from ethanol.

Fluffy golden yellow needle crystals; yield (90%); m.p.:138°C. IR [KBr, cm^{-1}]: 3370 (NH); 3079 (CH aromatic); 2912 (CH aliphatic); 1691 (C=O); 1522 (C=C); 1108 (p-Cl-phenyl). ^{1}H NMR(DMSO-d_6, δ ppm): 4.76 (s, 2 H, CH_2CO); 6.39 (s, 1 H, NH, D_2O exchangeable); 6.82 (s, 1 H, 3-CF_3-C_6H_4-C_2-H); 6.94 (d, 1 H, J = 7.2 Hz, 3-CF_3-C_6H_4-C_6-H); 6.97-7.00 (m, 1 H, 3-CF_3-C_6H_4-C_4-H); 7.25-7.30 (m, 1 H, 3-CF_3-C_6H_4-C_5-H); 7.64 (d, 2 H, J = 8.1 Hz, 4-Cl-C_6H_4-$C_{2,6}$-H); 8.11 (d, 2 H, J = 8.1 Hz, 4-Cl-C_6H_4-$C_{3,5}$-H). Anal. Calc. (%) for $C_{15}H_{11}ClF_3NO$ (313.7): C, 57.43; H, 3.53; N, 4.46. Found (%): C, 57.52; H, 3.55; N, 4.48.

3.1.2. 1-(4-Chlorophenyl)-2-(Pyridin-3-Ylamino)Ethanone; 3b

An equimolar mixture of 4-chlorophenacyl bromide (0.38 g, 2 mmol.) and 3-aminopyridine (0.19 g, 2 mmol.) were stirred for 2 h. at room temperature in (20 mL) diethyl ether. The solid product was collected by filtration and recrystallized from the ethanol.

Buff needle crystals; yield (95%); m.p.: 205 °C. IR [KBr, cm^{-1}]: 3332, 3220 (NH); 3060 (CH aromatic); 2929 (CH aliphatic); 1693 (C=O); 1636 (C=N); 1585 (C=C); 1084 (p-Cl-phenyl); MS m/z (relative intensity %): 248 ($M^{+\bullet}$+2, 3); 246 ($M^{+\bullet}$, 11); 139 (100). Anal. Calc. (%) for $C_{13}H_{11}ClN_2O$ (246.7): C, 63.29; H, 4.49; N, 11.36. Found (%): C, 63.32; H, 4.57; N, 11.33.

3.1.3. General Procedure for Synthesis of Compounds 4a-d

The selected 2-(substitutedamino)-1-(4-chlorophenyl)ethanones 3a,b (0.01 mol.) was stirred at room temperature with the appropriate nitrile (0.01 mol.) namely; malononitrile and ethylcyanoacetate in (40 mL) sodium ethoxide [prepared by dissolving sodium metal (0.23 g, 0.01 mol.) in absolute ethanol (40 mL)] overnight. The obtained product was filtered, washed with water then recrystallized from glacial acetic acid.

1) 2-Amino-4-(4-chlorophenyl)-1-(3-trifluoromethylphenyl)-1H-pyrrole-3-carbonitrile; 4a.

Yellow needle crystals; yield (63%); m.p.: >300°C. IR [KBr, cm^{-1}]: 3427, 3300 (NH_2); 2900 (CH aliphatic); 2224 (C≡N); 1561 (C=C); 1023 (p-Cl-phenyl). ^{1}H NMR(DMSO-d_6, δ ppm): 7.07 (s, 1 H, CH-pyrrole); 7.22 (d, 2 H, J = 8.1 Hz, 4-Cl-C_6H_4-$C_{2,6}$-H); 7.31-7.38 (m, 3 H, 3-CF_3-C_6H_4-$C_{4,5,6}$-H); 7.64 (s, 1 H, 3-CF_3-C_6H_4-C_2-H); 7.86 (d, 2 H, J = 8.1 Hz, 4-Cl-C_6H_4-$C_{3,5}$-H); 8.45 (s, 2 H, NH_2, D_2O exchangeable). ^{13}C NMR (DMSO-d_6, δ ppm): 108 (pyrrole-C_3); 112 (3-CF_3-C_6H_4-C_2); 115 (C≡N); 118 (pyrrole-C_5); 120 (pyrrole-C_2); 122 (3-CF_3-C_6H_4-C_4); 125.7 (CF_3); 127.5 (3-CF_3-C_6H_4-C_6); 127.6 (pyrrole-C_4); 127.9 (4-Cl-C_6H_4-$C_{2,6}$); 128.1 (4-Cl-C_6H_4- $C_{3,5}$); 128.5 (3-CF_3-C_6H_4-C_5); 131.4 (3-CF_3-C_6H_4-C_3); 132.2 (4-Cl-C_6H_4-C_4); 132.5 (4-Cl-C_6H_4-C_1); 142 (3-CF_3-C_6H_4-C_1); MS m/z (relative intensity %): 363 ($M^{+\bullet}$+2) (0.3); 361 ($M^{+\bullet}$) (0.2); 76 (100). Anal. Calc. (%) for $C_{18}H_{11}ClF_3N_3$ (361.7): C, 59.76; H, 3.06; N, 11.62. Found (%): C, 59.80; H, 3.14; N, 11.68.

2) 2-Amino-4-(4-chlorophenyl)-1-(pyridin-3-yl)-1H-pyrrole-3-carbonitrile; 4b.

Dark grey crystals; yield (77%); m.p.: >300°C. IR [KBr, cm^{-1}]: 3445, 3200 (NH_2); 3080 (CH aromatic); 2880 (CH aliphatic); 2191 (C≡N); 1638 (C=N); 1598 (C=C); 1091 (p-Cl-phenyl); MS m/z (relative intensity %): 294 ($M^{+\bullet}$, 3); 112 (100). Anal. Calc. (%) for $C_{16}H_{11}ClN_4$ (294.7): C, 65.20; H, 3.76; N, 19.01. Found (%): C, 65.83; H, 4.01; N, 19.54.

3) Ethyl 2-amino-4-(4-chlorophenyl)-1-(3-trifluoromethylphenyl)-1H-pyrrole-3-carboxylate; 4c

Orange needle crystals; yield (57%); m.p.: 250°C - 252°C C. IR [KBr, cm^{-1}]: 3442, 3200 (NH_2); 3060 (CH aromatic); 2960 (CH aliphatic); 1728 (C=O); 1280, 1060 (C-O-C); 1121 (p-Cl-phenyl). MS m/z (relative intensity %): 410

($M^{+\bullet}$+2, 2); 409 ($M^{+\bullet}$+1, 0.4); 161 (100). Anal. Calc. (%) for $C_{20}H_{16}ClF_3N_2O_2$ (408.8): C, 58.76; H, 3.94; N, 6.85. Found (%): C, 58.80; H, 4.01; N, 6.82.

4) Ethyl 2-amino-4-(4-chlorophenyl)-1-(pyridin-3-yl)-1H-pyrrole-3-carboxylate; 4d

Brown crystals; yield (69%); m.p.: >300°C. IR [KBr, cm^{-1}]: 3360, 3250 (NH_2); 3002 (CH aromatic); 2819 (CH aliphatic); 1720 (C=O); 1641 (C=N); 1530 (C=C); 1259, 1089 (C-O-C); 1147 (p-Cl-phenyl). ^1H NMR (DMSO-d_6, δ ppm): 1.25 - 1.30 (m, 3 H, CH_2CH_3); 3.50 - 3.59 (m, 2 H, CH_2CH_3); 6.72 (s, 2 H, NH_2, D_2O exchangeable); 7.30 (d, 2 H, J = 8.3 Hz, 4-Cl-C_6H_4-$C_{2,6}$-H); 7.55 - 7.58 (m, 1 H, pyridyl-C_5-H); 7.64 - 7.72 (m, 1 H, pyridyl-C_6-H); 7.85 (d, 2 H, J = 8.3 Hz, 4-Cl-C_6H_4-$C_{3,5}$-H); 7.92 (s, 1 H, CH-pyrrole); 8.03 - 8.06 (m, 1 H, pyridyl-C_4-H); 8.51 (s, 1 H, pyridyl-C_2-H). Anal. Calc. (%) for $C_{18}H_{16}ClN_3O_2$ (341.8): C, 63.25; H, 4.72; N, 12.29. Found (%): C, 63.75; H, 4.98; N, 12.78.

3.1.4. General Procedure for Synthesis of Compounds 5a-d

An equimolar mixture of the appropriate 1-(substitutedaryl)-2-amino-4-(4-chlorophenyl)-1H-pyrrole-3-carboni- trile 4a,b (2 mmol.) and the selected aromatic aldehyde namely; benzaldehyde and 2,4-dichlorobenzaldehyde (2 mmol.) was heated under reflux for 16 h in absolute ethanol (30 mL) containing glacial acetic acid (2 - 3 drops). The reaction mixture was allowed to cool and the product was filtered off, washed with ethanol and recrystallized from dioxane.

1) 2-(Benzylidenimino)-4-(4-chlorophenyl)-1-(3-trifluoromethylphenyl)-1H-pyrrole-3-carbonitrile; 5a

Faint yellow needle crystals; yield (55%); m.p.: 280°C - 282°C. IR [KBr, cm^{-1}]: 2927 (CH aromatic); 2863 (CH aliphatic); 2223 (C≡N); 1634 (C=N); 1560 (C=C); 1023 (p-Cl-phenyl). ^1H NMR(DMSO-d_6, δ ppm): 6.77 - 6.88 (m, 3 H, C_6H_5-$C_{3,4,5}$-H); 7.03 (d, 2 H, J = 8.1Hz, 4-Cl-C_6H_4-$C_{2,6}$-H); 7.25 (s, 1 H, CH-pyrrole); 7.30 (d, 2 H, J = 8.1 Hz, 4-Cl-C_6H_4-$C_{3,5}$-H); 7.34 - 7.37 (m, 1 H, 3-CF_3-C_6H_4-C_5-H); 7.45 - 7.50 (m, 2 H, 3-CF_3-C_6H_4-$C_{4,6}$-H); 7.85 (d, 2 H, J = 8.1 Hz,C_6H_5, $C_{2,6}$-H); 8.36 (s, 1 H, CH-benzylidinimine); 8.49 (s, 1 H, 3-CF_3-C_6H_4-C_2-H); MS m/z (relative intensity %): 450 ($M^{+\bullet}$+1) (0.4); 449 ($M^{+\bullet}$, 0.5); 448 ($M^{+\bullet}$-1, 0.5); 83 (100). Anal. Calc. (%) for $C_{25}H_{15}ClF_3N_3$ (449.9): C, 66.75; H, 3.36; N, 9.34. Found (%): C, 66.82; H, 3.58; N, 9.74.

2) 2-(Benzylidenimino)-4-(4-chlorophenyl)-1-(pyridin-3-yl)-1H-pyrrole-3-carbonitrile; 5b

Dark brown powder; yield (67%); m.p.: >300°C. IR [KBr, cm^{-1}]: 2926 (CH aromatic); 2864 (CH aliphatic); 2207 (C≡N); 1585 (C=N); 1562 (C=C); 1105 (p-Cl-phenyl); MS m/z (relative intensity %): 384 ($M^{+\bullet}$+2, 0.5); 383 ($M^{+\bullet}$+1, 0.1); 382 ($M^{+\bullet}$, 0.1); 67 (100). Anal. Calc. (%) for $C_{23}H_{15}ClN_4$ (382.8): C, 72.16; H, 3.95; N, 14.63. Found (%): C, 72.52; H, 4.01; N, 14.60.

3) 2-(2,4-Dichlorobenzylidenimino)-4-(4-chlorophenyl)-1-(3-trifluoromethylphenyl)-1H-pyrrole-3-carbonitrile; 5c

Yellow needle crystals; yield (49%); m.p.: 276°C - 278°C. IR [KBr, cm^{-1}]: 2933 (CH aromatic); 2850 (CH aliphatic); 2220 (C≡N); 1561 (C=N); 1024 (p-Cl-phenyl). ^1H NMR(DMSO-d_6, δ ppm): 7.30 (s, 1 H, CH-pyr- role); 7.33 - 7.36 (m, 1 H, 3-CF_3-C_6H_4-C_5-H); 7.42 - 7.51 (m, 3 H, 2,4-$(Cl)_2$-C_6H_3-$C_{3,5,6}$-H); 7.54 - 7.59 (m, 2 H, 3-CF_3-C_6H_4-$C_{4,6}$-H); 7.57 (d, 2 H, J = 8.4 Hz, 4-Cl-C_6H_4-$C_{2,6}$-H); 7.81 (d, 2 H, J = 8.4 Hz, 4-Cl-C_6H_4-$C_{3,5}$-H); 7.86 (s, 1 H, 3-CF_3-C_6H_4-C_2-H); 8.36 (s, 1 H, CH-benzylidinimine). Anal. Calc. (%) for $C_{25}H_{13}Cl_3F_3N_3$ (518.7): C, 57.88; H, 2.53; N, 8.10. Found (%): C, 57.65; H, 2.87; N, 8.59.

4) 2-(2,4-Dichlorobenzylidenimino)-4-(4-chlorophenyl)-1-(pyridin-3-yl)-1H-pyrrole-3-carbonitrile; 5d

Dark brown needle crystals; yield (85%); m.p.:240°C - 242°C. IR [KBr, cm^{-1}]: 3030 (CH aromatic); 2900 (CH aliphatic); 2199 (C≡N); 1620 (C=N); 1591 (C=C); 1094 (p-Cl-phenyl); MS m/z (relative intensity %): 452 ($M^{+\bullet}$+2, 1); 95 (100). Anal. Calc. (%) for $C_{23}H_{13}Cl_3N_4$ (451.7): C, 61.15; H, 2.90; N, 12.40. Found (%): C, 61.45; H, 3.02; N, 12.56.

3.1.5. General Procedure for Synthesis of Compounds 6a,b

An equimolar mixture of the selected 1-(substitutedaryl)-2-amino-4-(4-chlorophenyl)-1H-pyrrole-3-carbonitrile 4a,b (2 mmol.) and ethyl cyanoacetate (0.23 g, 0.1 mL, 2 mmol.) was heated under reflux in absolute ethanol (30 mL) for 8 h. The reaction mixture was allowed to cool and the product was collected and recrystallized from glacial acetic acid.

1) 4-Amino-3-(4-chlorophenyl)-6-oxo-1-(3-trifluoromethylphenyl)-6,7-dihydro-1H-pyrrolo[2,3-b]pyridine-5- carbonitrile; 6a

Pale yellow needle crystals; yield (77%); m.p.: >300°C. IR [KBr, cm^{-1}]: 3490, 3461 (br. OH tautomer); 3454, 3366, 3316 (NH_2, NH); 2960 (CH aromatic); 2200 (C≡N); 1660 (C=O); 1566 (C=C); 1022 (p-Cl-phenyl). ^1H NMR(DMSO-d_6, δ ppm): 4.93 (s, 2 H, NH_2, D_2O exchangeable); 7.08 (s, 1 H, CH-pyrrole); 7.34 - 7.38 (m, 1 H, 3-CF_3-C_6H_4-C_5-H); 7.48-7.62 (m, 2 H, 3-CF_3-C_6H_4-$C_{4,6}$-H); 7.64 (d, 2 H, J = 8.4 Hz, 4-Cl-C_6H_4-$C_{2,6}$-H); 7.86 (d, 2 H, J = 8.1 Hz, 4-Cl-C_6H_4-$C_{3,5}$-H); 7.90 (s, 1/2 H, NH, D_2O exchangeable); 8.45 (s, 1 H, 3-CF_3-C_6H_4-C_2-H); 10.28 (s, ½ H, OH tautomer, D_2O exchangeable). Anal. Calc. (%) for $C_{21}H_{12}ClF_3N_4O$ (428.8): C, 58.82; H, 2.82; N, 13.07. Found (%): C, 58.93; H, 2.88; N, 13.18.

2) 4-Amino-3-(4-chlorophenyl)-6-oxo-1-(pyridin-3-yl)-6,7-dihydro-1H-pyrrolo[2,3-b]pyridine-5-carbonitrile; 6b

Pale brown powder; yield (40%); m.p.: >300°C. IR [KBr, cm^{-1}]: 3462, 3412 (br. OH tautomer); 3380, 3330, 3292, 3225 (NH_2, NH); 2920 (CH aromatic);

2206 ($C\equiv N$); 1680 (C=O); 1634 (C=N); 1525 (C=C); 1092 (p- Cl-phenyl); MS m/z (relative intensity %): 361 ($M^{+\bullet}$, 2); 360 ($M^{+\bullet}$-1, 3); 57 (100). Anal. Calc. (%) for $C_{19}H_{12}ClN_5O$ (361.8): C, 63.08; H, 3.34; N, 19.36. Found (%): C, 63.17; H, 3.76; N, 19.51.

3.1.6. General Procedure for Synthesis of Compounds 7a,b

Equimolar amounts of the appropriate 1-(substitutedaryl)-2-amino-4-(4-chlorophenyl)-1H-pyrrole-3-carbonitrile 4a,b (2 mmol.) and malononitrile (0.13 g, 0.1 mL, 2 mmol.) were heated under reflux in absolute ethanol (20 mL) for 14 h. The reaction mixture was allowed to cool and the product was collected, washed with ethanol and recrystallized from dioxane.

1) 4,6-Diamino-3-(4-chlorophenyl)-1-(3-trifluoromethylphenyl)-1H-pyrrolo[2,3-b]pyridine-5-carbonitrile; 7a

Golden yellow needle crystals; yield (88%); m.p.: >300°C. IR [KBr, cm^{-1}]: 3375, 3325 (NH_2); 2928 (CH aromatic); 2200 ($C\equiv N$); 1646 (C=N); 1560 (C=C); 1024 (p-Cl-phenyl). ^1H NMR(DMSO-d_6, δ ppm): 4.18 (s, 2 H, pyridine-C_4-NH_2, D_2O exchangeable); 5.19 (s, 2 H, pyridine-C_2-NH_2, D_2O exchangeable); 6.83 (s, 1 H, CH- pyrrole); 7.10 - 7.13 (m, 1 H, 3-CF_3-C_6H_4-C_5-H); 7.31 (d, 2 H, J = 8.1 Hz, 4-Cl-C_6H_4-$C_{2,6}$-H); 7.34 - 7.47 (m, 2 H, 3-CF_3-C_6H_4-$C_{4,6}$-H); 7.82 (d, 2 H, J = 8.1 Hz, 4-Cl-C_6H_4-$C_{3,5}$-H); 8.38 (s, 1 H, 3-CF_3-C_6H_4-C_2-H). Anal. Calc. (%) for $C_{21}H_{15}ClF_3N_5$ (429.8): C, 58.96; H, 3.06; N, 16.37. Found (%): C, 58.74; H, 3.21; N, 16.34.

2) 4,6-Diamino-3-(4-chlorophenyl)-1-(pyridin-3-yl)-1H-pyrrolo-[2,3-b]pyridine-5-carbonitrile; 7b

Buff powder; yield (58%); m.p.: >300°C. IR [KBr, cm^{-1}]: 3428, 3364 (NH_2); 3066 (CH aromatic); 2198 ($C\equiv N$); 1638 (C=N); 1550 (C=C); 1092 (p-Cl-phenyl); MS m/z (relative intensity %): 363 ($M^{+\bullet}$+3, 2); 58 (100). Anal. Calc. (%) for $C_{19}H_{15}ClN_6$ (362.8): C, 63.25; H, 3.63; N, 23.29. Found (%): C, 63.92; H, 3.48; N, 23.20.

3.1.7. General Procedure for Synthesis of Compounds 8a,b

A mixture of the selected 1-(substitutedaryl)-2-amino-4-(4-chlorophenyl)-1H-pyrrole-3-carbonitrile 4a,b (2 mmol.) and hydroxylamine hydrochloride (0.14 g, 2 mmol.) in (30 mL) sodium ethoxide [prepared by dissolving sodium metal (0.06 g, 2 mmol.) in absolute ethanol (30 mL)] was refluxed for 12 h. Then the reaction was concentrated under reduced pressure and the residue was collected by filtration, washed thoroughly with water and recrystallized from ethanol/benzene mixture.

1) 4-(4-Chlorophenyl)-6-(3-trifluoromethylphenyl)-1,6-dihydropyrrolo[2,3-c]pyrazol-3-amine; 8a

Yellow powder; yield (75%); m.p.: 293°C - 295°C. IR [KBr, cm^{-1}]: 3418, 3290, 3240 (NH_2, NH); 3000, 2925 (CH aromatic); 1650 (C=N); 1560 (C=C); 1090 (p-Cl-phenyl); MS m/z (relative intensity %): 378 ($M^{+\bullet}$+2, 1); 377($M^{+\bullet}$+1, 1); 376 ($M^{+\bullet}$, 1); 58 (100). Anal. Calc. (%) for $C_{18}H_{12}ClF_3N_4$ (376.8): C, 57.38; H, 3.21; N, 14.87. Found (%): C, 57.45; H, 3.36; N, 14.92.

2) 4-(4-Chlorophenyl)-6-(pyridin-3-yl)-1,6-dihydropyrrolo[2,3-c]pyrazol-3-amine; 8b

Pale brown powder; yield (64%); m.p.: > 300°C. IR [KBr, cm^{-1}]: 3428, 3332, 3270, 3214, 3132 (NH_2, NH), 3040 (CH aromatic); 1636 (C=N); 1550 (C=C); 1094 (p-Cl-phenyl). ^{1}H NMR(DMSO-d_6, δ ppm): 5.27 (s, 2 H, NH_2, D_2O exchangeable); 6.77 (s, 1 H, CH-pyrrole); 7.33 (d, 2 H, J = 7.8 Hz, 4-Cl-C_6H_4-$C_{2,6}$-H); 7.56 - 7.70 (m, 3 H, pyridyl-$C_{4,5,6}$-H); 7.87 (d, 2 H, J = 7.8 Hz, 4-Cl-C_6H_4-$C_{3,5}$-H); 7.92 (s, 1 H, pyridyl-C_2-H); 8.06 (s, 1 H, NH, D_2O exchangeable). Anal. Calc. (%) for $C_{16}H_{12}ClN_5$ (309.8): C, 62.04; H, 3.90; N, 22.61 Found (%): C, 62.10; H, 3.93; N, 22.69.

3.1.8. General Procedure for Synthesis of Compounds 9a-d

The appropriate compound 4a,b (2 mmol.) was refluxed with equimolar amount of urea or thiourea (2 mmol.) in absolute ethanol (20 mL) containing sodium ethoxide [prepared by dissolving sodium metal (0.03 g, 2 mmol.) in absolute ethanol (20 mL)] for 10 h. The reaction was allowed to cool and the solid product was filtered and washed with ethanol.

1) 1-[4-(4-Chlorophenyl)-3-cyano-1-(3-trifluoromethylphenyl)-1H-pyrrol-2-yl]urea; 9a

Pale yellow crystals; gl. acetic; yield (80%); m.p.: >300°C. IR [KBr, cm^{-1}]: 3324, 3248, 3164 (NH_2, NH); 3092 (CH aromatic); 2225 (C≡N); 1700 (C=O); 1562 (C=C); 1024 (p-Cl-phenyl);); MS m/z (relative intensity %): 405 ($M^{+\bullet}$+1, 0.2); 404 ($M^{+\bullet}$, 0.2); 57 (100). Anal. Calc. (%) for $C_{19}H_{12}ClF_3N_4O$ (404.8): C, 56.38; H, 2.99; N, 13.84. Found (%): C, 56.44; H, 3.05; N, 13.96.

2) 1-[4-(4-Chlorophenyl)-3-cyano-1-(pyridin-3-yl)-1H-pyrrol-2-yl]urea; 9b

Buff powder; gl. acetic; yield (67%); m.p.: 324°C - 326°C. IR [KBr, cm^{-1}]: 3375, 3310, 3244, 3200 (NH_2, NH); 3025 (CH aromatic); 2225 (C≡N); 1680 (C=O); 1636 (C=N);1456 (C=C); 1100 (p-Cl phenyl). ^{1}H NMR(DMSO-d_6, δ ppm): 5.46 (s, 2 H, NH_2, D_2O exchangeable); 6.70 (s, 1 H, CH-pyrrole); 7.30 (d, 2 H, J = 8.1 Hz, 4-Cl-C_6H_4-$C_{2,6}$-H); 7.57 - 7.65 (m, 3 H, pyridyl-$C_{4,5,6}$-H); 7.84 (d, 2 H, J = 8.1 Hz, 4-Cl-C_6H_4-$C_{3,5}$-H); 8.04 (s, 1 H, NH, D_2O exchangeable); 8.48 (s, 1 H, pyridyl-C_2-H). ^{13}C NMR (DMSO-d_6, δ ppm): 106.2 (pyrrole-C_3); 114 (C≡N); 117 (pyrrole-C_5); 119 (pyrrole-C_2); 124

(pyrrole-C_4); 124.4 (pyridyl-C_5); 126.7 (pyridyl- C_6); 127.1 (4-Cl-C_6H_4-$C_{2,6}$); 127.6 (4-Cl-C_6H_4-$C_{3,5}$); 130.8 (4-Cl-C_6H_4-C_4); 131.6 (4-Cl-C_6H_4-C_1); 133.5 (pyridyl-C_1); 138.9 (pyridyl-C_2); 148.4 (pyridyl-C_4); 159.8 (C=O) of amide. Anal. Calc. (%) for $C_{17}H_{12}ClN_5O$ (337.8): C, 60.45; H, 3.58; N, 20.73. Found (%): C, 60.71; H, 3.63; N, 20.78.

3) 1-[4-(4-Chlorophenyl)-3-cyano-1-(3-trifluoromethylphenyl)-1H-pyrrol-2-yl]thiourea; 9c

Yellow powder; petroleum ether 60/80; yield (54%); m.p.: >300°C. IR [KBr, cm^{-1}]: 3413, 3250, 3183 (NH_2, NH); 2933 (CH aromatic); 2259 (C≡N); 1562 (C=C); 1562, 1416, 1100, 922 (I, II, III, IV bands N-C=S); 1025 (p-Cl-phenyl). ^1H NMR(DMSO-d_6, δ ppm): 5.47 (s, 2 H, NH_2, D_2O exchangeable); 7.05 (s, 1 H, CH-pyrrole); 7.12-7.15 (m, 1 H, 3-CF_3-C_6H_4-C_5-H); 7.30 (d, 2 H, J = 8.1 Hz, 4-Cl-C_6H_4-$C_{2,6}$-H); 7.34-7.47 (m, 2 H, 3-CF_3- C_6H_4-$C_{4,6}$-H); 7.82 (d, 2 H, J = 8.1 Hz, 4-Cl-C_6H_4-$C_{3,5}$-H); 8.40 (s, 1 H, 3-CF_3-C_6H_4-C_2-H); 10.62 (s, 1 H, NH, D_2O exchangeable). Anal. Calc. (%) for $C_{19}H_{12}ClF_3N_4S$ (420.8): C, 54.23; H, 2.87; N, 13.31. Found (%): C, 54.27; H, 2.93; N, 13.56.

4) 1-[4-(4-Chlorophenyl)-3-cyano-1-(pyridin-3-yl)-1H-pyrrol-2-yl]thiourea; 9d

Pale browncrystals; dioxane; yield (89%); m.p.: 334°C - 336°C. IR [KBr, cm^{-1}]: 3424, 3380, 3330 (NH_2, NH); 3020 (CH aromatic); 2196 (C≡N); 1628 (C=N); 1552 (C=C); 1500, 1280, 1098, 1006 (I, II, III, IV bands N-C=S); 1110 (p-Cl-phenyl); MS m/z (relative intensity %): 355 ($M^{+\bullet}$+2, 0.5); 353 ($M^{+\bullet}$, 1); 352 ($M^{+\bullet}$−1, 1); 57 (100). Anal. Calc. (%) for $C_{17}H_{12}ClN_5S$ (353.8): C, 57.71; H, 3.42; N, 19.79. Found (%): C, 57.80; H, 3.45; N, 19.88.

3.1.9. General Procedure for Synthesis of Compounds 10a-d

Method 1:

A mixture of the selected 1-(substitutedaryl)-2-amino-4-(4-chlorophenyl)-1H-pyrrole-3-carbonitrile 4a,b (2 mmol.) was refluxed and urea (0.12 g, 2 mmol.) and/or thiourea (0.15 g, 2 mmol.) was refluxed for 12 h. in a mixture of glacial acetic acid and HCl (20 mL) (3:1). The reaction was allowed to cool, filtered and washed with ethanol to yield the target compounds 10a-d in an average yield of 82%.

Method 2:

The selected urea or thiourea derivative 9a-d (2 mmol.) was refluxed in pyridine (10 mL) for 16 h. The solvent was evaporated under reduced pressure and the solid obtained was collected to yield the target compounds 10a-d in an average yield of 63%.

1) 4-Amino-5-(4-chlorophenyl)-7-(3-trifluoromethylphenyl)-1H-pyrrolo[2,3-d]pyrimidine-2(7H)-one; 10a

Light brown crystals; Ethanol/Benzene; yield (73%); m.p.: >300°C. IR [KBr, cm^{-1}]: 3450 (br. OH tautomer); 3356, 3211 (NH_2, NH); 3080 (CH aromatic); 1654 (C=O); 1458 (C=C); 1100 (p-Cl-phenyl); MS m/z (relative intensity %): 406 ($M^{+\bullet}$+2, 1); 405 ($M^{+\bullet}$+1, 1); 404 ($M^{+\bullet}$, 1); 61 (100). Anal. Calc. (%) for $C_{19}H_{12}ClF_3N_4O$ (404.8): C, 56.38; H, 2.99; N, 13.84. Found (%): C, 56.43; H, 3.07; N, 13.93.

2) 4-Amino-5-(4-chlorophenyl)-7-(pyridin-3-yl)-1H-pyrrolo[2,3-d]pyrimidine-2(7H)-one; 10b

Pale brownpowder; dioxane; yield (40%); m.p.: 286°C - 288°C. IR [KBr, cm^{-1}]: 3480 (br. OH tautomer); 3420, 3223 (NH_2, NH); 3100 (CH aromatic); 1690 (C=O); 1095 (p-Cl-phenyl). ^1H NMR (DMSO-d_6, δ ppm): 4.15 (s, 2 H, NH_2, D_2O exchangeable); 4.94 (s, 1/2 H, NH, D_2O exchangeable); 7.39 - 7.41 (m, 3 H, pyridyl- $C_{4,5,6}$-H); 7.42 (s, 1 H, CH-pyrrole); 7.50 (d, 2 H, J = 8.4 Hz, 4-Cl-C_6H_4-$C_{2,6}$-H); 7.83 (s, 1/2 H, OH, tautomer, D_2O exchangeable); 8.05 (s, 1 H, pyridyl-C_2-H); 8.17 (d, 2 H, J = 8.4 Hz, 4-Cl-C_6H_4-$C_{3,5}$-H). Anal. Calc. (%) for $C_{17}H_{12}ClN_5O$ (337.8): C, 60.45; H, 3.58; N, 20.73. Found (%): C, 60.62; H, 3.60; N, 20.81.

3) 4-Amino-5-(4-chlorophenyl)-7-(3-trifluoromethylphenyl)-1H-pyrrolo[2,3-d]pyrimidine-2(7H)-thione; 10c

Pale yellowpowder; Ethanol; yield (65%); m.p.: >360°C. IR [KBr, cm^{-1}]: 3367, 3265 (NH_2, NH); 3164 (CH aromatic); 1480, 1328, 1120, 1097 (I, II, III, IV bands N-C=S); 1080 (p-Cl-phenyl). ^1H NMR(DMSO-d_6, δ ppm): 4.76 (s, 1 H, NH, D_2O exchangeable); 6.37 (s, 2 H, NH_2, D_2O exchangeable); 6.85 (d, 1 H, J = 7.2 Hz, 3- CF_3-C_6H_4-C_6-H); 6.95 (d, 2 H, J = 8.3 Hz, 4-Cl-C_6H_4-$C_{2,6}$-H); 7.00 (s, 1 H, CH-pyrrole); 7.25 - 7.30 (m, 1 H, 3-CF_3-C_6H_4-C_5-H); 7.64 (d, 2 H, J = 8.3 Hz, 4-Cl-C_6H_4-$C_{3,5}$-H); 8.08 - 8.12 (m, 2 H, 3-CF_3-C_6H_4-$C_{2,4}$-H). ^{13}C NMR(DMSO-d_6, δ ppm): 107.8 (pyrrolopyrimidine C_{4a}); 113 (3-CF_3-C_6H_4-C_2); 116.6 (pyrrolopyrimidine C_6); 120 (pyrrolopyrimidine C_{7a}); 121.9 (3-CF_3-C_6H_4-C_4); 124.1 (CF_3); 124.8 (3-CF_3-C_6H_4-C_6); 126 (pyrrolopyrimidine C_5); 128.9 (4-Cl-C_6H_4-$C_{2,6}$); 129.8 (4-Cl-C_6H_4-$C_{3,5}$-H); 130.9 (3-CF_3-C_6H_4-C_5); 131.04 (3-CF_3-C_6H_4-C_3); 131.08 (4-Cl-C_6H_4-C_4); 135.8 (4-Cl-C_6H_4-C_1); 145 (3-CF_3-C_6H_4-C_1); 157 (pyrrolopyrimidine C_4); 185 (C=S). Anal. Calc. (%) for $C_{19}H_{12}ClF_3N_4S$ (420.8): C, 54.23; H, 2.87; N, 13.31. Found (%): C, 54.27; H, 2.91; N, 13.46.

4) 4-Amino-5-(4-chlorophenyl)-7-(pyridin-3-yl)-1H-pyrrolo[2,3-d]pyrimidine-2(7H)-thione; 10d

Pale brown powder; Ethanol; yield (83%); m.p.: >300°C. IR [KBr, cm^{-1}]: 3428, 3350, 3220 (NH_2, NH); 3080 (CH aromatic); 1624 (C=N); 1404, 1316, 1150, 1000 (I, II, III, IV bands N-C=S); 1100 (p-Cl-phenyl); MS m/z (relative intensity %): 355 ($M^{+\bullet}$+2, 0.6); 354 ($M^{+\bullet}$+1, 1); 353 ($M^{+\bullet}$, 2); 58 (100). Anal. Calc. (%) for $C_{17}H_{12}ClN_5S$ (353.8): C, 57.71; H, 3.42; N, 19.79. Found (%): C, 57.80; H, 3.51; N, 19.84.

3.1.10. General Procedure for Synthesis of Compounds 11a,b

A mixture of the appropriate 1-(substitutedaryl)-2-amino-4-(4-chlorophenyl)-1H-pyrrole-3-carbonitrile 4a,b (2 mmol.) and phenyl isothiocyanate (0.27 g, 0.24 mL, 2 mmol.) was refluxed for 7 h in absolute ethanol (30 mL) and the solid product was filtered and washed with ethanol.

1) 1-[4-(4-Chlorophenyl)-3-cyano-1-(3-trifluoromethylphenyl)-1H-pyrrol-2-yl]-3-phenylthiourea; 11a

Yellow needle crystals; gl.acetic; yield (57%); m.p.: >360°C. IR [KBr, cm^{-1}]: 3372, 3286 (NH); 3050 (CH aromatic); 2200 (C≡N); 1562 (C=C); 1418, 1282, 1170, 1020 (I, II, III, IV bands N-C=S); 1098 (p-Cl-phenyl). ^1H NMR (DMSO-d_6, δ ppm): 6.90 (s, 1 H, CH-pyrrole); 7.10 - 7.13 (m, 3 H, C_6H_5-$C_{3,4,5}$-H); 7.30 (d, 2 H, J = 8.4 Hz, 4-Cl-C_6H_4-$C_{2,6}$-H); 7.34-7.39 (m, 3 H, 3-CF_3-C_6H_4-$C_{4,5,6}$-H); 7.47 (d, 2 H, J = 8.4 Hz, 4-Cl-C_6H_4-$C_{3,5}$-H); 7.80 (d, 2 H, J = 7.2 Hz, C_6H_5-$C_{2,6}$-H); 7.83 (s, 1 H, 3-CF_3-C_6H_4-C_2-H); 8.70 (s, 1 H, pyrrole-C_2-NH, D_2O exchangeable); 9.59 (s, 1 H, NH, D_2O exchangeable). Anal. Calc. (%) for $C_{25}H_{16}ClF_3N_4S$ (496.9): C, 60.42; H, 3.25; N, 11.27. Found (%): C, 60.44; H, 3.35; N, 11.31.

2) 1-[4-(4-Chlorophenyl)-3-cyano-1-(pyridin-3-yl)-1H-pyrrol-2-yl]-3-phenylthiourea; 11b

Buff crystals; DMF; yield (70%); m.p.: 288°C - 290°C. IR [KBr, cm^{-1}]: 3374, 3326, 3260, 3200 (NH); 3090 (CH aromatic); 2206 (C≡N); 1640 (C=N); 1566 (C=C); 1420, 1396, 1250, 1110 (I, II, III, IV bands N-C=S); 1098 (p-Cl-phenyl); MS m/z (relative intensity %): 429 ($M^{+\bullet}$, 2); 83 (100). Anal. Calc. (%) for $C_{23}H_{16}ClN_5S$ (429.9): C, 64.25; H, 3.75; N, 16.29. Found (%): C, 64.29; H, 3.77; N, 16.32.

3.1.11. General Procedure for Synthesis of Compounds 12a,b

Method 1:

A mixture of the appropriate 1-(substitutedaryl)-2-amino-4-(4-chlorophenyl)-1H-pyrrole-3-carbonitrile 4a,b (2 mmol.) and phenyl isothiocyanate (0.27 g, 0.24 mL, 2 mmol.) was refluxed in pyridine (15 mL) for 10 h and the solid product was filtered, washed with ethanol to yield the target compounds 12a,b in an average yield of 53%.

Method 2:

The selected phenyl thiourea derivatives 11a,b was refluxed in pyridine (10 mL) for 18 h. The solvent was evaporated under reduced pressure to yield the target compounds 12a,b in an average yield of 48%.

1) 5-(4-Chlorophenyl)-4-imino-3-phenyl-7-(3-trifluoromethylphenyl)-3,4-dihydro-1H-pyrrolo[2,3-d]pyrimi dine- 2(7H)-thione; 12a

Light brown crystals; dioxane; yield (68%); m.p.: >300°C. IR [KBr, cm^{-1}]: 3190 (NH); 3050 (CH aromatic); 1546 (C=C); 1434, 1332, 1180, 1020 (I, II, III, IV bands N-C=S); 1100 (p-Cl-phenyl); MS m/z (relative intensity %): 496 ($M^{+\bullet}$, 1); 140 (100). Anal. Calc. (%) for $C_{25}H_{16}ClF_3N_4S$ (496.9): C, 60.42; H, 3.25; N, 11.27. Found (%): C, 60.51; H, 3.28; N, 11.29.

2) 5-(4-Chlorophenyl)-4-imino-3-phenyl-7-(pyridin-3-yl)-3,4-dihydro-1H-pyrrolo[2,3-d]pyrimidine-2(7H)-thione; 12b

Dark grey powder; dioxane; yield (59%); m.p.: > 300°C. IR [KBr, cm^{-1}]: 3317, 3191 (NH); 3020 (CH aromatic); 1625 (C=N); 1520 (C=C); 1406, 1250, 1134, 1026 (I, II, III, IV bands N-C=S); 1080 (p-Cl-phenyl). ^1H NMR (DMSO-d_6, δ ppm): 5.30 (s, 1 H, NH, D_2O exchangeable); 6.63 (s, 1 H, CH-pyrrole); 6.85 - 7.05 (m, 3 H, C_6H_5-$C_{3,4,5}$-H); 7.30 - 7.50 (m, 2 H, C_6H_5-$C_{2,6}$-H); 7.56 (d, 2 H, J = 8.4 Hz, 4-Cl-C_6H_4-$C_{2,6}$-H); 7.62 - 7.78 (m, 2 H, pyridyl-$C_{5,6}$-H); 7.94 (d, 2 H, J = 8.4 Hz, 4-Cl-C_6H_4-$C_{3,5}$-H); 7.99 - 8.03 (m, 2 H, pyridyl-$C_{2,4}$-H); 10.58 (s, 1 H, imine NH, D_2O exchangeable). Anal. Calc. (%) for $C_{23}H_{16}ClN_5S$ (429.9): C, 64.25; H, 3.75; N, 16.29. Found (%): C, 64.29; H, 3.80; N, 16.36.

3.1.12. General Procedure for Synthesis of Compounds 13a,b

A mixture of the appropriate 1-(substitutedaryl)-2-amino-4-(4-chlorophenyl)-1H-pyrrole-3-carbonitrile 4a,b (2 mmol.) and triethyl orthoformate (0.29 g, 0.33 mL, 2 mmol.) in acetic anhydride (25 mL) was heated under reflux for 6 h. The reaction mixture was concentrated to the minimum and left to cool. The obtained product was collected washed with ethanol and recrystallized from dimethylformamide.

1) Ethyl N-4-(4-chlorophenyl)-3-cyano-1-(3-trifluoromethylphenyl)-1H-pyrrol-2-ylformimidate; 13a

Pale yellow crystals; yield (67%); m.p.: 298°C - 300°C. IR [KBr, cm^{-1}]: 3100 (CH aromatic); 2900 (CH-aliphatic); 2190 (C≡N); 1620 (C=N); 1560 (C=C); 1270, 1025 (C-O-C); 1090 (p-Cl-phenyl). ^1H NMR (DMSO-d_6, δ ppm): 1.17 (t, 3 H, J = 6.8 Hz, CH_2CH_3); 4.05 - 4.10 (m, 2 H, CH_2CH_3); 6.80 (s, 1 H, CH-pyrrole); 7.15 - 7.18 (m, 1 H, 3-CF_3-C_6H_4-C_5-H); 7.30 - 7.33 (m, 2 H, 3-CF_3-C_6H_4-$C_{4,6}$-H); 7.35 (s, 1 H, N=CH); 7.51(d, 2 H, J = 8.4 Hz, 4-Cl-C_6H_4-$C_{2,6}$-H); 7.81 (d, 2 H, J = 8.4 Hz, 4-Cl-C_6H_4-$C_{3,5}$-H); 8.36 (s, 1 H, 3-CF_3-C_6H_4-C_2-H). Anal. Calc. (%) for $C_{21}H_{15}ClF_3N_3O$ (417.8): C, 60.37; H, 3.62; N, 10.06. Found (%): C, 60.41; H, 3.65; N, 10.12.

2) Ethyl N-4-(4-chlorophenyl)-3-cyano-1-(pyridin-3-yl)-1H-pyrrol-2-ylformimidate; 13b

Dark brown needle crystals; yield (49%); m.p.:< 300°C. IR [KBr, cm^{-1}]: 3178 (CH aromatic); 2923, 2860 (CH-aliphatic); 2198 (C≡N); 1630 (C=N); 1557 (C=C); 1280, 1023 (C-O-C); 1023 (p-Cl-phenyl); MS m/z (relative intensity %): 352 ($M^{+\bullet}$+2, 0.4); 351 ($M^{+\bullet}$+1, 0.5); 350 ($M^{+\bullet}$, 0.4); 349 ($M^{+\bullet}$−1, 0.4); 58 (100). Anal. Calc. (%) for $C_{19}H_{15}ClN_4O$ (350.8): C, 65.05; H, 4.31; N, 15.97. Found (%): C, 65.10; H, 4.37; N, 15.99.

3.1.13. General Procedure for Synthesis of Compounds 14a,b

Method 1:

The appropriate 1-(substitutedaryl)-2-amino-4-(4-chlorophenyl)-1H-pyrrole-3-carbonitrile 4a,b (2 mmol.) was refluxed in excess formamide (30 mL) for 10 h. The reaction mixture was concentrated to the minimum and left to cool then triturated with glacial acetic acid to yield the target compounds 14a,b in an average yield of 66%.

Method 2:

The selected ethyl N-1-(substitutedaryl)-4-(4-chlorophenyl)-3-cyano-1H-pyrrol-2-ylformimidate 13a,b (2 mmol.) was stirred in (100 mL) of methanolic ammonia solution [methanol: ammonia (2:1)] at room temperature for 7 h. The separated solid was then collected to yield the target compounds 14a,b in an average yield of 41%.

1) 5-(4-Chlorophenyl)-7-(3-trifluoromethylphenyl)-7H-pyrrolo[2,3-d]pyrimi-din-4-amine; 14a

Pale yellow powder; gl. acetic; yield (51%); m.p.:< 300°C. IR [KBr, cm^{-1}]: 3408, 3350, 3240 (NH_2); 3020 (CH aromatic); 1525 (C=C); 1098 (p-Cl-phenyl); MS m/z (relative intensity %): 390 ($M^{+\bullet}$+2, 2); 389 ($M^{+\bullet}$+1, 1); 388 ($M^{+\bullet}$, 1); 387 ($M^{+\bullet}$−1, 2); 57 (100). Anal. Calc. (%) for $C_{19}H_{12}ClF_3N_4$ (388.8): C, 58.70; H, 3.11; N, 14.41. Found (%): C, 58.73; H, 3.18; N, 14.49.

2) 5-(4-Chlorophenyl)-7-(pyridin-3-yl)-7H-pyrrolo[2,3-d]pyrimidin-4-amine; 14b

Dark brown powder; gl. acetic; yield (67%); m.p.: 320°C - 322°C. IR [KBr, cm^{-1}]: 3436, 3382 (NH_2); 3092 (CH aromatic); 1550 (C=N); 1450 (C=C); 1100 (p-Cl phenyl). ^1H NMR (DMSO-d_6, δ ppm): 7.38 (s, 2 H, NH_2, D_2O exchangeable); 7.41 - 7.50 (m, 2 H, pyridyl-$C_{4,5}$-H); 7.69 (s, 1 H, CH-pyrrole); 7.78 (d, 2 H, J = 8.4 Hz, 4-Cl-C_6H_4-$C_{2,6}$-H); 7.91-8.10 (m, 3 H, pyridyl-C_6-H & 4-Cl-C_6H_4-$C_{3,5}$-H); 8.87 (s, 1 H, pyridyl-C_2-H); 9.06 (s, 1 H, pyrimidine-C_2-H). Anal. Calc. (%) for $C_{17}H_{12}ClN_5$ (321.8): C, 63.46; H, 3.76; N, 21.77. Found (%): C, 63.53; H, 3.80; N, 21.80.

3.1.14. General Procedure for Synthesis of Compounds 15a,b

Method 1:

Equimolar amounts of the selected of 1-(substituted aryl)-2-amino-4-(4-chlorophenyl)-1H-pyrrole-3-carboni- trile 4a,b (2 mmol.) and N,N-dimethylformamide-dimethylacetal (0.24 g, 0.27 mL, 2 mmol.) in xylene (20 mL) were refluxed for 6 h. The reaction mixture was concentrated under reduced pressure and the obtained products were filtered, washed with ethanol to yield the target compounds 15a,b in an average yield of 45%.

Method 2:

The appropriate ethyl 1-(substitutedaryl)-N-4-(4-chlorophenyl)-3-cyano-1H-pyrrol-2-ylformimidate 13a,b (2 mmol.) and dimethylamine (0.09 g, 0.1 mL, 2 mmol.) were stirred for 2 h in absolute ethanol (50 mL). The obtained product was filtered off, washed with ethanol to yield the target products 15a,b in an average yield of 57%.

1) N'-[4-(4-Chlorophenyl)-3-cyano-1-(3-trifluoromethylphenyl)-1H-pyrrol-2-yl]-N,N-dimethylformimidamide; 15a

Faint yellow crystals; Ethanol/Benzene.; yield (67%); m.p.: <300°C. IR [KBr, cm^{-1}]: 3040 (CH aromatic); 2900, 2850 (CH-aliphatic); 2220 (C≡N); 1648 (C=N); 1560 (C=C); 1024 (p-Cl-phenyl); MS m/z (relative intensity %): 418 ($M^{+\bullet}$+2, 0.6); 415 ($M^{+\bullet}$−1, 0.4); 284 (100). Anal. Calc. (%) for $C_{21}H_{16}ClF_3N_4$ (416.8): C, 60.51; H, 3.87; N, 13.44. Found (%): C, 60.55; H, 3.93; N, 13.51.

2) N'-[4-(4-Chlorophenyl)-3-cyano-1-(pyridin-3-yl)-1H-pyrrol-2-yl]-N,N-dimethylformimidamide; 15b

Buff crystals; Ethanol/Benzene.; yield (59%); m.p.: 288°C - 290°C. IR [KBr, cm^{-1}]: 3030 (CH aromatic); 2930, 2840 (CH-aliphatic); 2202 (C≡N); 1648 (C=N); 1564 (C=C); 1090 (p-Cl-phenyl).^{1}H NMR (DMSO-d_6, δ ppm): 3.80 (s, 6 H, two CH_3); 7.30 (d, 2 H, J = 8.4 Hz, 4-Cl-C_6H_4-$C_{2,6}$-H); 7.45-7.65 (m, 3 H, CH-pyrrole, pyridyl-$C_{5,6}$-H); 7.82 (d, 2 H, J = 8.4 Hz, 4-Cl-C_6H_4-$C_{3,5}$-H); 7.98 (s, 1 H, N=CH); 8.21 (d, 1 H, J = 6.7 Hz, pyridyl-C_4-H); 8.41 (s, 1 H, pyridyl-C_2-H). Anal. Calc. (%) for $C_{19}H_{16}ClN_5$ (349.8): C, 65.24; H, 4.61; N, 20.02. Found (%): C, 65.28; H, 4.72; N, 20.10.

3.1.15. General Procedure for Synthesis of Compounds 16a,b

A mixture of the selected 1-(substitutedaryl)-2-amino-4-(4-chlorophenyl)-1H-pyrrole-3-carbonitrile 4a,b (2 mmol.) and p-toluene sulphonyl chloride (0.38 g, 2 mmol.) was refluxed for 7 h. in toluene (15 mL) containing drops of TEA. The solid obtained was filtered off, washed with ethanol and recrystallized from ethanol/benzene mixture.

1) N-[4-(4-Chlorophenyl)-3-cyano-1-(3-trifluoromethylphenyl)-1H-pyrrol-2-yl)-4-methylbenzenesulfonamide; 16a

Orange yellow crystals; yield (59%); m.p.: >300°C. IR [KBr, cm^{-1}]: 3230, 3198 (NH); 3000 (CH aromatic); 2921, 2858 (CH-aliphatic); 2260 (C≡N); 1522 (C=C); 1452, 1350, 1120 (SO_2); 1035 (p-Cl-phenyl); MS m/z (relative intensity %): 516 ($M^{+\bullet}$+1, 2); 57 (100). Anal. Calc. (%) for $C_{25}H_{17}ClF_3N_3O_2S$ (515.9): C, 58.20; H, 3.32; N, 8.14. Found (%): C, 58.26; H, 3.37; N, 8.17.

2) N-[4-(4-Chlorophenyl)-3-cyano-1-(pyridin-3-yl)-1H-pyrrol-2-yl]-4-methylbenzenesulfonamide; 16b

Buff powder; yield (63%); m.p.: >360°C. IR [KBr, cm^{-1}]: 3310, 3180 (NH); 3050 (CH aromatic); 2923, 2858 (CH-aliphatic); 2209 (C≡N); 1543 (C=C); 1420, 1300, 1177 (SO_2); 1026 (p-Cl phenyl).^{1}H NMR (DMSO- d_6, δ ppm): 2.29 (s, 3 H, CH_3); 7.11 (d, 2 H, J = 8.1 Hz, 4-CH_3-C_6H_4-$C_{3,5}$-H); 7.18 (s, 1 H, CH-pyrrole); 7.38 - 7.42 (m, 4 H, 4-Cl-C_6H_4-$C_{2,6}$-H & 4-CH_3-C_6H_4-$C_{2,6}$-H); 7.48 (d, 2 H, J = 8.4 Hz, 4-Cl-C_6H_4-$C_{3,5}$-H); 7.60 - 7.85 (m, 3 H, pyridyl-$C_{4,5,6}$-H); 8.07 (s, 1 H, pyridyl-C_2-H); 9.40 (s, 1 H, NH, D_2O exchangeable). ^{13}C (DMSO-d_6): 20.7 (CH_3); 109 (pyrrole-C_3); 114 (C≡N); 117 (pyrrole-C_5); 119 (pyrrole-C_2); 124 (pyrrole-C_4); 124.4 (pyridyl- C_5); 125.4 (pyridyl-C_6); 128 (4-CH_3-C_6H_4-$C_{2,6}$); 128.9 (4-Cl-C_6H_4-$C_{2,6}$); 131 (4-Cl-C_6H_4-$C_{3,5}$& 4-CH_3-C_6H_4-$C_{3,5}$); 133 (4-Cl-C_6H_4-C_6); 134 (4-Cl-C_6H_4-C_1); 135 (pyridyl-C_1); 137.8 (4-CH_3-C_6H_4-C_1); 139 (4-CH_3-C_6H_4-C_4); 143.9 (pyridyl-C_2); 145.2 (pyridyl-C_4). Anal. Calc. (%) for $C_{23}H_{17}ClN_4O_2S$ (448.9): C, 61.54; H, 3.82; N, 12.48. Found (%): C, 61.58; H, 3.88; N, 12.51.

3.1.16. General Procedure for Synthesis of Compounds 17a,b

The selected compound 4a,b (2 mmol.) was stirred at room temperature for 3 h. inconc. sulfuric acid (15 mL) then poured drop by drop on to crushed ice. The reaction mixture was neutralized with ammonium hydroxide and the obtained product was filtered, washed thoroughly with water, left to dry and recrystallized from ethanol.

1) 2-Amino-4-(4-chlorophenyl)-1-(3-trifluoromethylphenyl)-1H-pyrrole-3-carboxamide; 17a

Brown needle crystals; yield (43%); m.p.: >300°C. IR [KBr, cm^{-1}]: 3390, 3167 (NH_2); 3020 (CH aromatic); 1683 (C=O); 1110 (p-Cl-phenyl). ^{1}H NMR (DMSO-d_6, δ ppm): 6.37 (s, 2 H, NH_2, D_2O exchangeable); 6.94 (d, 1 H, J = 7.2 Hz, 3-CF_3-C_6H_4-C_6-H); 7.00 (s, 1 H, CH-pyrrole); 7.25 - 7.30 (m, 2 H, 3-CF_3-C_6H_4-$C_{4,5}$-H); 7.34 (d, 2 H, J = 8.1 Hz, 4-Cl-C_6H_4-$C_{2,6}$-H); 7.63 (d, 2 H, J = 8.1 Hz, 4-Cl-C_6H_4-$C_{3,5}$-H); 8.11 (s, 1 H, 3-CF_3-C_6H_4-C_2-H); 8.60 (s, 2 H, $CONH_2$, D_2O exchangeable). Anal. Calc. (%) for $C_{18}H_{13}ClF_3N_3O$ (379.8): C, 56.93; H, 3.45; N, 11.06. Found (%): C, 56.97; H, 3.52; N, 11.13.

2) 2-Amino-4-(4-chlorophenyl)-1-(pyridin-3-yl)-1H-pyrrol-3-carboxamide; 17b

Dark brown powder; yield (65%); m.p.: 305°C - 307°C. IR [KBr, cm^{-1}]: 3430, 3292, 3182 (NH_2); 3015 (CH aromatic); 1672 (C=O); 1620 (C=N); 1512 (C=C); 1069 (p-Cl-phenyl); MS m/z (relative intensity %): 314 ($M^{+\bullet}$+2, 1); 313 ($M^{+\bullet}$+1, 2); 312 ($M^{+\bullet}$, 0.3); 197 (100). Anal. Calc. (%) for $C_{16}H_{13}ClN_4O$ (312.8): C, 61.44; H, 4.19; N, 17.91. Found (%): C, 61.47; H, 4.23; N, 17.96.

3.1.17. General Procedure for Synthesis of Compounds 18a,b

Method 1:

The selected 1-(substitutedaryl)-2-amino-4-(4-chlorophenyl)-1H-pyrrole-3-carbonitrile 4a,b (2 mmol.) was refluxed for 9 h. in excess formic acid 80% (30 mL). The reaction mixture was then concentrated and the obtained product was filtered, washed with ethanol to yield the target compounds 18a,b in an average yield of 52%.

Method 2:

The selected 1-(substitutedaryl)-2-amino-4-(4-chlorophenyl)-1H-pyrrole-3-carboxamide 17a,b (2 mmol.) was refluxed in excess formamide (30 mL) for 8 h. The reaction mixture was concentrated to yield the target compounds 18a,b in an average yield of 67%.

1) 5-(4-Chlorophenyl)-7-(3-trifluoromethylphenyl)-3H-pyrrolo[2,3-d]pyrimidin-4(7H)-one; 18a

Faint yellow powder; Ethanol; yield (48%); m.p.: >300°C. IR [KBr, cm^{-1}]: 3386 (br. OH, tautomer); 3280 (NH); 3000 (CH aromatic); 1680 (C=O); 1470 (C=C); 1090 (p-Cl-phenyl). ^1H NMR(DMSO-d_6, δ ppm): 5.27 (s, 1/2 H, NH, D_2O exchangeable); 6.76 (s, 1 H, CH-pyrrole); 6.80 - 6.90 (m, 1 H, 3-CF_3-C_6H_4-C_5-H); 6.98 (d, 1 H, J = 7.2 Hz, 3-CF_3-C_6H_4-C_6-H); 7.33 (d, 2 H, J = 8.4 Hz, 4-Cl-C_6H_4-$C_{2,6}$-H); 7.54 (d, 1 H, J = 6.9 Hz, 3-CF_3-C_6H_4-C_4-H); 7.65 (d, 2 H, J = 8.4 Hz, 4-Cl-C_6H_4-$C_{3,5}$-H);7.88 (s, 1 H, CH-pyrimidine); 7.91 (s, 1 H, 3-CF_3-C_6H_4-C_2-H); 8.06 (s, 1/2 H, OH tautomer, D_2O exchangeable). Anal. Calc. (%) for $C_{19}H_{11}ClF_3N_3O$ (389.8): C, 58.55; H, 2.84; N, 10.78. Found (%): C, 58.58; H, 2.89; N, 10.81.

2) 5-(4-Chlorophenyl)-7-(pyridin-3-yl)-3H-pyrrolo[2,3-d]pyrimidin-4(7H)-one; 18b

Grey powder; Ethanol; yield (54%); m.p.: >300°C. IR [KBr, cm^{-1}]: 3429 (br. OH tautomer); 3200 (NH); 3048 (CH aromatic); 1670 (C=O); 1598 (C=C); 1098 (p-Cl-phenyl); MS m/z (relative intensity %): 324 ($M^{+\bullet}$+2, 0.1); 323 ($M^{+\bullet}$+1, 0.1); 78 (100). Anal. Calc. (%) for $C_{17}H_{11}ClN_4O$ (322.7): C, 63.26; H, 3.44; N, 17.36. Found (%): C, 63.31; H, 3.54; N, 17.41.

3.1.18. General Procedure for Synthesis of Compounds 19a,b

The selected 1-(substitutedaryl)-2-amino-4-(4-chlorophenyl)-1H-pyrrol-3-carboxamide 17a,b (2 mmol.) was refluxed in excess thionyl chloride (15 mL) for 6 h. The reaction mixture was evaporated under reduced pressure. The residue was triturated with ethanol, filtered then recrystallized from glacial acetic acid.

1) 7-(4-Chlorophenyl)-5-(3-trifluoromethylphenyl)-1H-pyrrolo[3,2-d]1,2,6-thiadiazine-2,4(1H,3H)-dione; 19a

Faint orange powder; yield (48%); m.p.: 294°C - 296°C. IR [KBr, cm^{-1}]: 3181 (NH); 3070 (CH aromatic); 1706 (C=O); 1520 (C=C); 1139 (S=O); 1050 (p-Cl-phenyl).^{1}H NMR(DMSO-d_6, δ ppm): 4.20 (s, 1 H, NH, D_2O exchangeable); 7.03 (s, 1 H, CH-pyrrole); 7.10 - 7.20 (m, 3 H, 3-CF_3-C_6H_4-$C_{4,5,6}$-H); 7.37 (d, 2 H, J = 8.1 Hz, 4-Cl-C_6H_4-$C_{2,6}$-H); 7.68 (d, 2 H, J = 8.1 Hz, 4-Cl-C_6H_4-$C_{3,5}$-H); 7.92 (s, 1 H, 3-CF_3-C_6H_4-C_2-H); 8.06 (s, 1 H, NH, D_2O exchangeable); MS m/z (relative intensity %): 427 ($M^{+\bullet}$+2, 0.1); 426 ($M^{+\bullet}$+1, 0.1); 425 ($M^{+\bullet}$, 0.2); 424 ($M^{+\bullet}$−1, 52); 229 (100). Anal. Calc. (%) for $C_{18}H_{11}ClF_3N_3O_2S$ (425.8): C, 50.77; H, 2.60; N, 9.87. Found (%): C, 50.81; H, 2.68; N, 9.91.

2) 7-(4-Chlorophenyl)-5-(pyridin-3-yl)-1H-pyrrolo[3,2-d]1,2,6-thiadiazin-2,4(1H,3H)-dione; 19b

Dark brown powder; yield (53%); m.p.: 310°C - 312°C. IR [KBr, cm^{-1}]: 3354 (NH); 3090 (CH aromatic); 1690 (C=O); 1650 (C=N); 1524 (C=C); 1112 (S=O); 1046 (p-Cl-phenyl); MS m/z (relative intensity %): 360 ($M^{+\bullet}$+2, 2); 358 ($M^{+\bullet}$, 2); 71 (100). Anal. Calc. (%) for$C_{16}H_{11}ClN_4O_2S$(358.8): C, 53.56; H, 3.09; N, 15.61. Found (%): C, 53.58; H, 3.12; N, 15.67.

3.1.19. General Procedure for Synthesis of Compounds 20a,b

The selected 1-(substitutedaryl)-2-amino-4-(4-chlorophenyl)-1H-pyrrole-3-carboxamide 17a,b (2 mmol.) was refluxed in excess acetic anhydride (15 mL) for 10 h. The reaction mixture was concentrated to the minimum. The solid product was collected, washed with ethanol then recrystallized from dioxane.

1) 5-(4-Chlorophenyl)-2-methyl-7-(3-trifluoromethylphenyl)-3H-pyrrolo[2,3-d]pyrimidin-4(7H)-one; 20a

Faint brown needle crystals; yield (51%); m.p.: >300°C. IR [KBr, cm^{-1}]: 3418 (br. OH tautomer); 3150 (NH); 3060 (CH aromatic); 1616 (C=N); 1539 (C=C); 1095 (p-Cl-phenyl). ^{1}H NMR (DMSO-d_6, δ ppm): 2.73 (s, 3 H, CH_3); 7.04 (s, 1 H, CH-pyrrole); 7.31 (d, 2 H, J = 8.1 Hz, 4-Cl-C_6H_4-$C_{2,6}$-H); 7.42-7.54 (m, 1 H, 3-CF_3-C_6H_4- C_5-H); 7.60 (d, 1 H, J = 7.2 Hz, 3-CF_3-C_6H_4-C_6-H); 7.69 (d, 1 H, J = 6.9 Hz, 3-CF_3-C_6H_4-C_4-H); 7.81 (d, 2 H, J = 8.1 Hz, 4-Cl-C_6H_4-$C_{3,5}$-H); 7.96 (s, 1 H, 3-CF_3-C_6H_4-C_2-H); 9.63 (s, 1/2 H, NH, D_2O exchangeable); 11.50

(s, 1/2 H, OH tautomer, D_2O exchangeable). Anal. Calc. (%) for $C_{20}H_{13}ClF_3N_3O$ (403.8): C, 59.49; H, 3.25; N, 10.41. Found (%): C, 59.53; H, 3.31; N, 10.48.

2) 5-(4-Chlorophenyl)-2-methyl-7-(pyridin-3-yl)-3H-pyrrolo[2,3-d]pyrimidin-4(7H)-one; 20b

Dark brown needle crystals; yield (47%); m.p.: 272°C - 274°C. IR [KBr, cm^{-1}]: 3140 (NH); 3030 (CH aromatic); 2890 (CH aliphatic); 1652 (C=O); 1610 (C=N); 1107 (p-Cl-phenyl). ^1H NMR (DMSO-d_6, δ ppm): 2.07 (s, 3 H, CH_3); 7.30 - 7.40 (m, 2 H, pyridyl-$C_{5,6}$-H); 7.58 (d, 2 H, J = 8.4 Hz, 4-Cl-C_6H_4-$C_{2,6}$-H); 7.64 (s, 1 H, CH-pyrrole); 7.95 (d, 2 H, J = 8.4 Hz, 4-Cl-C_6H_4-$C_{3,5}$-H); 7.99 - 8.16 (m, 2 H, pyridyl-$C_{2,4}$-H); 9.29 (s, 1/2 H, NH, D_2O exchangeable); 10.99 (s, 1/2 H, OH tautomer, D_2O exchangeable). Anal. Calc. (%) for $C_{18}H_{13}ClN_4O$ (336.8): C, 64.19; H, 3.89; N, 16.64. Found (%): C, 64.22; H, 3.91; N, 16.65.

3.1.20. General Procedure for Synthesis of Compounds 21a,b

The selected 7-(substitutedaryl)-5-(4-chlorophenyl)-3H-pyrrolo[2,3-d]pyrimidin-4(7H)-one 18a,b (2 mmol.) was refluxed in excess $POCl_3$ (15 mL) for 18 h. The reaction mixture was allowed to cool then poured on to crushed ice then alkalinized with sodium bicarbonate till pH 10. The obtained product was filtered, washed with water then recrystallized from glacial acetic acid.

1) 4-Chloro-5-(4-chlorophenyl)-7-(3-trifluoromethylphenyl)-7H-pyrrolo[2,3-d]pyrimidine; 21a

Yellow crystals;yield (71%); m.p.: >300°C. IR [KBr, cm^{-1}]: 2924 (CH aromatic); 1620 (C=N); 1542 (C=C); 1100 (p-Cl-phenyl); MS m/z (relative intensity %): 408 ($M^{+\bullet}$+1, 0.2); 407 ($M^{+\bullet}$, 0.2); 406 ($M^{+\bullet}$−1, 0.2); 405 ($M^{+\bullet}$−2) (0.2); 58 (100). Anal. Calc. (%) for $C_{19}H_{10}Cl_2F_3N_3$ (408.2): C, 55.90; H, 2.47; N, 10.29. Found (%): C, 55.93; H, 2.51; N, 10.32.

2) 4-Chloro-5-(4-chlorophenyl)-7-(pyridin-3-yl)-7H-pyrrolo[2,3-d]pyrimidine; 21b

Brown crystals; yield (83%); m.p.: >300°C. IR [KBr, cm^{-1}]: 3080 (CH aromatic); 1654 (C=N); 1524 (C=C); 1059 (p-Cl-phenyl). ^1H NMR (DMSO-d_6, δ ppm): 7.29 (d, 1 H, J = 7.8 Hz, pyridyl-C_6-H); 7.59 (s, 1 H, CH- pyrrole); 7.84 (d, 2 H, J = 8.4 Hz, 4-Cl-C_6H_4-$C_{2,6}$-H); 8.00 - 8.10 (m, 3 H, 4-Cl-C_6H_4-$C_{3,5}$-H & pyridyl-C_5-H); 8.22 (d, 1 H, J = 7.8 Hz, pyridyl-C_4-H); 8.52 (s, 1 H, pyridyl-C_2-H); 10.23 (s, 1 H, pyrimidine-C_2-H). Anal. Calc. (%) for $C_{17}H_{10}Cl_2N_4$ (341.2): C, 59.84; H, 2.95; N, 16.42. Found (%): C, 59.92; H, 2.97; N, 16.51.

3.1.21. General Procedure for Synthesis of Compounds 22a,b

An equimolar mixture of the appropriate 7-(substitutedaryl)-4-chloro-5-(4-chlorophenyl)-7H-pyrrolo[2,3-]py- rimidine 21a,b (2 mmol.) and morpholine (0.17 g, 0.2 mL, 2 mmol.) was refluxed in absolute ethanol for 9 h (30 mL) in presence of a catalytic amount of TEA (3 - 5 drops). The reaction mixture was allowed to cool then treated with 10% acetic acid. The obtained product was filtered, washed with ethanol then recrystallized from ethanol.

1) 7-(3-Trifluoromethylphenyl)-5-(4-chlorophenyl)-4-morpholino-7H-pyrrolo[2,3-d]pyrimidine; 22a

Yellow cubes; yield (76%); m.p.: >300°C. IR [KBr, cm^{-1}]: 3090 (CH aromatic); 2924, 2862 (CH aliphatic); 1660 (C=N); 1580 (C=C); 1250, 1046 (C-O-C); 1110 (p-Cl-phenyl). ^{1}H NMR (DMSO-d_6, δ ppm): 3.43 (t, 4 H, J = 4.5 Hz, morpholine-$C_{2,6}$-H); 3.93 (t, 4 H, J = 4.5 Hz, morpholine-$C_{3,5}$-H); 7.33 (d, 2 H, J = 8.1 Hz, 4-Cl-C_6H_4-$C_{2,6}$-H); 7.46 - 7.51 (m, 3 H, 3-CF_3-C_6H_4-$C_{5,6}$-H & CH-pyrrole); 7.60 - 7.64 (m, 2 H, 3-CF_3-C_6H_4- $C_{2,4}$-H); 7.79 (d, 2 H, J = 8.1 Hz, 4-Cl-C_6H_4-$C_{3,5}$-H); 8.30 (s, 1 H, pyrimidine-C_2-H). Anal. Calc. (%) for $C_{23}H_{18}ClF_3N_4O$ (458.9): C, 60.20; H, 3.95; N, 12.21. Found (%): C, 60.23; H, 3.97; N, 12.28.

2) 7-(Pyridine-3-yl)-5-(4-chlorophenyl)-4-morpholino-7H-pyrrolo[2,3-d]pyrimidine; 22b

Dark brown powder; yield (83%); m.p.: > 300°C. IR [KBr, cm^{-1}]: 3000 (CH aromatic); 2924, 2861 (CH aliphatic); 1620 (C=N); 1550 (C=C); 1270, 1040 (C-O-C); 1110 (p-Cl-phenyl). ^{1}H NMR(DMSO-d_6, δ ppm): 3.40 - 3.43 (m, 4 H, morpholine-$C_{2,6}$-H); 4.13 (t, 4 H, J = 4.8 Hz, morpholine-$C_{3,5}$-H); 7.04 (s, 1 H, CH-pyrrole); 7.32 (d, 2 H, J = 8.4 Hz, 4-Cl-C_6H_4-$C_{2,6}$-H); 7.54 - 7.70 (m, 3 H, pyridyl-$C_{4,5,6}$-H); 7.81 (d, 2 H, J = 8.4 Hz, 4-Cl- C_6H_4-$C_{3,5}$-H);7.88 (s, 1 H, pyridyl-C_2-H); 8.00 (s, 1 H, pyrimidine-C_2-H). Anal. Calc. (%) for $C_{21}H_{18}ClN_5O$ (391.9): C, 64.37; H, 4.63; N, 17.87. Found (%): C, 64.41; H, 4.67; N, 17.92.

3.1.22. General Procedure for Synthesis of Compounds 23a,b

A mixture of compounds 21a,b (2 mmol.) and hydrazine hydrate 98% (0.1 g, 0.1 mL, 2 mmol.) was refluxed in absolute ethanol (30 mL) in presence of TEA as a catalyst for 10 h. The reaction mixture was allowed to cool and poured on to crushed ice. The obtained solid was filtered, washed with water and recrystallized from dioxane to yield the target compounds 23a,b; respectively.

1) 5-(4-Chlorophenyl)-4-hydrazinyl-7-(3-trifluoromethylphenyl)-7H-pyrrolo[2,3-d]pyrimidine; 23a

Faint brown needle crystals; yield (45%); m.p.: > 300°C. IR [KBr, cm^{-1}]: 3400, 3200, 3110 (NH_2, NH); 2923 (CH aromatic); 1590 (C=N); 1573 (C=C); 1031 (p-Cl-phenyl). ^1H NMR (DMSO-d_6, δ ppm): 4.06 (s, 2 H, NH_2, D_2O exchangeable); 7.01 (s, 1 H, CH-pyrrole); 7.15 - 7.22 (m, 3 H, 3-CF_3-C_6H_4-$C_{4,5,6}$-H); 7.36 (d, 2 H, J = 8.1 Hz, 4-Cl-C_6H_4-$C_{2,6}$-H); 7.67 (d, 2 H, J=8.1 Hz, 4-Cl-C_6H_4-$C_{3,5}$-H); 8.10 (s, 1 H, 3-CF_3-C_6H_4-C_2-H); 9.20 (s, 1 H, pyrimidine-C_2-H); 10.45 (s, 1 H, NH, D_2O exchangeable). Anal. Calc. (%) for $C_{19}H_{13}ClF_3N_5$ (403.8): C, 56.52; H, 3.25; N, 17.34. Found (%): C, 56.54; H, 3.29; N, 17.37.

2) 5-(4-Chlorophenyl)-4-hydrazinyl-7-(pyridin-3-yl)-7H-pyrrolo[2,3-d]pyrimidine; 23b

Dark brown powder; yield (56%); m.p.: >300°C. IR [KBr, cm^{-1}]: 3382, 3230, 3120 (NH_2, NH); 2924 (CH aromatic); 1651 (C=N); 1527 (C=C); 1044 (p-Cl-phenyl). ^1H NMR (DMSO-d_6, δ ppm): 4.52 (s, 2 H, NH_2, D_2O exchangeable); 7.00 (s, 1 H, CH-pyrrole); 7.32 (d, 2 H, J = 8.4 Hz, 4-Cl-C_6H_4-$C_{2,6}$-H); 7.47-7.52 (m, 1 H, pyridyl-C_5-H); 7.56 (d, 1 H, J = 6.4 Hz, pyridyl-C_6-H); 7.76 (d, 2 H, J = 8.4 Hz, 4-Cl-C_6H_4-$C_{3,5}$-H); 7.83 - 7.86 (m, 2 H, pyridyl $C_{2,4}$-H); 7.88 (s, 1 H, pyrimidine-C_2-H); 8.00 (s, 1 H, NH, D_2O exchangeable). Anal. Calc. (%) for $C_{17}H_{13}ClN_6$ (336.8): C, 60.63; H, 3.89; N, 24.95. Found (%): C, 60.66; H, 3.91; N, 24.97.

3.1.23. General Procedure for Synthesis of Compounds 24a,b

To a stirred solution of the selected ethyl N-1-(substitutedaryl)-4-(4-chlorophenyl)-3-cyano-1H-pyrrol-2-ylfor- mimidate 13a,b (2 mmol.) in absolute ethanol (50 mL), hydrazine hydrate 98% (0.1 g, 0.1 mL, 2 mmol.) was added and stirring was continued for 5 h at room temperature. The solid product was filtered, washed with ethanol and recrystallized from ethanol/benzene.

1) 3-Amino-5-(4-chlorophenyl)-4-imino-7-(3-trifluoromethylphenyl)-4,7-dihydro-3H-pyrrolo[2,3-d]pyrimidines; 24a

Faint yellow powder; yield (42%); m.p.: >300°C. IR [KBr, cm^{-1}]: 3447, 3292, 3183 (NH_2, NH); 2928 (CH aromatic); 1643 (C=N); 1556 (C=C); 1025 (p-Cl-phenyl); MS m/z (relative intensity %): 403 ($M^{+\bullet}$, 1); 85 (100). Anal. Calc. (%) for $C_{19}H_{13}ClF_3N_5$ (403.8): C, 56.52; H, 3.25; N, 17.34. Found (%): C, 56.57; H, 3.31; N, 17.42.

2) 3-Amino-5-(4-chlorophenyl)-4-imino-7-(pyridin-3-yl)-4,7-dihydro-3H-pyrrolo[2,3-d]pyrimidines; 24b

Dark grey powder; yield (57%); m.p.: > 300°C. IR [KBr, cm^{-1}]: 3310, 3192 (NH_2, NH); 2921 (CH aromatic); 1650 (C=N); 1558 (C=C); 1091 (p-Cl-phenyl). ^{1}H NMR (DMSO-d_6, δ ppm): 3.48 (s, 2 H, NH_2, D_2O exchangeable); 6.70 (s, 1 H, CH-pyrrole); 7.24 - 7.31 (m, 2 H, 4-Cl-C_6H_4-$C_{2,6}$-H); 7.42 - 7.51 (m, 3 H, pyridyl-$C_{4,5,6}$-H); 7.82 (d, 2 H, J = 8.4 Hz, 4-Cl-C_6H_4-$C_{3,5}$-H); 8.28 (s, 1 H, pyridyl-C_2-H); 9.40 (s, 1 H, pyrimidine-C_2-H); 10.60 (s, 1 H, imino NH, D_2O exchangeable). Anal. Calc. (%) for $C_{17}H_{13}ClN_6$ (336.8): C, 60.63; H, 3.89; N, 24.95. Found (%): C, 60.68; H, 3.92; N, 24.97.

3.1.24. General Procedure for Synthesis of Compounds 25a,b

An equimolar mixture of the selected 7-(substitutedaryl)-5-(4-chlorophenyl)-4-imino-4,7-dihydro-3H-pyrro- lo[2,3-d]pyrimidin-3-amine 24a,b (2 mmol.) and ethyl cyanoacetate (0.23 g, 0.1 mL, 2 mmol.) was refluxed in absolute ethanol (30 mL) containing few drops of glacial acetic acid (3 - 5 drops) for 18 h. The reaction mixture was concentrated, then left to cool and the solid product was filtered, washed with ethanol then recrystallized from ethanol.

1) 2-[9-(4-Chlorophenyl)-7-(3-trifluoromethylphenyl)-7H-pyrrolo[3,2-e][1,2,4]triazolo[1,5-c]pyrimidin-2-yl]- acetonetrile; 25a

Faint brown powder; yield (47%); m.p.: 280°C - 282°C. IR [KBr, cm^{-1}]: 2934 (CH aromatic); 2850 (CH aliphatic); 2263 (C≡N); 1632 (C=N); 1569 (C=C); 1091 (p-Cl-phenyl); MS m/z (relative intensity %):454 ($M^{+\bullet}$+2, 0.3); 452 ($M^{+\bullet}$, 1); 58 (100). Anal. Calc. (%) for $C_{22}H_{12}ClF_3N_6$ (452.8): C, 58.35; H, 2.67; N, 18.56. Found (%): C, 58.41; H, 2.69; N, 18.62.

2) 2-[9-(4-Chlorophenyl)-7-(pyridin-3-yl)-7H-pyrrolo[3,2-e][1,2,4]triazolo[1,5-c]pyrimidin-2-yl]acetonitrile; 25b

Dark brown powder; yield (57%); m.p.: >300°C. IR [KBr, cm^{-1}]: 3000 (CH aromatic); 2924, 2868 (CH aliphatic); 2200 (C≡N); 1641 (C=N); 1533 (C=C); 1026 (p-Cl-phenyl). ^{1}H NMR (DMSO-d_6, δ ppm): 4.16 (s, 2 H, CH_2); 7.10 (s, 1 H, CH-pyrrole); 7.32 - 7.34 (m, 2 H, 4-Cl-C_6H_4-$C_{2,6}$-H); 7.42 - 7.62 (m, 3 H, pyridyl-$C_{4,5,6}$-H); 7.87 (d, 2 H, J = 8.4 Hz, 4-Cl-C_6H_4-$C_{3,5}$-H); 8.09 (s, 1 H, pyridyl-C_2-H); 9.70 (s, 1 H, pyrimidine-C_2-H). Anal. Calc. (%) for $C_{20}H_{12}ClN_7$ (385.8): C, 62.26; H, 3.14; N, 25.41. Found (%): C, 62.32; H, 3.18; N, 25.47.

3.1.25. General Procedure for Synthesis of Compounds 26a,b

An equimolar mixture of the appropriate 7-(substitutedaryl)-3-amino-5-(4-chlorophenyl)-4-imino-4,7-dihydro- 3H-pyrrolo[2,3-d]pyrimidines 24a,b (2 mmol.), carbon disulfide (0.15 g, 2 mmol.) and KOH (0.11 g, 2 mmol.) was refluxed for 9 h in absolute ethanol (30 mL). The reaction mixture was allowed to cool then poured on to crushed ice and neutralized with few drops of c.HCl. The obtained solid product was filtered off, washed thoroughly with water then recrystallized from dioxane.

1) 9-(4-Chlorophenyl)-7-(3-trifluoromethylphenyl)-3,7-dihydro-2H-pyrrolo[3,2-e][1,2,4]triazolo[1,5- c]pyrimidin- 2-thione; 26a

Yellow crystals; yield (57%); m.p.: 274°C - 276°C. IR [KBr, cm^{-1}]: 3230, 3140 (NH); 2922 (CH aromatic); 1650 (C=N); 1522 (C=C); 1464, 1320, 1170, 1002 (I, II, III, IV bands N-C=S); 1108 (p-Cl-phenyl); MS m/z (relative intensity %): 447 ($M^{+\bullet}$+2, 6); 446 ($M^{+\bullet}$+1, 7); 56 (100). Anal. Calc. (%) for $C_{20}H_{11}ClF_3N_5S$ (445.8): C, 53.88; H, 2.49; N, 15.71. Found (%): C, 53.91; H, 2.52; N, 15.76.

2) 9-(4-Chlorophenyl)-7-(pyridin-3-yl)-3,7-dihydro-2H-pyrrolo[3,2-e][1,2,4]triazolo[1,5-c]pyrimidin-2-thione; 26b

Brown needle crystals; yield (75%); m.p.: 310°C - 312°C. IR [KBr, cm^{-1}]: 3275, 3125 (NH); 3005 (CH aromatic); 1605 (C=N); 1542 (C=C); 1460, 1398, 1260, 1108 (I, II, III, IV bands N-C=S); 1020 (p-Cl-phenyl). ^{1}H NMR (DMSO-d_6, δ ppm): 3.40 (s, 1 H, NH, D_2O exchangeable); 7.00 (s, 1 H, CH-pyrrole); 7.34 (d, 2 H, J = 8.4 Hz, 4-Cl-C_6H_4-$C_{2,6}$-H); 7.48 - 7.51 (m, 1 H, pyridyl-C_5-H); 7.58 (d, 1 H, J = 6.4 Hz, pyridyl-C_6-H); 7.79 (d, 2 H, J = 8.4 Hz, 4-Cl-C_6H_4-$C_{3,5}$-H); 7.86 (d, 1 H, J = 6.4 Hz, pyridyl-C_4-H); 8.30 (s, 1 H, pyridyl-C_2-H); 8.47 (s, 1 H, pyrimidine-C_2-H). Anal. Calc. (%) for $C_{18}H_{11}ClN_6S$ (378.8): C, 57.07; H, 2.93; N, 22.18. Found (%): C, 57.12; H, 2.96; N, 22.21.

3.1.26. General Procedure for Synthesis of Compounds 27a,b

The appropriate 7-(substitutedaryl)-3-amino-5-(4-chlorophenyl)-4-imino-4,7-dihydro-3H-pyrrolo[2,3-d]pyrimi- dine 24a,b (2 mmol.) was refluxed in excess triethyl orthoformate (10 mL) for 12 h. The reaction mixture was left to cool and the solid product was filtered, washed with ethanol then recrystallized from glacial acetic acid.

1) 9-(4-Chlorophenyl)-7-(3-trifluoromethylphenyl)-7H-pyrrolo[3,2-e][1,2,4]triazolo[1,5-c]pyrimidine; 27a

Faint yellow crystals; yield (47%); m.p.: >300°C. IR [KBr, cm^{-1}]: 3000 (CH aromatic); 1620 (C=N); 1561 (C=C); 1025 (p-Cl-phenyl). ^{1}H NMR (DMSO-d_6, δ ppm): 6.40 (s, 1 H, CH-pyrrole); 6.80 - 6.85 (m, 3 H, 3-CF_3-C_6H_4-$C_{4,5,6}$-H); 6.92 - 6.97 (m, 2 H, 4-Cl-C_6H_4-$C_{2,6}$-H); 7.28 - 7.35 (m, 2 H, 4-Cl-C_6H_4-$C_{3,5}$-H); 7.45 (s, 1 H, 3-CF_3-C_6H_4-C_2-H); 7.48 (s, 1 H, triazole-C_3-H); 8.58 (s, 1 H, pyrimidine-C_2-H). Anal. Calc. (%) for $C_{20}H_{11}ClF_3N_5$ (413.8): C, 58.05; H, 2.68; N, 16.93. Found (%): C, 58.10; H, 2.71; N, 16.96.

2) 9-(4-Chlorophenyl)-7-(pyridin-3-yl)-7H-pyrrolo[3,2-e][1,2,4]triazolo[1,5-c]pyrimidine; 27b

Dark brown powder, yield (56%); m.p.: 278°C - 280°C. IR [KBr, cm^{-1}]: 2924 (CH aromatic); 1647 (C=N); 1562 (C=C); 1027 (p-Cl-phenyl). ^{1}H NMR (DMSO-d_6, δ ppm): 6.92 - 6.97 (m, 1 H, pyridyl-C_5-H); 7.26 (s, 1 H, CH-pyrrole); 7.30 (d, 1 H, J = 7.8 Hz, pyridyl-C_6-H); 7.46 (d, 2 H, J = 8.4 Hz, 4-Cl-C_6H_4-$C_{2,6}$-H); 7.78 (d, 2 H, J = 8.4 Hz, 4-Cl-C_6H_4-$C_{3,5}$-H); 8.18 - 8.22 (m, 1 H, pyridyl-C_4-H); 8.50 (s, 1 H, pyridyl-C_2-H); 8.63 (s, 1 H, triazole-C_3-H); 9.00 (s, 1 H, pyrimidine-C_2-H). Anal. Calc. (%) for $C_{18}H_{11}ClN_6$ (346.8): C, 62.34; H, 3.20; N, 24.23. Found (%): C, 62.38; H, 3.27; N, 24.28.

3.1.27. General Procedure for Synthesis of Compounds 28a,b

An equimolar mixture of the selected compound 24a,b (2 mmol.) and phenyl isothiocyanate (0.27 g, 0.24 mL, 2 mmol.) was refluxed for 7 h in absolute ethanol (30 mL) containing few drops TEA (2 - 4 drops). The reaction mixture was left to cool and the obtained product was filtered off, washed with ethanol then recrystallized from ethanol.

1) 2-Amino-9-(4-chlorophenyl)-N-phenyl-7-(3-trifluoromethylphenyl)-7H-pyrrolo[3,2-e][1,2,4]triazolo[1,5-c] pyrimidine; 28a

Pale yellowish needle crystals; yield (42%); m.p.: >300°C. IR [KBr, cm^{-1}]: 3220 (NH); 2924 (CH aromatic); 1627 (C=N); 1554 (C=C); 1091 (p-Cl-phenyl). ^1H NMR (DMSO-d_6, δ ppm): 7.00 (s, 1 H, CH-pyrrole); 7.12 - 7.15 (m, 3 H, C_6H_5-$C_{3,4,5}$-H); 7.32 (d, 2 H, J = 8.4 Hz, 4-Cl-C_6H_4-$C_{2,6}$-H); 7.36-7.40 (m, 3 H, 3-CF_3-C_6H_4-$C_{4,5,6}$- H); 7.47 (d, 2 H, J = 8.4 Hz, 4-Cl-C_6H_4-$C_{3,5}$-H); 7.80 (d, 2 H, J = 7.2 Hz, C_6H_5-$C_{2,6}$-H); 7.83 (s, 1 H, 3-CF_3- C_6H_4-C_2-H); 9.70 (s, 1 H, pyrimidine-C_2-H); 10.40 (s,1 H, NH, D_2O exchangeable). Anal. Calc. (%) for $C_{26}H_{16}ClF_3N_6$ (504.9): C, 61.85; H, 3.19; N, 16.65. Found (%): C, 61.88; H, 3.23; N, 16.71.

2) 2-Amino-9-(4-chlorophenyl)-N-phenyl-7-(pyridin-3-yl)-7H-pyrrolo[3,2-e][1,2,4]triazolo[1,5-c]pyrimidine; 28b

Grey crystals; yield (56%); m.p.: >300°C. IR [KBr, cm^{-1}]: 3189 (NH); 3002, 2924 (CH aromatic); 1610 (C=N); 1550 (C=C); 1087 (p-Cl phenyl). ^1H NMR(DMSO-d_6, δ ppm): 6.93 (t, 1 H, J = 7.5 Hz, C_6H_5-C_4-H); 7.27 (s, 1 H, CH-pyrrole); 7.31 (d, 2 H, J = 7.5 Hz, C_6H_5-$C_{3,5}$-H); 7.38 - 7.48 (m, 2 H, pyridyl-$C_{5,6}$-H); 7.55 (d, 2 H, J = 8.4 Hz, 4-Cl-C_6H_4-$C_{2,6}$-H); 7.65 (d, 2 H, J = 7.5 Hz, C_6H_5-$C_{2,6}$-H); 7.87 (d, 2 H, J = 8.4 Hz, 4-Cl- C_6H_4-$C_{3,5}$-H); 8.22 - 8.32 (m, 1 H, pyridyl-C_4-H); 8.67 (s, 1 H, pyridyl-C_2-H); 9.87 (s, 1 H, pyrimidine-C_2-H); 10.20 (s, 1 H, NH, D_2O exchangeable). ^{13}C (DMSO-d_6): 107.9 (pyrrolotriazolopyrimidine-C_{9a}); 116.5 (pyrrolotriazolopyrimidine-C_8); 116.8 (N-C_6H_5-$C_{2,6}$); 120.9 (N-C_6H_5-C_4); 123.6 (4-Cl-C_6H_4-C_1); 124 (pyridyl-C_5); 126 (pyrrolotriazolopyrimidine-C_9); 128 (pyridyl-C_6); 128.9 (4-Cl-C_6H_4-$C_{2,6}$); 129.7 (4-Cl-C_6H_4-$C_{3,5}$); 129.9 (N- C_6H_5-$C_{3,5}$); 131.5 (4-Cl-C_6H_4-C_6); 133.9 (pyridyl-C_1); 136 (N-C_6H_5-C_1); 139.4 (pyrrolotriazolopyrimidine-C_5); 141.2 (pyridyl-C_2); 144.6 (pyridyl-C_4); 146 (pyrrolotriazolopyrimidine-C_{9b}); 155.6 (pyrrolotriazolopyrimidine- C_{6a}); 157.9 (pyrrolotriazolopyrimidine-C_2). Anal. Calc. (%) for $C_{24}H_{16}ClN_7$ (437.9): C, 65.83; H, 3.68; N, 22.39. Found (%): C, 65.87; H, 3.72; N, 22.43.

3.1.28. General Procedure for Synthesis of Compounds 29a,b

A mixture of the appropriate 1-(substitutedaryl)-2-amino-4-(4-chlorophenyl)-1H-pyrrole-3-carbonitrile 4a,b (2 mmol.) and sodium azide (0.13 g, 2 mmol.) and ammonium chloride (0.22 g, 2 mmol.) was heated in water bath for 10 h. in excess DMF (15 mL). The reaction mixture was allowed to cool, then triturated with water. The solid product was filtered off, washed with ethanol then recrystallized from glacial acetic acid.

1) 2-Amino-4-(4-chlorophenyl)-1-(3-trifluoromethylphenyl)-3-(1H-tetrazol-5-yl)-1H-pyrrole; 29a

Pale orange crystals; yield (73%); m.p.: >300°C IR [KBr, cm^{-1}]: 3280, 3180 (NH_2, NH); 2923 (CH aromatic); 1559 (C=C); 1383 (N=N); 1081 (p-Cl-phenyl); MS m/z (relative intensity %): 405 ($M^{+\bullet}$+1, 1); 58 (100). Anal. Calc. (%) for $C_{18}H_{12}ClF_3N_6$ (404.8): C, 53.41; H, 2.99; N, 20.76. Found (%): C, 53.48; H, 2.97; N, 20.88.

2) 2-Amino-4-(4-chlorophenyl)-1-(pyridin-3-yl)-3-(1H-tetrazol-5-yl)-1H-pyrrole; 29b

Dark brown needle crystals; yield (92%); m.p.: >300°C. IR [KBr, cm^{-1}]: 3345, 3200 (NH_2, NH); 2926 (CH aromatic); 1520 (C=C); 1419 (N=N); 1096 (p-Cl-phenyl). ^1H NMR (DMSO-d_6, δ ppm): 5.20 (s, 2 H, NH_2, D_2O exchangeable); 6.80 (s, 1 H, CH-pyrrole); 7.23 - 7.28 (m, 1 H, pyridyl-C_5-H); 7.45 (d, 1 H, J = 6.5 Hz, pyridyl-C_6-H); 7.53 (d, 2 H, J = 8.4 Hz, 4-Cl-C_6H_4-$C_{2,6}$-H); 7.61 (d, 1 H, J = 6.5 Hz, pyridyl-C_4-H); 7.92 (d, 2 H, J = 8.4 Hz, 4-Cl-C_6H_4-$C_{3,5}$-H); 8.20 (s, 1 H, pyridyl-C_2-H); 9.92 (s, 1 H, NH, D_2O exchangeable). Anal. Calc. (%) for $C_{16}H_{12}ClN_7$ (337.8): C, 56.89; H, 3.58; N, 29.03. Found (%): C, 56.91; H, 3.54; N, 29.08.

3.1.29. General Procedure for Synthesis of Compounds 30a,b

A mixture of the appropriate 1-(substitutedaryl)-2-amino-4-(4-chlorophenyl)-1H-pyrrole-3-carbonitrile 4a,b (2 mmol.), CS_2 (0.15 g, 0.13 mL, 2 mmol.) and excess ethylene diamine (5 mL) was refluxed in a water bath for 7 h. The reaction mixture was allowed to cool then triturated with ethanol. The obtained product was filtered off, washed with ethanol then recrystallized from dioxane.

1) 2-Amino-4-(4-chlorophenyl)-3-(4,5-dihydro-1H-imidazol-2-yl)-1-(3-trifluoromethylphenyl)-1H-pyrrole; 30a

Yellow powder; yield (67%); m.p.: >300°C. IR [KBr, cm^{-1}]: 3250, 3120 (NH_2, NH); 3000 (CH aromatic); 2950, 2850 (CH aliphatic); 1514 (C=C); 1100 (p-Cl-phenyl); MS m/z (relative intensity %): 406 ($M^{+\bullet}$+2, 0.1); 405 ($M^{+\bullet}$+1, 0.1); 404 ($M^{+\bullet}$, 0.1); 56 (100). Anal. Calc. (%) for $C_{20}H_{16}ClF_3N_4$ (404.8): C, 59.34; H, 3.98; N, 13.84. Found (%): C, 59.39; H, 3.92; N, 13.87.

2) 2-Amino-4-(4-chlorophenyl)-3-(4,5-dihydro-1H-imidazol-2-yl)-1-(pyridin-3-yl)-1H-pyrrole; 30b

Grey crystals;yield (83%); m.p.: >300°C. IR [KBr, cm^{-1}]: 3360; 3145 (NH_2, NH); 3050 (CH aromatic); 2890 (CH aliphatic); 1654 (C=N); 1520 (C=C); 1090 (p-Cl-phenyl). ^1H NMR (DMSO-d_6, δ ppm): 4.75 (s, 1 H, NH, D_2O exchangeable); 5.64 - 5.67 (m, 4 H, imidazolidine-$C_{4,5}$-H); 7.37 (s, 1 H, CH-pyrrole); 7.40 (s, 2 H, NH_2, D_2O exchangeable); 7.51 - 7.54 (m, 2 H, 4-Cl-C_6H_4-$C_{2,6}$-H); 7.57 - 7.62 (m, 1 H, pyridyl-C_5-H); 7.65 - 7.66 (m, 1 H, pyridyl-C_6-H); 8.01 - 8.03 (m, 2 H, 4-Cl-C_6H_4-$C_{3,5}$-H); 8.04 - 8.05 (m, 1 H, pyridyl-C_4-H); 8.06 (s, 1 H, pyridyl-C_2-H). Anal. Calc. (%) for $C_{18}H_{16}ClN_5$ (337.8): C, 64.00; H, 4.77; N, 20.73. Found (%): C, 64.16; H, 4.81; N, 20.79.

3.1.30. General Procedure for Synthesis of Compounds 31a,b

A mixture of the appropriate 1-(substitutedaryl)-2-amino-4-(4-chlorophenyl)-1H-pyrrole-3-carbonitrile 4a,b (2 mmol.) and CS_2 (0.13 mL, 0.15 g, 2 mmol.) was heated in excess pyridine (10 mL) in a water bath for 10 h, then left to cool. The solid product collected, washed with ethanol then recrystallized from ethanol.

1) 5-(4-Chlorophenyl)-7-(3-trifluoromethylphenyl)-1H-pyrrolo[2,3-d]pyrimidin-2,4(3H,7H)-dithione; 31a

Faint yellow needle crystals;yield (62%); m.p.: >300°C. IR [KBr, cm^{-1}]: 3236 (NH); 3050 (CH aromatic); 1622 (C=N); 1560 (C=C); 1424, 1328, 1129, 1005 (I, II, III, IV bands N-C=S); 1129 (p-Cl-phenyl). ^1H NMR (DMSO-d_6, δ ppm): 8.10 (s, 1 H, pyrimidine-N_3-H, D_2O exchangeable); 7.35 - 7.41 (m, 1H, 3-CF_3-C_6H_4-C_5-H); 7.50 (s, 1H, CH-pyrrole); 7.56 (d, 2 H, J = 8.4 Hz, 4-Cl-C_6H_4-$C_{2,6}$-H); 7.61 - 7.63 (m, 1 H, 3-CF_3-C_6H_4-C_6-H); 7.78 - 7.82 (m, 1 H, 3-CF_3-C_6H_4-C_4-H); 7.93 (d, 2 H, J = 8.4 Hz, 4-Cl-C_6H_4-$C_{3,5}$-H); 8.58 (s, 1 H, 3-CF_3-C_6H_4-C_2-H); 8.76 (s, 1 H, pyrimidine-N_1-H, D_2O exchangeable). Anal. Calc. (%) for $C_{19}H_{11}ClF_3N_3S_2$ (437.9): C, 52.11; H, 2.53; N, 9.60. Found (%): C, 52.17; H, 2.63; N, 9.71.

2) 5-(4-Chlorophenyl)-7-(pyridin-3-yl)-1H-pyrrolo[2,3-d]pyrimidin-2,4(3H,7H)-dithione; 31b

Brown needle crystals; yield (78%); m.p.: > 300°C. IR [KBr, cm^{-1}]: 3306, 3207 (NH); 2923 (CH aromatic); 1641 (C=N); 1534 (C=C); 1480, 1390, 1128, 1000 (I, II, III, IV bands N-C=S); 1128 (p-Cl-phenyl); MS m/z (relative intensity %): 372 ($M^{+\bullet}$+2, 1); 371 ($M^{+\bullet}$+1, 1); 58 (100). Anal. Calc. (%) for $C_{17}H_{11}ClN_4S_2$ (370.9): C, 55.05; H, 2.99; N, 15.11. Found (%): C, 55.01; H, 2.86; N, 15.16.

3.1.31. General Procedure for Synthesis of Compounds 32a,b

The selected 1-(substitutedaryl)-2-amino-4-(4-chlorophenyl)-1H-pyrrole-3-carbonitrile 4a,b (2 mmol.) was dissolved in a mixture of glacial acetic acid/c.HCl (10:7) (20 mL). To such solution, a solution of sodium nitrite (0.14 g, 2 mmol.) in (1 mL) water was added drop wise while stirring at 0°C - 5°C then stirring was continued for 3 h while maintaining the reaction temperature at 5°C. The solid product was then collected, washed with ethanol then recrystallized from glacial acetic acid.

1) 5-(4-Chlorophenyl)-7-(3-trifluoromethylphenyl)-3H-pyrrolo[2,3-d][1,2,3]triazine-4(7H)one; 32a

Yellow powder; yield (62%); m.p.:287°C - 289°C. IR [KBr, cm^{-1}]: 3100 (NH); 3050 (CH aromatic); 1691 (C=O); 1528 (C=C); 1411 (N=N); 1112 (p-Cl-phenyl).^1H NMR (DMSO-d_6, δ ppm): 7.25 (s, 1 H, CH-pyrrole); 7.41 (s, 1 H, NH, D_2O exchangeable); 7.54 (d, 2 H, J = 8.4 Hz, 4-Cl-C_6H_4-$C_{2,6}$-H); 7.62 - 7.86 (m, 4 H, 3-CF_3- C_6H_4-$C_{2,4,5,6}$-H); 7.92 (d, 2 H, J = 8.4 Hz, 4-Cl-C_6H_4-$C_{3,5}$-H). Anal. Calc. (%) for $C_{18}H_{10}ClF_3N_4O$ (390.7): C, 55.33; H, 2.58; N, 14.34. Found (%): C, 55.33; H, 2.58; N, 14.34.

2) 5-(4-Chlorophenyl)-7-(pyridin-3-yl)-3H-pyrrolo[2,3-d][1,2,3]triazin-4(7H)-one; 32b

Dark grey crystals; yield (78%); m.p.: 267°C - 269°C. IR [KBr, cm^{-1}]: 3114 (NH); 3010 (CH aromatic); 1690 (C=O); 1531 (C=C); 1407 (N=N); 1088 (p-Cl-phenyl). ^1H NMR(DMSO-d_6, δ ppm): 7.28 (s, 1 H, CH-pyrrole); 7.45 (s, 1 H, NH, D_2O exchangeable); 7.52 (d, 2 H, J = 8.1 Hz, 4-Cl-C_6H_4-$C_{2,6}$-H); 7.55 - 7.62 (m, 4 H, pyridyl-$C_{2,4,5,6}$-H); 7.91 (d, 2 H, J = 8.1 Hz, 4-Cl-C_6H_4-$C_{3,5}$-H). Anal. Calc. (%) for $C_{16}H_{10}ClN_5O$ (323.7): C, 59.36; H, 3.11; N, 21.63. Found (%): C, 59.36; H, 3.11; N, 21.63.

3.1.32. General Procedure for Synthesis of Compounds 33a,b

To a stirred solution of the selected ethyl 1-(substitutedaryl)-2-amino-4-(4-chlorophenyl)-1H-pyrrol-3-carbox- ylate 4c, d (2 mmol.) in absolute ethanol (50 mL), hydrazine hydrate 98% (0.1 g, 0.12 mL, 2 mmol.) was added and the reaction was refluxed for 6 h. The solid product was isolated by filtration, washed with ethanol and recrystallized from ethanol.

1) 2-Amino-4-(4-chlorophenyl)-1-(3-trifluoromethylphenyl)-1H-pyrrol-3-carbohydrazide; 33a

Pale yellow crystals; yield (67%); m.p.: >300°C. IR [KBr, cm^{-1}]: 3450, 3401, 3200 (NH_2, NH); 2934 (CH aromatic); 1690 (C=O); 1520 (C=C); 1021 (p-Cl-phenyl).^1H NMR (DMSO-d_6, δ ppm): 5.30 (s, 2 H, NH-NH_2, D_2O exchangeable); 6.74 (s, 1 H, CH-pyrrole); 7.31 (d, 2 H, J = 8.1 Hz, 4-Cl-C_6H_4-$C_{2,6}$-H); 7.38 - 7.45 (m, 1 H, 3-CF_3-C_6H_4-C_5-H); 7.50 - 7.60 (m, 2 H, 3-CF_3-

C_6H_4-$C_{4,6}$-H); 7.87 (d, 2 H, J = 8.1 Hz, 4-Cl-C_6H_4-$C_{3,5}$-H), 8.35 (s, 1 H, 3-CF_3-C_6H_4-C_2-H); 8.50 (s, 2 H, pyrrole-C_2-NH_2, D_2O exchangeable); 8.59 (s, 1 H, NH, D_2O exchangeable). Anal. Calc. (%) for $C_{18}H_{14}ClF_3N_4O$ (394.8): C, 54.76; H, 3.57; N, 14.19. Found (%): C, 54.65; H, 3.43; N, 14.23.

2) 2-Amino-4-(4-chlorophenyl)-1-(pyridin-3-yl)-1H-pyrrole-3-carbohydrazide; 33b

Faint brown crystals; yield (72%); m.p.: >300°C. IR [KBr, cm^{-1}]: 3390, 3213 (NH_2, NH); 2980 (CH aromatic); 1680 (C=O), 1522 (C=C); 1031 (p-Cl-phenyl); MS m/z (relative intensity %): 329 ($M^{+\bullet}$+2, 1); 327 ($M^{+\bullet}$, 9); 140 (100). Anal. Calc. (%) for $C_{16}H_{14}ClN_5O$ (327.8): C, 58.63; H, 4.31; N, 21.37. Found (%): C, 58.71; H, 4.42; N, 21.43.

3.1.33. General Procedure for Synthesis of Compounds 34a,b

The selected ethyl 1-(substitutedaryl)-2-amino-4-(4-chlorophenyl)-1H-pyrrol-3-carboxylate 4c, d (2 mmol.) was refluxed for 7 h in excess acetic anhydride (30 mL). The reaction mixture was concentrated under reduced pressure and the residue was collected by filtration, washed with ethanol and recrystallized from ethanol.

1) Ethyl 2-acetamido-4-(4-chlorophenyl)-1-(3-trifluoromethylphenyl)-1H-pyrrol-3-carboxylate; 34a

Faint orange powder; yield (59%); m.p.: > 300°C. IR [KBr, cm^{-1}]: 3293, 3180 (NH); 2927 (CH aromatic); 2860 (CH-aliphatic); 1689, 1650 (C=O); 1421 (C=C); 1250, 1026 (C-O-C); 1120 (p-Cl-phenyl). ^{1}H NMR (DMSO-d_6, δ ppm): 1.17 - 1.23 (m, 3 H, CH_2CH_3); 2.05 (s, 3 H, $COCH_3$); 3.91 - 4.01 (m, 2 H, CH_2CH_3); 7.30 (d, 1 H, J = 7.8 Hz, 3-CF_3-C_6H_4-C_6-H); 7.37 (d, 1 H, J = 7.8 Hz, 3-CF_3-C_6H_4-C_4-H); 7.50 - 7.55 (m, 3 H, 4-Cl- C_6H_4-$C_{2,6}$-H & 3-CF_3-C_6H_4-C_5-H); 7.75 (s, 1 H, CH-pyrrole); 7.84 (d, 2 H, J = 8.4 Hz, 4-Cl-C_6H_4-$C_{3,5}$-H); 8.08 (s, 1 H, 3-CF_3-C_6H_4-C_2-H); 10.45 (s, 1 H, NH, D_2O exchangeable). Anal. Calc. (%) for $C_{22}H_{18}ClF_3N_2O_3$ (450.8): C, 58.61; H, 4.02; N, 6.21. Found (%): C, 58.64; H, 4.10; N, 6.28.

2) Ethyl 2-acetamido-4-(4-chlorophenyl)-1-(pyridin-3-yl)-1H-pyrrole-3-carboxylate; 34b

Dark brown needle crystals; yield (71%); m.p.: > 300°C. IR [KBr, cm^{-1}]: 3220 (NH); 3096 (CH aromatic); 2921, 2858 (CH-aliphatic); 1693 (C=O); 1544 (C=C); 1245, 1024 (C-O-C); 1110 (p-Cl-phenyl). MS m/z (relative intensity %): 385 ($M^{+\bullet}$+2, 0.6); 384 ($M^{+\bullet}$+1, 0.2); 383 ($M^{+\bullet}$, 0.2); 58 (100). Anal. Calc. (%) for $C_{20}H_{18}ClN_3O_3$ (383.8): C, 62.58; H, 4.73; N, 10.95. Found (%): C, 62.55; H, 4.81; N, 10.97.

3.1.34. General Procedure for Synthesis of Compounds 35a,b

An equimolar mixture of the selected 1-(substitutedaryl)-2-amino-4-(4-chlorophenyl)-1H-pyrrole-3-carbohy- drazide 33a,b (2 mmol.) and glucose (0.22 g, 2 mmol.) was refluxed in absolute ethanol (30 mL) for 12 h, then left to cool. The obtained product was filtered off, washed with ethanol then recrystallized from ethanol.

1) 2-Amino-4-(4-chlorophenyl)-N'-2,3,4,5,6-pentahydroxy-1-hexylidene)-1-(3-trifluoromethylphenyl)-1H-pyr-rol- 3-carbohydrazide; 35a

Yellow needle crystals; yield (89%); m.p.: 296°C - 298°C. IR [KBr, cm^{-1}]: multiple absorption bands at 3496, 3438 (OH); 3360, 3310, 3250 (NH_2, NH); 3000 (CH aromatic); 2924, 2860 (CH-aliphatic); 1710 (C=O); 1644 (C=N); 1542 (C=C); 1036 (p-Cl-phenyl). MS m/z (relative intensity %): 558 ($M^{+\bullet}$+2, 0.1); 57 (100). Anal. Calc. (%) for $C_{24}H_{24}ClF_3N_4O_6$ (556.9): C, 51.76; H, 4.34; N, 10.06. Found (%): C, 51.81; H, 4.31; N, 10.15.

2) 2-Amino-4-(4-chlorophenyl)-N'-2,3,4,5,6-pentahydroxy-1-hexylidene)-1-(pyridin-3-yl)-1H-pyrrol-3-car-bohy- drazide; 35b

Buff needle crystals; yield (91%); m.p.: 326°C - 328°C. IR [KBr, cm^{-1}]: 3400, 3386 (OH); 3300, 3210(NH_2, NH); 3000 (CH aromatic); 2920 (CH-aliphatic); 1700 (C=O), 1640 (C=N); 1470 (C=C); 1090 (p-Cl-phenyl). MS m/z (relative intensity %): 490 ($M^{+\bullet}$+1, 0.2); 57 (100). Anal. Calc. (%) for $C_{22}H_{24}ClN_5O_6$ (489.9): C, 53.94; H, 4.94; N, 14.30. Found (%): C, 53.97; H, 4.91; N, 14.38.

3.1.35. General Procedure for Synthesis of Compounds 36a,b

The selected ethyl 1-(substitutedaryl)-2-acetamido-4-(4-chlorophenyl)-1H-pyrrol-3-carboxylates 34a,b (2 mmol.) was refluxed with hydrazine hydrate 98% (0.1 g, 0.12 mL, 2 mmol.) in absolute ethanol (30 mL) for 6 h. The reaction mixture was allowed to cool and the solid product was filtered, washed with ethanol then recrystallized from dioxane.

1) 3-Amino-5-(4-chlorophenyl)-2-methyl-7-(3-trifluoromethylphenyl)-3H-pyrrolo[2,3-d]pyrimidin-4(7H)-one; 36a

Pale orange crystals; yield (71%); m.p.: >300°C. IR [KBr, cm^{-1}]: 3280, 3200 (NH_2); 2924 (CH aromatic); 2861 (CH-aliphatic); 1680 (C=O); 1428 (C=C); 1105 (p-Cl-phenyl). ^1H NMR (DMSO-d_6, δ ppm): 2.07 (s, 3 H, CH_3); 4.91 (s, 2 H, NH_2, D_2O exchangeable); 7.30 (d, 2 H, J = 8.4 Hz, 4-Cl-C_6H_4-$C_{2,6}$-H); 7.33 (s, 1 H, CH- pyrrole); 7.38 - 7.48 (m, 2 H, 3-CF_3-C_6H_4-$C_{5,6}$-H); 7.50 - 7.60 (m, 1 H,3-CF_3-C_6H_4-C_4-H); 7.85 (d, 2 H, J = 8.4 Hz, 4-Cl-C_6H_4-$C_{3,5}$-H); 8.09 (s, 1 H, 3-CF_3-C_6H_4-C_2-H). Anal. Calc. (%) for $C_{20}H_{14}ClF_3N_4O$(418.8): C, 57.36; H, 3.37; N, 13.38. Found (%): C, 57.44; H, 3.31; N, 13.42.

2) 3-Amino-5-(4-chlorophenyl)-2-methyl-7-(pyridin-3-yl)-3H-pyrrolo[2,3-d]pyrimidin-4(7H)-one; 36b

Dark brown crystals; yield (87%); m.p.: >300°C. IR [KBr, cm^{-1}]: 3391, 3185 (NH_2); 2929 (CH aromatic); 2860 (CH-aliphatic); 1690 (C=O), 1543 (C=C); 1033 (p-Cl-phenyl). MS m/z (relative intensity %): 353 ($M^{+\bullet}$+2, 0.2); 351 ($M^{+\bullet}$, 1); 58 (100). Anal. Calc. (%) for $C_{18}H_{14}ClN_5O$ (351.8): C, 61.46; H, 4.01; N, 19.91. Found (%): C, 61.51; H, 4.07; N, 19.96.

3.1.36. General Procedure for Synthesis of Compounds 37a,b

A solution of the appropriate ethyl 1-(substitutedaryl)-2-amino-4-(4-chlorophenyl)-1H-pyrrol-3-carboxylate 4c, d (2 mmol.) in (30 mL) ethanolic sodium hydroxide solution [prepared by dissolving sodium hydroxide (0.1 g, 2 mmol.) in (30 mL) absolute ethanol] was heated under reflux for 9 h. The reaction mixture was allowed to cool, then acidified with dilute acetic acid (1.3 mL) and extracted with ether. The obtained product from ethereal solution was collected and recrystallized from ethanol.

1) 2-Amino-4-(4-chlorophenyl)-1-(3-trifluoromethylphenyl)-1H-pyrrol-3-carboxylic acid; 37a

Pale yellow powder; yield (57%); m.p.: >300°C. IR [KBr, cm^{-1}]: 3350 (br. OH); 3119 (NH_2); 2996 (CH aromatic); 1696 (C=O); 1522 (C=C); 1108 (p-Cl-phenyl). MS m/z (relative intensity %): 382 ($M^{+\bullet}$+2, 1); 380 ($M^{+\bullet}$, 2); 139 (100). Anal. Calc. (%) for $C_{18}H_{12}ClF_3N_2O_2$ (380.7): C, 56.78; H, 3.18; N, 7.36. Found (%): C, 56.82; H, 3.21; N, 7.47.

2) 2-Amino-4-(4-chlorophenyl)-1-(pyridin-3-yl)-1H-pyrrol-3-carboxylic acid; 37b

Grey powder; yield (67%); m.p.: > 300°C. IR [KBr, cm^{-1}]: 3391 (br. OH); 3219 (NH_2); 3050 (CH aromatic); 1684 (C=O); 1517 (C=C); 1089 (p-Cl phenyl). ^1H NMR (DMSO-d_6, δ ppm): 5.37 (s, 2 H, NH_2, D_2O exchangeable); 7.38 (d, 2 H, J = 8.4 Hz, 4-Cl-C_6H_4-$C_{2,6}$-H); 7.52 - 7.58 (m, 1 H, pyridyl-C_5-H); 7.59 - 7.68 (m, 1 H, pyridyl-C_6-H); 7.89 (s, 1 H, CH-pyrrole); 7.92 - 7.96 (m, 2 H, 4-Cl-C_6H_4-$C_{3,5}$-H); 8.01 - 8.06 (m, 1 H, pyridyl- C_4-H); 8.07 (s, 1 H, pyridyl-C_2-H); 9.47 (s, 1 H, OH, D_2O exchangeable). Anal. Calc. (%) for $C_{16}H_{12}ClN_3O_2$ (313.7): C, 61.25; H, 3.86; N, 13.39. Found (%): C, 61.27; H, 3.89; N, 13.43.

3.1.37. General Procedure for Synthesis of Compounds 38a,b

An equimolar mixture of the selected ethyl 1-(substitutedaryl)-2-amino-4-(4-chlorophenyl)-1H-pyrrol-3-car- boxylate 4c,d (2 mmol.) and phenyl isothiocyanate (0.27 g, 0.24 mL, 2 mmol.) was refluxed in for 12 h in absolute ethanol (20 mL). The reaction mixture was left to cool and the obtained product was filtered, washed with ethanol then recrystallized from glacial acetic acid.

1) Ethyl 4-(4-chlorophenyl)-2-(3-phenylthioureido)-1-(3-trifluoromethylphenyl)-1H-pyrrol-3-carboxylate; 38a

Faint orange needle crystals; yield (62%); m.p.: >300°C. IR [KBr, cm^{-1}]: 3200 (NH); 2924 (CH aromatic); 2861 (CH-aliphatic); 1720 (C=O); 1560 (C=C); 1457, 1380, 1140, 998 (I, II, III, IV bands N-C=S); 1250, 1031 (C-O-C); 1031 (p-Cl-phenyl). ^1H NMR (DMSO-d_6, δ ppm): 1.12 - 1.20 (m, 3 H, CH_2CH_3); 4.38 (q, 2 H, J = 6 Hz, CH_2CH_3); 6.90 (s, 1 H, CH-pyrrole); 7.10 - 7.17 (m, 3 H, C_6H_5-$C_{3,4,5}$-H); 7.20 - 7.28 (m, 1 H, 3-CF_3-C_6H_4- C_5-H); 7.31 - 7.38 (m, 1 H, 3-CF_3-C_6H_4-C_6-H); 7.51 - 7.58 (m, 3 H, 4-Cl-C_6H_4-$C_{2,6}$-H & 3-CF_3-C_6H_4-C_4-H); 7.70 - 7.78 (m, 2 H, C_6H_5-$C_{2,6}$-H); 7.81 (d, 2 H, J = 8.1 Hz, 4-Cl-C_6H_4-$C_{3,5}$-H); 8.01 (s, 1 H, 3-CF_3-C_6H_4-C_2-H); 11.40 (s, 1 H, pyrrole-C_2-NH, D_2O exchangeable); 11.83 (s, 1 H, C_6H_5-NH, D_2O exchangeable). Anal. Calc. (%) for $C_{27}H_{21}ClF_3N_3O_2S$ (544): C, 59.61; H, 3.89; N, 7.72. Found (%): C, 59.67; H, 3.93; N, 7.81.

2) Ethyl 4-(4-chlorophenyl)-2-(3-phenylthioureido)-1-(pyridin-3-yl)-1H-pyrrole-3-carboxylate; 38b

Dark brown needle crystals; yield (76%); m.p.: >300°C. IR [KBr, cm^{-1}]: 3390, 3213 (NH); 3080 (CH aromatic); 2927, 2820 (CH-aliphatic); 1680 (C=O); 1525 (C=C); 1411, 1360, 1134, 1033 (I, II, III, IV bands N-C=S); 1250, 1090 (C-O-C); 1040 (p-Cl-phenyl). MS m/z (relative intensity %): 478 ($M^{+\bullet}$+2, 1); 94 (100). Anal. Calc. (%) for $C_{25}H_{21}ClN_4O_2S$ (477): C, 62.95; H, 4.44; N, 11.75. Found (%): C, 62.98; H, 4.51; N, 11.78.

3.1.38. General Procedure for Synthesis of Compounds 39a,b

A mixture of the appropriate 1-(substitutedaryl)-2-amino-4-(4-chlorophenyl)-1H-pyrrole-3-carboxylic acid 37a,b (2 mmol.) and excess acetic anhydride (20 mL) was heated under reflux for 10 h. The reaction was left to cool and the solid product which was formed, was filtered off, washed with ethanol then recrystallized from glacial acetic acid.

1) 5-(4-Chlorophenyl)-2-methyl-7-(3-trifluoromethylphenyl)pyrrolo[2,3-d][1,3]oxazin-4(7H)-one; 39a

Yellow powder; yield (73%); m.p.: >300°C. IR [KBr, cm^{-1}]: 3040 (CH aromatic); 2922, 2852 (CH-aliphatic); 1686 (C=O); 1550 (C=C); 1290, 1088 (C-O-C); 1025 (p-Cl-phenyl). MS m/z (relative intensity %):404 ($M^{+\bullet}$, 0.2); 57 (100). Anal. Calc. (%) for $C_{20}H_{12}ClF_3N_2O_2$ (404.8): C, 59.35; H, 2.99; N, 6.92. Found (%): C, 59.41; H, 2.95; N, 6.97.

2) 5-(4-Chlorophenyl)-2-methyl-7-(pyridin-3-yl)pyrrolo[2,3-d][1,3]oxazin-4(7H)-one; 39b

Dark brown crystals; yield (83%); m.p.: >300°C. IR [KBr, cm^{-1}]: 3060 (CH aromatic); 2925, 2850 (CH-ali- phatic); 1688 (C=O); 1426 (C=C); 1236,

1084 (C-O-C); 1020 (p-Cl-phenyl). MS m/z (relative intensity %): 338 ($M^{+\bullet}$+1, 0.2); 337 ($M^{+\bullet}$, 0.2); 57 (100). Anal. Calc. (%) for $C_{18}H_{12}ClN_3O_2$ (337.8): C, 64.01; H, 3.58; N, 12.44. Found (%): C, 64.07; H, 3.62; N, 12.51.

3.1.39. General Procedure for Synthesis of Compounds 40a,b

A solution of the selected ethyl 1-(substituted aryl)-4-(4-chlorophenyl)-2-(3-phenylthioureido)-1H-pyrrol-3- carboxylate 38a,b (2 mmol.) in sodium ethoxide [prepared by dissolving sodium metal (0.1 g, 2 mmol.) in absolute ethanol (30 mL)] was heated under reflux for 8 h. The reaction mixture was acidified with 10% HCl and the obtained product was collected, washed with ethanol and recrystallized from ethanol.

1) 5-(4-Chlorophenyl)-3-phenyl-7-(3-trifluoromethylphenyl)-2-thioxo-2,3-dihydro-1H-pyrrolo[2,3-d]pyrimi- din-4(7H)-one; 40a

Faint orange needle crystals; yield (69%); m.p.: >300°C. IR [KBr, cm^{-1}]: 3180 (NH); 2922 (CH aromatic); 1700 (C=O); 1550 (C=C); 1460, 1350, 1134, 1036 (I, II, III, IV bands N-C=S); 1036 (p-Cl-phenyl). ^{1}H NMR (DMSO-d_6, δ ppm): 5.02 (s,1 H, NH, D_2O exchangeable); 6.60 - 6.78 (m, 5 H, C_6H_5); 7.31 (d, 2 H, J = 8.4 Hz, 4-Cl-C_6H_4-$C_{2,6}$-H); 7.40 (s, 1 H, CH-pyrrole); 7.48 - 7.61 (m, 2 H, 3-CF_3-C_6H_4-$C_{5,6}$-H); 7.85 (d, 2 H, J = 8.4 Hz, 4-Cl-C_6H_4-$C_{3,5}$-H); 7.98 - 8.10 (m, 2 H, 3-CF_3-C_6H_4-$C_{2,4}$-H). Anal. Calc. (%) for $C_{25}H_{15}ClF_3N_3OS$(497.9): C, 60.30; H, 3.04; N, 8.44. Found (%): C, 60.38; H, 3.10; N, 8.48.

2) 5-(4-Chlorophenyl)-3-phenyl-7-(pyridin-3-yl)-2-thioxo-2,3-dihydro-1H-pyrrolo[2,3-d]pyrimidin-4(7H)-one; 40 b

Dark brown needle crystals; yield (69%); m.p.: >300°C. IR [KBr, cm^{-1}]: 3200 (NH); 2924 (CH aromatic); 1690 (C=O); 1541 (C=C); 1455, 1250, 1122, 1037 (I, II, III, IV bands N-C=S); 1037 (p-Cl-phenyl). MS m/z (relative intensity %): 432 ($M^{+\bullet}$+2, 0.4); 57 (100). Anal. Calc. (%) for $C_{23}H_{15}ClN_4OS$ (430.9): C, 64.11; H, 3.51; N, 13.00. Found (%): C, 64.17; H, 3.61; N, 13.08.

3.2. Biology

3.2.1. Anti-Inflammatory Screening

1) Animals

The screening for anti-inflammatory activity for all the newly synthesized compounds 4a-40b was carried out by using adult albino rats of both sexes weighing 120 - 150 g which were obtained from animal house laboratory of Nile company, Cairo, Egypt. Rats were divided into eighty four groups; each group consists of five rats per cage in the Department of Pharmacology, Faculty of Medicine, Al-Azhar University.

The rats were kept under constant temperature 30°C and 12 hours light/dark cycle. All animals were acclimatized in the animal facility for at least two weeks prior the experiments.

The animals were kept fastened for 24 hours prior to the experiment, but they were allowed free access to water [37] . The animal experiments described below comply with the ethical principles and guidelines for the care and use of laboratory animals adopted by the National Egyptian Community.

The equipment used was Dial micrometer model (120 - 1206 Baty, Sussex, England).

2) Anti-inflammatory activity:

Rat paw edema assay was carried out according to Winter et al. [38] . Prepared compounds (equimolar to the referencedrug) were dissolved in DMSO and administrated subcutaneously.

One hour after drug administration acute inflammation was induced by injection of 0.05 mL of 1% of carrageenan sodium (Sigma-Aldrich, St. Louis, USA) subcutaneously into the sub planter region of the right hind paw.

The thickness of the injected paw was measured (from dorsal to ventral surfaces) immediately after carrageenan injection and after (1, 2, 3, 4, 5 and 6 hours) by using a micrometer. The size of edema was expressed as the increase in the thickness in mm after carrageenan injection.

The percentage inhibition of edema thickness at each time interval was calculated from the mean effect in control and treated animals according to the equation [39] [40] .

$$\% \text{ Inhibition of edema thickness} = \left[\left(Tc - Tt \right) / Tc \right] \times 100$$

where, Tc and Tt are the mean increase in thickness of the carrageenan injected paw of the control group and drug treated groups; respectively.

Control group: received the excipients (water mixed with few drops of tween 80) followed by carrageenan after 1 hour. Indomethacin (Indocin®) (5 mg/kg) was used as the referencedrug [41] . Then the potencies of compounds were calculated after 6 hours of carrageenan injection where the % edema inhibition reached maximum.

3) Statistical analysis

Results are expressed as (mean ± standard deviation) statistically analyzed using two way analysis of variance (ANOVA) followed by Bonferroni test [42] .

3.2.2. Ulcerogenicity

All animals subjected to this experimental test were sacrificed immediately after the last measurement (6 hour), by diethyl ether and stomachs were separated. An opening at the great curvature was made and the stomachs were washed with distilled water and cleaned gently by dipping in normal saline. The mucosal damage was inspected with a 3× magnifying lens for any evidence of hyperemia, hemorrhage or ulcer. For each stomach the mucosal damage was assessed [43] .

The percentage ulceration for each group was calculated as follows:

$$\%\ \text{Ulceration} = \frac{\text{Number of animals bearing ulcerin a group}}{\text{Total number of animals in the same group}} \times 100$$

3.2.3. Analgesic Screening

1) Animals:

The screening for analgesic activity for all synthesized compounds 4a-40b was carried out by using mices of both sexes weighing 25 - 30 g which were obtained from animal house laboratory, Nile company, Cairo, Egypt.

Mices were divided into eighty four groups; each group consists of four mices per cage in the animal facility of Faculty of Medicine, Al-Azhar University. The mices were kept under constant temperature 25°C and 12 hours light/dark cycle. All animals were acclimatized in the animal facility for at least two weeks prior the experiments.The animals were kept fastened for 24 hours prior to the experiment, but they were allowed free access to water.

2) Assessment of analgesic screening:

The analgesic activity was evaluated according to writhing test reported by Koster et al. [44] . The newly synthesized compounds (equimolar to the reference drug) were dissolved in DMSO and administrated to the groups orally (using intragastric tube) followed by injection of 0.6% acetic acid solution (10 mL/kg) after 1 hour [45] . Indomethacin (Indocin®) (2.5 mg/kg) was used as the reference drug.

Stretching movements (arching of the back, developments of the tension in the abdominal muscles, elongation of the body and extension of the forelimbs) were counted as a writhing response. The number of writhes was counted for 15 minutes immediately after the acetic acid injection. The percentage of inhibition of writhes number was calculated as follows:

$$\% \text{ of Inhibition} = \frac{Nc - Nt}{Nc} \times 100$$

where Nc and Nt are number of writhes in the control group and drug groups; respectively.

3.2.4. Antimicrobial Activity

1) Materials and methods

Antimicrobial activity was examined by the cup-diffusion method [46] . The in-vitro antimicrobialactivity of the synthesized compounds was investigated against several pathogenic representatives; Gram-negative bacteria; Pseudomonas aeruginosa (ATCC 27853) and Escherichia coli (ATCC 8739) and Gram-positive bacteria; Staphylococcus aureus (ATCC 25923) and Candida albicans (ATCC 10231) as a representative for fungi. All microorganisms used were obtained from the culture collection of the Department of Microbiology and Immunology, Faculty of Pharmacy (boys), Al-Azhar University, Cairo, Egypt.

Media for disc sensitivity tests were the nutrient agar and Muller-Hinton agar (MHA) purchased from Difco (USA). Non-sterile powder of the tested compounds was dissolved in dimethylformamide in a concentration of 1 mg/mL. Each 100 mL of sterile molten agar (at 45°C) received 1 mL of 6 h cultured broth with the microorganism, then the seeded agar was poured into sterile petri dishes. Cups (8 mm in diameter) were cut in the agar. Each cup received 0.1 mL of the 1 mg/mL solution of the test compounds. Plates were then incubated at 37°C for 24 h. for bacteria and 48 h. for fungi. Ampicillin (Bioanalyses Turkey) and Fluconazole (Sigma-Aldrich, USA) were used as reference substances.

3.3. Computer Aided Docking

3.3.1. Materials

All the molecular studies were carried out on an Intel Pentium 1.6 GHz processor, 512 MB memory with windows XP operating system using Molecular Operating Environment (MOE 10.2008) software provided by chemical computing group, Montreal, Canada.

All the minimizations were performed with MOE until a RMSD gradient of 0.05 $Kcal{\cdot}mol^{-1}{\cdot}A^{0-1}$ with MMFF94X force field and the partial charges were automatically calculated.

3.3.2. General Methodology

The coordinates of the X-ray crystallographic structure of the COX-2 complex with its co-crystallized ligand (Diclofenac) in the file (PDP ID: 1CX2) was obtained from the protein data bank (PDB). Enzyme structures were checked for missing atoms, bonds and contacts. The ligand molecules were constructed using the builder molecule and were energy minimized. The enzyme was prepared for docking studies where Ligand molecule was removed from the enzyme active site. Also, hydrogen atoms were added to the structure with their standard geometry. However, MOE Alpha site Finder was used for the active sites search in the enzyme structure and dummy atoms were created from the obtained alpha spheres.

4. CONCLUSIONS

It can be concluded that, among the newly synthesized compounds, compounds 5d, 9d, 11b, 12a, 13b and 32a showed multiple activities; anti-inflammatory, analgesic and anti-bacterial activities. Furthermore, it is to be noted that, some functions exerted multiple activities among these functions are 2,4-dichlorobenzylidine imino function in 2-position of pyrrole as in compound 5d. Also, the thiourea and phenylurea function as in compounds 9d and 11b, respectively. In addition to, imino function such as iminopyrrolopyrimidine thione and ethoxymethyleneimino functions as in compounds 12a and 13b, respectively, besides to, the pyrrolo[1,2,3]tri-azine derivative 32a.

In addition to, the molecular docking for the twenty most active anti-inflammatory compounds was performed on the active site of COX-2 enzyme in a trial to predict their mode of action as anti-inflammatory drugs, in which the compounds showed several interactions leading to the conclusion that they might exert their action through inhibition of COX-2 enzyme.

ACKNOWLEDGEMENTS

The biological anti-inflammatory screening was performed in the Histology Department, Faculty of Medicine (Girls), Al-Azhar University. While, the biological anti-microbial screening was performed in the Microbiology Department, Faculty of Pharmacy (Girls), Al-Azhar University. The Molecular docking study was performed in the pharmaceutical Chemistry Department, Faculty of Pharmacy, Alexandria University.

CITE THIS PAPER

M. T. Sarg, M. M. Koraa, A. H. Bayoumi, S. M. Abd El Gilil (2015) Synthesis of Pyrroles and Condensed Pyrroles as Anti-Inflammatory Agents with Multiple Activities and Their Molecular Docking Study. *Open Journal of Medicinal Chemistry*,**05**,49-96. doi: 10.4236/ojmc.2015.54005

REFERENCES

1. Hatamjafari, F. and Montazeri, N. (2009) Three-Component Process for the Synthesis of Some Pyrrole Derivatives under Microwave Irradiation. Turkish Journal of Chemistry, 33,797-802.
2. Gajera, R. V. (2010) Synthesis and Physicochemical Studies of Some Molecules of Meducinal Interest. PhD. Thesis, Saurashtra University, Rajkot, p. 23.
3. Idhayadhulla, A., Kumar, R. S. and Abdul Nasser, A. J. (2011) Synthesis, Characterization and Antimicrobial Activity of New Pyrrole Derivatives. Journal of the Mexican Chemical Society, 55, 218-223.
4. Mohamed, M. S., Mostafa, A. G. and Abd El-hameed, R. H. (2012) Evaluation of the Anti-Inflammatory Activity of Novel Synthesized Prrole, Pyrrolopyrimidine and Spiropyrrolopyrimidne Derivatives. Pharmacophore, 3, 44-54.
5. Dannhardt, G., Kiefer, W., Kramer, G., Maehrlein, S., Nowe, U. and Fiebich, B. (2000) The Pyrrole Moiety as a Template for COX-1/COX-2 Inhibitors. European Journal of Medicinal Chemistry, 35, 499-510.
6. Bijander, K., Vinit, R., Arvind, K., and Vaishali, S. (2012) Anti-Inflammatory Activity of 1,3,4-Oxadiazole Derivatives Compound. International Journal of Current Pharmaceutical Research, 4, 9-14.
7. Rachel, C., Monica, K. (2015) In Vitro Anti-Oxidant and Anti-Inflammatory Activity of Newly Synthesized Schiff Bases Derived from 2-Aminothiazole Derivatives. Indo American Journal of Pharmaceutical Research, 5, 2078-2088.
8. Grillo, M. P. and Hua, F. (2003) Identification of Zomepirac-S-acyl-glutathione in Incubations with Rat Hepatocytes and in Vivo in Rat Bile. Drug Metabolism and Disposition, 31, 1429-1436.
9. Lessigiarska, I., Nankov, A., Bocheva, A., Pajeva, I. and Bijev, A. (2005) 3D-QSAR and Preliminary Evaluation of Anti-Inflammatory Activity of Series of N-Pyrrolylcarboxylic Acids. IL Farmaco, 60, 209-218.
10. Harrak, Y., Rosell, G., Daidone, G., Plescia, S., Schillaci, D. and Pujol, M.D. (2007) Synthesis and Biological Activity of New Anti-Inflammatory Compounds Containing the 1,4-Benzodioxine and/or Pyrrole System. Bioorganic & Medicinal Chemistry, 15, 4876.
11. Ushiyama, S., Yamada, T., Murakami, Y., Kumakura, S., Inoue, S., Suzuki, K., Nakao, A., Kawara, A. and Kimura, T. (2008) Preclinical Pharmacology Profile of CS-706, a Novel Cyclooxygenase-2 Selective Inhibitor, with Potent Antinociceptive and Anti-Inflammatory Effects. European Journal of Pharmacology, 578, 76.
12. Mohamed, M.S., Kamel, R. and Fatahala, S.S. (2011) Synthesis of New Pyrroles of Potential Anti-Inflammatory Activity. Archiv der Pharmazie, 344, 830-839.
13. Pham, V.C., Shin, J.S., Choi, M.J., Kim, T.W., Lee, K., Kim, K.J., Huh, G., Kim, J., Choo, D.J., Lee, K.T. and Lee, J.Y. (2012) Biological Evaluation and Molecular Docking Study of 3-(4-Sulfamoylphenyl)-4-phenyl-1H-pyrrole- 2,5-dione as COX-2 Inhibitor. Bulletin of the Korean Chemical Society, 33, 721-724.

14. Parmar, K., Sutariya, S., Shukla, M. and Goswami, K. (2012) Synthesis and Antimicrobial Activity of Substituted 2H-Pyrrole-2-ones Derivatives Based on 1-N-phenyl-3-phenyl-4-formyl Pyrazole (PFP). Journal of Chemical and Pharmaceutical Research, 4, 3478-3482.
15. Joshi, S.D., More, U.A., Pansuriya, K., Aminabhavi, T.M. and Gadad, A.K. (2013) Synthesis and Molecular Modeling Studies of Novel Pyrrole Analogues as Antimycobacterial Agents. Journal of Saudi Chemical Society.
16. Jana, G.H., Jain, S., Arora, S.K. and Sinha, N. (2006) Corrigendum to "Synthesis of Some Diguanidino 1-Methyl-2,5- Diaryl-1H-Pyrroles as Antifungal Agents" [Bioorg. Med. Chem. Lett. 15 (2005) 3592-3595]. Bioorganic & Medicinal Chemistry Letters, 16, 751.
17. Raimondi, M.V., Cascioferro, S., Schillaci, D. and Petruso, S. (2006) Synthesis and Antimicrobial Activity of New Bromine-Rich Pyrrole Derivatives Related to Monodeoxypyoluteorin. European Journal of Medicinal Chemistry, 41, 1439-1445.
18. Biava, M., Porretta, G.C., Poce, G., Logu, A.D., Meleddu, R., Rossi, E.D., Manettic, F. and Botta, M. (2009) 1,5-Di- aryl-2-ethyl Pyrrole Derivatives as Antimycobacterial Agents: Design, Synthesis, and Microbiological Evaluation. European Journal of Medicinal Chemistry, 44, 4734-4738.
19. Mehta, S. (2013) Synthesis and Biological Activity of Pyrrole and Pyrrolidine Compounds from 4-Chloro-2-hydroxy- benzoic Acid Hydrazide. International Journal of Pharmaceutical Research and Bio-Science, IJPRBS, 2, 382.
20. Ghorab, M.M., Heiba, H.I., Hassan, A.A., Abd El-Aziz, A.B. and El-Gazzar, M.G. (2011) Antimicrobial Evaluation of Novel Pyrrole, Pyrazole, Pyrimidine and Pyrrolo [2,3-d]-Pyrimidine Derivatives Bearing Sulfonamide Moiety. The Journal of American Science, 7, 1063-1073.
21. Nicolaou, K.C., Simmons, N.L., Chen, J.S., Haste, N.M. and Nizet, V. (2011) Total Synthesis and Biological Evaluation of Marinopyrrole A and Analogs. Tetrahedron Letters, 52, 2041-2043.
22. Idhayadhulla, A., Kumar, R.S., Abdul Nasser, A.J. and Manilal, A. (2012) Synthesis and Antimicrobial Activity of Some New Pyrrole Derivatives. Bulletin of the Chemical Society of Ethiopia, 26, 429-435.
23. Varaprasad, C.V., Ramasamy, K.S., Girardet, J.L., Gunic, E., Lai, V., Zhong, W., An, H. and Hong, Z. (2007) Synthesis of Pyrrolo[2,3-d]pyrimidine Nucleoside Derivatives as Potential Anti-HCV Agents. Bioorganic Chemistry, 35, 25- 34.
24. Dodonova, J., Uogintaite, I., Masevicius, V. and Tumkevicius, S. (2010) Palladium-Catalyzed Reaction of Methyl 5-Amino-4-chloro-2-methylthiopyrrolo[2,3-d]-pyrimidine-6-carboxylate with Arylboronic Acids. Synthesis of 1,3,4,6- Tetraazadibenzo[cd,f]-azulene Heterocyclic System. Chemistry of Heterocyclic Compounds, 46, 1122-1126.
25. Hilmy, K.M.H., Soliman, D.H., Shahin, E.B.A. and Abd Alhameed, R. (2012) Synthesis and Molecular Modeling Study of Novel Pyrrole Schiff Bases as Anti-HSV-1 Agents. Life Science Journal, 9, 736-745.
26. de Courcy, B., Piquemal, J., Garbay, C. and Gresh, N. (2010) Polarizable Water Molecules in Ligand-Macromolecule Recognition. Impact on the Relative Affinities of Competing Pyrrolopyrimidine Inhibitors for FAK Kinase. Journal of the American Chemical Society, 132, 3312.
27. Shamsuzzaman, Siddiqui, T., Alam, M.G. and Dar, A.M. (2015) Synthesis, Characterization and Anticancer Studies of New Steroidal Oxadiazole, Pyrrole and Pyrazole Derivatives. Journal of Saudi Chemical Society, 19, 387-391
28. Idhayadhulla, A., Kumar, R.S., Abdul Nasser, A.J. and Manilal, A. (2013) Synthesis of Some New Pyrrole and Pyridine Derivatives and their Antimicrobial, Anticancer Activities. International Journal of Biological Chemistry, 7, 15- 26.
29. Kalgutkar, A.S., Crews, B.C., Rowlinson, S.W., Marnett, A.B., Kozak, K.R., Remmel, R.P. and Marnett, L.J. (2000) Biochemically Based Design of Cyclooxygenase-2 (COX-2) Inhibitors: Facile Conversion of Nonsteroidal Antiinflammatory Drugs to Potent and Highly Selective COX-2 Inhibitors. Proceedings of the National Academy of Sciences of the United States of America, 97, 925-930.

30. Zarghi, A., Javid, F.S., Ghodsi, R., Dadress, O.G., Daraie, B. and Hedayati, M. (2011) Design, Synthesis and Biological Evaluation of New 5,5-Diarylhydantoin Derivatives as Selective Cyclooxygenase-2 Inhibitors. Scienta Pharmaceutica, 79, 449-460.
31. Etcheverry, S.B., Barrio, D.A., Cortizo, A.M. and Williams, P.A. (2002) Three New Vanadyl(IV) Complexes with Non-Steroidal Anti-Inflammatory Drugs (Ibuprofen, Naproxen and Tolmetin). Bioactivity on Osteoblast-Like Cells in Culture. Journal of Inorganic Biochemistry, 88, 94-100.
32. Fernandes, E., Costa, D., Toste, S.A., Lima, J.L.F.C. and Reis, S. (2004) In Vitro Scavenging Activity for Reactive Oxygen and Nitrogen Species by Nonsteroidal Anti-Inflammatory Indole, Pyrrole, and Oxazole Derivative Drugs. Free Radical Biology and Medicine, 37, 1895-1905.
33. Ghoneim, A.A. (2009) A Clean Procedure for Synthesis of Phenylquinoline Derivatives. Oriental Journal of Chemistry, 25, 449-504.
34. Kotb, E.R., El-Hashash, M.A., Salama, M.A., Kalf, H.S., Abdel Wahed, N.A.M. (2009) Synthesis and Reactions of Some Novel Nicotinonitrile Derivatives for Anticancer and Antimicrobial Evaluation. Acta Chimica Slovenica, 56, 908-919.
35. Carpenter, C.A., Kenar, J.A. and Price, N.P.J. (2010) Preparation of Saturated and Unsaturated Fatty Acid Hydrazides and Long Chain C-Glycoside Ketohydrazones. Green Chemistry, 12, 2012-2018.
36. Chavan, H.V., Bandgar, B.P., Adsul, L.K., Dhakane, V.D., Bhale, P.S., Thakare, V.N. and Masand, V. (2013) Design, Synthesis, Characterization and Anti-Inflammatory Evaluation of Novel Pyrazole Amalgamated Flavones. Bioorganic & Medicinal Chemistry Letters, 23, 1315-1321.
37. Kumar, S.T., Kanti, BT., Liaquat, A. and Biswapati, M. (2003) Anti-Inflammatory and Anti-Platelet Aggregation Activity of Human Placental Extract. Acta Pharmacologica Sinica, 24, 187-192.
38. Winter, C.A., Risley, E.A. and Nuss, G.W. (1962) Carrageenin-Induced Edema in Hind Paw of the Rat as an Assay for Antiinflammatory Drugs. Experimental Biology and Medicine, 111, 544-547.
39. Sammour, O.A., Al-Zuhair, H.H. and El-Sayed, M. I. (1998) Inhibitory Effect of Liposome-Encapsulated Piroxicam on Inflammation and Gastric Mucosal Damage. Pharmazeutische Industrie, 60, 1084-1087.
40. Khalifa, M.M. and Abdelbaky, N.A. (2007) Synthesis of New Imidazolyl Acetic Acid Derivatives as Potential Anti- Inflammatory Agents. Az. J. Pharm. Sci., 35, 22-30.
41. Hemamalini, K., Naik, K.O. and Ashok, P. (2010) Anti-Inflammatory and Analgesic Effect of Methanolic Extract of Anogeissus acuminata Leaf. International Journal of Pharmaceutical and Biomedical Research, 1, 98-101.
42. Curfman, G.D., Morrissey, S. and Drazen, J.M. (2005) Expression of Concern: Bombardier et al., "Comparison of Upper Gastrointestinal Toxicity of Rofecoxib and Naproxen in Patients with Rheumatoid Arthritis," N Engl J Med 2000;343:1520-8. The New England Journal of Medicine, 353, 2813-2814.
43. Metwally, K.A., Yaseen, S.H., Lashine, E.M., El-Fayomi, H.M. and El-Sadek, M.E. (2007) Non-Carboxylic Analogues of Arylpropionic Acids: Synthesis, Anti-Inflammatory Activity and Ulcerogenic Potential. European Journal of Medicinal Chemistry, 42, 152-160.
44. Koster, R., Anderson, M. and Beer, D. (1959) Acetic Acid for Analgesic Screening. Federation Proceedings, 18, 412.
45. Narayana, B., Raj, K.K.V., Ashalatha, B.V. and Kumari, N.S. (2005) Synthesis of Some New 2-(6-Methoxy-2-Naph- thyl)-5-Aryl-1,3,4-Oxadiazoles as Possible Non-steroidal Anti-inflammatory and Analgesic Agents. Archiv der Pharmazie, 338, 373-377.
46. Sharma, K.K., Saikia, R., Kotoky, J., Kalita, J.C. and Das, J. (2011) Evaluation of Antidermatophytic Activity of Piper betle, Allamanda cathertica and Their Combination: An in Vitro and in Vivo Study. International Journal of PharmTech Research, 3, 644-651.

CHAPTER 8

Synthesis of Novel Acid Dyes with Coumarin Moiety and Their Utilization for Dyeing Wool and Silk Fabrics

Mahmoud S. Bashandy[1,2*], Fatma A. Mohamed[3,4], Mohamed M. El-Molla[3,5], Mahmoud B. Sheier[1], Ahmed H. Bedair[1]

[1]*Chemistry Department, Faculty of Science (Boys), Al-Azhar University, Cairo, Egypt*

[2]*Chemistry Department, University College in Al-Jamoum, Umm Al-Qura University, Makkah, Saudi Arabia*

[3]*Textile Research Division, National Research Centre, Giza, Egypt*

[4]*Balguenvzh University College, Department of Female Student, Umm Al-Qura University, Makkah, Saudi Arabia*

[5]*Chemistry Department, College of Sciences & Art, Al Jouf University, El-Quraiat, Saudi Arabia*

ABSTRACT

This article describes the synthesis of some novel coumarin compounds to use as acid dyes by using compounds 1 - 4 as starting materials, which were prepared by interaction of 2-hydroxyben- zaldehyde with ethyl 3-oxobutanoate, diethylmalonate, 4-nitrobenzenediazonium chloride and 4-sulfobenzene-diazonium chloride, respectively. Compound 1 reacted with bromine and 2-cyano- acetohydrazide to give phenacyl bromide derivative 5 and 2-cyanoacetohydrazone derivative 6, respectively. Coupling of compound 6 with equimolar amount of 2-sulfo-4-((4-sulfophenyl) diazenyl)benzenediazonium chloride gave coumarin acid dye 8. Phenacyl bromide derivative 5 reacted with potassium cyanide in refluxing ethanol to produce compound 7, which on coupling with equimolar amount of 8-hydroxy-6-sulfonaphthalene-2-diazonium chloride and 8-hydroxy-3,6-disulfonaphthalene-1-diazonium chloride gave coumarin acid dyes 9 and 10, respectively. Interaction of compound 2 with 2-amino-5-((4-sulfophenyl)diazenyl)benzenesulfonic acid, benzene-1,4- diamine and 3,3'-dimethoxy-[1,1'-biphenyl]-4,4'-diamine in refluxing ethanol afforded compounds 11, 12 and 14, respectively. Diazonium sulphate of compounds 12 and 14 coupling with 4-amino- 5-hydroxynaphthalene-2,7-disulfonic acid gave compounds 13 and 15, respectively. Cyclocondensation of compound 3 with ethyl 3-oxobutanoate, diethyl malonate and malononitrile afforded derivatives of 3-acetyl-2H-chromen-2-one 16, ethyl 2-oxo-2H-chromene-3-carboxylate 17 and 2- imino-2H-chromene-3-carbonitrile 18, respectively. Reaction of sodium benzenesulfonate deriva- tive 4 with ethyl 3-oxobutanoate and hydrazine hydrate gave compounds 19 and 20, respectively. The structures of the newly synthesized compounds were confirmed by elemental analysis, UV/ VIS, IR, ^{1}H NMR and Ms spectral data. The suitability of the prepared dyestuffs for dyeing of wool and silk fabrics has been investigated. The dyed fabric shows good light fastness, very good rubbing, perspiration, washing and excellent sublimation fastness. These dyes have been color shade from blue to violet with very good depth and levelness on fabrics. The dye bath exhaustion and fixation on fabric has been found to be very good.

Keywords: Synthesis, Coumarin, Acid Dye, Dyeing, Wool Fabric

1. INTRODUCTION

The considerable innovation has been witnessed in past three decades in the field of azo dye chemistry based on heterocyclic systems and studies in the synthesis of such derivatives have been reported [1] - [5] . Most of the recent research has focused on structural variations of existing types, for example, variations in substituent, especially on the side chains of the coupling components. The use of heterocyclic coupling component and diazo components in the synthesis of azo dyes is well established, and the resultant dyes exhibit better tinctorial strength and brighter dyeing than those derived from aniline-based components. Most heterocyclic dyes of technical interest for application to textiles are derived from diazo components consisting of five-membered rings containing one sulphur heteroatom and to which a diazotisable amino group is directly attached.

The ring may also possess one or more nitrogen heteroatoms and be fused to another aromatic ring. These diazo components are capable of providing red to blue color dyes that meet the rigorous technical and economical requirements demanded of them by both manufacturer and user. Intensive efforts have been made in the investigation of monoazo dyes in which a heterocyclic system replaces one of the usual carboxylic systems. Many different heterocyclic diazo components have been studied, especially derivatives of thiazole, imidazole, benzimidazole owing to the marked bathochromic effect of such groups [6] - [12] . A majority of acid dyestuffs are sulphonic acid derivatives of azo dyes. The free dye acids are difficult to isolate and are hydroscopic in nature making it difficult to pack and store them. These dyes are invariably isolated as sodium salts. Coumarins are attractive and versatile molecules that find applications in various fields like medicine, perfumery, dyes, pigments, optical brighteners, lasers, optical data storage devices, solar cells [13] - [19] . The coumarin is not fluorescent, but the introduction of an electron-withdrawing group such as a diazotized aromatic amine or an acetyl group makes it highly fluorescent. Coumarin establishes a family of dyes [20] - [24] that are applicable in different fields of science and technology [25] - [27] . The present work was carried out with the following objectives, synthesis and identification of some newly acid dyes based on coumarin derivatives, and the possibility of its use in dyeing of wool and silk fabrics.

2. RESULTS AND DISCUSSION

2.1 Chemistry

The present investigation deals with the synthesis of novel coumarin compounds to use as acid dyes by using compounds 1-4 as a starting materials, which were prepared by interaction of 2-hydroxybenzaldehyde with ethyl 3-oxobutanoate, diethylmalonate, 4-nitrobenzenediazonium chloride and 4-sulfobenzenediazonium chloride, respectively (Scheme 1). Structure of compound 3 was established on the basis of its elemental analysis and spectral data. Thus, IR spectrum of 3indicated absorption bands at ν_{max} = 3370 cm^{-1} for hydroxyl group, 1653 cm^{-1} for carbonyl group and 1456, 1290 cm^{-1} for nitro group. ^{1}H NMR spectrum showed three singlet signals at δ = 8.05, 10.30 and 11.78 ppm corresponding to H_6 of hyroxyphenyl ring, hydroxyl and formyl protons, respectively, two doublet signals, each doublet for two protons at δ = 7.95 and 8.32 ppm for $H_{2,6}$and $H_{3,5}$of nitrophenyl ring, respectively, two doublet signals, each doublet for one proton at δ = 7.17 and 8.14 ppm due to H_3 and H_4 of hyroxyphenyl ring, respectively. IR spectrum of compound 4 showed absorption bands at ν_{max}

= 3462 cm^{-1} for hydroxyl group, ν_{max} = 1658 cm^{-1} for carbonyl group and ν_{max} = 1386, 1150 cm^{-1} for sulphate group. ^{1}H NMR spectrum showed three singlet signals at δ = 7.19, 10.33 and 11.56 ppm corresponding to H_6 of hyroxyphenyl- ring, hydroxyl and formyl protons, respectively, two doublet signals, each doublet for two protons at δ = 7.76 - 7.79 and 8.15 ppm for $H_{2,6}$ and$H_{3,5}$ of benzenesulfonate ring, respectively, two doublet signals, each doublet for one proton at δ = 7.17 and 8.06 - 8.08 ppm due to H_3 and H_4 of hyroxyphenyl ring, respectively.

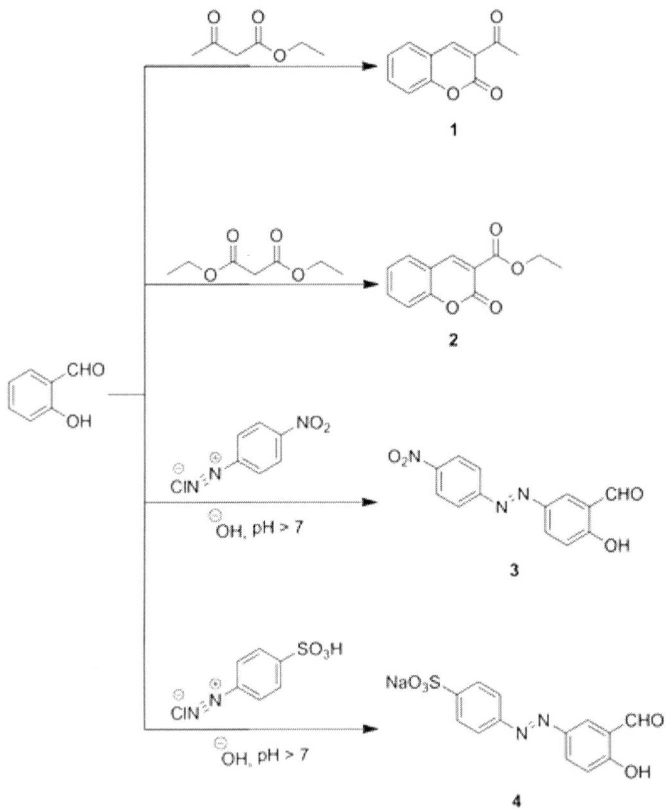

Scheme 1. Synthetic pathways for compounds 1-4.

3-Acetyl-2H-chromen-2-one (1) reacted with bromine and 2-cyanoaceto-hydrazide to give phenacyl bromide derivative 5and 2-cyanoacetohydrazone derivative 6, respectively. Phenacyl bromide derivative 5 reacted with potassium cyanide in refluxing ethanol to produce 3-oxo propanenitrile derivative 7 (Scheme 2). The structure of compound 7 was established on the basis of their elemental analysis and spectral data. Thus, IR spectrum of compound 7 revealed absorption bands at v_{max} = 2207 cm^{-1} for cyano group, v_{max} = 1703 and 1636 cm^{-1} for cyclic carbonyl of coumarin and acyclic carbonyl, respectively. The mass spectrum of compound 7 showed a molecular ion peak at m/z = 213 and a base peak at m/z = 101.The methylene group in compound 6 proved to be highly reactive. Thus, 2-cyanoacetohydrazone derivative 6 underwent coupling with equimolar amount of 2- sulfo-4-((4-sulfophenyl)-diazenyl) benzenediazonium chloride to give coumarin acid dye 8 (Scheme 2). The IR spectrum of coumarin acid dye 8 showed, two bi-forked characteristic absorption bands at v_{max} = 3421 and 3350 cm^{-1} assignable to 2NH groups, another absorption bands at v_{max} = 2212 cm^{-1} for cyano group and at v_{max} = 1755 and 1657 cm^{-1} for two carbonyl groups of coumarin ring and amide group, respectively. Mass spectrum of compound 8 showed a molecular ion peak at m/z = 681 and a base peak at m/z = 50.

3-Oxo-3-(2-oxo-2H-chromen-3-yl)propanenitrile (7) underwent coupling with equimolar amount of 8-hy- droxy-6-sulfonaphthalene-2-diazonium chloride and 8-hydroxy-3,6-disulfonaphthalene-1-diazonium chloride to give coumarin acid dyes 9 and 10,respectively (Scheme 3). The structure of coumarin acid dyes 9 and 10 were established on the basis of their elemental analysis and spectral data. Thus, IR spectrum of compound 9 revealed absorption bands at v_{max} = 3424 and 3320 $cm^{-1}$for OH and NH groups, respectively, v_{max} = 2248 cm^{-1} for cyano group and at v_{max} = 1710 and 1632 cm^{-1} for cyclic carbonyl of coumarin and acyclic carbonyl, respectively.

^{1}H NMR spectrum of compound 9 showed two singlet signals at δ = 12.71 and 12.80 ppm corresponding to protons of OH and NH groups. IR spectrum of compound 10 revealed broad absorption band from v_{max} = 3486 to 3380 cm^{-1} for OH and NH groups, v_{max} = 2225 cm^{-1} for cyano group and at v_{max} = 1710 and 1632 cm^{-1} for cyclic carbonyl of coumarin and acyclic carbonyl, respectively. Mass spectrum of compound 10 showed a molecular ion peak at m/z = 587 and a base peak at m/z = 219.

Interaction of ethyl 2-oxo-2H-chromene-3-carboxylate (2) with 2-amino-5-((4-sulfophenyl)diazenyl) benzene-sulfonic acid in refluxing ethanol afforded coumarin acid dye 11 (Equation 1). The structure of coumarin acid dye 11 was established on the basis of their elemental analysis and spectral data. Thus, IR spectrum of compound 11 revealed absorption bands at v_{max}= 3325 cm^{-1} for NH group, v_{max} = 1736 and 1617 cm^{-1} for two carbonyl groups of coumarin ring and amide group, respectively. Mass spectrum of compound 11 showed a molecular ion peak at m/z = 573 and a base peak at m/z = 394.

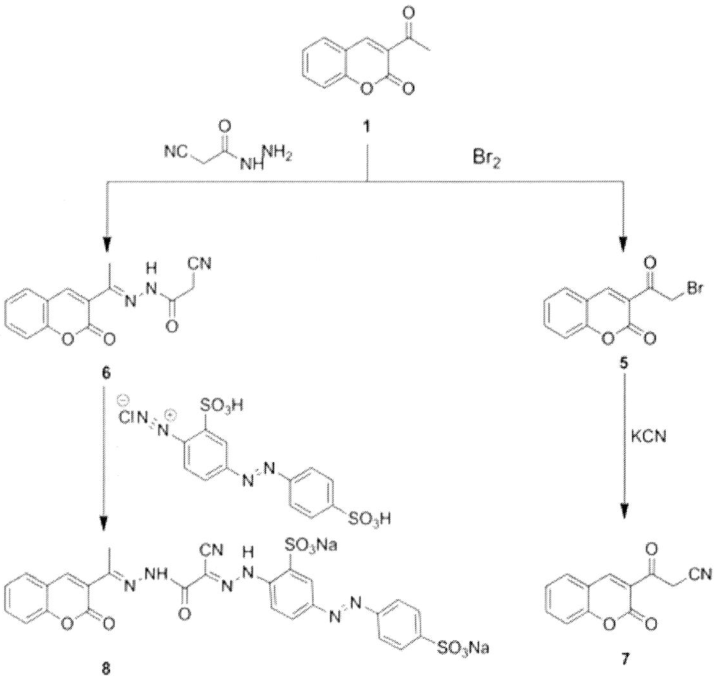

Scheme 2. Synthetic pathways for compounds 5-8.

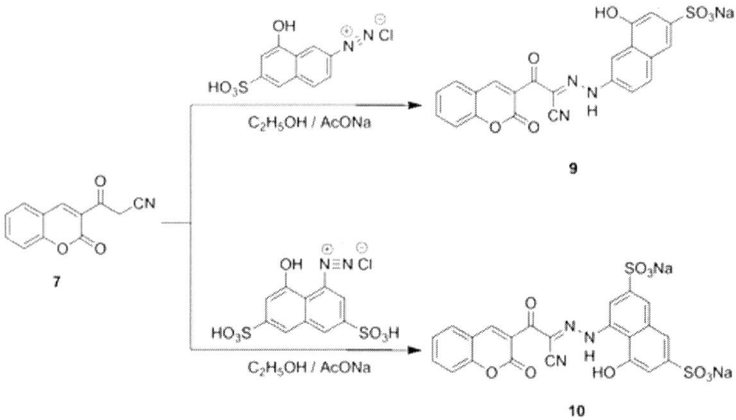

Scheme 3. Synthetic pathways for compounds 9 and 10.

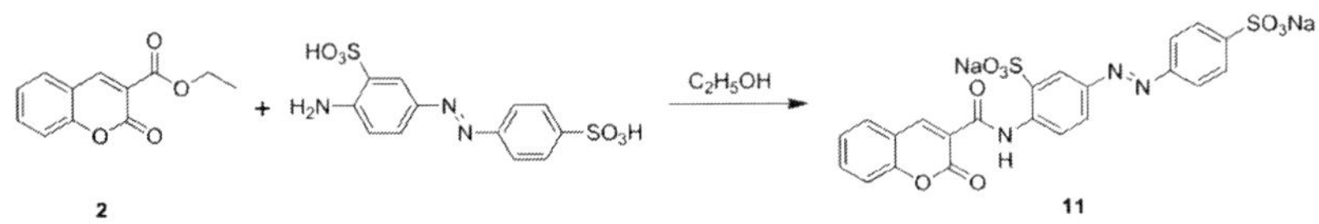

Equation 1. Synthetic pathway for compound 11.

Interaction of compound 2 with benzene-1,4-diamine in refluxing ethanol afforded a single product identified as N-(4-aminophenyl)-2-oxo-2H-chromene-3-carboxamide (12) on the basis of elemental analysis and spectral data. Thus, IR spectrum of compound 12 showed, two bi-forked characteristic absorption bands at ν_{max} = 3458 and 3360 cm^{-1} assignable to amino group, ν_{max} = 3320 cm^{-1} for NH group and ν_{max} = 1702 and 1650 cm^{-1} for two carbonyl groups of coumarin ring and amide group, respectively. Mass spectrum of compound 12 showed a molecular ion peak at m/z = 280 and showed other peaks at m/z = 173 corresponding to fragment (M^+-$C_6H_7N_2$), m/z = 107as a base peak for fragment ($C_6H_7N_2^+$) (Chart 1).

Compounds 12 was suspended with stirring in concentrated sulfuric acid and cooled to 0°C - 5°C then diazotized by adding sodium nitrite. After stirring at 0°C for 1 h, the diazonium sulphate solution of compound 12 was added to dissolved coupler compound of 4-amino-5-hydroxynaphthalene-2,7-disulfonic acid in 20% sodium hydroxide solution at 0°C to give compound 13 (Scheme 4). IR spectrum of compound 13 revealed broad absorption band from ν_{max} = 3505 to 3273 cm^{-1} for hydroxyl, amino and NH groups and at ν_{max} = 1700 and 1648 cm^{-1} for two carbonyl groups of coumarin ring and amide group, respectively. Mass spectrum of compound 13 showed a molecular ion peak at m/z = 654 and a base peak at m/z = 107.

In the same manner, compound 2 underwent condensation with 3,3'-dimethoxy-[1,1'-biphenyl]-4,4'-diamine to produce N-(4'-amino-3,3'-dimethoxy-[1,1'-biphenyl]-4-yl)-2-oxo-2H-chromene-3-carboxamide (14). The elemental analysis and spectral data of the latter structure were in agreement with its assigned structure. Thus, IR spectrum of compound 14 revealed a broad absorption band from ν_{max} = 3400 to 3270 cm^{-1} due to amino and NH groups and ν_{max} = 1718 and 1658 cm^{-1} for two carbonyl groups of coumarin ring and amide group, respectively. Besides, the mass spectrum was compatible with the molecular formula $C_{24}H_{20}N_2O_5$, m/z =416 confirmed structure 14. Diazonium sulphate of compound 14 coupling with 4-amino-5-hydroxynaphthalene-2,7-disulfonic acid to give compound 15 (Scheme 5). IR spectrum of compound 15 revealed a broad absorption band from ν_{max} = 3509 to 3380 cm^{-1} due to hydroxyl, amino and NH groups, ν_{max} = 1726 and 1661 cm^{-1} for two carbonyl groups of coumarin ring and amide group, respectively and ν_{max} = 1449 cm^{-1} for N=N group. Mass spectrum of compound 15 showed a molecular ion peak at m/z = 791 and a base peak at m/z = 64.

Cyclocondensation of 2-hydroxy-5-((4-nitrophenyl)diazenyl)benzaldehyde (3) with ethyl 3-oxobutanoate, diethyl malonate and/or malononitrile afforded derivatives of 3-acetyl-2H-chromen-2-one 16, ethyl 2-oxo-2H-chromene-3-carboxylate17 and 2-imino-2H-chromene-3-carbonitrile 18, respectively (Scheme 6). The structures of compounds 16-18 were established on the basis of their elemental analysis and spectral data. Thus, IR spectrum of compound 16 revealed an absorption bands atν_{max} = 1745 and 1676 cm^{-1} for two carbonyl groups of coumarin ring and acetyl group, respectively, ν_{max} = 1522 cm^{-1} corresponding to N=N group and two absorption bands at ν_{max} = 1338 and 1231 cm^{-1} for nitro group.^{1}H NMR spectrum of 16 revealed three singlet signals at δ= 2.53, 8.42 and 8.80 ppm for methyl, $H_5$and H_4 of coumarin, respectively, two doublet signals each doublet signal for one proton at δ= 7.65, 8.25 ppm corresponding to H_8 and H_7 of coumarin, respectively, besides two doublet signals each doublet signal for two protons at δ= 8.08 and 8.59 ppm corresponding to $H_{2,6}$ and $H_{3,5}$ of nitrophenyl ring, respectively.IR spectrum of compound 17 revealed an absorption bands at ν_{max} = 1741 and 1705 cm^{-1} for two carbonyl groups of coumarin ring and ester group, respectively, ν_{max} = 1521 cm^{-1} corresponding to N=N group and two absorption bands at ν_{max} = 1338 and 1251 cm^{-1} for nitro group. ^{1}H NMR spectrum of 17 revealed triplet and quartet signals at δ= 1.30 and 4.30 ppm, respectively corresponding to protons of methyl and methylene of ester group, respectively, two singlet signals at δ= 8.40 and 8.89 ppm for H_5 and H_4 of coumarin, respectively, two doublet signals each doublet signal for one proton at δ = 7.60, 8.19 ppm corresponding to H_8 and H_7 of coumarin, respectively, besides two doublet signals each doublet signal for two protons at δ = 8.03 and 8.52 ppm corresponding to $H_{2,6}$ and $H_{3,5}$ of nitrophenyl ring, respectively.

IR spectrum of compound 18 revealed an absorption band at ν_{max} = 3337 for NH group, ν_{max} = 2205 for cyano group, ν_{max} = 1556 cm^{-1} corresponding to N=N group and two absorption bands at ν_{max} = 1338 and 1251 cm^{-1} for nitro group.Mass spectrum of compound 18 showed a molecular ion peak at m/z = 321 (M^{+}+2) and showed other peaks at m/z = 319 (M^{+}), m/z = 170 corresponding to fragment (M^{+}-$C_6H_3N_3O_2$), m/z = 150 for fragment (M^{+}-$C_{10}H_5N_2O$), m/z = 106 for fragment ($C_7H_6O^{+}$), m/z = 77 for fragment ($C_6H_5^{+}$), m/z = 65 for fragment ($C_5H_5^{+}$) and m/z = 52 for fragment ($C_4H_4^{+}$) (Chart 2).

Reaction of sodium benzenesulfonate derivative 4 with ethyl 3-oxobutanoate and/or hydrazine hydrate gave compounds 19 and 20, respectively (Scheme 7). IR spectrum of compound 19 revealed an absorption bands at ν_{max} = 1754 and 1672 cm^{-1} for two carbonyl groups of coumarin ring and acetyl group, respectively, ν_{max} = 1564 cm^{-1} corresponding to N=N group and two absorption bands at ν_{max} = 1382 and

1122 cm^{-1} for SO_2 group. ^1H NMR spectrum of 19 revealed triplet signal at δ = 2.57 ppm corresponding to protons of methyl group, two singlet signals at δ = 8.19 and 8.80 ppm for H_5 and H_4 of coumarin, respectively, two doublet signals each doublet signal for one proton at δ = 7.62, 7.85 ppm corresponding to H_8 and H_7 of coumarin, respectively, besides two doublet signals each doublet signal for two protons at δ = 7.78 and 8.52 ppm corresponding to $H_{2,6}$ and $H_{3,5}$ of benzenesulfonate ring, respectively.

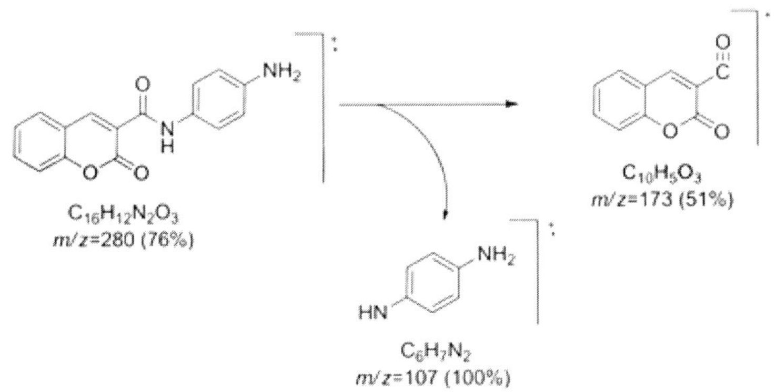

Chart 1. Fragmentation pattern for compound 12.

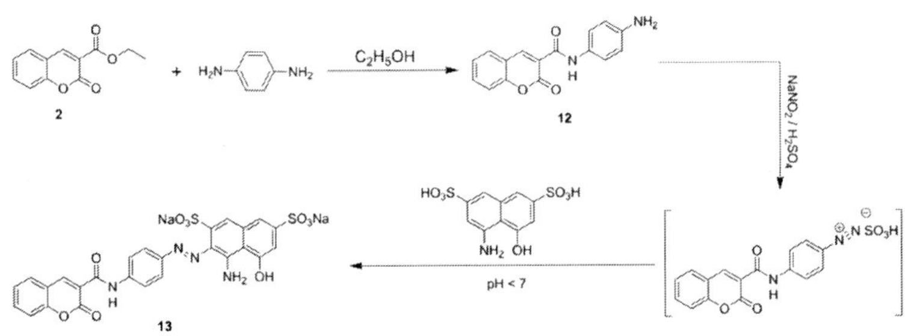

Scheme 4. Synthetic pathways for compounds 12 and 13.

IR spectrum of compound 20 revealed an absorption broad band at v_{max} = 3384 for OH groups, v_{max} = 1486 cm^{-1} corresponding to N=N group and two absorption bands at v_{max} = 1384 and 1132 cm^{-1} for SO_2 group. Mass spectrum of compound 20 showed a molecular ion peak at m/z = 552 (M^{+}-$2N_2$) and showed other peaks at m/z = 450 corresponding to fragment ($C_{20}H_{15}N_6O_5S^+$) and m/z = 295 for fragment ($C_{14}H_{10}N_6O_2^+$) (Chart 3).

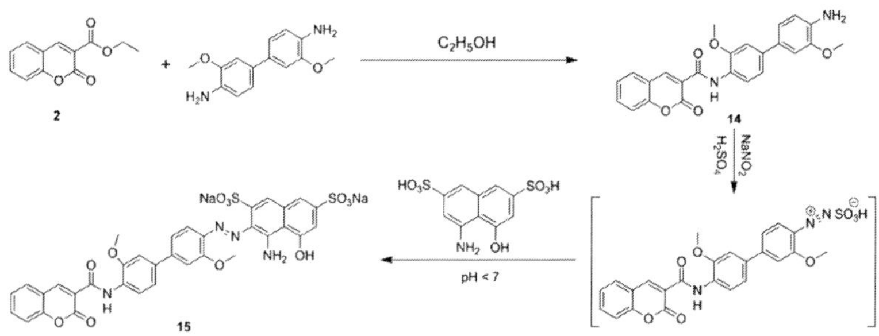

Scheme 5. Synthetic pathways for compounds 14 and 15.

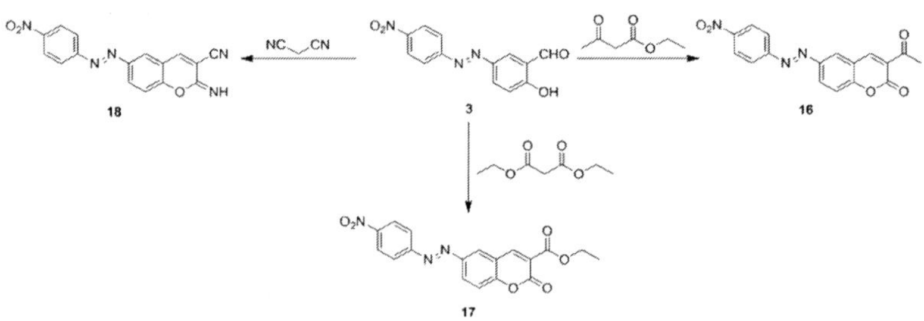

Scheme 6. Synthetic pathways for compounds 16-18.

3. EXPERIMENTAL

3.1. Methods and Materials

3.1.1. Wool Fabric

Wool fabric of 310 g/m^2, supplied by Golden Tex Co., Tenth of Ramadan-Egypt, was initially treated in an aqueous solution with a liquor ratio 50:1 containing 0.5 g/L sodium carbonate and 2 g/L nonionic detergent at 60°C for 30 min, then thoroughly washed, and air dried at room temperature.

3.1.2. Silk Fabric

Degummed and bleached silk fabric (El-Khateib Co., Egypt) weighing 90 g/m^2 was used throughout this work. Before dyeing, the fabric was treated in an aqueous solution containing 2 g/l non-ionic detergent for 1 h at 90°C and at a liquor ratio 50:1, then washed thoroughly in water and air dried at room temperature.

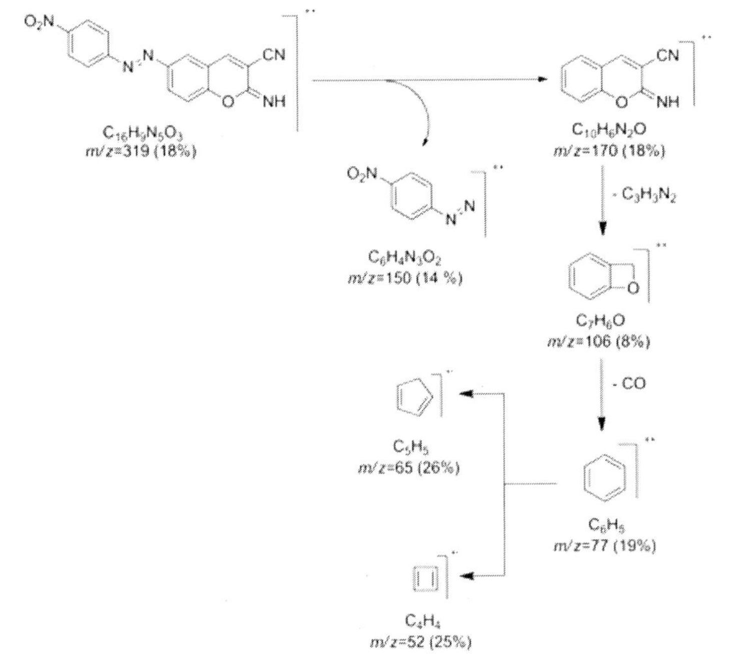

Chart 2. Fragmentation pattern for compound 18.

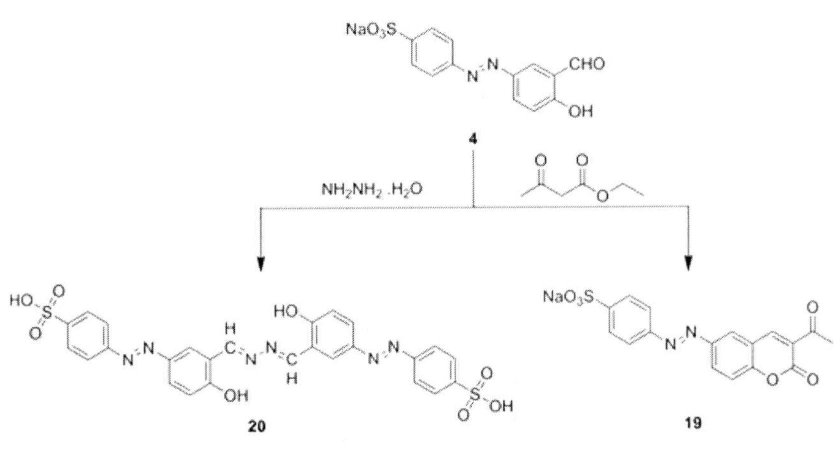

Scheme 7. Synthetic pathways for compounds 19 and 20.

3.1.3. Chemicals

H-acid, γ-acid and 4-aminoazobenzene-3,4'-disulphonic acid, 1-amino-3-bromo-5,10-dioxoanthracene-2-sul- phonic acid were obtained from Fluka Chemie AG. All other chemicals used in the study were of reagent gradeand applied without further purification.

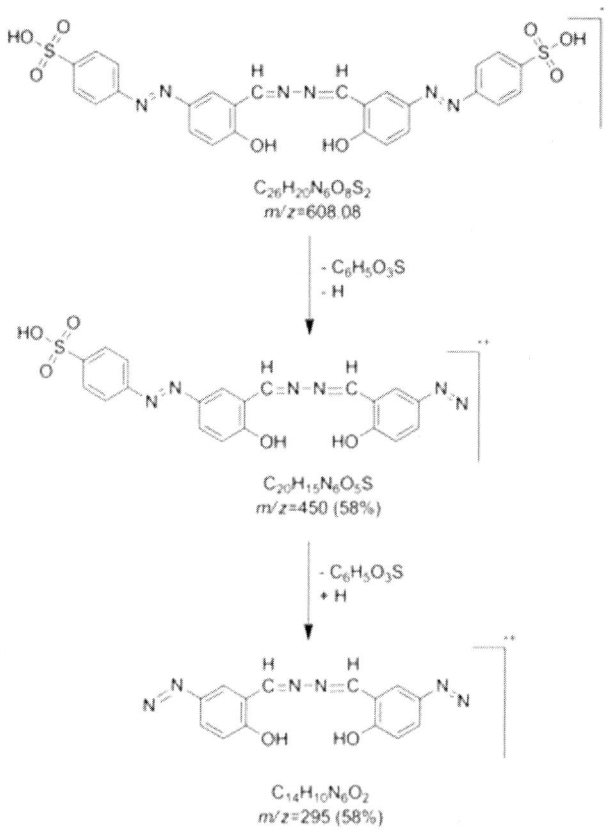

Chart 3. Fragmentation pattern for compound 20.

3.2. Chemistry

Melting points (°C, uncorrected) were determined in open capillaries on a Gallen Kemp melting point apparatus (Sanyo Gallen Kemp, Southborough, UK). IR spectra (KBr) were recorded on FT-IR 5300 spectrometer and

Perkin Elmer spectrum RXIFT-IR system (ν, cm^{-1}). Pre-coated silica gel plates (silica gel 0.25 mm, 60 G F 254; Merck, Germany) were used for thin layer chromatography. The NMR spectra in (DMSO-d_6) were recorded at 400 MHz on a Varian Gemini NMR spectrometer (δ, ppm). Mass spectra were obtained on GC Ms-QP 1000 EX mass spectrometer at 70 ev. Elemental analyses were performed on Carlo Erba 1108 Elemental Analyzer (Heraeus, Hanau, Germany). All compounds were within ±0.4% of the theoretical values. Analyses were carried out by the Micro analytical Research Center, Faculty of Science, Cairo University and Al-Azhar University.

3.2.1. 3-Acetyl-2H-Chromen-2-One (1)

Compound 1 was synthesized according to the literature procedure [28] .

3.2.2. Ethyl 2-Oxo-2H-Chromene-3-Carboxylate (2)

Compound 2 was synthesized according to the literature procedure [29] .

3.2.3. 2-Hydroxy-5-((4-Nitrophenyl)Diazenyl)Benzaldehyde (3)

2-Hydroxybenzaldehyde (1.22 g, 0.01 mol) was dissolved in water (20 mL) containing (0.4 g, 0.01 mol) of sodium hydroxide and (4.24 g, 0.04 mol) of sodium carbonate during the period of 30 min at 0°C. The resulting solution was added slowly to a solution of diazonium chloride of 4-nitroaniline (1.38 g, 0.01 mol) in water at 0°C - 5°C. The reaction mixture was stirred for 1 h at 0°C and then allowed to warm slowly to room temperature. The product was collected by filtration and washed with 100 mL of NaCl solution (10%) under vacuum. The obtained solid was dried under vacuum at 80°C overnight to give 3. Brown crystals, yield, 92%; mp 185-188°C. IR (KBr, cm^{-1}): $ν_{max}$ = 3370 (OH), 1653 (C=O), 1578 (C=C), 1526 (N=N), 1456, 1290 cm^{-1} (NO_2). ^1H NMR (DMSO-d_6): δ = 7.17 (1H, d, Ar-H_3), 7.95 (2H, d, nitrophenyl-$H_{2,6}$), 8.05 (1H, s, Ar-H_6), 8.14 (1H, d, Ar-H_4), 8.32 (2H, d, nitrophenyl-$H_{3,5}$), 10.30 ppm (1H, s, OH), 11.78 (1H, s, CHO). Anal. Calcd. for $C_{13}H_9N_3O_4$(271.23): C, 57.57; H, 3.34; N, 15.49. Found: C, 57.29; H, 3.56; N, 15.28%.

3.2.4. Sodium 4-((3-Formyl-4-Hydroxyphenyl)Diazenyl)Benzenesulfonate (4)

2-Hydroxybenzaldehyde (1.22 g, 0.01 mol) was dissolved in water (20 mL) containing (0.40 g, 0.01 mol) of sodium hydroxide and (4.24 g, 0.04 mol) of sodium carbonate during the period of 30 min at 0°C. The resulting solution was added slowly to a solution of 4-sulfobenzenediazonium chloride (2.2 g, 0.01 mol) in water at 0°C - 5°C. The reaction mixture was stirred for 1 h at 0°C and then allowed to warm slowly to room temperature. The product was collected by filtration and washed with 100 mL of NaCl solution (10%)

under vacuum. The obtained solid was dried under vacuum at 80°C overnight to give 4. Yellow crystals, yield: 61%; mp 350°C - 351°C (dec.). IR (KBr, cm^{-1}): v_{max} = 3462 (OH), 1658 (C=O), 1558 (C=C), 1478 (N=N), 1386, 1150 cm^{-1} (SO_2). ^1H NMR (DMSO-d_6): δ =7.17 (1H, d, Ar-H_3), 7.19 (1H, s, Ar-H_6), 7.76 - 7.79 (2H, d, benzenesulfonate-$H_{2,6}$), 8.06-8.08 (1H, d, Ar-H_4), 8.15 (2H, d, benzenesulfonate-$H_{3,5}$), 10.33 (1H, s, OH), 11.56 ppm (1H, s, CHO). Anal. Calcd. for $C_{13}H_9N_2NaO_5S$ (328.28): C, 47.56; H, 2.76; N, 8.53; S, 9.77. Found: C, 47.21; H, 3.03; N, 8.18; S, 10.02%.

3.2.5. 3-(2-Bromoacetyl)-2H-Chromen-2-One (5)

Compound 1 was synthesized according to the literature procedure [30] .

3.2.6. 2-Cyano-N'-(1-(2-Oxo-2H-Chromen-3-Yl)Ethylidene)Acetohydrazide (6)

Compound 1 was synthesized according to the literature procedure [31] .

3.2.7. 3-Oxo-3-(2-Oxo-2H-Chromen-3-yl)Propanenitrile (7)

A mixture of 3-(2-bromoacetyl)-2H-chromen-2-one (5; 2.66 g, 0.01 mol) and potassium cyanide (0.65 g, 0.01 mol) in ethanol (20 mL) was heated under reflux for 4 h. during the reflux period, a brown crystalline solid was separated. The separated solid was filtered off, washed with ethanol and recrystallized from ethanol to give 7. Brown crystals, yield 62%; mp 180°C - 182°C. IR (KBr, cm^{-1}): v_{max} = 2931 (CH-aliphatic), 2207 (C≡N), 1703 (O-C=O), 1636 (C=O). MS m/z (%): 213 [M^+] (99), 175 (27), 134 (47), 101 (100). Anal. Calcd. for $C_{12}H_7NO_3$ (213.19): C, 67.61; H, 3.31; N, 6.57. Found: C, 67.09; H, 3.13; N, 7.02%.

3.2.8. Sodium 2-(2-(1-Cyano-2-Oxo-2-(2-(1-(2-oxo-2H-Chromen-3-yl)Ethylidene)Hydr-Azinyl) Ethylidene)Hydrazinyl)-5-((4-Sulfonatophenyl)Diazenyl)Benzenesulfonate (8)

To a stirred solution of (6; 2.69 g, 0.01 mol) in ethanol (50 mL) containing, sodium acetate (3 g), 2-sulfo-4- ((4-sulfophenyl)diazenyl) benzenediazonium chloride (prepared by adding sodium nitrite (0.69 g, 0.01 mol) to 2-amino-5-((4-sulfophenyl)diazenyl)benzenesulfonic acid (3.57 g, 0.01 mol) in conc. HCl (6 mL) at (0°C - 5°C) under stirring) was added drop wise while cooling to (0°C -5°C) and stirring. The reaction mixture was then left at room temperature for 2 h and the solid product formed was collected by filtration and recrystallized from DMF/acetone to give 8.Orange crystals, yield, 82%; mp >300°C. l_{max} (H_2O) 518 nm. IR (KBr, cm^{-1}): v_{max} = 3421, 3350 (2NH), 2929 (CH-aliphatic), 2212 (C≡N), 1755 (O-C=O), 1657 cm^{-1} (C=O). MS m/z (%): 681 [M^+] (5), 356 (77), 310 (32), 267 (78), 50 (100). Anal.

Calcd. for $C_{26}H_{17}N_7Na_2O_9S_2$ (681.56): C, 45.82; H, 2.51; N, 14.39; S, 9.41. Found: C, 45.25; H, 2.69; N, 15.35; Na, 6.09; S, 8.98, O, 21.60%.

3.2.9. General Procedure for Preparation of Compounds 9 and 10

To a stirred solution of (7; 2.13 g, 0.01 mol) in ethanol (50 mL) containing, sodium acetate (3 g) 8-hydroxy- 6-sulfonaphthalene-2-diazonium chloride and/or 8-hydroxy-3,6-disulfonaphthalene-1-diazonium chloride (prepared by adding sodium nitrite (0.69 g, 0.01 mol) to 6-amino-4-hydroxynaphthalene-2-sulfonic acid and/or 4-amino-5-hydroxynaphthalene-2,7-disulfonic acid (0.01 mol) in conc. HCl (6 mL) at (0°C - 5°C) under stirring) was added drop wise while cooling to (0°C - 5°C) and stirring. The reaction mixture was then left at room temperature for 2 h and the solid product formed was collected by filtration and recrystallized from the appropriate solvents to afford colored products9and 10,respectively.

Violet crystals, yield, 88%; mp >300°C (DMF/acetone), $l_{max}(H_2O)$ 545 nm. IR (KBr, cm^{-1}): v_{max} = 3424(OH), 3320 (NH), 2907 (CH-aliphatic), 2248 (C≡N), 1710 (O-C=O), 1632 cm^{-1} (C=O). ^1H NMR (DMSO-d_6): δ = 7.22 - 8.81 (10H, m, 2Ar-H + coumarin-H_{4-8}), 12.71, 12.80 ppm (2H, 2s, OH, NH). Anal. Calcd. for $C_{22}H_{12}N_3NaO_7S$ (485.40): C, 54.44; H, 2.49; N, 8.66; S, 6.61. Found: C, 54.84; H, 2.56; N, 8.32; S, 6.07%.

Violet crystals, yield, 88%; mp >300°C (DMF/acetone), $l_{max}(H_2O)$ 520 nm. IR (KBr, cm^{-1}): v_{max} = 3486 - 3380 (OH, NH), 2225 (C≡N), 1710 (O-C=O), 1632 cm^{-1} (C=O). MS m/z (%): 587 [M^+] (23), 537 (52), 506 (18), 358 (71), 219 (100). Anal. Calcd. for $C_{22}H_{11}N_3Na_2O_{10}S_2$ (587.45): C, 44.98; H, 1.89; N, 7.15; S, 10.92. Found: C, 45.04; H, 2.02; N, 6.96; S, 11.06%.

3.2.10. General Procedure for Preparation of Compounds 11, 12, 14

Interaction of ethyl 2-oxo-2H-chromene-3-carboxylate(2; 2.18 g, 0.01 mol) with aromatic aryl amine derivatives namely, 2-amino-5-((4-sulfophenyl)diazenyl)-benzenesulfonic acid, benzene-1,4-diamine and/or 3,3'-dime- thoxy-[1,1'-biphenyl]-4,4'-diamine (0.01 mol) in ethanol (20 mL) was heated under reflux for 4 h. The separated solid wazxzs filtered off, washed with ethanol and recrystallized from the appropriate solvents to give the compounds 11, 12 and 14, respectively.

Yellow crystals, yield, 91%; mp >300°C (DMF/acetone), $l_{max}(H_2O)$ 440 nm. IR (KBr, cm^{-1}): v_{max}= 3325 (NH), 3060 (CH-aromatic), 1736 (O-C=O), 1617 (C=O), 1509 cm^{-1} (N=N). MS m/z (%): 573 [M^+] (2), 448 (99), 394 (100), 358 (78), 246 (86). Anal. Calcd. for $C_{22}H_{13}N_3Na_2O_9S_2$(573.46): C, 46.08; H, 2.28; N, 7.33; S, 11.18. Found: C, 45.89; H, 2.12; N, 7.96; S, 11.36%.

Bright orange crystals, yield, 65%; mp 233°C - 235°C (AcOH). IR (KBr, cm^{-1}): ν_{max} = 3458, 3360 (NH_2), 3320 (NH), 3020 (CH-aromatic), 1702 (O-C=O), 1650 cm^{-1} (C=O). MS m/z (%): 280 [M^+] (76), 173 [M^+-$C_6H_7N_2$] (51), 107 [$C_6H_7N_2^+$] (100) (Chart 1). Anal. Calcd. for $C_{16}H_{12}N_2O_3$(280.28): C, 68.56; H, 4.32; N, 9.99. Found: C, 68.25; H, 4.51; N, 10.06%.

Brownish-yellow crystals, yield, 70%; mp 110°C - 112°C (DMF/acetone). IR (KBr, cm^{-1}): ν_{max} = 3400-3270 (NH_2, NH), 3183 (CH-aromatic), 2935 (CH-aliphatic), 1718 (O-C=O), 1658 cm^{-1}(C=O). MS m/z (%): 416 [M^+] (1), 244 (74), 201 (100), 186 (35), 158 (36). Anal. Calcd. for $C_{24}H_{20}N_2O_5$ (416.43): C, 69.22; H, 4.84; N, 6.73. Found: C, 68.97; H, 5.01; N, 6.86%.

3.2.11. Sodium 4-Amino-5-Hydroxy-3-((4-(2-Oxo-2h-Chromene-3-Carboxamido)Phenyl)-Diazenyl) Naphthalene-2,7-Disulfonate (13)

Compounds 12 (2.80 g, 0.01 mol) was suspended with stirring in concentrated H_2SO_4 (3 mL) and cooled to 0°C - 5°C then diazotized by adding $NaNO_2$ (0.9 g, 0.013 mol) drop wise at 0°C - 5°C. After stirring at 0°C for 1h, the diazonium sulphate solution of compound 12 were checked with starch paper for the presence of HNO_2. There were a colour change from white to brown, sulfamic acid was added to destroy excess HNO_2. Coupler compound 4-amino-5-hydroxynaphthalene-2,7-disulfonic acid (3.19 g, 0.01 mol) was dissolved in NaOH solution (3 mL, 20%) at 0°C. The cold solution of diazo compound 12 that were prepared above were added to this solution, keeping the temperature at 0°C - 5°C and maintaining pH 4 by adding AcONa. After stirring for 1 h at 0°C, the dissert 13dyes was collected by filtration then recrystallized from DMF/acetone. Violet crystals, yield, 89%; mp >300°C, l_{max}(H_2O) 570 nm. IR (KBr, cm^{-1}): ν_{max} = 3505 - 3273 (OH, NH_2, NH), 3073 (CH-aromatic), 1700 (O-C=O), 1648 (C=O), 1502 cm^{-1} (N=N). MS m/z (%): 654 [M^+] (4), 486 (5), 280 (71), 173 (50), 107 (100). Anal. Calcd. for $C_{26}H_{16}N_4Na_2O_{10}S_2$ (654.54): C, 47.71; H, 2.46; N, 8.56; S, 9.80. Found: C, 47.29; H, 3.01; N, 8.86; S, 10.16%.

3.2.12. Sodium 4-Amino-3-((3,3'-Dimethoxy-4'-(2-Oxo-2h-Chromene-3-Carboxamido)-[1,1'-Bi- Phenyl]-4-Yl)Diazenyl)-5-Hydroxynaphthalene-2,7-Disulfonate (15)

Compounds 14 (4.16 g, 0.01 mol) was suspended with stirring in concentrated H_2SO_4 (3 mL) and cooled to 0°C - 5°C then diazotized by adding $NaNO_2$ (0.9 g, 0.013 mol) drop wise at 0°C - 5°C. After stirring at 0°C for 1h, the diazonium sulphate solution of compound 14 were checked with starch paper for the presence of HNO_2. There were a colour change from white to brown, sulfamic acid was added to destroy excess HNO_2. Coupler compound 4-amino-5-hydroxynaphthalene-2,7-disulfonic acid (3.19 g, 0.01 mol) was dissolved in NaOH solution (3 mL, 20%) at 0°C. The cold solution of diazo compound 14 that were prepared above were added to this solution, keeping the temperature at 0-5°C and maintaining pH 4 by adding AcONa. After stirring for 1h at 0°C, the dissert dye 15 was collected by filtration then recrystallized from DMF/acetone. Violet crystals, yield, 89%; mp >300°C, $l_{max}(H_2O)$ 405 nm. IR (KBr, cm^{-1}): v_{max} = 3509 - 3380 (OH, NH_2, NH), 1726 (O-C=O), 1661 (C=O), 1449 cm^{-1} (N=N). MS m/z (%): 791 [M^++1] (5), 744 (8), 472 (9), 146 (52), 118 (63), 64 (100). Anal. Calcd. for $C_{34}H_{24}N_4Na_2O_{12}S_2$ (790.68): C, 51.65; H, 3.06; N, 7.09; S, 8.11. Found: C, 51.29; H, 2.96; N, 6.86; S, 7.92%.

3.2.13. General Procedure for Preparation of Compounds 16 - 18

To a mixture of 2-hydroxy-5-((4-nitrophenyl)diazenyl)benzaldehyde (3, 2.71 g, 0.01 mol) and ethyl 3-oxobuta- noate, diethyl malonate and/or malononitrile (0.01 mol) in ethanol (20 mL), glacial AcOH (0.3 mL) and piperidine (0.3 mL) were added under rapid stirring. The reaction mixture was heated under reflux for 5 h and left to cool, the solid formed after cooling filtered off and recrystallized from a suitable solvent to afford the pure products 16-18, respectively.

Brown crystals, yield, 59%; mp 225-227°C (DMF).IR (KBr, cm^{-1}): v_{max} = 3050 (CH-aromatic), 2928 (CH-ali- phatic), 1745 (O-C=O), 1676 (C=O), 1600 (C=C), 1522 (N=N), 1338, 1231 cm^{-1}(NO_2). ^1H NMR (DMSO-d_6): δ = 2.53 (3H, s, CH_3), 7.65(1H, d, coumarin-H_8), 8.08 (2H, d, nitrophenyl-$H_{2,6}$), 8.25 (1H, d, coumarin-H_7), 8.42 (1H, s, coumarin-H_5), 8.59 (2H, d, nitrophenyl-$H_{3,5}$), 8.80 ppm (1H, s, coumarin-H_4). Anal. Calcd. for $C_{17}H_{11}N_3O_5$ (337.29): C, 60.54; H, 3.29; N, 12.46. Found: C, 60.30; H, 3.48; N, 12.28%.

Brown crystals, yield, 56%; mp 245°C - 246°C (DMF). IR (KBr, cm^{-1}): v_{max} = 3097 (CH-aromatic), 2960 (CH-aliphatic), 1741 (coumarin C=O), 1705 (ester C=O), 1521(N=N), 1338, 1251 cm^{-1} (NO_2). ^1H NMR (DMSO-d_6): δ = 1.30 (3H, t, CH_3), 4.30 (2H, q, CH_2), 7.60 (1H, d, coumarin-H_8), 8.03 (2H, d, nitrophenyl-$H_{2,6}$), 8.19 (1H, d, coumarin-H_7), 8.40 (1H, s, coumarin-H_5), 8.52 (2H, d, nitrophenyl-$H_{3,5}$), 8.89 ppm (1H, s, coumarin-H_4). Anal. Calcd. for $C_{18}H_{13}N_3O_6$ (367.31): C, 58.86; H, 3.57; N, 11.44. Found: C, 60.01; H, 3.34; N, 11.68%.

Brown crystals, yield, 56%; mp 250-252°C (dec.) (DMF). IR (KBr, cm^{-1}): v_{max} = 3337 (NH), 2205 (C≡N), 1605 (C=N), 1556 (N=N), 1336, 1246 cm^{-1} (NO_2). MS m/z (%): 321 [M^++2] (11), 319 [M^+] (18), 300 (100), 170 [M^+-$C_6H_3N_3O_2$] (18), 150 [M^+-$C_{10}H_5N_2O$] (14), 106 [$C_7H_6O^+$] (8), 77 [$C_6H_5^+$] (19), 65 [$C_5H_5^+$] (26), 52 [$C_4H_4^+$] (25) (Chart 2). Anal. Calcd. for $C_{16}H_9N_5O_3$ (319.27): C, 60.19; H, 2.84; N, 21.94. Found: C, 59.94; H, 3.01; N, 22.05%.

3.2.14. Sodium 4-((3-Acetyl-2-Oxo-2h-Chromen-6-yl)Diazenyl)Benzenesulfonate (19)

To a mixture of sodium 4-((3-formyl-4-hydroxyphenyl)diazenyl)benzene-sulfonate (4, 3.28 g, 0.01 mol) and ethyl 3-oxobutanoate (1.30 g, 0.01 mol) in ethanol (20 mL), glacial AcOH (0.3 mL) and piperidine (0.3 mL) were added under rapid stirring. The reaction mixture was heated under reflux for 5 h and, after cooling, the solid was filtered and recrystallized from DMF/acetone to afford the pure product 19. Bright yellow crystals, yield, 63%; mp > 360°C (dec.). IR (KBr, cm^{-1}): ν_{max} = 3058 (CH-aromatic), 2922 (CH-aliphatic), 1754 (O-C=O), 1672 (C=O), 1620 (C=C), 1564 (N=N), 1382, 1122 cm^{-1} (SO_2). ^1H NMR (DMSO-d_6): δ = 2.57 (3H, t, CH_3), 7.62(1H, d, coumarin-H_8), 7.78 (2H, d, Ar-$H_{2,6}$), 7.85 (1H, d, coumarin-H_7), 8.19 (1H, s, coumarin-H_5), 8.52 (2H, d, Ar-$H_{3,5}$), 8.80 ppm (1H, s, coumarin-H_4). Anal. Calcd. for $C_{17}H_{11}N_2NaO_6S$ (394.33): C, 51.78; H, 2.81; N, 7.10; S, 8.13. Found: C, 52.01; H, 3.05; N, 6.94; S, 7.95%.

3.2.15. 4,4'-(((Hydrazine-1,2-Diylidenebis(Methanylylidene))Bis(4-Hydroxy-3,1-Phenyl-Ene)) Bis(Diazene-2,1-diyl))Dibenzenesulfonic Acid (20)

A mixture of sodium 4-((3-formyl-4-hydroxyphenyl)diazenyl) benzene-sulfonate (4, 6.56 g, 0.02 mol) and hydrazine hydrate (0.5 g, 0.01 mol) in ethanol (30 mL) were refluxed for 2 h. The reaction mixture was concentrated and left to cool then recrystallized from DMF/acetone to give 20. Bright yellow crystals, yield, 63%; mp 320°C - 322°C (dec.). IR (KBr, cm^{-1}): ν_{max} = 3384 (OH), 3108 (CH-aromatic), 2934 (CH-aliphatic), 1600 (C=N), 1486 (N=N), 1384, 1132 cm^{-1} (SO_2). MS m/z (%): 552 [M^{+}-$2N_2$] (60), 450 [$C_{20}H_{15}N_6O_5S^{+}$] (58), 295 [$C_{14}H_{10}N_6O_2^{+}$] (58), 80 (100) (Chart 3). Anal. Calcd. for $C_{26}H_{20}N_6O_8S_2$ (608.60): C, 51.31; H, 3.31; N, 13.81; S, 10.54. Found: C, 51.61; H, 2.95; N, 14.02; S, 10.15%.

4. DYEING OF COUMARIN COMPOUNDS 8-11

4.1. Dyeing of Coumarin Compounds 8-11 on Wool

4.1.1. Effect of pH and Dyeing Temperature on Percent of Dyeing Exhaustion

Table 1 and Figure 1 show the effect of pH on the exhaustion of acid dyes on wool fibre. The data revealed that at lower dyeing pH values (pH 4), the substantivity of the acid dyes on wool was virtually high. From these results, we can deduce that the dye 9 shows the most exhaustion value because this dye has lower molecular weight than others. The low substantivety of acid dye 8 is due to high molecular weight Although the dye 8 has higher molecular weight than dye 11, it is more substantivety, due to it contains diazo component.

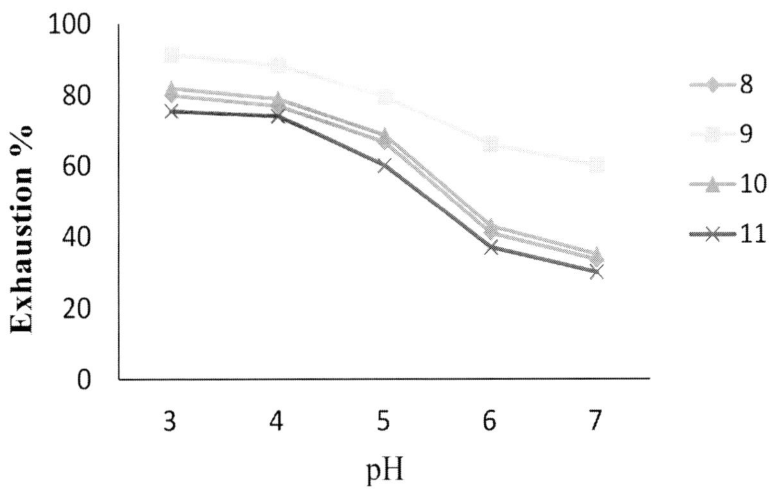

Figure 1. Exhaustion of acid dyes 8-11 on wool at various pH values at 2% o.w.f.

Table 1. Exhaustion of acid dyes 8-11 on wool at various pH values.

Comp. No.	Exhaustion (%)				
	pH 3	pH 4	pH 5	pH 6	pH 7
8	80.0	77.0	66.5	41	33.5
9	91.5	88.5	79.5	66	60.0
10	82.0	79.0	68.5	43	35.0
11	75.5	74.0	60.0	37	30.0

4.1.2. Effect of Dye Concentration on Percent of Dyeing Exhaustion

Figure 2 shows the exhaustion of the dyes on wool using different depth of shades (1% - 5% o.w.f.) at pH 4 and 100°C. Increasing the dye concentration reduces the exhaustion on wool. From these results, we can deduce that the dye 9 exhibits the most exhaustion value while the dye 11 shows the lowest one.

4.1.3. Effect of Dyeing Time on Percent of Dyeing Exhaustion

Figure 3 indicates the variation of dyeing time with temperature at pH 4 and (2% o.w.f.) dye concentration. It is apparent that the dye-fibre reaction is characterized by fast initial rate followed by slower rate, which levels off within the last 30 min of dyeing process. From these results, we can deduce that the dye 9 shows the most exhaustion value while the dye 11 shows the lowest one.

4.2. Dyeing of Coumarin Compounds 8-11 on Silk

4.2.1. Effect of pH on Percent of Dyeing Exhaustion

Table 2 and Figure 4 show the effect of pH on the exhaustion of acid dyes on silk fibre. The data revealed that at lower dyeing pH values (pH 5), the substantivity of the acid dyes on silk is virtually high. From these results, we can deduce the acid dye 9 exhibits the most exhaustion value because this dye has lower molecular weight than others. The low substantivety of acid dye 10 is due to high molecular weight, which decreases substantivety. Although the dye 10 has higher molecular weight than dye 11, it is more substantivety, because it contains diazo component.

4.2.2. Effect of Dye Concentration on Percent of Dyeing Exhaustion

Figure 5 shows the exhaustion of the dyes on silk using different depth of shades (1% - 5% o.w.f.) at pH 5 and 100°C. Increasing the dye concentration reduces the exhaustion on silk. From these results, we can deduce the dye 9 is the most exhaustion value while the dye 11 is the lowest one.

4.2.3. Effect of Dyeing Time on Percent of Dyeing Exhaustion

Figure 6 indicates the variation of dyeing time with temperature at pH 5 and (2% o.w.f.) dye concentration. It is apparent that the dye-fibre reaction is characterized by fast initial rate followed by slower rate, which levels off within the last 30 min of dyeing process. From these results, we can deduce the acid dye 9 shows the most exhaustion value while the dye 11 is the lowest one i.e. increasing the time of dying, this leads to the increasing of the exhaustion values.

4.3. Colorimetric and Fastness Properties for All Dyeings

The colorimetric CIE L * a * b * C * h° data of the dyed wool and silk using the dye are shown in Table 3. The colour parameters were evaluated by means of the Cielab system and the modified CIEL * C * H° (D65/10°) system. The following colour parameters for the dyed samples were obtained by the digital Cielab system: L*— lightness, a*—redness if positive coordinate, or greenness if negative coordinate, b*—yellowness if positive coordinate, or blueness if negative coordinate, C*—chromaticity, Ho—hue of the colour, X—coordinate x, Y—coordinate y, Z—coordinate z [32] . As shown in Table 4, the fastness to washing, rubbing and perspiration of all samples dyed with the dye was excellent to very good irrespective to the fabric used. Chromatic parameters were determined in comparison for samples dyed with all dyes. The light fastness of the dyes was found to depend on the mobility of electrons through conjugated system from conjugated system from thiazole ring with coupler compounds afforded a good value of light fastness. The visible absorption spectra of some dyes showed that colours of dyes in the range are blue-purple, in addition, others in range are yellow-brown. The fastness of dyed fabrics to water, washing, alkaline and acid perspirations and rubbing was found to be very high irrespective of degree of sulphonation in the coupling component.

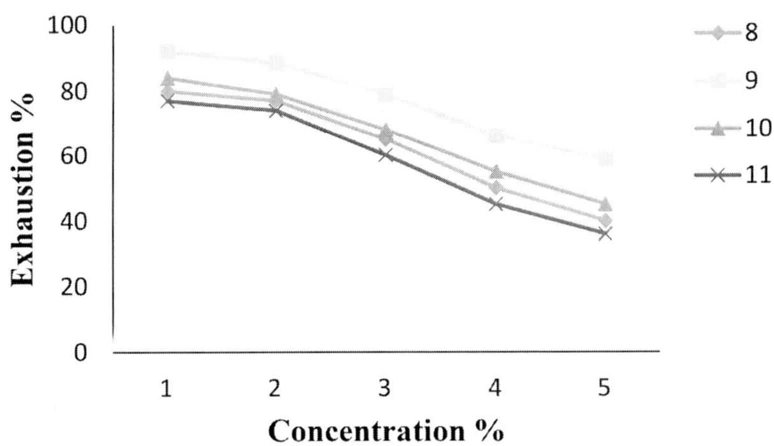

Figure 2. Exhaustion of acid dyes 8-11 on wool at various concentrations at pH 4.

5. CONCLUSION

This article shows that: the synthesis of some novel coumarin compounds is to be used as acid dyes. Dye 9 named sodium 6-(2-(1-cyano-2-oxo-2-(2-oxo-2H-chromen-3-yl)ethylidene)hydrazinyl)-4-hydroxynaphthalene- 2-sulfonate shows the most exhaustion value on wool and silk fibres.

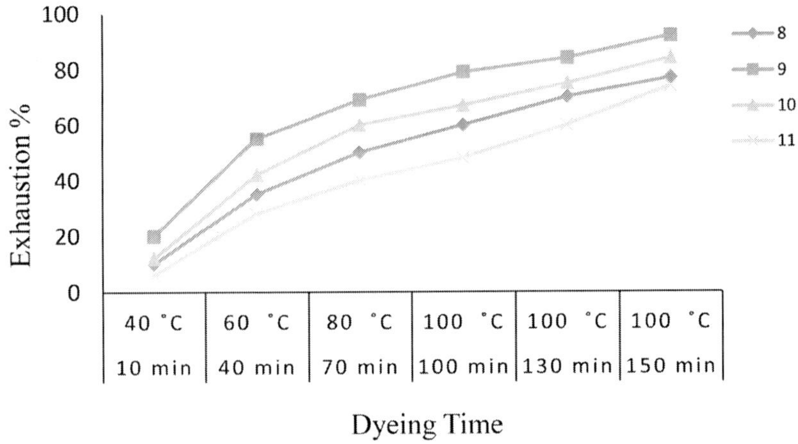

Figure 3. Exhaustion of acid dyes 8-11 on wool at pH 4 (2% o.w.f.) with varying of dyeing time.

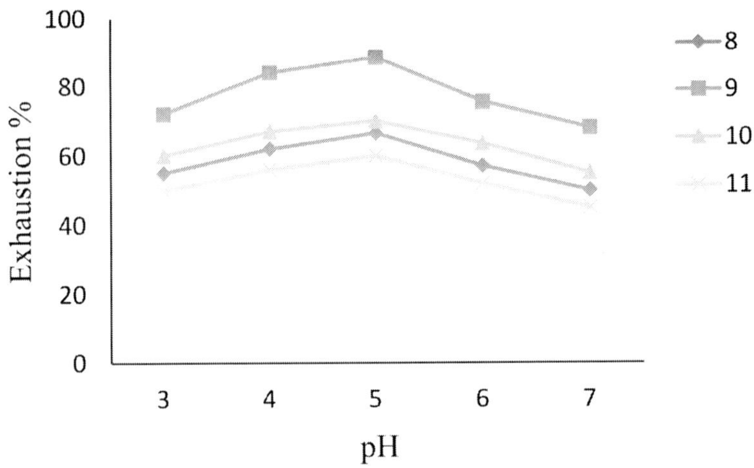

Figure 4. Exhaustion of acid dyes 8-11 on silk at various pH values.

Table 2. Exhaustion of acid dyes 8-11 on silk at various pH values.

Comp. No.	Exhaustion (%)				
	pH 3	pH 4	pH 5	pH 6	pH 7
8	55	62	66.5	57.0	50
9	72	84	88.5	75.5	68
10	60	67	70.0	63.5	55
11	50	56	60.0	52.0	45

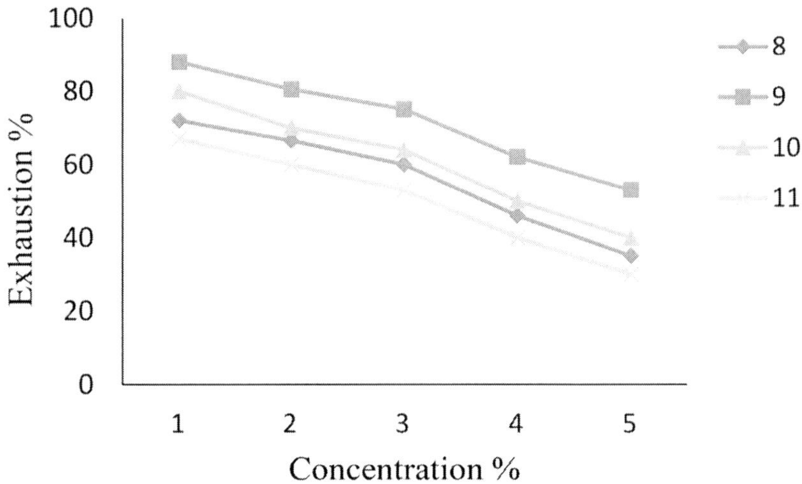

Figure 5. Exhaustion of acid dyes 8-11 on silk at various concentrations at pH 5.

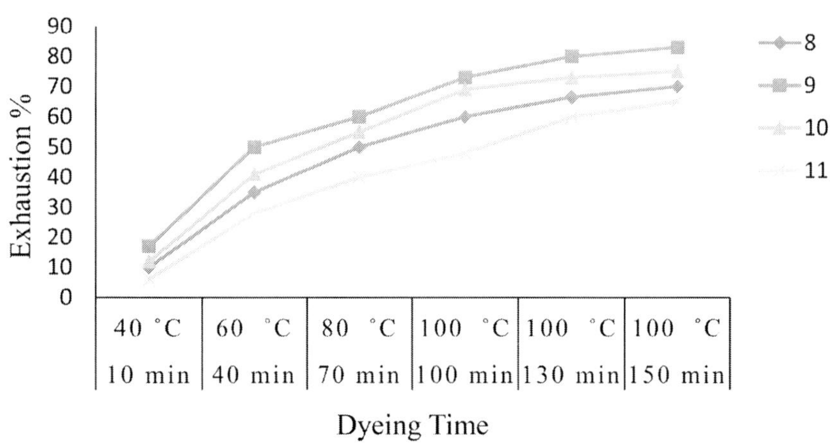

Figure 6. Exhaustion of acid dyes 8-11 on silk at pH 5 (2% o.w.f.) with varying of dyeing time.

Table 3. Colorimetric data of the dyed wool and silk fabrics using acid dyes 8-11 (2% o.w.f.) at 100°C and at pH 4.

Comp. No.	Fabric	K/S	L^*	a^*	b^*	C^*	Ho	E^*
8	Wool	22.67	45.81	45.08	54.94	71.06	50.63	84.55
	Silk	4.73	60.13	31.93	51.68	60.74	58.29	85.47
9	Wool	20.71	19.95	2.790	−9.940	10.32	74.32	22.46
	Silk	8.93	30.13	13.81	−8.150	16.03	30.54	34.13
10	Wool	16.20	23.50	12.10	−4.200	12.80	19.14	26.70
	Silk	4.900	34.00	5.800	−10.10	11.64	60.13	35.93
11	Wool	65.11	22.84	24.10	28.75	37.51	50.02	43.92
	Silk	75.10	−1.140	34.89	1.670	34.92	2.740	34.94

K/S: colour strength; L^*: lightness; C^*: chromaticity; Ho: hue of the colour; E^*: total colour; a^*: redness if positive coordinate, or greenness if negative coordinate; b^*: yellowness if positive coordinate, or blueness if negative coordinate.

Table 4. Fastness properties of dyed wool and silk fabrics using acid dyes 8-11 (2% o.w.f.) at 100°C and at pH 4.

Comp. No.	Fabric	Fastness to Rubbing		Wash Fastness			Fastness to Perspiration						Light Fastness
							Alkaline			Acidic			
		Dry	Wet	Alt	SC	SW	Alt	SC	SW	Alt	SC	SW	
8	Wool	4-5	4-5	4-5	4-5	4-5	4-5	4-5	4-5	4-5	4-5	4-5	5
	Silk	4-5	4-5	4-5	4-5	4-5	4-5	4-5	4-5	4-5	4-5	4-5	4
9	Wool	4-5	4-5	4-5	4-5	4-5	4-5	4-5	4-5	4-5	4-5	4-5	5
	Silk	4-5	4-5	4-5	4-5	4-5	4-5	4-5	4-5	4-5	4-5	4-5	4
10	Wool	4-5	4-5	4-5	4-5	4-5	4-5	4-5	4-5	4-5	4-5	4-5	5
	Silk	4-5	4-5	4-5	4-5	4-5	4-5	4-5	4-5	4-5	4-5	4-5	4
11	Wool	4-5	4-5	4-5	4-5	4-5	4-5	4-5	4-5	4-5	4-5	4-5	5
	Silk	4-5	4-5	4-5	4-5	4-5	4-5	4-5	4-5	4-5	4-5	4-5	4

Alt: colour change of dyed sample; SC: staining on cotton; SW: staining on wool, wash-scale (1-5) and light-scale (1-8).

CITE THIS PAPER

Mahmoud S. Bashandy, Fatma A. Mohamed,, Mohamed M. El-Molla,, Mahmoud B. Sheier, Ahmed H. Bedair (2016) Synthesis of Novel Acid Dyes with Coumarin Moiety and Their Utilization for Dyeing Wool and Silk Fabrics. *Open Journal of Medicinal Chemistry*,**06**,18-35.

REFERENCES

1. AL-Adilee, K.J., Abass, A.K. and Taher, A.M. (2016) Synthesis of Some Transition Metal Complexes with New Heterocyclic Thiazolyl Azo Dye and Their Uses as Sensitizers in Photoreactions. Journal of Molecular Structure, 1108, 378-397.
2. Ozkutuk, M., Ipek, E., Aydiner, B., Mamas, S. and Seferoglu, Z. (2016) Synthesis, Spectroscopic, Thermal and Electrochemical Studies on Thiazolyl Azo Based Disperse Dyes Bearing Coumarin. Journal of Molecular Structure, 1108, 521-532.
3. Rufchahi, E.O.M., Gilani, A.G., Taghvaei, V., Karimi, R. and Ramezanzade, N. (2016) Synthesis, Structural Elucidation, Solvatochromism and Spectroscopic Properties of Some Azo Dyes Derived from 6-Chloro-4-hydroxyquino-line-2(1H)-one. Journal of Molecular Structure, 1108, 623-630.
4. Geng, J., Dai, Y., Qian, H., Wang, N. and Huang, W. (2015) 2-Amino-4-chloro-5-formylthiophene-3-carbonitrile Derived Azo Dyes. Dyes and Pigments, 117, 133-140.
5. Megarajan, S., Ahmed, K.B.A., Reddy, G.R.K., Kumar, P.S. and Anbazhagan, V. (2016) Phytoproteins in Green Leaves as Building Blocks for Photosynthesis of Gold Nanoparticles: An Efficient Electrocatalyst towards the Oxidation of Ascorbic Acid and the Reduction of Hydrogen Peroxide. Journal of Photochemistry and Photobiology B: Biology, 155, 7-12.
6. Raposo, M.M., Castro, M.C., Belsley, M. and Fonseca, A.M. (2011) Push-Pull Bithiophene Azo-Chromophores Bearing Thiazole and Benzothiazole Acceptor Moieties: Synthesis and Evaluation of Their Redox and Nonlinear Optical Properties. Dyes and Pigments, 91, 454-465.
7. Kaur, N., Dhaka, G. and Singh, J. (2015) Simple Naked-Eye Ratiometric and Colorimetric Receptor for Anions Based on Azo Dye Featuring with Benzimidazole Unit. Tetrahedron Letters, 56, 1162-1165.
8. Kaur, N. and Kumar, S. (2011) Colorimetric Metal Ion Sensors. Tetrahedron, 67, 9233-9264.
9. Patel, D.R., Patel, N.B., Patel, B.M. and Patel, K.C. (2014) Synthesis and Dyeing Properties of Some New Monoazo Disperse Dyes Derived from 2-Amino-4-(2',4'-dichlorophenyl)-1,3 Thiazole. Journal of Saudi Chemical Society, 18, 902-913.
10. Mohammadi, A., Khalili, B. and Tahavor, M. (2015) Novel Push-Pull Heterocyclic Azo Disperse Dyes Containing Piperazine Moiety: Synthesis, Spectral Properties, Antioxidant Activity and Dyeing Performance on Polyester Fibers. Spectrochimica Acta Part A: Molecular and Biomolecular Spectroscopy, 150, 799-805.
11. Saylam, A., Seferoglu, Z. and Ertan, N. (2014) Azo-8-hydroxyquinoline Dyes: The Synthesis, Characterizations and Determination of Tautomeric Properties of Some New Phenyl-and Heteroarylazo-8-hydroxyquinolines. Journal of Molecular Liquids, 195, 267-276.
12. Sahoo, J., Mekap, S.K. and Kumar, P.S. (2015) Synthesis, Spectral Characterization of Some New 3-Heteroaryl Azo 4-Hydroxy Coumarin Derivatives and Their Antimicrobial Evaluation. Journal of Taibah University for Science, 9, 187-195.

13. Sunanda, B.P., Krishna, J.J. and Ganapati, S.S. (2013) Greener Protocol for One Pot Synthesis of Coumarin Styryl Dyes. Dyes and Pigments, 97, 105-112.
14. Maryam, A. and Farahnaz, N. (2015) Producing Fluorescent Digital Printing Ink: Investigating the Effect of Type and Amount of Coumarin Derivative Dyes on the Quality of Ink. Journal of Luminescence, 167, 254-260.
15. Simon, B., Azzam, C., Miroslav, M., Danie, T.G. and Denis, J. (2016) Optical Properties of V-Shaped Bis-Coumarins: Ab Initio Insights. Computational and Theoretical Chemistry, 1076, 57-64.
16. Anita, K.S., Kailas, K.S. and Ganapati, S.S. (2015) Synthesis and Photophysical Study of Novel Coumarin Based Styryl Dyes. Dyes and Pigments, 120, 190-199.
17. Krishnaiah, V., Rajesh, K.K., Karnewar, S., Rajeswar, R.V., Srigiridhar, K. and Murali, K.T. (2015) Synthesis, Biological Activity Evaluation and Molecular Docking Studies of Novel Coumarin Substituted Thiazolyl-3-aryl-pyrazole-4-carbaldehydes. Bioorganic & Medicinal Chemistry Letters, 25, 5797-5803.
18. Mahmoud, A.A. and Ali, E. (2015) Bismuth Triflate: A Highly Efficient Catalyst for the Synthesis of Bio-Active Coumarin Compounds via One-Pot Multi-Component Reaction. Chinese Journal of Catalysis, 36, 1124-1130.
19. Raghavendra, U.P., Basanagouda, M., Melavanki, R.M., Fattepur, R.H. and Thipperudrappa, J. (2015) Solvatochromic Studies of Biologically Active Iodinated 4-Aryloxymethyl Coumarins and Estimation of Dipole Moments. Journal of Molecular Liquids, 202, 9-16.
20. Palita, K., Narid, P., Thitiya, S., Ruangchai, T., Supawadee, N., Taweesak, S., Tinnagon, K., Siriporn, J. and Vinich, P. (2015) N-Coumarin Derivatives as Hole-Transporting Emitters for High Efficiency Solution-Processed Pure Green Electroluminescent Devices. Dyes and Pigments, 112, 227-235.
21. Tolga, E., Mustafa, B. and Meryem, C. (2015) Novel Phthalocyanines Bearing 7-Oxy-3-(3,5-difluorophenyl)coumarin Moieties: Synthesis, Characterization, Photophysical and Photochemical Properties. Journal of Photochemistry and Photobiology A: Chemistry, 300, 6-14.
22. He, X.W., Shang, Y.J., Zhou, Y., Yu, Z.Y., Han, G., Jin, W.J. and Chen, J.J. (2015) Synthesis of Coumarin-3-carboxylic Esters via FeCl3-Catalyzed Multicomponent Reaction of Salicylaldehydes, Meldrum's Acid and Alcohols. Tetrahedron, 71, 863-868.
23. Palita, K., Thitiya, S., Narid, P., Supawadee, N., Tinnagon, K., Siriporn, J., Taweesak, S. and Vinich, P. (2014) Synthesis, Characterization, and Properties of Novel Bis(aryl)carbazole-containing N-Coumarin Derivatives. Tetrahedron Letters, 55, 6689-6693.
24. Mei, M.P., Pushparaj, H., Mani, G., Muthiahpillai, P. and Hyun, T.J. (2014) Solvent Free Synthesis of Coumarin Derivative by the Use of AlSBA-1 Molecular Sieves. Journal of Industrial and Engineering Chemistry, 20, 953-960.
25. Margaret, S., Tarik, J.D. and Okenwa, O. (2016) Progress towards Self-Healing Polymers for Composite Structural Applications. Polymer, 83, 260-282.
26. Ashis, M., Kotni, S., Ajoy, B. and Sumanta, B. (2014) Role of Charge Transfer Interaction and the Chemical Physics behind Effective Fulleropyrrolidine/Porphyrin Non-Covalent Interaction in Solution. Spectrochimica Acta Part A: Molecular and Biomolecular Spectroscopy, 121, 559-568.
27. Matthew, V.D., Samer, D., Yi, H. and Ulrich, J.K. (2014) Lanthanide Upconversion Nanoparticles and Applications in Bioassays and Bioimaging. Analytica Chimica Acta, 832, 1-33.
28. Gursoy, A. and Karali, N. (2003) Synthesis, Characterization and Primary Antituberculosis Activity Evaluation of 4-(3-Coumarinyl)-3-benzyl-4-thiazolin-2-one Benzylidenehydrazones. Turkish Journal of Chemistry, 27, 545-551.
29. Bhat, M.A., Siddiqui, N. and Khan, S.A. (2008) Synthesis of Novel 3-(4-Acetyl-5H/methyl-5-substituted phenyl-4,5-dihydro-1,3,4-oxadiazol-2-yl)-2H-chromen-2-ones as Potential Anticonvulsant Agents. Acta Poloniae Pharmaceutica. Drug Research, 65, 235-239.
30. Siddiqui, N., Arshad, M.F. and Khan, S.A. (2009) Synthesis of Some New Coumarin Incorporated Thiazolyl Semicarbazones as Anticonvulsants. Acta Poloniae Pharmaceutica, 66, 161-167.

31. Gomha, S.M. and Khalil, K.D. (2012)A Convenient Ultrasound-Promoted Synthesis of Some New Thiazole Derivatives Bearing a Coumarin Nucleus and Their Cytotoxic Activity. Molecules, 17, 9335-9347.
32. McDonald, R. (1997) Color Physics for Industry. 2nd Edition, Society of Dyes and Colorists, Bradford.

CHAPTER 9

Utility of 2-Methyl-quinazolin-4(3H)-one in the Synthesis of Heterocyclic Compounds with Anticancer Activity

Aisha Youssif Hassan Helali[1], Marwa Taha Mostafa Sarg[2*], Makarem Mohamed Said Koraa[3], Mona Said Fathy El-Zoghbi[2]

[1]*Department of Chemistry, Faculty of Science (Girls), Al-Azhar University, Cairo, Egypt*

[2]*Department of Organic Chemistry, Faculty of Pharmacy (Girls), Al-Azhar University, Cairo, Egypt*

[3]*Department of Organic Chemistry, Faculty of Pharmacy, Cairo University, Giza, Egypt*

ABSTRACT

Quinolino[2,1-b]quinazolines 3 and 4, pyrrolo[2,1-b]quinazoline 5 and various substituted 2-(4- chlorostyryl)quinazoline derivatives including: 4-amino derivative 8, 4-hydrazino derivative 9, thiourea derivative 10, thiosemicarbazide derivative 19, 4-benzylidene hydrazinyl derivative 21, 4-amino thiazolidene derivatives 11, 12, 13, 22, imidazoquinazolines 15, 16, quinazolinium chloride 14, triazino[4,3-c]quinazolines 17, 18, tetrazino[1,6-c]quinazoline 20, 4-amino azetidinyl derivative 23, triazolo[4,3-c]quinazoline 24, 4-amino substituted quinazolines 25, 26, 27, 29 and quinazolino quinazoline 28 were synthesized through different chemical reactions. The obtained compounds were evaluated for their in vitro antitumor activity against HEPG2 and MCF-7 cell lines compared to the reference drug (doxorubicin). Compounds 18, 19, 20, 23 and 24 were found to be the most active against both cell lines exhibiting IC_{50} values ranging from 10.82 - 29.46 μM/L and 7.09 - 31.85 μM/L against Hep-G2 and MCF-7 cell lines, respectively, in addition to docking study of these five compounds against thymidylate synthase and dihydrofolate reductase enzymes active sites.

Keywords: Quinazoline; Anticancer; DHFR; TS

1. INTRODUCTION

Quinazoline-4(3H)-one ring system has been consistently regarded as promising privileged structural icon owing to its pharmacodynamic versatility in many of its synthetic derivatives as well as in several naturally occurring alkaloids isolated from animals, families of plant kingdoms and micro-organisms [1] -[3] .

A systematic perpusal of literature, reveals that quinazoline derivatives encompass a broad spectrum of pharmaceutical activity profile such as antitumor [4] -[10] , anti-HIV [11] -[13] , antimicrobial [14] -[16] , antibacterial, anti-inflammatory [17] -[20] , CNS activity [21] -[23] and cardiovascular activity [24] [25] . The quinazoline nucleus is the scaffold of many antitumor drugs mainly acting as inhibitors of tyrosine kinase receptor (TKR) [26] . Over expression of the receptors has been observed in a number of cancers such as breast, ovarian, colon and prostate cancer. Therefore, blocking of tyrosine kinase activity represents a rational approach to cancer therapy [27] -[34] . Also quinazoline derivatives have a therapeutic potential as an anti-invasive agent with potential activity in early and advanced solid tumors, metastatic bone disease and leukemia [35] . So the quinazoline ring is considered as one of the major classes of benzoheterocyclic compounds that have drawn much attention in the field of cancer chemotherapy besides other pharmacological activities [27] -[29] . Moreover, many 4-quinazoline derivatives revealed potent anticancer activity mediated through the inhibition of dihydrofolate reductase enzyme (DHFR) [30] [31] or through inhibition of chki kinase [32] .

2. RESULTS AND DISCUSSION

2.1. Chemistry

2-Methyl-quinazolin-4-one was synthesized from the fusion of anthranilic acid with thioacetamide [36] which was utilized as a building unit for novel quinazoline and fused quinazolinone compounds "(Scheme 1)". The extent of the pharmacological effect of a quinazolinone derivative depends on the active group which is attached to it. Recently, several scientists have elucidated that in quinazolinone system sites like position 2 and 3, can be suitably modified by the introduction of various heterocyclic moieties to show excellent pharmacological results [37] [38] .

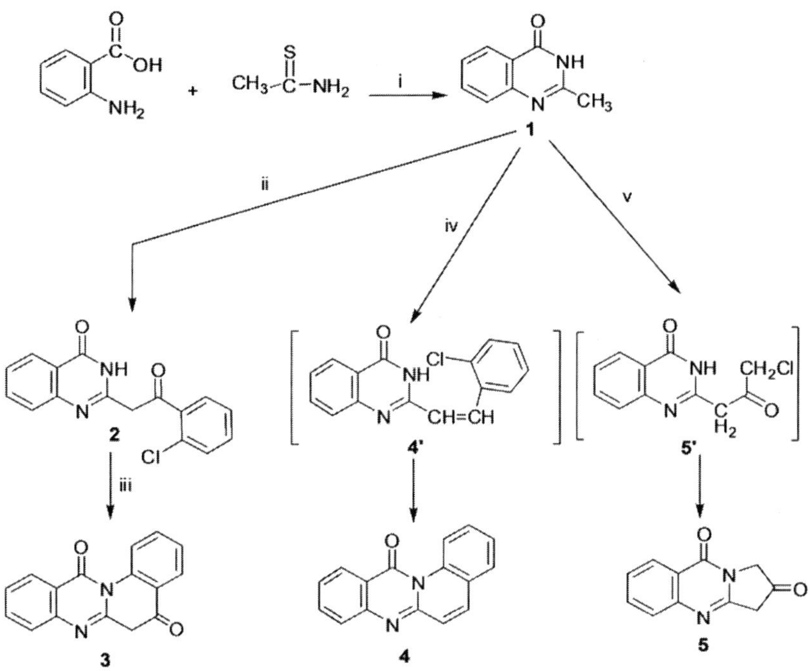

Reagents & conditions: i: Fusion; ii: 2-Cl-C_6H_4COCl/ THF/ 0°C; iii: Reflux; iv: 2-Cl-C_6H_4-CHO/Na Ac/CH_3COOH/Reflux; v: $ClCH_2COCl$/THF/R.T.

Scheme 1. Synthetic pathways for compounds 1, 2, 3, 4, and 5.

A known procedure for building up such heterocyclic system is based on cyclization of o-chlorobenzoyl chloride with 2-methyl-quinazolin-4(3H)-one 1 in dry THF. The reaction stopped at the stage of formation of 2-[2- (2-

chlorophenyl)-2-oxoethyl]quinazolin-4(3H)-one 2. This could be explained in the first place by steric hindrance of the o-substituent for optimal position of the reacting groups. We attempted to decrease the effect of the latter by heating compound 2 in THF for 2h. However under this reaction condition, cyclization occurred readily to yield 5H-quinolino[2,1-b]quinazolin-5, 12(6H)-dione 3.

The ^{1}H-NMR spectrum of compound 2 revealed a singlet at δ 4.33 ppm attributed to CH_2 protons and a deuterium oxide exchangeable singlet at δ 7.73 ppm due to the NH proton. While, ^{1}H-NMR spectrum of compound 3 showed a singlet at δ 2.49 ppm assigned to CH_2 protons and lacked any deuterium oxide exchangeable signals of its precursor.

Attempts to carry out a similar conversion of compound 1 into 4 by refluxing in THF are unsuccessful. However, the cyclic quinolono [2,1-b]quinazoline 4 was obtained through the reaction of compound 1 with 2-chlorobenzaldehyde in glacial acetic acid/sodium acetate. In this case, isolation of the intermediate 4' was difficult because of its high reactivity which led directly to the cyclic product 4.

Pyrrolo[2,1-b]quinazolindione derivative 5 was prepared by the reaction of compound 1 with chloroacetyl chloride in dry THF. The reaction is assumed to proceed through chloroacetylation of compound 1 at the 2-position followed by intramolecular cyclization on N_3-NH function with elimination of a hydrochloride molecule to give the target compound 5. ^{1}H-NMR spectrum of compound 5 showed two singlets at δ 2.08, 2.49 ppm each integrated for two protons attributed to two pyrrolo methylene protons.

Our attention was drawn to the earlier discovery by scientists that modification of methyl group present at the second position of 4(3H)-quinazolinone into other chemical intities yielded structural analogues with appreciable biological activity [39] [40] "(Scheme 2)". Therefore, compound 1 was condensed with 4-chlorobenzaldehyde in glacial acetic acid/sodium acetate to afford the chalcone derivative 6 [41] . Chlorination of compound 6 with phosphorous oxychloride was reported to give [42] poor yields of impure chlorinated product. However, the addition of triethylamine to the reaction mixture accelerated the reaction speed [43] and afforded 4-chloroquinazoline derivative 7 [41] . Furthermore, amination of compound 7 was carried out by refluxing in pyridine and ammonium acetate to afford the corresponding 4-aminoquinazoline derivative 8. Moreover, hydrazinolysis of compound 7 was accomplished by the use of hydrazine hydrate 99% in ethanol to afford 4-hydrazinylquinazoline derivative 9. The ^{1}H-NMR spectrum of compound 9 showed two deuterium oxide exchangeable singlets at δ 4.15 and 12.26 ppm due to NH_2 and NH protons; respectively.

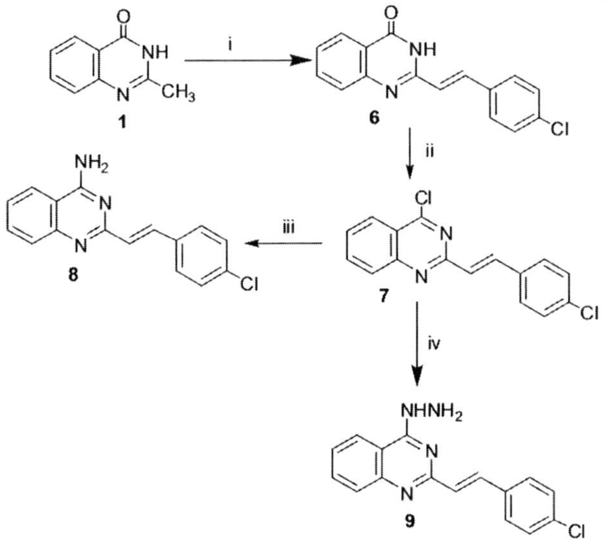

Reagents & conditions: i 4-Cl-C_6H_4-CHO/ NaAc/ CH_3COOH/Reflux; ii: $POCl_3$/T.E.A./Reflux; iii: CH_3COONH_4/pyridine/Reflux; iv: NH_2NH_2/ EtOH/Reflux.

Scheme 2. Synthetic pathways for compounds 6, 7, 8 and 9.

The work was extended to shed more light on the activity and synthetic potential of the amino group in compound 8 "(Scheme 3)". Thus, compound 8 was reacted with phenyl isothiocyanate in boiling pyridine to give the corresponding thiourea derivative 10. The ^{1}H-NMR spectrum of compound 10 revealed two deuterium oxide exchangeable singlets at δ 7.96 and 12.13 ppm due to thiourea N_3 and N_1-NH protons; respectively. Furthermore, the thiourea derivative was used as a good starting material for preparation of several polyfunctional thiazolyl derivatives with expected biological activities as the thiazolyl moieties were reported to possess a wide range of pharmacological actions especially anti cancer activity [44] . Therefore, compound 10 was reacted with ethyl chloroacetate as well as phenacyl chloride in boiling ethanol/sodium acetate to yield 3-phenylthiazolidin-4-one derivative 11 and 3,4-diphenylthiazolylidene derivative 12; respectively. It was also fused with chloroacetone to give 4-methyl-3-phenyl thiazol-ylidene dreivative 13.

The reaction mechanisms are expected to proceed via elimination of hydrochloride molecules followed by cyclization through ethanol elimination or dehydration. The ^{1}H-NMR spectrum of compound 11 revealed a singlet at δ 3.46 ppm due to SCH_2 protons. The ^{1}H-NMR spectrum of compounds 12 revealed a singlet at δ 7.49 ppm due to CH-thiazole proton.

However, tricyclic systems in which the quinazoline ring was condensed with imidazole ring have not been adequately studied "(Scheme 4)". Therefore our goal was directed to synthesize imidazoquinazoline derivatives through condensation of compound 8 with phenacyl chloride in ethanol, which afforded quinazolin-3-ium chloride derivative 14 and imidazo[1,2-c]quinazoline derivative 15 according to the mechanism suggested in the literature [45] . The first step of this process involved the addition of a phenacyl chloride molecule on N-3 position to form the corresponding quinazolinium derivative 14 which underwent cyclization under the same experimental conditions to give the imidazoquinazoline derivative 15. The ^{1}H-NMR spectrum of compound 14 showed a singlet at δ 3.63 ppm due to CH_2protons, in addition to two deuterium oxide exchangeable singlets at δ 5.18 and 12.33 ppm due to NH_2 protons and OH tautomer proton; respectively. While the ^{1}H-NMR spectrum of compound 15 revealed a singlet integrated for one proton at δ 7.96 ppm due to CH-imidazole.

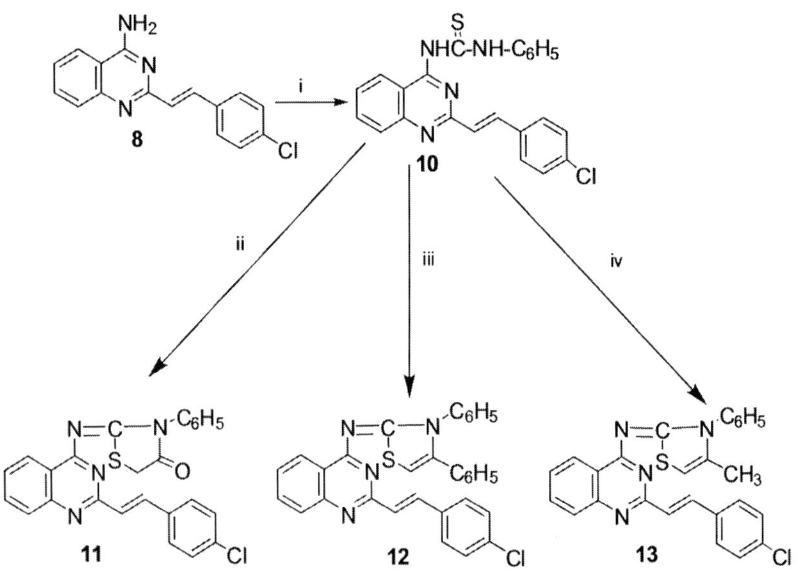

Reagents & conditions: i: C_6H_5NCS/pyridine/Reflux; ii: $ClCH_2COOC_2H_5$/NaAc/EtOH/Reflux; iii: $C_6H_5COCH_2Cl$/ NaAc/EtOH/Reflux; iv:$ClCH_2COCH_3$/Fusion

Scheme 3. Synthetic pathways for compounds 10, 11, 12 and 13.

Moreover, acylation of the amino group of compound 8 with chloroacetyl chloride under anhydrous conditions which underwent intramolecular cyclization on the ring to yield the cyclized product imidazo[1,2-

c]quinazolinone derivative 16. In the case of chloroacetyl chloride, the isolation of the intermediate 16' was difficult due to its high reactivity which led directly to the cyclic product 16. The ^{1}H-NMR spectrum of compound 16 revealed a singlet at δ 3.55 ppm due to CH_2 protons of imidazole moiety and another singlet at δ 7.50 ppm attributed to the imidazole = CH-tautomer. In addition to a deuterium oxide exchangeable singlet at δ 8.51 ppm corresponding to OH tautomer.

Recently, several biological activities have been reported for [1,2,4]triazino[4,3-c]quinazolines [46] , so our aim was extended to the development of different synthetic methods for novel [1,2,4]triazino[4,3-c]quinazolines 17 and 18 and investigation of their cytotoxicity and antitumor activity "(Scheme 5)". Therefore, condensation of compound 9 with benzoin and with chloroacetyl chloride were applied to furnish the target compounds 17 and 18; respectively. The ^{1}H-NMR spectrum of compound 17 revealed a singlet at δ 6.02 ppm due to the triazine C_5-proton. While, the ^{1}H-NMR spectrum of compound 18 showed a singlet at δ 4.25 ppm due to the triazine CH_2protons, in addition to a deuterium oxide exchangeable singlet at δ 12.20 ppm attributed to NH proton. Furthermore, from our survey in the literature, it was found that, a new series of quinazoline compounds was designed in such a way to accommodate the isothiocyanates, Schiff's bases and amine bridges at positions 2, 3 or 4 of the quinazoline ring. Isothiocyanate, Schiff's base and amide functions are known to contribute in the enhancement of the antitumor activity [4] . The reaction of phenyl isothiocyanate with compound 9 afforded hydrazine carbothioamide derivative 19. The ^{1}H-NMR spectrum of compound 19 showed a deuterium oxide exchangeable singlet at δ 3.98 ppm due to two NH protons that are attached to C=S function and a deuterium oxide exchangeable singlet at δ 11.08 ppm due to NH proton attached to the quinazoline ring which is more deshielded by the nitrogens of the quinazoline ring. The thiosemicarbazide derivative 19 underwent oxidative cyclization by stirring with bromine to yield [1,2,4,5]tetrazino[1,6-c]quinazoline thione derivative 20. The ^{1}H-NMR spectrum of compound 20 showed a deuterium oxide exchangeable singlet at δ 4.00 ppm due to NH proton.

Moreover, the reaction of compound 9 with 4-chlorobenzaldehyde in presence of piperidine furnished the Schiff's base 21 "(Scheme 6)". The Schiff's base 21 was used as a good starting material for preparation of several quinazoline derivatives. Among which its reaction with thioglycolic acid in presence of anhydrous zinc chloride to furnish the thiazolidinone derivative 22. Furthermore, the reaction of the Schiff's base 21 with chloroacetyl chloride in presence of triethylamine as a base afforded azetidinone derivative 23. The ^{1}H-NMR spectrum of compound 23 revealed two doublets at δ 4.06 and 4.30 ppm attributed to CH-C_6H_4-Cl and CH-Cl of the azetidine ring; respectively. In addition to a deuterium oxide exchangeable singlet at δ 10.68 ppm attributed to NH proton. Moreover, a group of Slovak's scientists reported that [1,2,4]triazolo[4,3-c]quinazoline derivatives possess considerable in vitro antitumor activity [46] . So we synthesized [1,2,4]triazolo[4,3-c]quinazoline compound 24 through the cyclization of the Schiff's base 21 in glacial acetic acid/sodium acetate.

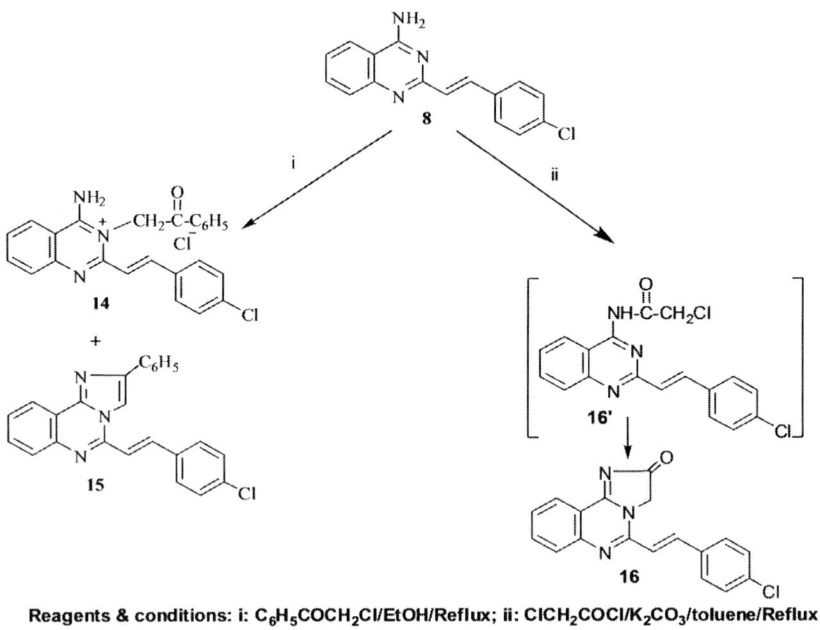

Scheme 4. Synthetic pathways for compounds 14, 15 and 16.

Finally, our work was extended to react 4-chloroquinazoline derivative 7 with several primary and secondary amines preparing a series of promising derivatives in order to compare their biological activities "(Scheme 7)". Focusing on setting up a simple, cheap, fast and efficient synthesis; the reaction of compound 7 with the primary amine, 2-aminoethanol was accomplished by fusion with excess amine reducing as much as possible both reaction time and temperature so as to preserve the styryl group from reacting. This rapid reaction protocol resulted in the formation of the secondary aminoquinazoline derivative 25 in good yields. The ^{1}H-NMR spectrum of compound 25 showed a multiplet at δ 3.66 - 3.71 ppm integrated for four protons corresponding to two CH_2 protons and two deuterium oxide exchangeable singlets at δ 4.40 and δ 12.25 ppm attributed to OH and NH protons, respectively. Moreover, the substrate 7 was also reacted very easily with secondary amines such as dimethylamine and pyrrole by the same manner as aminoethanol to yield tertiary amino quinazolines 26 and 27; respectively in very good yields.

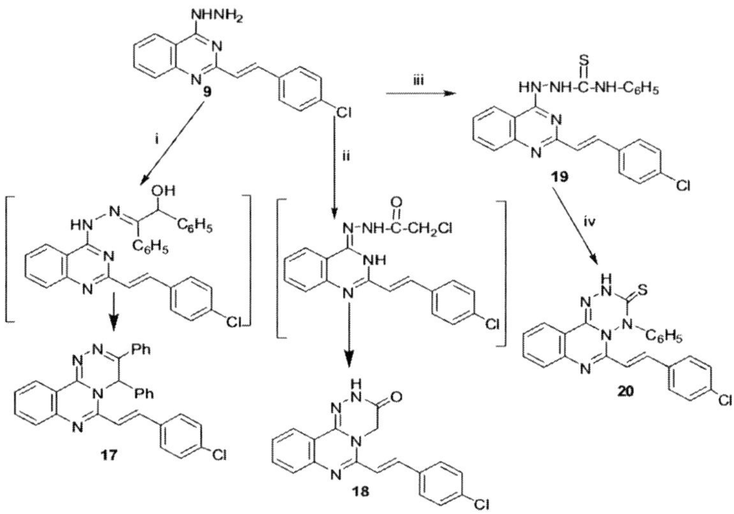

Reagents & conditions: i: Benzoin/EtOH/Reflux; ii: $ClCH_2COCl$/DMF/Reflux; iii: C_6H_5NCS/MeOH/Reflux; iv: Br_2/CCl_4/R.T.

Scheme 5. Synthetic pathways for compounds 17, 18, 19 and 20.

Furthermore, compound 7 was reacted with anthranilic acid in glacial acetic acid to afford the polycyclic quinazoline derivative 28 while hot, while the reaction filtrate yielded the open chain secondary aminoquinazoline derivative 29 that bears a free carboxylic group which could possibly alter the biological properties including cytotoxic activity [47] . However, the reaction is expected to proceed through formation of the secondary aminobenzoic acid derivative 29 first which carried out intramolecular dehydration reaction leading to the cyclic quinazolinoquinazoline derivative 28. The ^{1}H-NMR spectrum of compound 29 showed two deuterium oxide exchangeable singlets at δ 8.76 and δ 12.18 ppm attributed to NH and carboxylic OH protons; respectively. While the ^{1}H-NMR spectrum of compound 28 lacked any deuterium oxide exchangeable singlets due to NH and OH protons.

2.2. Biology

The six dose growth inhibition percent and the IC_{50} values of the tested compounds (2 - 29) against liver HepG2 and breast MCF-7 cell lines are represented in Tables 1 and 2; respectively.

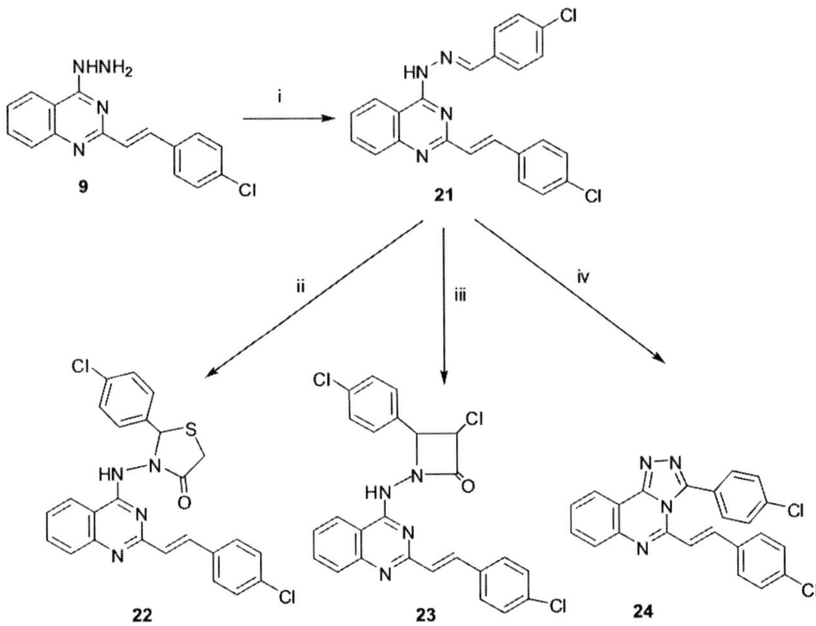

Reagents & conditions: i: 4-Cl-C_6H_4CHO/EtOH/piperidine/Reflux;
ii: $HSCH_2COOH/ZnCl_2$/benzene//Reflux;
iii: $ClCH_2COCl$/T.E.A./DMF/Reflux; iv: CH_3COOH/NaAc/Reflux.

Scheme 6. Synthetic pathways for compounds 21, 22, 23 and 24.

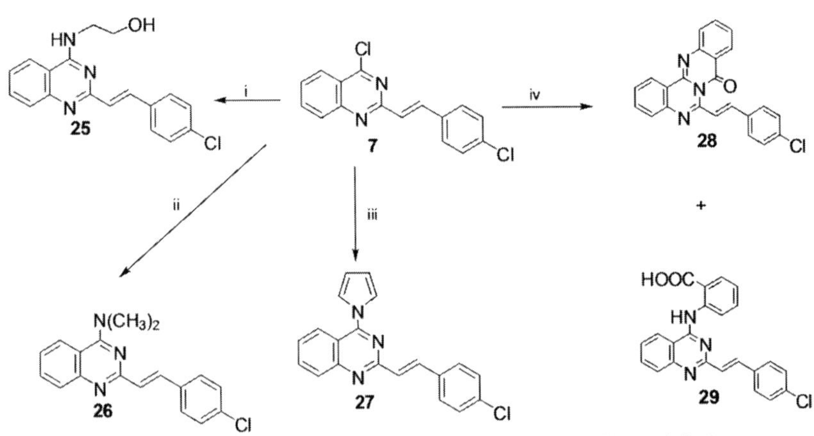

Reagents & conditions: i: $H_2NCH_2CH_2OH$/Fusion; ii: $(CH_3)_2NH$/Fusion; iii: pyrrole/Fusion;
iv: Anthranilic acid/CH_3COOH/Fusion

Scheme 7. Synthetic pathways for compounds 25, 26, 27, 28 and 29.

Table 1. Six dose growth inhibition percent and IC_{50} values of the test compounds against HepG2 cell line.

Sample conc. (μg/mL.) Compound No.	Growth inhibition % 50	25	12.5	6.25	3.13	1.6	IC_{50} (μM/L)
2	81.73	65.02	39.08	21.46	10.59	3.20	59.73
3	64.81	56.58	43.37	34.70	23.03	11.54	71.76
4	73.61	57.42	40.81	33.50	24.62	15.84	78.86
5	57.48	48.74	35.82	27.14	20.38	15.55	>100
6	87.08	66.57	41.14	25.41	9.19	1.38	59.93
7	85.04	72.69	50.38	28.67	9.75	1.27	41.33
8	83.71	61.58	33.25	12.68	4.22	0.00	70.82
9	83.46	52.65	33.96	17.07	8.26	2.11	78.38
10	90.38	87.53	77.16	47.12	31.39	12.58	16.35
11	75.41	53.45	42.52	30.35	16.69	8.86	46.27
12	71.28	58.09	26.91	18.02	11.68	4.86	42.25
13	72.66	39.07	18.28	9.13	1.84	0.00	72.91
14	74.81	37.46	17.07	6.94	3.12	0.00	76.78
15	88.97	77.44	61.57	45.44	26.13	14.39	20.99
16	84.36	50.68	21.46	8.30	2.17	0.00	76.01
17	83.76	73.57	39.05	30.43	20.19	13.94	34.96
18	83.65	70.57	59.45	27.08	10.52	4.84	31.85
19	92.11	86.69	77.52	38.72	20.64	10.36	18.79
20	84.07	73.89	53.36	29.14	16.87	7.53	27.04
21	85.88	74.64	48.76	31.68	16.04	7.46	31.34
22	80.15	66.32	28.25	16.48	5.81	2.42	39.84
23	91.03	83.09	74.41	60.58	48.53	25.08	7.09
24	92.14	86.26	78.62	63.85	40.19	23.47	10.58
25	73.37	46.82	30.46	16.22	5.85	1.27	86.15
26	67.52	55.94	26.38	20.59	13.84	6.39	72.81
27	72.18	38.42	14.26	3.52	1.04	0.00	>100
28	70.26	31.38	20.41	12.84	4.78	1.06	96.61
29	91.26	80.92	66.71	43.36	21.89	10.33	19.95
Doxorubicin	89.05	85.71	83.10	78.97	69.68	51.75	8.55

The thiourea derivative 10 exhibited strong anticancer activity against liver cancer cell line about half the activity of doxorubicin (IC_{50} = 8.55 μM/L) with IC_{50} 16.35 μM/L. However, it showed moderate activity against breast cancer MCF-7 with IC_{50} 68.75 μM/L. Fusion of [1,2,4]triazine or [1,2,4,5]tetrazine to the quinazoline back bone as in compounds 18 and 20 resulted in marked increase in activity against both HepG2 and MCF-7 cell lines exerting IC_{50} 31.85 and 27.04 μM/L against HepG2 cell line and 29.46 and 20.98 μM/L against MCF-7 cell line; respectively. It is to be noted that, the presence of a phenyl thiosemicarbazide side chain attached to the C_4-quinazoline ring as in compound 19 exhibited marked anticancer activity against both HepG2 and MCF-7 cell lines exerting IC_{50} 18.79 and 13.46 μM/L; respectively. However, incorporating N-azetidinone- 4-amino moiety to the quinazoline back bone as in compound 23 resulted in improvement in activity against both HepG2 and MCF-7 cell lines which exhibited slightly better anticancer activity against HepG2 (IC_{50} 7.09 µM/L) than the reference drug doxorubicin (IC_{50} 8.55 µM/L). While, it showed comparable activity to that of doxorubicin (IC_{50} 8.90 µM/L) against MCF-7 exerting IC_{50} 11.94 µM/L.

Table 2. Six dose growth inhibition percent and IC_{50} values of the test compounds against MCF-7 cell line.

Sample conc. (µg/mL)	Growth inhibition %						IC_{50} (µM/L)
Compound No.	**50**	**25**	**12.5**	**6.25**	**3.13**	**1.6**	
2	69.06	52.75	30.68	16.86	7.14	1.53	78.52
3	58.28	47.59	30.12	21.88	10.66	4.22	>100
4	62.47	46.84	37.45	25.22	14.08	8.56	>100
5	53.68	40.22	28.46	13.64	5.78	1.87	>100
6	72.05	45.17	21.46	8.28	2.62	100.00	>100
7	71.61	50.54	36.18	27.03	20.66	9.49	81.66
8	71.41	50.76	36.27	14.04	5.18	1.85	86.48
9	74.68	48.14	29.06	10.44	3.82	0.00	90.54
10	76.22	45.61	30.19	12.66	4.78	1.83	68.75
11	70.92	47.86	31.05	21.84	10.47	5.62	59.87
12	75.42	50.25	27.47	18.64	9.43	3.76	48.26
13	65.02	37.83	20.46	8.22	2.68	0.00	79.96
14	68.66	33.21	9.35	2.86	0.00	0.00	84.59
15	79.02	65.77	41.36	20.85	10.28	3.54	44.37
16	84.06	63.33	37.77	20.52	12.47	5.81	57.63
17	62.71	53.62	37.46	18.81	6.17	2.35	47.03
18	86.18	71.67	58.64	37.86	28.08	12.14	29.46
19	84.06	73.17	60.24	52.42	36.83	11.66	13.46
20	89.46	74.83	61.25	41.18	30.19	11.27	20.98
21	72.82	60.69	63.29	24.16	17.25	10.88	46.41
22	82.66	70.58	51.37	21.07	9.14	1.87	24.80
23	89.14	72.82	61.28	52.69	30.26	10.50	11.94
24	93.03	81.77	72.32	59.05	43.28	28.46	10.82
25	68.46	42.54	28.79	14.28	5.32	1.87	99.08
26	59.46	47.67	38.31	25.22	13.96	7.64	96.76
27	61.06	36.97	15.82	3.15	0.00	0.00	>100
28	65.06	51.89	32.54	18.76	9.47	2.62	62.14
29	78.22	61.86	45.08	19.22	10.47	3.84	40.40
Doxorubicin	90.76	88.45	84.26	77.78	70.82	55.16	8.90

Furthermore, fusion of triazole ring to quinazoline nucleus as in compound 24 resulted in marked increase in activity against both HepG2 and MCF-7 cell lines exerting IC_{50} 10.58 and 10.82 µM/L; respectively which represent a strong anticancer agent against these two cell lines (Figures 1 and 2).

Moreover, the presence of 2-amino benzoic acid moiety on the C_4-quinazoline ring as in compound 29 exhibited remarkable increase in activity towards HepG2 cell line showing IC_{50} 19.95 μM/L.

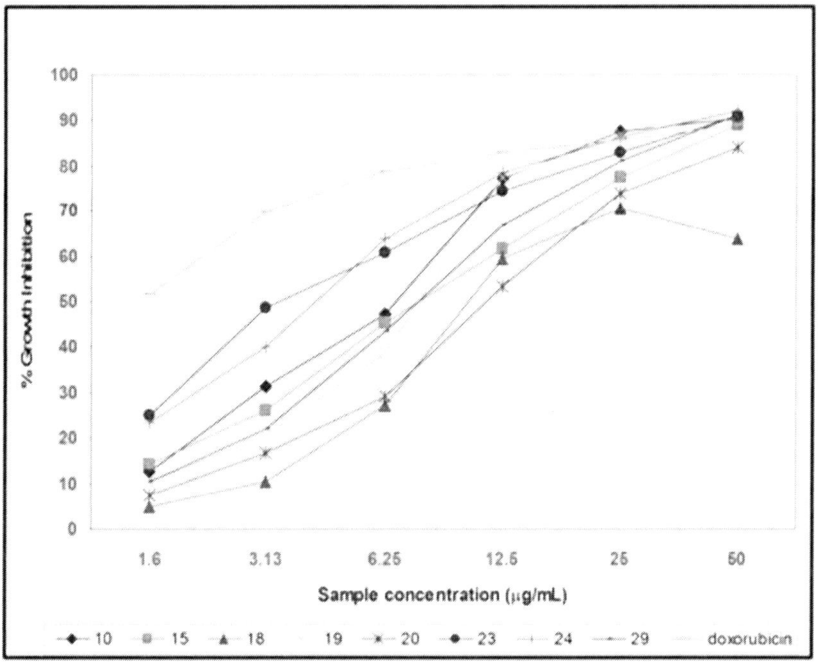

Figure 1. Growth inhibition curves of most active compounds against HepG2 cell line.

2.3. Computer Aided Docking

Five representative active anticancer compounds 18, 19, 20, 23 and 24 were subjected to docking using Molecular Operating Environment (MOE) program [48] on the 3D structure of two enzymes, namely; thymidylate synthase (TS) and dihydrofolate reductase (DHFR). The study also shows the enzyme interactions with their substrates; 2'-deoxyuridine-5'-monophosphate (DUMP) and methotrexate (MTX); respectively.

2.3.1. Docking on the Active Site of Thymidylate Synthase

MOE docking studies of the inhibitors were performed using thymidylate synthase co-crystallized with the substrate DUMP(PDB ID: IBID) as a template.

1) Docking of DUMP into TS active site revealed that:

Several interactions were considered to be responsible for the observed affinity of the compound. As it acts as a hydrogen bond acceptor with the backbone Asp 169 residue and the side chain residues; Arg 166, Ser 167, Asn 177, His 207 and Tyr 209 as well as hydrogen bond donor with the side chain residues; Asn 177, His 207 and Tyr 209. In addition to hydrophobic interactions with: Arg 21, Cys 146, His 147, Gln 165, Arg 166, Ser 167, Cys 168, Asp 169, Gly 173, Asn 177, His 207 and Tyr 209 as shown in Figure 3(a).

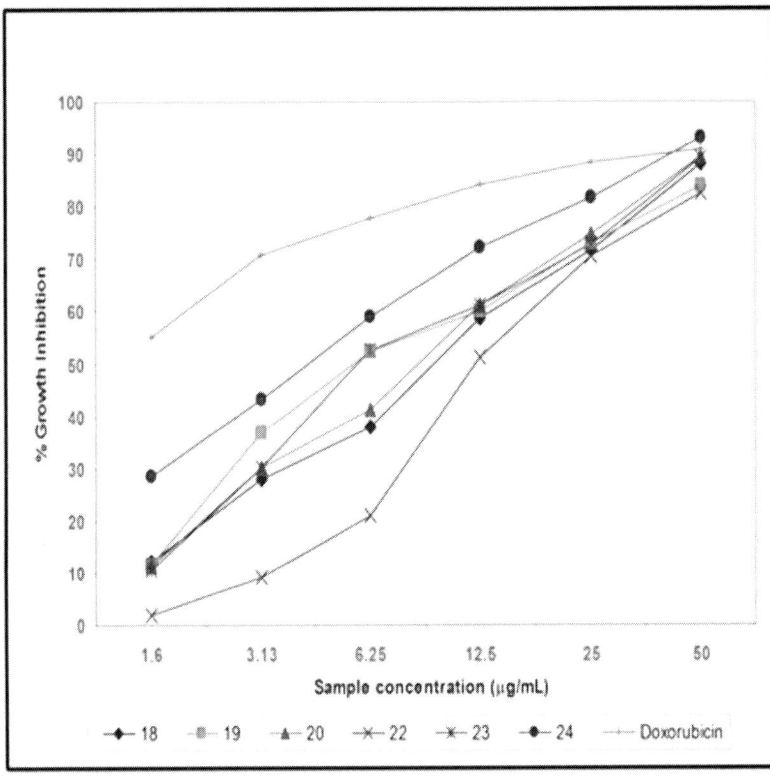

Figure 2. Growth inhibition curves of most active compounds against MCF-7 cell line.

2) Docking of compound 18 into TS active site revealed:

The presence of hydrophobic interactions between the chlorine atom and the styryl moiety as well as the triazine ring C_5 carbon and C_6 carbonyl function with the following amino acid residues: Glu 87, Ile 108, Trp 109, Asn 112, His 196, Leu 221, Phe 225, Asn 226, Met 311 and Ump 314 as shown in Figure 3(b).

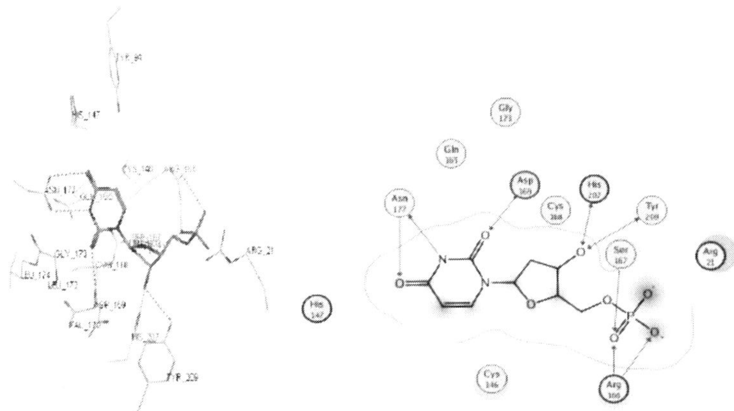

(a)

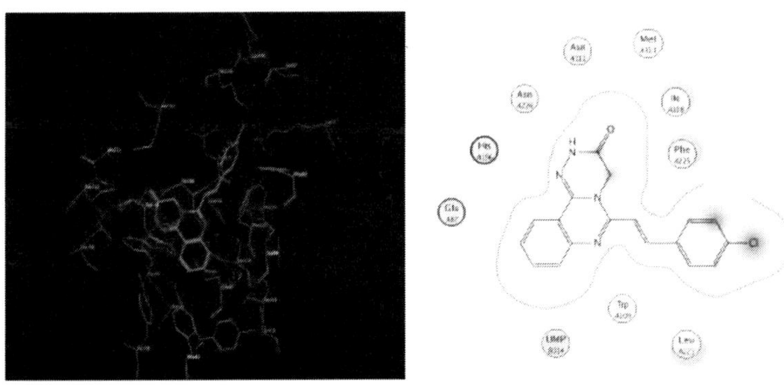

(b)

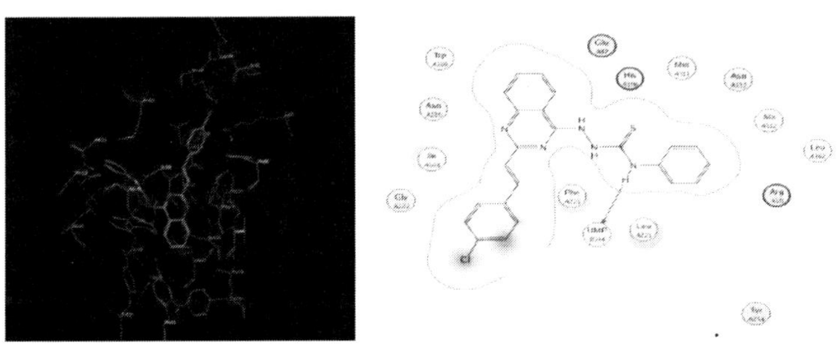

(c)

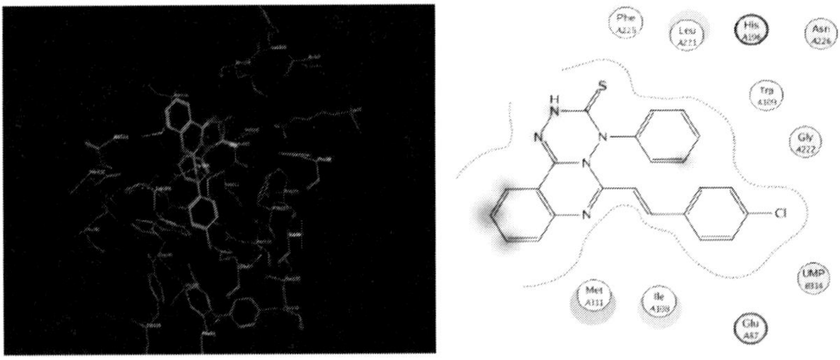

(d)

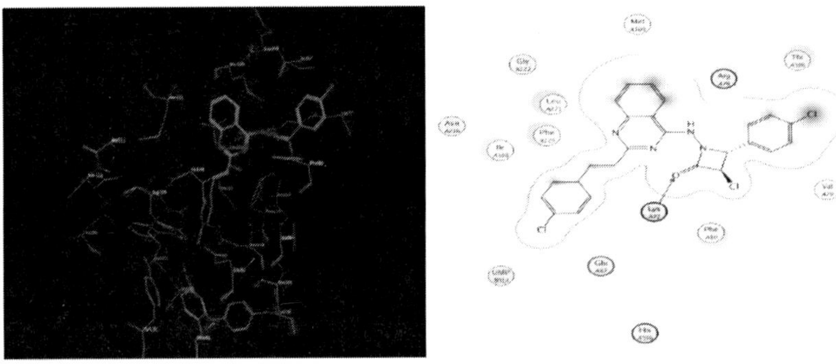

(e)

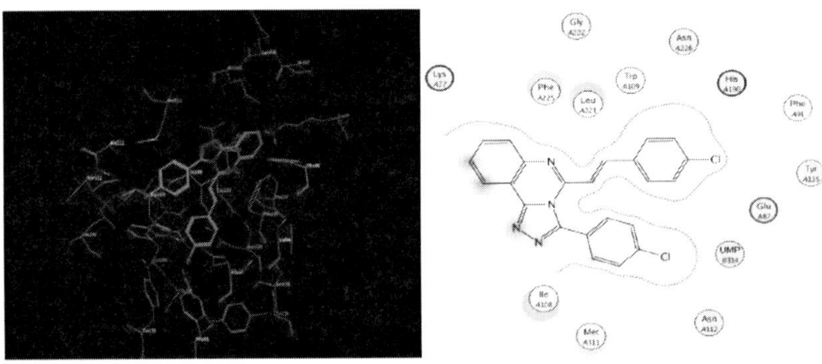

(f)

Figure 3. (a) Docking of DUMP in the active site of TS; (b) Docking of compound 18 in the active site of TS; (c) Docking of compound 19 in the active site of TS; (d) Docking of compound 20 in the active site of TS; (e) Docking of compound 23 in the active site of TS; (f) Docking of compound 24 in the active site of TS.

3) Docking of compound 19 into TS active site revealed:

The interaction of the NH group that is attached to the phenyl ring as a hydrogen bond donor with the side chain residue Ump 314 (2.89 Å) at a strength of 4.7%. This beside many hydrophobic interactions concerning the chlorine atom, styryl moiety and C=S function with the following amino acid residues: Arg 50, Glu 87, Ile 108, Trp 109, Asn 112, Leu 192, His 196, Leu 221, Gly 222, Phe 225, Asn 226, Tyr 258, Met 311, Ala 312 and Ump 314 as shown in Figure 3(c).

4) Docking of compound 20 into TS active site:

Displayed many hydrophobic interactions between quinazoline-$C_{5,6,7}$ carbons, tetrazine NH, tetrazine N_2 and C=S function, in addition to phenyl ring with many amino acid residues which are; Glu 87, Ile 108, Trp 109, His 196, Leu 221, Gly 222, Phe 225, Asn 226, Met 311 and Ump 314 as shown in Figure 3(d).

5) Docking of compound 23 into TS active site illustrated:

The interaction of carbonyl oxygen atom as a hydrogen bond acceptor with the side chain residue Lys77 (2.90 Å) at a strength of 46.9% as well as hydrophobic interactions involving azetidine-C_4-4-chlorophenyl moiety, styryl moiety, quinazoline-C_4-NH function and quinazoline-$C_{5,6,7,8}$ carbons with the following amino acid residues: Lys 77, Arg 78, Val 79, Phe 80, Glu 87, Ile 108, His 196, Leu 221, Gly 222, Phe 225, Asn 226, Thr 306, Met 309 and Ump 314 as shown in Figure 3(e).

6) Docking of compound 24 into TS active site showed:

Hydrophobic interactions between triazole-N_1 and N_2, quinazoline-C_5 and C_6, triazole-C_3-4-chlorophenyl moiety and styryl moiety with the following amino acid residues: Lys 77, Glu 87, Phe 91, Ile 108, Trp 109, Asn 112, Tyr 135, His 196, Leu 221, Gly 222, Phe 225, Asn 226, Met 311 and Ump 314 as shown in Figure 3(f).

2.3.2. Docking on the Active Site of Dihydrofolate Reductase (DHFR)

MOE docking studies of the inhibitors were performed using dihydrofolate reductase co-crystallized with me thotrexate (PDB ID: 4DFR) as a template.

1) Docking of MTX into DHFR active site revealed that:

Hydrogen bond interactions beside hydrophobic interactions were considered to be responsible for the observed affinity as it acts as a hydrogen bond donor to the backbone Ile 5 and Ile 94 residues and the side chain Asp 27 residue. It also acts as a hydrogen bond accepter to Arg 52 and Arg 57 residues. This beside many hydrophobic interactions with various amino acid residues: ILe 5, Ala 6, Ala 7, Asp 27, Leu 28, Phe 31, Lys 32, Ser

49, Ile 50, Arg 52, Leu 54, Arg 57, Ile 94, Tyr 100 and Thr 113 as shown in Figure 4(a).

2) Docking of compound 18 into DHFR active site:

Revealed the interaction of quinazoline-N_1 atom as a hydrogen bond acceptor with the side chain residue Ser 59 (2.93 Å) at a strength of 43.7%, in addition to the interaction of carbonyl group oxygen as a hydrogen bond acceptor with the side chain residue Tyr 121 (2.86 Å) at a strength of 18.9% beside many hydrophobic interactions among quinazoline-C_6 atom, chlorine atom, styryl moiety and triazine-C_5 with the following amino acid residues: Val 8, Ile 16, Lys 55, Thr 56, Ser 59, Ile 60, Pro 61, Val 115, Gly 117 and Tyr 121 as shown in Figure 4(b).

3) Docking of compound 19 into DHFR active site:

Illustrated the interactions of the two NH functions on both sides of C=S function as hydrogen bond donors with the same side chain residue Ser 59 (2.49 Å, 1.71 Å) at a strength of 7.7% and 90% as well as hydrophobic interactions with the chlorine atom, styryl moiety, phenyl group, quinazoline ring and quinazoline-C_4-NH moiety with the following amino acid residues: Ala 9, Ile 16, Asp 21, Leu 22, Phe 31, Phe 34, Lys 55, Thr 56, Ser 59, Asn 64, Leu 67, Arg 70 and Thr 146 as shown in Figure 4(c).

4) Docking of compound 20 into DHFR active site showed:

The interaction of the tetrazine-NH function as a hydrogen bond donor with the side chain residue Asp 21 (1.32 Å) at a strength of 50.8%. In addition to hydrophobic interactions concerning the sulphur atom, phenyl group, quinazoline-$C_{7,8}$ and styryl moiety with the following amino acid residues: Ile 7, Val 8, Ala 9, Ile 16, Gly 20, Asp 21, Leu 22, Glu 30, Phe 31, Phe 34, Thr 56, Ser 59, Ile 60 and Thr 146 as shown in Figure 4(d).

5) Docking of compound 23 into DHFR active site displayed:

The interaction of the carbonyl group oxygen as a hydrogen bond acceptor with the side chain residue Ala 9 (3.40 Å) at a strength of 2.7%, in addition to interaction of quinazoline-C_4-NH function as a hydrogen bond donor with the side chain residue Val 115 (2.65 Å) at a strength of 2.0%, besides arene hydrophobic interaction between the center of quinazoline ring and amino acid residue Phe 34, in addition to many hydrophobic interactions among the 4-chlorostyryl moiety, azetidine-C_4-4-chlorophenyl group as well as the quinazoline ring and the following amino acid residues: Val 8, Ala 9, Ile 16, Gly 20, Leu 22, Glu 30, Phe 31, Phe 34, Lys 55, Thr 56, Ser 59, Asn 64, Leu 67, Val 115 and Tyr 121 as shown in Figure 4(e).

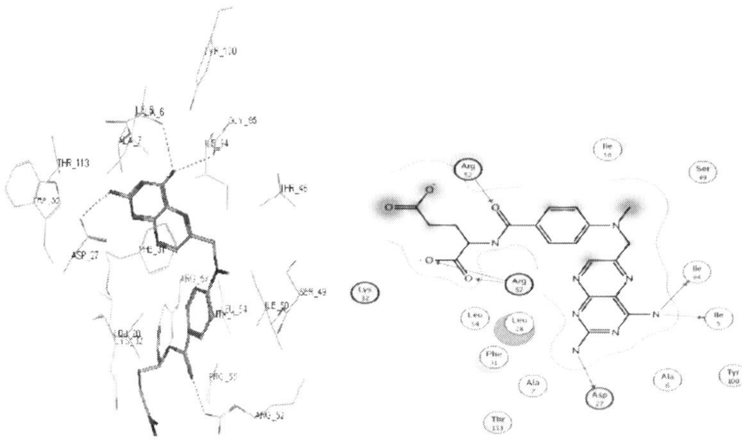

(a)

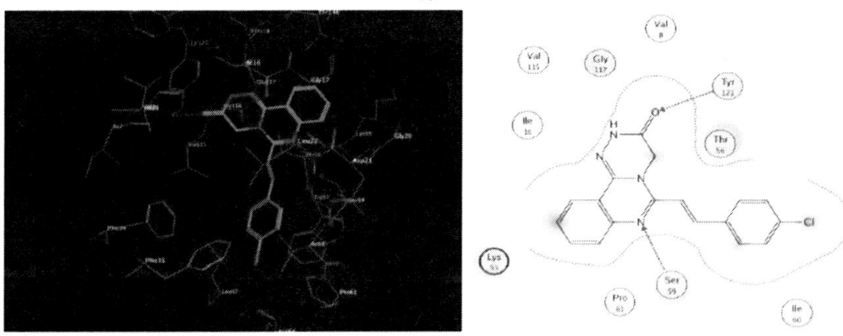

(b)

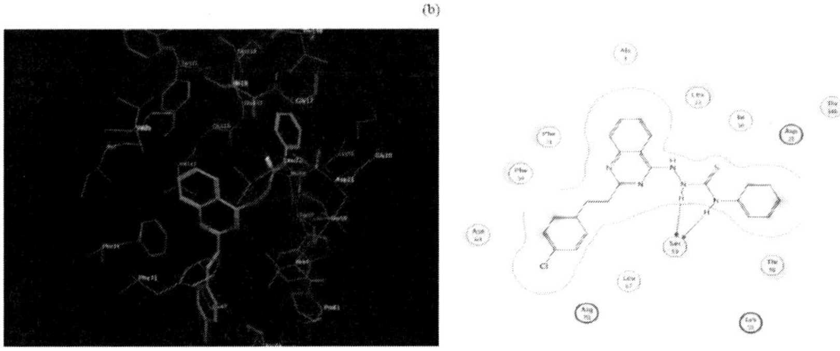

(c)

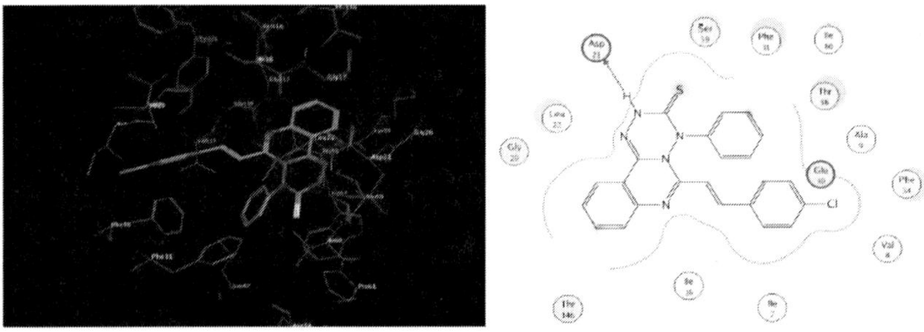

(d)

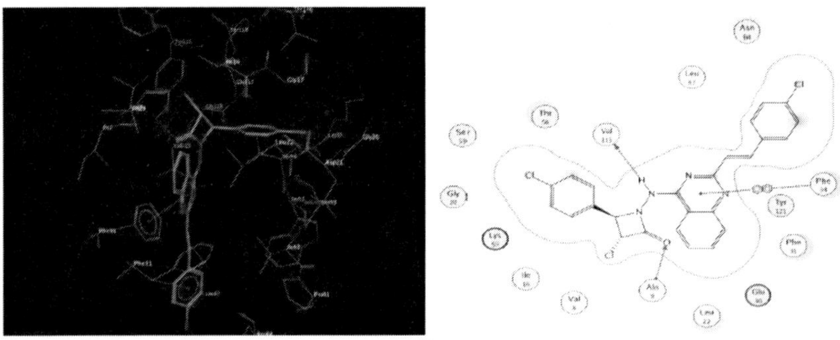

(e)

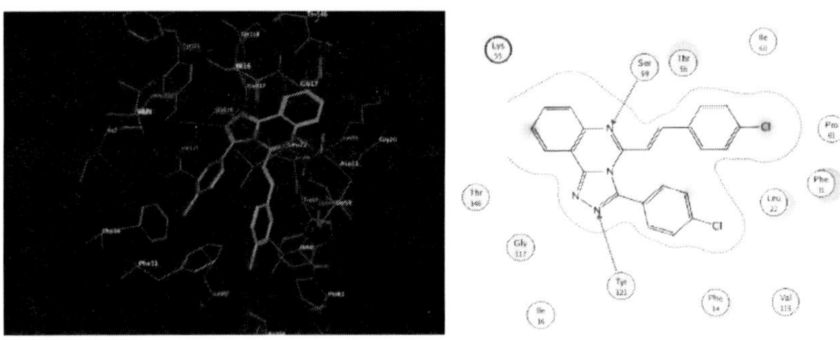

(f)

Figure 4. (a) Docking of MTX in the active site of DHFR; (b) Docking of compound 18 in the active site of DHFR; (c) Docking of compound 19 in the active site of DHFR; (d) Docking of compound 20 in the active site of DHFR; (e) Docking of compound 23 in the active site of DHFR; (f) Docking of compound 24 in the active site of DHFR.

6) Docking of compound 24 into DHFR active site revealed:

The interaction of quinazoline-N_1 atom as a hydrogen bond acceptor with the side chain residue Ser 59 (3.00 Å) at a strength of 5.6% as well as the interaction of triazole-N_2 atom as a hydrogen bond acceptor with the side chain residue Tyr 121 (3.66 Å) at a strength of 3.9 %, besides many hydrophobic interactions among the chlorine atom, quinazoline-C_6 carbon, triazole-C_3 carbon, in addition to triazole-C_3-4-chlorophenyl moiety and the following amino acid residues: Ile 16, Leu 22, Phe 31, Phe 34, Lys 55, Thr 56, Ser 59, Ile 60, Pro 61, Val 115, Gly 117, and Tyr 146 as shown in Figure 4(f).

3. EXPERIMENTAL

3.1. Chemistry

All melting points were measured on Electro thermal LA 9000 SERIS, Digital Melting point Apparatus and are uncorrected. IR spectra (KBr) were recorded on FT-IR 200 spectrophotometer (ύ cm^{-1}), pharmaceutical analytical unit, Faculty of Pharmacy, Al-Azhar University. ^{1}H-NMR spectra were recorded in (DMSO-d_6) at 300 MHz on a Varian Gemini NMR spectrometer (δ ppm) using TMS as an internal standard, Resarch Service Unit, Faculty of Science, Cairo University. Mass spectra were recorded on GC Ms-QP 5050A mass spectrometer at 70 eV and microanalytical data were performed on Elementar Vario EI III CHN analyzer at the microanalytical unit, in Regional center for Mycology and Biotechnology, Al-Azhar University. Thin layer chromatography was performed on precoated (0.25 mm) silica gel GF_{254} plates (E. Merck, Germany). Compounds were detected with 254 nm UV lamp.

3.1.1. 2-[2-(2-Chlorophenyl)-2-oxoethyl]quinazolin-4(3H)-one; 2

A solution of compound 1 [36] (1.6 g, 0.01 mol) in dry THF (20 mL) was added dropwise while stirring to a solution of 2-chlorobenzoyl chloride (3.5 g , 2.8 mL, 0.02 mol.) in dry THF (10 mL) at 0°C and stirring was continued for 2 h. The reaction mixture was further stirred for 4h at room temperature then it was poured onto crushed ice. The solid obtained was filtered off, washed with water, dried and recrystallized from ethanol to yield compound 2.

White crystals, yield%: 87%, m.p.: 153°C - 155°C, IR (KBr, cm^{-1}): 3550 (br. OH); 3071 (C-H aromatic); 2973, 2817 (C-H aliphatic); 1693 (C=O); 1574 (C=N); 1473 (C=C). ^{1}H-NMR (DMSO-d_6, d ppm): 4.33 (s, 2H, CH_2); 7.51 - 7.59 (m, 4H, quinazoline-$C_{5,6,7,8}$-H); 7.73 (s, 1H, NH, D_2O exchangeable); 7.74 - 7.86 (m, 4H, 2-Cl-C_6H_4-$C_{3,4,5,6}$-H); MS m/z (relative intensity %): 298 ($M^{+\cdot}$, 2.90); 48 (100). Anal. Form: $C_{16}H_{11}ClN_2O_2$ (298.5). Calcd. (%): C, 64.32; H, 3.68; N, 9.38; Found (%): C, 64.37; H, 3.75; N, 9.30.

3.1.2. 5H-Quinolino[2,1-b]quinazolin-5,12-(6H)-dione; 3

A solution of compound 1 (1.6 g, 0.01 mol) in dry THF (20 mL) was added dropwise while stirring to a solution of 2-chlorobenzoyl chloride (3.5 g, 2.8 mL, 0.02 mol) in dry THF (10 mL) at 0°C and stirring was continued for 2 h. The reaction mixture was refluxed for another 2h and the obtained product was filtered, washed with THF, dried and recrystallized from ethanol to give compound 3.

White crystals, yield %: 75% , m.p.: 300°C - 302°C, IR (KBr, cm^{-1}): 3000 (C-H aromatic); 1708, 1690 (two C=O); 1569 (C=N); 1478 (C=C). ^1H-NMR (DMSO-d_6, d ppm): 2.49 (s, 2H, CH_2-C=O); 7.51 (d, 2H, J = 8.3 Hz, quinazolin-$C_{6,7}$-H); 7.61-7.69 (m, 2H, quinoline-$C_{6,7}$-H); 7.81 (d, 1H, J = 8.3 Hz, quinazoline-C_5-H); 7.91 (d, 1H, J = 7.2 Hz, quinoline-C_5-H); 7.96 (d, 1H, J = 7.2 Hz, quinoline-C_8-H); 8.15 (d, 1H, J = 8.3 Hz, quinazoline -C_8-H). MS m/z (relative intensity %): 262 (M^+ , 3.04); 74 (100). Anal. Form: $C_{16}H_{10}N_2O_2$ (262). Calcd. (%): C, 73.27; H, 3.84, N, 10.68; Found (%): C, 73.25; H, 3.89; N, 10.60.

3.1.3. 12H-Quinolino[2,1-b]quinazolin-12-one; 4

Equimolar amounts of compound 1 (1.6 g, 0.01 mol) and 2-chlorobenzaldehyde (1.4 g, 1.1 mL, 0.0l mol) were refluxed in glacial acetic acid (30 mL) in presence of anhydrous sodium acetate (2 g, 0.02 mol) for 8 h. The reaction mixture was allowed to cool then poured onto crushed ice. The obtained solid was filtered off, washed with water and recrystallized from ethanol to yield compound 4.

White crystals, yield %: 88%, m.p.: 253°C - 255°C, IR (KBr, cm^{-1}): 3058 (C-H aromatic); 1681 (C=O); 1568 (C=N); 1511 (C=C). ^1H-NMR (DMSO-d_6, d ppm): 7.01 (s, 1H, quinoline-C_4-H); 7.07 (s, 1H, quinoline-C_3-H); 7.43 - 7.58 (m, 3H, quinoline-$C_{5,6,7}$-H); 7.71 (d, 2H, J = 7.8 Hz, quinazoline-$C_{6,7}$-H); 7.80 - 7.86 (m, 2H, quinoline-C_8-H&quinazoline-C_5-H); 8.11 (d, 1H, J = 7.8 Hz, quinazoline-C_8-H). MS m/z (relative intensity %): 246 (M^+, 100). Anal. Form: $C_{16}H_{10}N_2O$ (246). Calcd. (%): C, 78.03; H, 4.09; N, 11.38; Found(%): C, 78.11; H, 4.12; N, 11.30.

3.1.4. Pyrrolo[2,1-b]quinazolin-2,9(1H, 3H)-dione; 5

A solution of compound 1 (1.6 g, 0.01 mol) in dry THF (20 mL) was added dropwise while stirring to a solution of chloroacetyl chloride (2.26 g, 1.6 mL, 0.02 mol) in dry THF (10 mL) at 0°C and stirring was continued for 2 h. The reaction mixture was further stirred for 4h at room temperature then the excess solvent was evaporated. The reaction mixture was cooled and poured onto crushed ice. The solid obtained was filtered off, washed with water, dried and recrystallized from ethanol.

White crystals, yield %: 65%, m.p.: 313°C - 315°C, IR (KBr, cm^{-1}): 3034 (C-H aromatic); 2819 (C-H aliphatic); 1704, 1669 (two C=O); 1561 (C=N); 1479 (C=C). ^{1}H-NMR(DMSO-d_6, d ppm): 2.08 (s, 2H, pyrroloC_3-H); 2.49 (s, 2H, pyrrolo-C_1-H); 7.60 - 7.65 (m, 1H, C_7-H); 7.76 (d, 1H, J = 8 Hz, C_6-H); 7.91 - 7.96 (m, 1H, C_5-H); 8.15 (d, 1H, J = 8Hz, C_8-H). MS m/z (relative intensity %): 200 (M^+, 1.99); 62 (100). Anal. Form: $C_{11}H_8N_2O_2$ (200). Calcd. (%): C, 66.00; H, 4.03. Found (%): C, 65.98; H, 4.08.

3.1.5. 2-(4-Chlorostyryl)quinazolin-4(3H)-one; 6

Equimolar amounts of compound 1 (1.6 g, 1 mol) and 4-chlorobenzaldehyde (1.4 g, 1 mol) were refluxed in glacial acetic acid (30 mL) for 8 h in presence of anhydrous sodium acetate (2 g, 0.02 mol). The reaction mixture was allowed to cool then poured onto crushed ice. The obtained product was filtered, washed with water and recrystallized from ethanol.

White crystals, yield %: 85%, m.p: 302°C - 305°C [41] .

3.1.6. 4-Chloro-2-(4-chlorostyryl)quinazoline; 7

Compound 6 (3.5 g, 0.123 mol) was refluxed in excess phosphorous oxychloride (5 mL) for 14 h and in presence of a catalytic amount of triethylamine (3 - 5 drops). The reaction mixture was allowed to cool then poured onto crushed ice. The obtained solid was collected.

Yellowish brown, yield %: 82%, m.p.: 268°C - 270°C [41] .

3.1.7. 4-Amino-2-(4-chlorostyryl)quinazoline; 8

A mixture of compound 7 (3.2 g, 0.1 mol) and ammonium acetate (2.3 g, 0.3 mol) was refluxed in pyridine (30 mL) for 24 h. The reaction mixture was allowed to cool then poured onto crushed ice. The obtained solid was filtered off, dried and recrystallized from acetone.

Brown crystals, yield%: 58%, m.p.: 318°C - 320°C, IR (KBr, cm^{-1}): 3251,3142 (NH_2); 3055 (C-H aromatic); 1600 (C=N); 1596 (C=C). ^{1}H-NMR (DMSO-d_6, d ppm): 4.20 (s, 2H, NH_2, D_2O exchangeable); 7.01 (s, 1H, CH=CH-4-Cl-C_6H_4); 7.06 (s, 1H, CH=CH-4-Cl-C_6H_4); 7.52 (d, 1H, J = 6.9 Hz, quinazoline-C_5-H); 7.55 (d, 1H, J = 6.9 Hz, quinazoline-C_8-H); 7.71 (d, 2H, J = 8.1 Hz, 4-Cl-C_6H_4-$C_{2,6}$-H); 7.83 - 7.86 (m, 1H, quinazoline-C_6-H); 7.87-7.97 (m, 1H, quinazoline-C_7-H); 8.14 (d, 2H, J = 8.1 Hz, 4-Cl-C_6H_4-$C_{3,5}$-H). Anal. Form: $C_{16}H_{12}ClN_3$ (281.5). Calcd. (%): C, 68.20; H, 4.26; N, 14.92 Found(%): C, 68.19; H, 4.43; N, 14.64.

3.1.8. 2-(4-Chlorostyryl)-4-hydrazinylquinazoline; 9

A mixture of compound 7 (3.2 g, 1mol) and hydrazine hydrate 99% (0.5 g, 0.5 mL, 5 mol) (1:5) was refluxed in absolute ethanol (30 mL) for 8 h. The reaction mixture was allowed to cool and poured on crushed ice. The obtained solid was filtered, dried and recrystallized from acetone/ethanol.

Yellow crystals, yield %: 42%, m.p.: 180°C - 182°C, IR (KBr, cm^{-1}): 3430, 3310, 3200 (NH_2, NH); 3050 (C-H aromatic); 2920 (C-H aliphatic); 1550 (C=N); 1495 (C=C). ^{1}H-NMR (DMSO-d_6, d ppm): 4.15 (s, 2H, NH_2, D_2O exchangeable); 6.99 (s, 1H, CH=CH-4-Cl-C_6H_4); 7.04 (s, 1H, CH=CH-4-Cl-C_6H_4); 7.25 - 7.36 (m, 1H, quinazoline-C_5-H); 7.44 - 7.54 (m, 1H, quinazoline-C_6-H); 7.60 - 7.70 (m, 1H, quinazoline-C_7-H); 7.74 (d, 1H, J = 6.9 Hz, quinazoline-C_8-H); 7.80 (d, 2H, J = 8.4 Hz, 4-Cl-C_6H_4-$C_{2,6}$-H); 8.11 (d, 2H, J = 8.4 Hz, 4-Cl-C_6H_4-$C_{3,5}$-H); 12.26 (s, 1H, NH, D_2O exchangeable). Anal. Form: $C_{16}H_{13}ClN_4$ (296.5) Calcd. (%): C, 64.75; H, 4.38; N, 18.86; Found (%): C, 64.80; H, 4.41; N, 18.64.

3.1.9. 1-[2-(4-Chlorostyryl)quinazolin-4-yl]-3-phenylthiourea; 10

A solution of compound 8 (2.8 g, 0.01 mol) in pyridine (30 mL) was treated with an equimolar amount of phenyl isothiocyanate (1.4 g, 1.2 mL, 0.01 mol). The reaction mixture was heated under reflux for 18 h, then cooled, poured onto crushed ice and acidified with 10% hydrochloric acid. The solid product obtained was filtered off, washed with water and recrystallized from ethanol.

Brown crystals, yield %: 78%, m.p.: 243°C - 245°C, IR (KBr, cm^{-1}): 3433, 3184 (NH); 3046 (C-H aromatic); 2921 (C-H aliphatic); 1603 (C=N); 1493 (C=C); 1572, 1293, 1141, 964 (N-C=S). ^{1}H-NMR (DMSO-d_6, d ppm): 6.99 (s, 1H, CH=CH-4-Cl-C_6H_4); 7.04 (s, 1H, CH=CH-4-Cl-C_6H_4); 7.48 (d, 2H, J = 7.5 Hz, quinazoline-$C_{6,7}$-H); 7.52 (d, 2H, J = 8.6 Hz, 4-Cl-C_6H_4-$C_{2,6}$-H); 7.69 (d, 2H, J = 8.6Hz, 4-Cl-C_6H_4-$C_{3,5}$-H); 7.79 - 7.83 (m, 5H, C_6H_5); 7.96 (s, 1H, NH-C_6H_5, D_2O exchangeable); 8.11 (d, 2H, J = 7.5Hz, quinazoline-$C_{5,8}$-H); 12.31 (s, 1H, NH-CS-NH-C_6H_5, D_2O exchangeable). Anal. Form: $C_{23}H_{17}ClN_4S$ (416.5) Calcd. (%): C, 66.26; H, 4.08; N, 13.44; S, 7.68. Found (%): C, 66.28; H, 4.16; N, 13.62; S, 7.74.

3.1.10. General Procedure for Synthesis of Compounds 11 & 12

A solution of compound 10 (0.4 g, 0.01 mol) in ethanol (30 mL) and in presence of a catalytic amount of sodium acetate (1 g, 0.01 mol) was heated under reflux for 6 h with an equimolar amount of the appropriate chloro compound namely; ethyl chloroacetate (0.6 g, 0.5 mL, 0.01mol) or phenacyl chloride (0.5 g, 0.01 mol). The solid obtained after cooling the reaction mixture was filtered off and washed with water to yield compounds 11 and 12 respectively.

1) 2-[2-(4-Chlorostyryl)quinazolin-4-ylimino]-3-phenylthiazolidin-4-one; 11

Recrystallized from ethanol, buff crystals, yield %: 43%, m.p.: 283°C - 285°C, IR (KBr, cm^{-1}): 3045 (C-H aromatic); 2923 (C-H aliphatic); 1665 (C=O); 1603 (C=N); 1571(C=C); 1342, 1084 (C-S-C). ^{1}H-NMR (DMSO-d$_6$, d ppm): 3.46 (s, 2H, S-CH$_2$); 6.96 (s, 1H, CH=CH-4-Cl-C$_6$H$_4$); 7.01 (s, 1H, CH=CH-4-Cl-C$_6$H$_4$); 7.49 (d, 2H, J = 8.4 Hz, 4-Cl-C$_6$H$_4$-C$_{2,6}$-H); 7.69 (d, 2H, J = 8.4 Hz, 4-Cl-C$_6$H$_4$-C$_{3,5}$-H); 7.77 - 7.82 (m, 5H, C$_6$H$_5$); 7.85 - 7.90 (m, 3H, quinazoline-C$_{5,6,7}$-H); 8.09 (d, 1H, J = 7.5 Hz, quinazoline-C$_8$-H). ms m/z (relative intensity %): 456 (M$^+$, 1.92); 62 (100). Anal. Form: $C_{25}H_{17}ClN_4OS$ (456.5) Calcd. (%): C, 65.71; H, 3.72; N, 12.26; S, 7.00. Found (%): C, 65.74; H, 3.82; N, 12.39; S, 7.10.

2) 2-(4-Chlorostyryl)-N-(3,4-diphenylthiazol-2(3H)ylidene)quinazolin-4-amine; 12

Recrystallized from ethanol, brown crystals, yield %: 45%, m.p.: 273°C - 275°C, IR (KBr, cm^{-1}): 3000 (C-H aromatic); 2927 (C-H aliphatic); 1615 (C=N); 1548 (C=C); 1380, 1040 (C-S-C). ^{1}H-NMR (DMSO-d$_6$, d ppm): 6.99 (s, 1H, CH=CH-4-Cl-C$_6$H$_4$); 7.05 (s, 1H, CH=CH-4-Cl-C$_6$H$_4$); 7.49 (s, 1H, CH-thiazole); 7.52 (d, 2H, J = 8.4Hz, 4-Cl-C$_6$H$_4$-C$_{2,6}$-H); 7.68 - 7.79 (m, 10H, two C$_6$H$_5$); 7.83 (d, 2H, J = 7.8Hz, quinazoline-C$_{6,7}$-H); 7.91 - 7.96 (m, 2H, 4-Cl-C$_6$H$_4$-C$_{3,5}$-H); 8.12 (d, 2H, J = 7.8 Hz, quinazoline-C$_{5,8}$-H). MS m/z (relative intensity %): 516 (M$^+$, 1.26), 117 (100). Anal. Form: $C_{31}H_{21}ClN_4S$ (516.5). Calcd. (%): C,72.02; H, 4.06; N, 10.84; S, 6.19. Found (%): C, 72.07; H, 4.13; N, 10.72; S, 6.18.

3.1.11. 2-(4-Chlorostyryl)-N-(4-methyl-3-phenylthiazol-2(3H)-ylidene)quinazolin-4-amine; 13

An equimolar mixture of compound 10 (0.4 g, 0.01 mol) and chloroacetone (0.09 g, 0.1 mL, 0.01 mol) was fused at 120°C for 24 h. The reaction mixture was triturated with ethanol. The obtained product was filtered off then washed with several solvents.

Black crystals, yield%: 52%, m.p.; >360°C, IR (KBr, cm^{-1}): 3070 (C-H aromatic); 2925 (C-H aliphatic); 1600 (C=N); 1490 (C=C); 1375, 1186 (C-S-C). MS m/z (relative intensity %): 454 (M$^+$, 1.06), 53 (100). Anal. Form: $C_{26}H_{19}ClN_4S$ (454.5). Calcd. (%): C, 68.64; H, 4.18; N, 12.32; S, 7.04. Found (%): C, 68.70; H, 4.23; N, 12.39; S, 7.12.

3.1.12. General Procedure for Synthesis of Compounds 14 & 15

An equimolar mixture of compound 8 (1.4 g, 0.01 mol) and phenacyl chloride (0.8 g, 0.01 mol) in ethanol (30 mL) was refluxed for 18 h. The reaction mixture was then concentrated, allowed to cool and the solid product obtained was filtered off to yield compound 14. The filtrate was poured onto crushed ice and the obtained solid was collected to give compound 15.

1) 4-Amino-2-(4-chlorostyryl)-3-(2-oxo-2-phenylethyl)quinazolin-3-ium chloride; 14

Recrystallized from ethanol, brown crystals, yield%: 64%, m.p.: 278°C - 280°C, IR (KBr, cm^{-1}): 3433 (NH_2); 3048 (C-H aromatic); 2921 (C-H aliphatic); 1672 (C=O); 1582 (C=N); 1460 (C=C). ^{1}H-NMR (DMSO-d_6, d ppm): 3.63 (s, 2H, CH_2); 5.18 (s, 2H, NH_2, D_2O exchangeable); 7.00 (s, 1H, CH=CH-4-Cl-C_6H_4); 7.05 (s, 1H, CH=CH-4-Cl-C_6H_4); 7.52 (d, 1H, J = 7.2 Hz, quinazoline-C_5-H); 7.59 - 7.74 (m, 5H, C_6H_5); 7.77 - 7.84 (m, 2H, quinazoline-$C_{6,7}$-H); 8.06 (d, 2H, J = 7.5 Hz, 4-Cl-C_6H_4-$C_{2,6}$-H); 8.07 - 8.10 (m, 2H, 4-Cl-C_6-H_4-$C_{3,5}$-H); 8.13 (s, 1H, quinazoline-C_8-H); 12.33 (s, 1H, OH, D_2Oexchangeable). Anal.Form: $C_{24}H_{19}Cl_2N_3O$ (436) Calcd. (%): C, 66.05; H, 4.35; N, 9.63. Found (%): C, 66.04; H, 4.31; N, 9.62.

2) 5-(4-Chlorostyryl)-2-phenylimidazo[1,2-c] quinazoline; 15

Recrystallized from DMF/methanol, yellowish orange crystals, yield%: 63%, m.p.: 193°C - 195°C, IR (KBr, cm^{-1}): 3048 (C-H aromatic); 2924 (C-H aliphatic), 1604 (C=N); 1587 (C=C). ^{1}H-NMR (DMSO-d_6, d ppm): 6.98 (s, 1H, CH=CH-4-Cl-C_6H_4), 7.04 (s, 1H, CH=CH-4-Cl-C_6H_4); 7.48 (d, 1H, J = 8.1 Hz, quinazolineC_5-H); 7.53 (d, 2H, J = 8.3 Hz, 4-Cl-C_6H_4-$C_{2,6}$-H); 7.66 - 7.70 (m, 5H, C_6H_5); 7.79-7.85 (m, 3H, quinazoline$C_{6,7,8}$-H); 7.96 (s, 1H, CH-imidazole), 8.01 (d, 2H, J = 8.3 Hz, 4-Cl-C_6H_4-$C_{3,5}$-H). MS m/z (relative intensity %): 381 (M^+, 1.09); 76 (100). Anal. Form: $C_{24}H_{16}ClN_3$ (381.5) Calcd. (%): C, 75.49; H, 4.19; N, 11.00. Found (%): C, 75.57; H, 4.28; N, 11.09.

3.1.13. 5-(4-Chlorostyryl)imidazo[1,2-c]quinazolin-2(3H)-One; 16

A mixture of compound 8 (2.8 g, 0.1 mol) and chloroacetyl chloride (1.7 g, 1.2 mL, 0.15 mol) was refluxed in dry toluene (30 mL) in presence of anhydrous potassium carbonate (1.37 g, 0.1 mol) for 8 h. The reaction mixture was then filtered while hot and the filtrate was poured onto crushed ice to yield the desired product which was filtered, washed with water, dried and recrystallized from ethanol.

White crystals, yield%: 41%, m.p., >360°C, IR (KBr, cm^{-1}): 3096 (C-H aromatic); 2952 (C-H aliphatic); 1680 (C=O); 1617 (C=N), 1484 (C=C). ^{1}H-NMR (DMSO-d_6, d ppm): 3.55 (s, 2H, imidazole-C_5-CH_2 tautomer); 7.07 (m, 1H, CH=CH-4-Cl-C_6H_4); 7.26 - 7.27 (m, 1H, CH=CH-4-Cl-C_6H_4); 7.50 (s, 1H, imidazole-C_5, =CH tautomer); 7.68 - 7.86 (m, 8H, two C_6H_4); 8.51 (s, 1H, OH tautomer, D_2Oexchangeable). MS m/z (relative intensity %): 323 ($M^{+\bullet}$+2, 0.53); 321 (M^+, 1.03); 53 (100). Anal. Form: $C_{18}H_{12}ClN_3O$ (321.5). Calcd. (%): C, 67.18; H, 3.73; N, 13.06. Found (%): C, 67.24; H, 3.81; N, 13.19.

3.1.14. 6-(4-Chlorostyryl)-3,4-diphenyl-4H-[1,2,4]triazino[4,3-c]quinazoline; 17

An equimolar mixture of compound 9 (3 g, 0.1mol) and benzoin (2.1 g, 1.6 mL, 0.1 mol) in ethanol (30 mL) containing few drops of glacial acetic acid (2 - 3 drops) was refluxed for 8 h. The reaction mixture was cooled and the obtained solid was filtered, dried and recrystallized from acetone.

Yellowish white crystals, yield%: 43%, m.p.: 100°C - 102°C, IR (KBr, cm^{-1}): 3044 (C-H aromatic); 1600 (C=N); 1569 (C=C). ^{1}H-NMR (DMSO-d_6, d ppm): 6.02 (s, 1H, triazine-C_5-H); 7.22 (s,1H, CH=CH-4-Cl-C_6H_4); 7.23 (s, 1H, CH=CH-4-Cl-C_6H_4); 7.24-7.29 (m, 5H, triazine-C_5-C_6H_5); 7.31 - 7.37 (m, 5H, triazine-C_6-C_6H_5); 7.41 (d, 2H, J = 8.3 Hz, 4-Cl-C_6H_4-$C_{2,6}$-H); 7.44 - 7.48 (m, 2H, quinazoline-$C_{6,7}$-H); 7.53 (d, 1H, J = 8.1 Hz, quinazoline-C_5-H); 7.57 - 7.60 (m, 1H, quinazoline-C_8-H); 7.99 (d, 2H, J = 8.3 Hz, 4-Cl-C_6H_4-$C_{3,5}$-H). MS m/z (relative intensity %): 472 (M^+, 1.04); 76 (100). Anal. Form: $C_{30}H_{21}ClN_4$ (472.5) Calcd. (%): C, 76.19; H, 4.44; N, 11.85. Found (%): C, 76.22; H, 4.52; N, 11.97.

3.1.15. 6-(4-Chlorostyryl)-2H-[1,2,4]triazino[4,3-c]quinazolin-3(4H)-one; 18

Compound 9 (3 g, 0.1 mol) and chloroacetyl chloride (1.7 g, 1.2 mL, 0.15 mol) were stirred in dry DMF (30 mL) for 2 h at room temperature, then refluxed for another 5 h. The reaction mixture was allowed to cool then poured onto crushed ice. The solid separated was filtered, washed with water, dried and recrystallized from ethanol.

Black crystals, yield %: 52%, m.p.: 150°C - 152°C, IR (KBr, cm^{-1}): 3416 (NH); 3039 (C-H aromatic); 2922 (C-H aliphatic); 1674 (C=O); 1618 (C=N); 1461 (C=C). ^{1}H-NMR (DMSO-d_6, d ppm): 4.25 (s, 2H, CH_2); 7.30 (s, 1H, CH=CH-4-Cl-C_6H_4); 7.31 (s, 1H, CH=CH-4-Cl-C_6H_4); 7.32 - 7.46 (m, 2H, quinazoline-$C_{6,7}$-H); 7.51 (d, 2H, J = 8.1 Hz, 4-Cl-C_6H_4-$C_{2,6}$-H); 7.61 (d, 1H, J = 8.1 Hz, quinazoline-C_5-H); 7.77 (d, 2H, J = 8.1 Hz, 4-Cl-C_6H_4-$C_{3,5}$-H); 8.08 (d, 1H, J = 8.1 Hz , quinazoline-C_8-H); 12.20 (s, 1H, NH, D_2O exchangeable). MS m/z (relative intensity %): 336 (M^+, 1.13); 64(100). Anal. Form: $C_{18}H_{13}ClN_4O$ (336.5). Calcd. (%): C, 64.19; H, 3.86; N, 16.64. Found (%): C, 64.22; H, 3.53; N, 16.74.

3.1.16. 2-[2-(4-Chlorostyryl)quinazolin-4-yl]-N-phenylhydrazine Carbothioamide; 19

A mixture of compound 9 (3 g, 0.01 mol) and phenyl isothiocyanate (1.4 g, 1.2 mL, 0.01 mol) in methanol (25 mL) was refluxed for 12 h. The solvent was evaporated. The obtained compound was filtered off, washed with methanol and recrystallized from ethanol.

Yellow crystals, yield%: 75%, m.p:, 108°C - 110°C, IR (KBr, cm^{-1}): 3210, 3183 (NH); 3014 (C-H aromatic); 1534, 1364, 1130, 905 (N-C=S); 1450 (C=C). ^{1}H-NMR (DMSO-d_6, d ppm): 3.98 (s, 2H, NH-CS-NH, D_2O exchangeable); 7.13 (d, 1H, J = 7.5 Hz, CH=CH-4-Cl-C_6H_4); 7.16 (d, 1H, J = 7.5 Hz, CH=CH-4-Cl-C_6H_4); 7.31 - 7.46 (m, 13H, aromatic-CH); 11.08 (s, 1H, NH, quinazoline-C_4-NH, D_2O exchangeable). MS m/z (relative intensity %): 431 (M^+, 1.11); 54 (100). Anal. Form: $C_{23}H_{18}ClN_5S$ (431.5) Calcd. (%): C, 63.96; H, 4.17; N, 16.22. Found (%): C, 64.04; H, 4.24; N, 16.32.

3.1.17. 6-(4-Chlorostyryl)-4-phenyl-2H-[1,2,4,5]tetrazino[1,6-c]quinazolin-3(4H)-thione; 20

To a solution of compound 19 (2.2 g, 0.01 mol) in carbon tetrachloride (30 mL), bromine (2.4 g, 0.8 mL, 0.02 mol) was added drop wise while stirring. After complete addition, the reaction mixture was stirred at room temperature for 12 h. The obtained crystals were separated and washed with carbon tetrachloride.

Yellow needle crystals, yield%: 55%, m.p.: 153°C - 155°C, IR (KBr, cm^{-1}): 3310 (NH); 3090, 3000 (C-H aromatic); 2924 (C-H aliphatic); 1573 (C=N); 1554, 1372, 1115, 1041 (N-C=S); 1508 (C=C). ^{1}H-NMR (DMSO-d_6, d ppm): 4.00 (s, 1H, NH, D_2O exchangeable); 7.12 (s, 1H, CH=CH-4-Cl-C_6H_4); 7.15 (s, 1H, CH=CH-4-Cl-C_6H_4); 7.53 - 7.77 (m, 13H, aromatic-CH). MS m/z (relative intensity%): 429 (M^+, 2.02); 54 (100). Anal. Form: $C_{23}H_{16}ClN_5S$ (429.5) Calcd. (%): C, 64.26; H, 3.72; N, 16.29. Found (%): C, 64.28; H, 3.70; N, 16.37.

3.1.18. 4-[2-(4-Chlorobenzylidene)hydrazinyl]-2-(4-chlorostyryl)quinazoline; 21

Equimolar amounts of compound 9 (3 g, 0.01 mol) and 4-chlorobenzaldehyde (1.4 g, 0.01 mol) were refluxed for 8 h in absolute ethanol (30 mL) and in presence of few drops of piperidine as catalyst (2 - 3 drops). The reaction mixture was concentrated and the obtained solid residue was collected and recrystallized from ethanol.

Transparent fine crystals, yield%: 92%, m.p.: 233°C - 235°C, IR (KBr, cm^{-1}): 3422 (NH); 3010 (C-H aromatic); 2952 (C-H aliphatic); 1593 (C=N); 1457 (C=C). MS m/z (relative intensity %): 419 (M^+, 2.01); 85 (100). Anal. Form: $C_{23}H_{16}Cl_2N_4$(419) Calcd. (%): C, 65.87; H, 3.81; N, 13.36. Found (%): C, 65.92; H, 3.84; N, 13.42.

3.1.19. 2-(4-Chlorophenyl)-3-[2-(4-chlorostyryl)quinazolin-4-yl amino]thiazolidin-4-one; 22

To a solution of compound 21 (2.1 g, 0.01 mol) and anhydrous $ZnCl_2$ (2.7 g, 0.02 mol) in dry benzene (50 mL), thioglycolic acid (1.8 g, 1.4 mL, 0.02 mol) was added drop wise while stirring at ambient temperature. The reaction mixture was kept for 3 days at room temperature and then refluxed for 14 h. The reaction mixture was filtered. The filtrate was concentrated and poured onto crushed ice. The obtained solid was collected and washed with several solvents.

Black crystals, yield%: 82%, m.p., >360°C, IR (KBr, cm^{-1}): 3428 (NH); 3000 (C-H aromatic); 2927 (C-H aliphatic); 1690 (C=O); 1610 (C=N), 1410 (C=C); 1290, 1163 (C-S-C). MS m/z (relative intensity %): 493 (M^+, 2.02); 50 (100). Anal. Form: $C_{25}H_{18}Cl_2N_4OS$ (493) Calcd. (%): C, 60.85; H, 3.65; N, 11.35; S, 6.49. Found (%): C, 60.94; H, 3.72; N, 11.51; S, 6.47.

3.1.20. 3-Chloro-4-(4-chlorophenyl)-1-[2-(4-chlorostyryl)quinazolin-4-yl amino]azetidin-2-one; 23

To a solution of compound 21 (2.1 g, 0.01 mol) in dry benzene (50 mL) containing triethyl amine (1 mL) as a base, chloroacetyl chloride (1.7 g, 1.2 mL, 0.015 mol) was added drop wise at room temperature. The reaction mixture was stirred for 40 minutes then refluxed for 32 h. The reaction mixture was filtered while hot to remove the triethyl amine hydrogen chloride salt and the filtrate was poured onto crushed ice with continuous stirring. The solid obtained was maintained in the refrigerator for 7 days.

Solid with low melting point, yield %: 84%, IR (KBr, cm^{-1}): 3402 (OH); 3290, 3240 (NH); 3004 (C-H aromatic), 2804 (C-H aliphatic); 1720 (C=O); 1635 (C=N); 1465 (C=C). ^{1}H-NMR (DMSO-d_6, d ppm): 4.06 (d, 1H, J = 9 Hz, CH-C_6H_4-Cl); 4.30 (d, 1H, J = 9 Hz, CH-Cl); 7.90 (s, 1H, CH=CH-4-Cl-C_6H_4); 8.06 (s, 1H, CH=CH-4-Cl-C_6H_4); 8.68 - 9.42 (m, 12H, two 4-Cl-C_6H_4 & quinazoline-$C_{5,6,7,8}$-H); 10.68 (s, 1H, NH, D_2O exchangeable). MS m/z (relative intensity %): 495 (M^+, 0.82); 62 (100). Anal. Form: $C_{25}H_{17}Cl_3N_4O$ (495.5) Calcd. (%): C, 60.54; H, 3.43; N, 11.30. Found (%): C, 60.62; H, 3.44; N, 11.39.

3.1.21. 3-(4-Chlorophenyl)-5-(4-chlorostyryl)-[1,2,4]triazolo[4,3-c]quinazoline; 24

Compound 21 (2.1 g, 0.01 mol) was refluxed in glacial acetic acid (15 mL) for 14 h in presence of anhydrous sodium acetate (1.5 g, 0.02 mol.). The reaction mixture was then filtered while hot, the filtrate was poured onto crushed ice. The obtained solid was filtered, washed with water and recrystallized from ethanol.

Black crystals, yield%: 52%, m.p.: 90°C - 92°C, IR (KBr, cm^{-1}): 3010 (C-H aromatic); 2900 (C-H aliphatic); 1641 (C=N); 1569 (C=C). MS m/z (relative intensity %): 417 (M^+, 28.57); 81 (100). Anal. Form: $C_{23}H_{14}Cl_2N_4$ (417) Calcd. (%): C, 66.18; H, 3.35; N, 13.42. Found (%): C, 66.19; H, 3.42; N, 13.58.

3.1.22. General Procedure for Synthesis of Compounds 25, 26 and 27

An equimolar mixture of compound 7 (1.6 g, 0.01 mol) and the appropriate amine namely; 2-aminoethanol (0.3 g, 0.3 mL, 0.01 mol), dimethyl amine (0.5 g, 0.6 mL, 0.01 mol) or pyrrole (0.3 g, 0.3 mL, 0.01 mol) was fused for 18 h. The reaction mixture was allowed to cool then triturated with ethanol. The obtained solid was filtered off to yield compounds 25, 26 and 27; respectively.

1) 2[2-(4-Chlorostyryl)quinazolin-4-ylamino]ethanol; 25

Recrystallized from ethanol, yellowish white crystals, yield%: 42%, m.p.: >360°C, IR (KBr, cm^{-1}): 3435 (OH broad band); 3184, 3126 (NH); 3046 (C-H aromatic); 2923, 2856 (C-H aliphatic); 1604 (C=N); 1571 (C=C). ^1H-NMR (DMSO-d_6, d ppm): 3.66 - 3.71 (m, 4H, HN-CH_2-CH_2-OH); 4.40 (s, 1H, OH, D_2O exchangeable); 7.19 (s, 1H, CH=CH-4-Cl-C_6H_4); 7.22 (s, 1H, CH=CH-4-Cl-C_6H_4); 7.38 (d, 1H, J = 6.5 Hz, quinazoline-C_5-H); 7.40 - 7.48 (m, 2H, quinazoline-$C_{6,7}$-H); 7.53 (d, 2H, J = 7.5 Hz, 4-Cl-C_6H_4-$C_{2,6}$-H); 7.67 (d, 1H, J = 6.5 Hz, quinazoline-C_8-H); 8.11 (d, 2H, J = 7.5 Hz, 4-Cl-C_6H_4-$C_{3,5}$-H); 12.35 (s, 1H, NH, D_2O exchangeable). MS m/z (relative intensity %): 327 ($M^{+\bullet}$+2, 1.34), 325 (M^+, 2.35); 78 (100). Anal. Form: $C_{18}H_{16}ClN_3O$ (325.5) Calcd. (%): C, 66.35; H, 4.91; N, 12.90. Found (%): C, 66.41; H, 5.03; N, 13.04.

2) 2-(4-Chlorostyryl)-N,N-dimethylquinazoline-4-amine; 26

Recrystalized from ethanol, white crystals, yield%: 53%, m.p.: 298°C - 300°C, IR (KBr, cm^{-1}): 3055 (C-H aromatic); 2938 (C-H aliphatic); 1572 (C=N); 1504 (C=C). ^1H-NMR (DMSO-d_6, d ppm): 2.09 (s, 6H, $N(CH_3)_2$); 6.99 (s, 1H, CH=CH-4-Cl-C_6H_4); 7.04 (s, 1H, CH=CH-4-Cl-C_6H_4); 7.48 (d, 2H, J = 6.9Hz, quinazoline-$C_{6,7}$-H); 7.53 (d, 2H, J = 8.7Hz, 4-Cl-C_6H_4-$C_{2,6}$-H); 7.69 (d, 2H, J = 8.7 Hz, 4-Cl-C_6H_4-$C_{3,5}$-H); 7.83 (d, 1H, J = 6.9 Hz, quinazoline -C_5-H); 8.11 (d, 1H, J = 7.8 Hz, quinazoline-C_8-H). MS m/z (relative intensity %): 309 ($M^{+\bullet}$, 0.02); 62 (100). Anal. Form: $C_{18}H_{16}ClN_3$ (309.5) Calcd. (%): C, 69.78; H, 5.16; N, 13.57. Found (%): C, 69.86; H, 5.24; N, 13.62.

3) 2-(4-Chlorostyryl)-4-(1H-pyrrol-1-yl)quinazoline; 27

Recrystallized from DMF, black crystals, yield%: 67%, m.p.: >360°C, IR (KBr, cm^{-1}): 3124 (C-H aromatic); 1657 (C=N); 1530 (C=C). ^1H-NMR (DMSO-d_6, d ppm): 6.55 (s, 2H, CH=CH); 7.69 (s, 2H, pyrrole-$C_{3,4}$-H); 7.70 (s, 2H, pyrrole-$C_{2,5}$-H); 7.92 - 8.00 (m, 8H, 4-Cl-C_6H_4 & quinazoline-$C_{5,6,7,8}$-H). MS m/z (relative intensity %): 331 ($M^{+\bullet}$, 1.22); 54 (100). Anal. Form: $C_{20}H_{14}ClN_3$(331.5) Calcd. (%): C, 72.39; H, 4.22; N, 12.66. Found (%): C, 72.45; H, 4.23; N, 12.72.

3.1.23. General Procedure for Synthesis of Compounds 28 and 29

A mixture of compound 7 (1.6 g, 0.01 mol) and anthranilic acid (1.4 g, 0.02 mol) was refluxed for 12 h in glacial acetic acid (30 mL). The reaction mixture was concentrated and filtered while hot to yield compound 28. The filtrate was poured onto crushed ice and the obtained solid was collected to give compound 29.

1) 8H-6-(4-Chlorostyryl)quinazolino[4,3-b]quinazolin-8-one; 28

Recrystallized from ethanol, yellow crystals, yield%: 44%, m.p.: >360°C, IR (KBr, cm^{-1}): 3057 (C-H aromatic), 2922 (C-H aliphatic); 1690 (C=O); 1637 (C=N); 1570 (C=C). ^{1}H-NMR (DMSO-d_6, d ppm): 7.30 (s, 1H, CH=CH-4-Cl-C_6H_4); 7.35 (s, 1H, CH=CH-4-Cl-C_6H_4); 7.50 (d, 2H, J = 8.5 Hz, 4-Cl-C_6H_4-$C_{2,6}$-H); 7.82 (d, 2H, J=8.5Hz, 4-Cl-C_6H_4-$C_{3,5}$-H); 7.87 - 7.96 (m, 8H, quinazoline-C-H). MS m/z (relative intensity %): 386 (M^++3, 0.22); 382 ($M^{+\bullet}$-1, 0.22); 54 (100). Anal. Form: $C_{23}H_{14}ClN_3O$ (383). Calcd. (%): C, 71.96; H, 3.65; N, 10.95. Found (%): C, 71.81; H, 3.71; N, 11.03.

2) 2-[2-(4-Chlorostyryl)quinazolin-4-amino]benzoic acid; 29

Recrystallized from ethanol, brown crystals, yield%: 85%, m.p., 203°C - 205°C, IR (KBr, cm^{-1}): 3179 (NH); 2920 (broad OH, Hydrogen bonded); 1683 (C=O); 1603 (C=N); 1519 (C=C). ^{1}H-NMR (DMSO-d_6, d ppm): 6.99 (s, 1H, CH=CH-4-Cl-C_6H_4); 7.04 (s, 1H, CH=CH-4-Cl-C_6H_4); 7.24 - 7.33 (m, 1H, aminophenyl-C_4-H); 7.37 (d, 1H, J = 7.8Hz, aminophenyl-C_6-H); 7.47 - 7.55 (m, 1H, aminophenyl-C_5-H); 7.69 (d, 2H, J = 8.4 Hz, 4-Cl-C_6H_4-$C_{2,6}$-H); 7.78 (d, 1H, J = 8.1 Hz, quinazoline-C_5-H); 7.83 (d, 2H, J = 8.4 Hz, 4-Cl-C_6H_4-$C_{3,5}$-H); 7.87 - 8.01 (m, 2H, quinazoline$C_{6,7}$-H); 8.12 (d, 1H, J = 8.1 Hz, quinazolineC_8-H); 8.34 (d, 1H, J = 7.8 Hz, aminophenyl-C_3-H); 8.76 (s, 1H, NH, D_2O exchangeable); 12.18 (s, 1H, OH, D_2O exchangeable) . MS m/z (relative intensity %): 401 (M^+, 0.50); 64 (100). Anal. Form: $C_{23}H_{16}ClN_3O_2$ (401) Calcd. (%): C, 68.79; H, 3.98; N, 10.46. Found (%): C, 68.80; H, 3.97; N, 10.52.

3.2. Biology

3.2.1. Anticancer Screening

The synthesized compounds were screened for their in vitro cytotoxic activity against human hepatocellular liver carcinoma (Hep-G2) and human breast cancer cell line (MCF-7). Doxorubicin was used as the reference drug.

Mammalian cell lines MCF-7 cells (human breast cancer cell line) were obtained from VACSERA Tissue culture unit. Hep-G2 cells (human cell line of a well differentiated hepatocellular carcinoma isolated from a liver biopsy of a male Caucasian aged 15 years) were obtained from the American type culture collection (ATCC).

3.2.2. Cell line Propagation

The cells were propagated in Dulbecco's modified Eagle's medium (DMEM) supplemented with 10% heat-inactivated fetal bovine serum, 1% L-glutamine, HEPES buffer and 50 µg/mL gentamycin. All cells were maintained at 37°C in a humidified atmosphere with 5% CO_2 and were subcultured two times a week.

3.2.3. Cytotoxicity Evaluation Using Viability Assay

For cytotoxicity assay, the cells were seeded in 96-well plate at a cell concentration of 1×10^4 cells per well in 100 µL of growth medium. Fresh medium containing different concentrations of the test sample was added after 24 h of seeding. Serial two-fold dilutions of the tested chemical compound were added to confluent cell monolayers dispensed into 96-well, flat-bottomed microtiter plates (Falcon, NJ, USA) using a multichannel pipette. The microtiter plates were incubated at 37°C in a humidified incubator with 5% CO_2 for a period of 48 h. Three wells were used for each concentration of the test sample. Control cells were incubated without test sample and with or without DMSO. The little percentage of DMSO present in the wells (maximal 0.1%) was found not to affect the experiment. After incubation of the cells for 24 h at 37°C, various concentrations of sample (50, 25, 12.5, 6.25, 3.125 & 1.56 µg/L) were added, and the incubation was continued for 48 h and viable cells yield was determined by a colorimetric method. After the end of the incubation period, media were aspirated and the crystal violet solution (1%) was added to each well for at least 30 minutes. The stain was removed and the plates were rinsed using tap water until all excess stain is removed. Glacial acetic acid (30%) was then added to all wells and mixed thoroughly, and then the absorbance of the plates were measured after gently shaken on Microplate reader (TECAN, Inc.), using a test wavelength of 490 nm. All results were corrected for background absorbance detected in wells without added stain. Treated samples were compared with the cell control in the absence of the tested compounds. All experiments were carried out in triplicate. The cell cytotoxic effect of each tested compound was calculated [49] [50] . The cytotoxicity of the tested compounds was estimated in terms of percent growth inhibition compared to untreated control cells and their IC_{50} in µM/L which is the concentration of the compound that inhibits the tumor cell growth by 50%.

3.3. Computer Aided Docking

3.3.1. Materials

All the molecular studies were carried out on an Intel pentium 1.6 GHz processor, 512 MB memory with windows XP operating system using Molecular Operating Environment (MOE 2005.06; Chemical Computing Group, Montreal, Canada) as the computational software. All the minimizations were performed with MOE until a RMSD gradient of 0.05 Kcal mol^{-1} A^{0-1} with MMFF94X forcefield and the partial charges were automatically calculated.

3.3.2. General Methodology

The coordinates of the X-ray crystal structure of 2'-deoxyuridine-5'-monophosphate (DUMP) bound to thymidylate synthase enzyme were obtained from Protein Data Bank (PDB ID: 1BID) as well as the coordinates of the X-ray crystal structure of methotrexate (MTX) bound to dihydrofolate reductase (DHFR) enzyme (PDB ID: 4DFR). Enzyme structures were checked for missing atoms, bonds and contacts. Hydrogen atoms were added to the enzyme structure. Water molecules and bound ligands were manually deleted. The ligand molecules were constructed using the builder molecule and were energy minimized. The active site was generated using the MOE-Alpha site finder. Dummy atoms were created from the obtained alpha spheres. Ligands were docked within the thymidylate synthase and dihydrofolate reductase active sites using the MOE-Dock with simulated annealing used as the search protocol and MMFF94X molecular mechanics forcefield for 8000 interactions. The lowest energy conformation was selected and subjected to an energy Minimization using MMFF94X force field.

4. CONCLUSION

Compounds 18, 19, 20, 23 and 24 exhibited very potent anticancer activity against both Hep-G2 and MCF-7 cell lines. However, compounds 10, 15 and 29, showed remarkable anticancer activity against liver Hep-G2 cell line, while compound 22 exhibited potent anticancer activity against MCF-7 cell line. Docking was performed for the five most active anticancer compounds 18, 19, 20, 23 and 24 on the two enzymes; thymidylate synthase and dihydrofolate reductase in a trial to predict their mode of action as anticancer drugs. Also, the compounds show several interactions with both enzymes but they exhibit strong interactions with dihydrofolate reductase enzyme, mainly compounds 18, 19, 23 and 24, giving rise to the conclusion that they might exert their action through inhibition of both enzymes but mainly DHFR enzyme.

ACKNOWLEGEMENTS

Aisha Youssif Hassan Helali, Marwa Taha Mostafa Sarg, Makarem Mohamed Said Koraa, Mona Said Fathy El-Zoghbi The biological anticancer screening was performed in the Regional Center for Mycology and Biotechnology, Al-Azhar University. The Molecular docking study was performed in the pharmaceutical Chemistry Department, Faculty of Pharmacy, Alexandria University.

REFERENCES

1. Witt, A. and Bergman, J. (2003) Recent Developments in the Field of Quinazoline Chemistry. Current Organic Chemistry, 7, 659-677.
2. Wong, H. and Gansan, A. (2003) Total Synthesis of the Fumiquinazoline Alkaloids: Solution-Phase Studies. Journal of Organic Chemistry, 65, 1022-1030.
3. Micheal, J.P. (2003) Quinoline, Quinazoline and Acridone Alkaloids. Natural Product Reports, 18, 543-559.
4. Al-Obaid, A.M., Abdel-Hamide, S.G., El-kashef, H.A., Abdel-Aziz, A.A., El-Azab, A.S., Al-khamees, H.A. and El-Subbagh, H.I. (2009) Synthesis, in Vitro Antitumor Activity and Molecular Modeling Study of Certain 2-Thieno-4(3H)-quinazolinone Analogs. European Journal of Medicinal Chemistry, 44, 2379-2391.
5. Al-Omary, F.A., Abou-Zeid, L.A., Nagi, M.N., Habib, S., Abdel-Aziz, A.A., El-Azab, A.S., Abdel-Hamide, S.G., Al-Omar, M.A., Al-Obaid, A.M. and El-Subbag, H.I. (2010) Non-Classical Antifoliates Part 2: Synthesis, Biological Evaluation and Molecular Modeling Study of Some New 2,6-Substituted Quinazolin-4-Ones. Bioorganic & Medicinal Chemistry, 18, 2849-2863.
6. El-Azab, A.S., Al-Omar, M.A., Abdel-Aziz, A.A., Abdel-Aziz, N.I., El-Sayed, M.A., Aleisa, A.M., Sayed-Ahmad, M.M. and Abdel-Hamide, S.G. (2010) Design, Synthesis and Biological Evaluation of Novel Quinazoline Derivatives as Potential Antitumor Agents: Molecular Docking Study. European Journal of Medicinal Chemistry, 45, 4188-4198.
7. Berest, G.G., Voskoboynik, O.Y., Kovalenko, S.I., Antypenko, O.M., Nosulenko, I.S., Katsev, A.M. and Shandrovskaya, O.S. (2011) Synthesis and Biological Activity of Novel N-Cycloalkyl-(cycloalkylaryl)-2-[(3-R-2-oxo-2H- [1,2,]">4]triazino[2,3-c]quinazoline-6-yl)thio]acetamides. European Journal of Medicinal Chemistry, 46, 6066-6074.
8. Yadav, M.R., Grande, F., Chouhan, B.S., Naik, P.P., Giridhar, R., Garofalo, A. and Neamati, N. (2012) Cytotoxic Potential of Novel 6,7-Dimethoxyquinazolines. European Journal of Medicinal Chemistry, 24, 231-243.
9. Li, H., Li, D., Lu, X., Xu, Y. and Zhu, H. (2012) Design and Synthesis of 4,6-Substituted-(diphenylamino)quinazolines as Potent EGFR Inhibitors with Antitumor Activity. Bioorganic & Medicinal Chemistry, 20, 317-323.
10. Mulakayala, N., Kandagatle, B., Ismail, Rapolu, R.K., Rao, P., Mulakayala, C., Kumar, C.S., Igbal, J. and Oruganti, S. (2012) $InCl_3$-Catalysed Synthesis of 2-Arylquinazolin-4(3H)-ones and 5-Arylpyrazolo[4,3-d]pyrimidin-7(6H)-ones and Their Evaluation as Potential Anticancer Agents. Bioorganic & Medicinal Chemistry, 22, 5063-5066.

11. Zeng, Z., He, Q., Liang, Y., Feng, X., Chen, F., Clercq, E.D., Balzarini, J. and Couque, C.P. (2010) Hybrid Diarylbenzopyrimidine Non-Nucleoside Reverse Transcriptase Inhibitors as Promising New Leads for Improved Anti-HIV-1 Chemotherapy. Bioorganic & Medicinal Chemistry, 18, 5039-5047.
12. Mohamed, Y.A., Amr, A., Mohamed, S.F., Abdalla, M.M., Al-Omar, M.A. and Shafik, S.H. (2012) Cytotoxicity and AntiHIV Evaluation of Some New Synthesized Quinazoline and Thiooxopyrimidine Derivatives Using 4-(thiophen-2-yl)-3,4,5,6- Tetrahydrobenzo[h] quinazolin-2(1H)-thione as Synthone. Journal of Chemical Sciences, 124, 693-702.
13. Vijayakumar, K., Ahamed, A.G. and Thiruneelakandan, G. (2013) Synthesis, Antimicrobial and Anti-HIV1 Activity of Quinazoline-4(3H)-one Derivatives. Journal of Applied Chemistry, 5, 387191.
14. Sharma, P. Kumar, A., Kumar, P., Singh, J. and Kaushik, M.P. (2011) QSAR Modeling of Synthesized 3-(1,3-benzothiazol- 2-yl)2-phenyl Quinazolin-4(3H)-ones as Potent Antibacterial Agents. Medicinal Chemistry Research, 21, 1136-1148.
15. Anand, R.V., Narasimhan, B., Chandran, R.V.P. and Jayaveera, K.N. (2011) 7-Chloro-3-(substituted benzylidene/phenylethylidene amino)-2-phenyl Quinazolin-4(3H)-ones: Synthesis, Antimicrobial and Antitubercular Evaluation. Medicinal Chemistry Research, 21, 2831-2836.
16. Ryu, C., Kim, Y.H., Im, A.A., Kim, J.Y., Yoon, J.H. and Kim, A. (2012) Synthesis and Antifungal Activity of 6,7- bis(arylthio)-quinazoline-5,8-diones and Furo[2,3-f]quinazolin-5-ols. Bioorganic & Medicinal Chemistry Letters, 22, 500-503.
17. Yadav, M.R., Shirude, S.T., Parmar, A., Balaraman, R. and Giridhar, R. (2006) Synthesis and Anti-Inflammatory Activity of 2,3-Diaryl-4(3H)-quinazolines. Chemistry of Heterocyclic Compounds, 42, 1038.
18. Chandria, P.M., Yakaiah, T., Rao, A.R.R., Narsaiah, B., Reddy, N.C., Sridhar, V. and Rao, J.V. (2008) Synthesis of Novel 4,6-Disubstituted Quinazoline Derivatives, Their Anti-Inflammatory and Anticancer Activity against U937 Leukemia Cell Lines. European Journal of Medicinal Chemistry, 43, 846-852.
19. Alafeefy, A.M., Kadi, A.A., El-Azab, A.S., Abdel-Hamide, S.G. and Daba, M.H. (2008) Synthesis, Analgesic and AntiInflammatory Evaluation of Some New 3H-Quinazolin-4-One Derivatives. Archiv der Pharmazie (wein heim), 341, 377-385.
20. Manivannan, E. and Chaturvedi, S.C. (2011) Analogue-Based Design, Synthesis and Molecular Docking Analysis of 2,3-Diarylquinazolinones as Non-Ulcerogenic Anti-Inflammatory Agents. Bioorganic & Medicinal Chemistry, 19, 4520- 4588.
21. Letourneau, J.J., Riviello, C.M., Li, H., Cole, A.G., Ho, K., Zonetakos, H.A., Desai, H., Zhao, J., Auld, D.S., Napier, S.E., Thomson, F.J., Goan, K.A., Morphy, J.R., Ohlmeyer, M.H.J. and Webb, M.L. (2010) Idendification and Optimization of Novel 2-(4-oxo-2-aryl-quinazolin-3(4H)-yl)acetamide Vasopressin V3 (V1b) Receptor Antagonists. Bioorganic & Medicinal Chemistry Letters, 20, 5394-5397.
22. Gupta, A., Kashaw, S.K., Jain, N., Jajak, H., Soni, A. and Stable, J.P. (2011) Design and Synthesis of Some Novel 3- [5-(4-Substituted)phenyl-1,3,4-oxadiazole-2-yl]-2-phenylquinazolin-4(3H)-ones as Possible Anticonvulsant Agent. Medicinal Chemistry Research, 20, 1638-1942.
23. Kumar, S., Kaur, H. and Kumar, A. (2012) Synthesis of New Azetidinonyl/Thiazolidinonyl Quinazolinone Derivatives as Antiparkinsonian Agents. Arabian Journal of Chemistry, 5, 475-484.
24. Laddha, S.S., Wadodkav, S.G. and Meghal, S.K. (2009) cAMP-Dependend Phosphodiesterase Inhibition and SAR Studies on Novel 6,8-Disubstituted-2-phenyl-3-(substituted benzothiazole-2-yl)-4(3H)-quinazolinone. Medicinal Chemistry Research, 18, 268-276.
25. Abou Seri, S.M., Abouzid, K. and Abou-El Ella, D.A. (2011) Molecular Modeling Study and Synthesis of QuinazolinoneArylpiperazine Derivatives as α_1-Adrenoreceptor Antagonists. European Journal of Medicinal Chemistry, 46, 647-658.

26. Abouzid, K. and Shouman, S. (2008) Design, Synthesis and in Vitro Antitumor Activity of 4-Aminoquinoline and 4-Aminoquinazoline Derivatives Targeting EGRF Tyrosine Kinase. Bioorganic & Medicinal Chemistry, 16, 7543-7551.
27. Morin, M.J. (2000) From Oncogene to Drug: Development of Small Molecule Tyrosine Kinase Inhibitors as Antitumor and Anti-Angiogenic Agents. Oncogene, 19, 6574-6583.
28. Bridges, A.J. (2001) Chemical Inhibitors of Protein Kinases. Chemical Reviews, 101, 2541-2572.
29. Xia, Y., Yang, S.Y., Hou, M.J., Kuo, S.C., Xia, P., Bastow, K.F., Nakanishi, Y., Nampoothiri, P., Hackl, T., Hamel, E. and Lee, K.H. (2001) Antitumor Agents. Part 204: Synthesis and Biological Evaluation of Substituted 2-Aryl Quinazolinones. Bioorganic & Medicinal Chemistry Letters, 11, 1193-1196.
30. Al-Rashood, S.T., Aboldaha, I.A., Nagi, M.N., Abou-Zeid, L.A., Abdel-Aziz, A.A., Abdel-Hamide, S.G., Youssef, K.M., Al-Obaid, A.M. and Subbagh, H.I. (2006) Synthesis, Dihydrofolate Reductase Inhibition, Antitumor Testing and Molecular Modeling Study of Some New 4(3H)-Quinazolinone Analogs. Bioorganic & Medicinal Chemistry, 14, 8608-8621.
31. Converso, A., Hartingh, T., Garbaccio, R.M., Tasber, E., Richert, K., Fraley, M.E., Youwei, Y., Kreatsoulas, C., Stirdivant, S., Drakas, B., Walsh, E.S., Kelly, H., Carolyn, B.A., Xianzhi, M., Marc, A.T., Stephen, B.C., Weikang, T., Rob, L., Loura, S.L., Joan, Z.M., Vinod, S., Sanjeev, M.K., Sylvie, J.S., Poul, Z.D. and Hartman, G.D. (2009) Development of Thioquinazolinones, Allosteric Chk1 Kinase Inhibitors. Bioorganic & Medicinal Chemistry Letters, 19, 1240-1244.
32. Raffa, D., Edler, M.C., Daidone, G., Maggio, B., Merickech, M., Plescia, S., Schillaci, D., Bai, R. and Hamel, E. (2004) Synthesis, Cytotoxicity and Inhibitory Effect on Tubulin Polymerization of 3-Heterocyclosubstituted 2-Styrylquinazolinone. European Journal of Medicinal Chemistry, 39, 299-304.
33. Alagarsamy, V., Revathi, S., Kalaiselvi, R., Phuvaneswari, S., Revathi, R., Amuthalakshmi, S., Vijay, S.K., Kumar, S.M.S., Angayarkanni, T., Sarathadevi, M., Saravan, K.S., Thaugatiruppathy, A., Venkatnarayanan, R. and Vankatesatermal, R. (2003) Analgesic, Anti-Inflammatory and Antibacterial Activity of Some Novel 2-Phenyl-3-(substituted methylamino)quinazolin-4(3H)-ones. Indian Journal of Pharmaceutical Sciences, 534-536.
34. Gupta, A., Mishra, P., Kashaw, S.K., Jatav, V. and Stables, J.P. (2008) Synthesis and Anticonvulsant Activity of Some Novel 3-Arylamino/Amino-4-aryl-5-imino-Delta(2)-1,2,4-Thiadiazoline. European Journal of Medicinal Chemistry, 43, 749-754.
35. Ashcroft, A.J., Davies, E.E. and Morgan, G.J. (2003)Aetiology of Bone Disease and the Role of Bisphosphonates in Multiple Myeloma. Lancet Oncology, 4, 284-292.
36. Neiementowski, V. (1895) Neiementowski Quinazoline Synthesis. Journal für Praktische Chemie, 51, 564.
37. Selvam, P., Breitenbach, J.M., Borysko, K.Z. and Drach, J.C. (2010) Synthesis, Antiviral Activity and Cytotoxicity of Some Novel 2-phenyl-3-Disubstituted Quinazolin-4(3H)-Ones. International Journal of Drug Design and Discovery, 1, 149-154.
38. Hemalatha, K. and Girija, K. (2011) Synthesis of Some Novel 2,3-Disubstituted Quinazolinones Derivatives as Analgesic and Anti-Inflammatory Agents. International Journal of Pharmacy and Pharmaceutical Sciences, 3, 103-106.
39. Raffa, D., Daidone, G., Maggio, B., Cascioferro, S., Plescia, F. and Schillaci, D. (2004) Synthesis, Antileukemic Activity of New 3-(5-Methylisoxazol-3-yl) and 3-(Pyrimidin-2-yl)-2-styrylquinazolin-4(3H)-ones. IL Farmaco, 59, 451- 455.
40. Jatav, V., Kashwa, S. and Mishra, P. (2008) Synthesis, Antibacterial and Antifungal Activity of Some Novel 3-[5-(4- Substituted phenyl)1,3,4-thiadiazol-2-yl]-2-styrylquinazolin-4(3H)-ones. Medicinal Chemistry Research, 17, 169-181.
41. Okano, M., Mito, J., Maruyama, Y., Masuda, H., Niwa, T., Nakagawa, S., Nakamura, Y. and Matsceura, A. (2009) Discovery and Structure-Activity Relationships of 4-Aminoquinazoline, a Novel Class of Opioid Receptor Like-1(ORL1) Antagonists. Bioorganic & Medicinal Chemistry, 17, 119-132.

42. Amin, K.M., Georgey, H.H. and Awadallah, F.M. (2011) EGFR Tyrosine Kinase Targeted Compounds: Synthesis, Docking Study and in Vitro Antitumor Activity of Some Newquinazoline and Benzo[d]isothiazole Derivatives. Medicinal Chemistry Research, 20, 1042-1053.
43. Stadlbauer, W., Fiala, W., Fisher, M. and Hojas, J. (2000) Thermal Cyclization of 4-Azido-3-nitropyridine-2-suroxanes. Journal of Heterocyclic Chemistry, 37, 1253-1256.
44. Chandrappa, S., Prasad, S.B.B., Vinaya, K., Kumar, C.S.A., Thimmegowda, N.R. and Rangappa, K.S. (2008) Synthesis and in Vitro Antiproliferative Activity against Human Cancer Cell Lines of Novel 5-(4-Methyl-benzylidine)-thiazolidino-2,4- diones. Investigational New Drugs, 26, 437-444.
45. Yurchenko, R.I., Ponomarenko, A.D., Savina, N.S. and Tolmachev, A.A. (2004) 2-(1-Adamantyl)-7-methylimidazo[1,2-a]pyridine and Its Reactions with N-Bromosuccinimide. Chemistry of Heterocyclic Compounds, 40, 1543-1545.
46. Berest, G.G., Voskoboynik, O.Y., Kovalenko, S.I., Antypenko, O.M., Nosulenko, I.S., Katsev, A.M. and Shandrovskaya, O.S. (2011) Synthesis and Biological Activity of Novel N-Cycloalkyl-(cycloalkylaryl)-2-[(3-R-2-oxo-2H- [1,2,]">4]triazino[2,3-c]quinazolin-6-yl)thio]acetamides. European Journal of Medicinal Chemistry, 46, 6066-6074.
47. Abouzid, K. and Shouman, S. (2008) Design, Synthesis and in Vitro Antitumor Activity of 4-Aminoquinoline and 4- Aminoquinazoline Derivatives Targeting EGFR Tyrosine Kinase. Bioorganic & Medicinal Chemistry, 16, 7543-7551.
48. Chemical Computing Group (2005) Molecular Operating Enviroment (MOE). Version 2005.06, Chemical Computing Group, Inc. Montreal. http://www.chemcomp.com
49. Mosmann, T. (1983) Rapid Colorimetric Assay for Cellular Growth and Survival Application to Proliferation and Cytotoxicity Assays. Journal of Immunological Methods, 65, 55-63.
50. Vijayan, P., Raghu, C., Ashok, G., Dhanaraj, S.A. and Suresh, B. (2004) Antiviral Activity of Medicinal Plants of Nilgiris. Indian Journal of Medical Research, 120, 24-29.

CHAPTER 10

Synthesis and Evaluation of Folate-Conjugated Phenanthraquinones for Tumor-Targeted Oxidative Chemotherapy

Ajay Kumar[1,2*], Venkatesh Chelvam[3,4], Mahalingam Sakkarapalayam[3], Guo Li[3], Pedro Sanchez-Cruz[2], Natasha S. Piñero[2], Philip S. Low[3*], Antonio E. Alegria[2*]

[1]*International Center for Trans-disciplinary Research, School of Environmental Affairs, Universidad Metropolitana, San Juan, Puerto Rico*

[2]*Department of Chemistry, University of Puerto Rico, Humacao, Puerto Rico*

[3]*Department of Chemistry, Purdue University, West Lafayette, Indiana*

[4]*Department of Chemistry, Centre for Biosciences and Biomedical Engineering, Indian Institute of Technology, Indore, Simrol Campus, Madhya Pradesh, India*

ABSTRACT

Almost all cells are easily killed by exposure to potent oxidants. Indeed, major pathogen defense mechanisms in both animal and plant kingdoms involve production of an oxidative burst, where host defense cells show an invading pathogen with reactive oxygen species (ROS). Although cancer cells can be similarly killed by ROS, development of oxidant-producing chemotherapies has been limited by their inherent nonspecificity and potential toxicity to healthy cells. In this paper, we describe the targeting of an ROS-generating molecule selectively to tumor cells using folate as the tumor-targeting ligand. For this purpose, we exploit the ability of 9,10-phenanthraquinone (PHQ) to enhance the continuous generation of H_2O_2 in the presence of ascorbic acid to establish a constitutive source of ROS within the tumor mass. We report here that incubation of folate receptor-expressing KB cells in culture with folate-PHQ plus ascorbate results in the death of the cancer cells with an IC_{50} of ~10 nM (folate-PHQ). We also demonstrate that a cleavable spacer linking folate to PHQ is significantly inferior to a noncleavable spacer, in contrast to most other folate-tar- geted therapeutic agents. Unfortunately, no evidence for folate-PHQ mediated tumor regression in murine tumor models is obtained, suggesting that unanticipated impediments to generation of cytotoxic quantities of ROS in vivo are encountered. Possible mechanisms and potential solutions to these unanticipated results are offered.

Keywords: Cancer, Folate Receptor, Reactive Oxygen Species

1. INTRODUCTION

Because cancer cells are rapidly mutating, they can develop resistance to almost any chemotherapeutic agent, resulting in recurrence of malignant disease. Thus, drug resistance can arise from mutations in the binding site of the drug on its therapeutic target, upregulation of multidrug resistance pumps, induction of immunosuppressive mechanisms, over-expression of the drug target, upregulation of catabolic pathways for drug degradation, or over-expression of alternate signaling pathways [1] . To avoid emergence of such resistance mechanisms, novel therapeutic agents will have to be designed that are immune to the above classes of escape mutations.

One class of drugs that could conceivably avoid escape mutations constitutes those molecules that continuously generate reactive oxygen species (ROS). Evidence that constant exposure to ROS does indeed constitute a mutation resistant cytotoxic mechanism derives from the fact that both the animal and plant kingdoms have exploited this strategy to kill invading pathogens since early evolutionary times [2] [3] . Indeed, oxidants and bleaches are still used in hospitals to sanitize areas exposed to potentially hazardous human body fluids [3] . Moreover, ROS generators need not enter cells to kill their targets and are consequently not subject to inactivation by multidrug resistance pumps. Although upregulation of antioxidant activity can confer some ROS resistance on certain cell types, cancer cells have been reported to have inherently poor antioxidant protection, and therefore, would be expected to be susceptible to killing by sustained oxidative stress [4] - [7] .

Because autoxidation of ascorbate produces H_2O_2 as a final product [8] [9] , ascorbate has been frequently proposed as a possible anticancer agent. Indeed, several recent reports have revisited the old controversy regarding the anticancer activity of ascorbate [10] [11] and have found significant cytotoxicity towards such cancers as prostate cancer [12] - [15] , neuroblastoma [16] , bladder carcinoma [15] [17] , malignant mesothelioma [18] [19] , chronic lymphocytic leukemia [20] , hepatocellular carcinoma [15] , mammary carcinoma [15] and cancer of the cervix [15] . In these studies, intravenous, but not oral, administration of ascorbate has been shown to generate both ascorbate radical and hydrogen peroxide in the extravascular space of both normal tissues [8] [21] and tumor xenografts [21] . These conditions not only result in decreased growth of tumor implants in mice [14] [15] [19] [22] - [26] , but also enhanced toxicity when administered together with menadione (1), a redox-cycling quinone that can continuously generate ROS in the presence of ascorbate [22] [27] [28] . The rationale for supplementing ascorbate therapy with 1 is that quinones such as 1 catalyze ascorbate-mediated production of H_2O_2 [9] by first accepting an electron from ascorbate to form a semiquinone and then reacting with oxygen to generate superoxide anion radical, $O_2^{\bullet-}$. The instability of this superoxide anion radical then results in its rapid disproportionation in water to hydrogen peroxide [29] , which in turn can react via an iron-catalyzed Haber-Weiss reaction to form the hydroxyl radical [30] . Because of the nonspecific reactivity of the hydroxyl radical, it can damage a variety of biological systems via induction of lipid and protein peroxidation [31] , degradation of deoxyribose [32] and promotion of DNA strand breaks [33] [34] , leading to cell death.

There is evidence that the rate limiting step in $O_2^{\bullet-}$ formation during redox cycling of simple quinones lies in the rate of quinone reduction and not reaction with oxygen [35] - [37] . Thus, rates of oxygen consumption by quinones in the presence of reducing agents are found to increase with quinone redox potential within a certain range of redox potentials, in accordance with Marcus electron transfer theory [35] [38] . We, therefore, hypothesized that if a more redox-active quinone than 1 was used in combination with ascorbate, more efficient anti- cancer activity should be observed.

Unfortunately, a major drawback of most oxidant-generating drugs is their lack of selectivity for malignant cells. Thus, when administered systemically, like other chemotherapeutic agents they distribute indiscriminately into all tissues, causing toxicity to malignant cells and healthy cells alike. For example, the LD_{50} value of 1 is 75 mg/rat Kg when administered intraperitoneally [39] . To circumvent this problem, we have exploited the use of folic acid to target attached drugs to folate receptors (FR) that are often over-expressed on cancer cells. Thus, human cancers known to upregulate FR relative to normal tissues include cancers of the ovary, lung, breast, kidney, brain, endometrium, colon, and hematopoietic cells of myelogenous origin [40]

Following binding to the FR, folate-drug conjugates are internalized by receptor-mediated endocytosis [41] and trafficked into recycling endosomes, where disulfide linkers connecting folate to its therapeutic payload can be cleaved, releasing the cytotoxic drug [42] . When desired, noncleavable linkers can also be designed that retain the drug in the targeting conjugate, preventing its release and diffusion from the cell. Folic acid (FA) conjugates of proteins, nucleotides, radiopharmaceuticals, imaging compounds, chemotherapeutic agents and a variety of nanoparticles have all been prepared and tested [43] - [46] . In most cases, FA conjugation to cytotoxic drugs has been shown to decrease their unwanted side effects and/or improve drug efficacy [43] [44] .

In this paper, we have exploited folic acid to target redox cycling derivatives of 9,10-phenanthraquinone (PHQ) to cancer cells in an attempt to facilitate ascorbate-mediated oxidation and killing of malignant cells. We selected PHQ as the redox cycling catalyst because its more positive redox potential was found to facilitate oxidation of ascorbate ~20 times faster than menadione [9] . Moreover, unlike menadione, PHQ does not undergo Michael addition with free thiols, resulting in improved stability in vivo where free thiols can be common. In addition, we have recently observed that PHQ is only 14% less active in producing OH radicals from H_2O_2 + ascorbate than the classic Fenton reagent cation, Fe^{2+}, at the same concentration (unpublished results). Thus, hydroxyl radical could also be produced where H_2O_2, ascorbate and PHQ are all coexisting.

2. MATERIALS AND METHODS

2.1. Chemicals

Protected amino acids were purchased from AAPTEC or CHEMIMPEX.N^{10}-(trifluoroacetyl)pteroic acid was purchased from Schirks. Other reagents are available from Sigma-Aldrich Chemicals, unless otherwise stated. Some of the important reagents are such as 3-acetylphenanthrene, mono-Boc-1,2-ethyldiamine. HCl, MeOCO(O)SCl, 2-mercaptopyridine, Fmoc-Lys(mtt)-Wang resin, Fmoc-Cys(4-methoxytrityl)-Wang resin, N^{10}TFA- pteroic acid, δ-protected Fmoc-Glu(O^{t}Bu)OH [where Fmoc = fluorenylmethy-loxycarbonyl & ^{t}Bu = tertiarybutyl], diisopropylethylamine (DIPEA), α-protected Fmoc-Glu(O^{t}Bu)OH, TFA, benzotriazol-1-yl-oxytripyrro- lidino-phosphonium hexafluorophosphate (PyBOP), isopropanol (i-PrOH), piperidine, triisopropyl silane (TIPS), ethanedithiol (EDT), hydrazine, dichloromethane (DCM), dimethyl formamide (DMF), and diethyl ether.

2.2. Equipment

^{1}H and ^{13}C NMR spectra were recorded using a Bruker 400 MHz NMR spectrometer. LC-ESI-MS spectra were obtained using a Micromass Quattro Micro API triple quadrupole mass spectrometer with an ESI source. A Waters 1500 HPLC with a 20.0 mL loop injector was used for semi-preparative chromatography. An Agilent 1100 UV analytical HPLC chromatograph was used to identify compound purity.

2.3. Cell Culture and Animal Husbandry

KB, A-549 and MDA-MB-231 cells were obtained from American Type Culture Collection. MDA-MB-231 cells expressing high levels of FR were generated by passaging the cells for 14 weeks in folate-free cell culture medium. The cell lines were grown continuously as a monolayer in folate-free RPMI medium containing 10% fetal bovine serum and 1% penicillin/streptomycin antibiotic cocktail in a 5% CO_2: 95% air-humidified atmosphere at 37°C.

All animal procedures were approved by the Purdue Animal Care and Use Committee in accordance with NIH guidelines. Normal rodent diets were not used, since they contain excessive amounts of folic acid which elevates serum folate levels significantly above normal physiological concentrations. Rather, all animals were maintained on a folate deficient diet (Harlan Teklad laboratories, WI) for at least 3 weeks prior to each study to lower their serum folate levels into the physiological range. Control animals were also maintained on a folate deficient diet.

2.4. Synthesis

2.4.1. 3-Carboxyphenanthrenequinone

This compound was synthesized and purified as described by Jacobsen et al. by CrO_3 oxidation of 3-acetylphe- nanthrene [47] .

3-Carboxyllic phenanthrenequinone was synthesized and purified as described by Jacobsen et al. by CrO_3 oxidation of 3-acetylphenanthrene.[60] 3-Acetylphenanthrene (5 g, 0.23 mmol) was dissolved in warm acetic acid (100 ml, 60°C) and chrom(VI)oxide (30 g, 0.6 mol) was added in small portions (Scheme 1). During this procedure the temperature rose to the boiling point. After addition of all chrom(VI)oxide the solution was diluted with water (500 ml) and the yellow precipitate was filtered off, washed with acetic acid/H_2O (1:1), cold acetic acid and finally with diethyl ether. Yield: 2.1 g (37% based on 3-acetylphenanthrene); MP: 280°C ^1H NMR (400 MHz, DMSO-d_6) δ: 7.577 (t, 1H, J = 7.55 Hz), 7.808 (t, 1H, J = 7.55 Hz), 8.041 (m, 2H), 8.120 (d, 1H, J = 8.12 Hz), 8.339 (d, 1H, J = 8.08 Hz), 8.706 (s, 1H), 13.636 (bs, 1H). ^{13}C NMR (100 MHz, DMSO-d_6) : ppm 125.00, 125.27, 129.64, 129.80, 129.92, 130.16, 132.00, 134.56, 134.94, 135.86, 135.95,136.74, 178.81, 178.87. ESI-MS: m/z 251 (calcd.) 252 $(M + H)^+$ found.

2.4.2. Non-Releasable FA-Conjugate of PHQ (3)

The synthetic protocol for 3 appears in Scheme 2. The use of a lysine-containing resin enabled the coupling of both the quinone and folate-linker to the resin followed by detachment and purification.

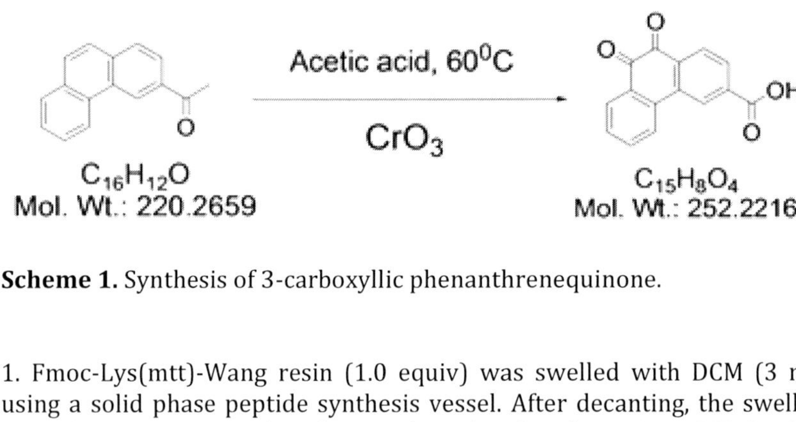

Scheme 1. Synthesis of 3-carboxyllic phenanthrenequinone.

1. Fmoc-Lys(mtt)-Wang resin (1.0 equiv) was swelled with DCM (3 mL) using a solid phase peptide synthesis vessel. After decanting, the swelling procedure was repeated with DMF (3 mL). After decanting DMF, 3 mL of 1% TFA in DMF was added to the resin and bubbled with Ar to promote homogeneous mixing for 5 min. This procedure was repeated 3 times and the resin was then washed with DMF (3 × 3 mL) and i-PrOH (3 × 3 mL). Formation of free amine was assessed by the Kaiser/ninhydrin test, where blue color shows the de- protection of amine and no more absorption at 304 nm in the de-protection waste showed complete deprotection.

2. The resin was swelled again in DMF. A solution of 3-carboxyllic phenanthraquinone (2.5 equiv), PyBOP (2.5 equiv), and DIPEA (4.0 equiv) in DMF was added and Ar was bubbled for 2 h. The resin was then washed with DMF (3 × 3 ml) and i-PrOH (3 × 3 mL). The coupling efficiency was assessed using the Kaiser test (absence of blue color indicates complete loading of protected glutamic acid).

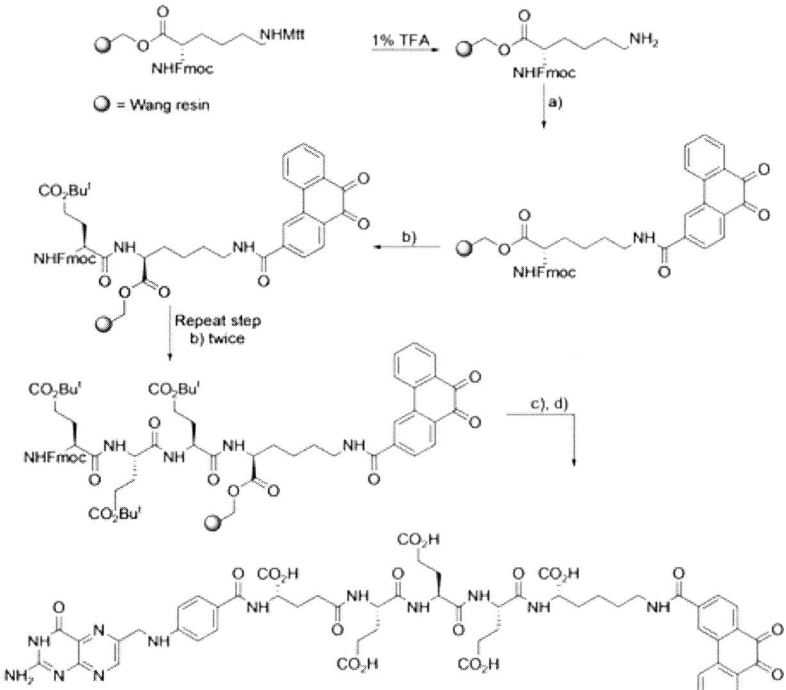

3

Chemical Formula: $C_{55}H_{58}N_{12}O_{19}$
Molecular Weight: 1191.1180

Scheme 2. Synthesis of 3: reagents and conditions: (a) 3-carboxy-9,10-phenanthraquinone, PyBOP, DIPEA, DMF; (b) (i) 20% piperidine, DMF, (ii) Fmoc-Glu(OtBu)-OH, PyBOP, DIPEA, DMF; (c) (i) 20% piperidine, DMF, (ii) N(10)-TFA-pte- roic acid, PyBOP, DIPEA, DMF; d) (i) TFA:TIPS:water (95:2.5:2.5:), (ii) sat.Na_2CO_3, pH = 10 - 11, 30 min.

3. The resin was swelled again in DMF and the solvent decanted. Three mLs of 20% piperidine in DMF was added to the resin and Ar bubbled for 5 min. This de-protection procedure was repeated three times and the resin was washed with DMF (3 × 3 mL) and i-PrOH (3 × 3 mL). Free amine formation was assessed by the Kaiser test.

4. The resin was again swelled in DMF and a solution of δ-protected Fmoc-Glu(O^tBu)-OH (2.5 equiv), PyBOP (2.5 equiv), and DIPEA (4.0 equiv) in DMF was added. Ar was bubbled for 2 h. The resin was then washed with DMF (3 × 3 ml) and i-PrOH (3 × 3 mL). The coupling efficiency was assessed using the Kaiser Test. Steps 3 and 4 were repeated to load two units of δ-protected Fmoc-Glu(O^tBu)-OH.

5. The resin was swelled in DMF and the solvent decanted. Three mLs of 20% piperidine in DMF were then added to the resin and Ar bubbled for 5 min. This procedure was repeated three times and the resin was washed with DMF (3 × 3 mL) and i-PrOH (3 × 3 mL). Formation of free amine was assessed by the Kaiser Test.

6. The resin was again swelled in DMF and a solution of α-protected Fmoc-Glu(O^{t}Bu)-OH (2.5 equiv), PyBOP (2.5 equiv), and DIPEA (4.0 equiv) in DMF was added. Ar was bubbled for 2 h. The resin was then washed with DMF (3 × 3 ml) and i-PrOH (3 × 3 mL). The coupling efficiency was assessed using the Kaiser Test.

7. The resin was again swelled in DMF. A solution of N^{10}-TFA-pteroic acid (1.25 equiv), PyBOP (2.5 equiv), and DIPEA (4.0 equiv) in DMF (and or DMSO) was then added. Ar was then bubbled for 8 h and the resin washed with DMF (3 × 3 ml) and i-PrOH (3 × 3 mL). The coupling efficiency was assessed using the Kaiser Test.

8. The resin was then washed with DMF (3 × 3 mL) and i-PrOH (3 × 3 mL). The final compound was cleaved from the resin using 3mL TFA:H_2O:TIPS (95:2.5:2.5) cocktail (30 min × 3) and concentrated under vacuum. The concentrated product (3) was then precipitated in ice cold diethyl ether, centrifuged and dried under vacuum.

9. The crude product (3) was dispersed in deoxygenated Millipore water (5 mL) and the solution pH was adjusted to 10 - 11 using a deoxygenated saturated aqueous solution of Na_2CO_3 with continuous bubbling of argon through the solution for 30-min to de-protect N^{10}-TFA on the folate moiety.

10. The crude product was purified by preparative RP-HPLC using a Waters XTerra MS C18 HPLC column 19 × 250 mm, 5 um, with detection at λ = 285 nm using the gradient 1%B to 50%B for 30 min, and washing with 80% B, for 5 min, where A = ammonium acetate buffer 20 mM, pH = 7 and B = ACN. Fractions were analyzed using analytical RP-HPLC and LC/MS. The combined fractions containing pure 3 were rotoevaporated under vacuum to remove ACN and lyophilized for 36 h to yield the final product as yellow solid. LC-MS: m/z 1190.4 (calcd.); found 1191.6 [M+H].

2.4.3. Releasable FA-Conjugate of PHQ (4)

In order to synthesize this conjugate, several precursors were prepared. First, 2 and the pyridyldisulfide-substi- tuted phenanthraquinone (5) were synthesized, Scheme 2. A folate-linker terminating with a cysteine (6) was also synthesized prior to assembly of the final product (4) (Schemes 3-6).

Synthesis of 2

Procedure. 3-Carboxylic phenanthrenequinone (186 mg, 0.7 mmol), DCC (178 mg, 0.84 mmol), and HOBt (117 mg, 0.84 mmol) were dissolved in dry dioxane (50 ml), and the mixture was stirred overnight at room temp. The dioxane was evaporated in vacuo, and the residue was dissolved again in anhydrous DMF (25 ml). Mono-Boc-1,2-ethyldiamine.HCl (122 ul, 0.77 mmol) was added to the suspension/solution followed by excess triethylamine (1 ml). After 1 hour, DCU (dicyclohexyl urea) was filtered off and water was added (150 ml). The yellow precipitate was collected by filtration, washed several times with cold water. Yield: 230 mg (83.3%). De-protection of the amine: PHQ-3-EDABoc (47.4 mg) was dissolved slightly in warmed acetic acid (1.3ml, 50°C), and 1 M HCl in acetic acid was added (1.3 ml). After 5 minutes, cold ether was added (10 ml), and the precipitated 2 was collected by filtration and washed several times with ether followed by a single wash with acetonitrile. Yield: 39 mg (92%). ^{1}H NMR (400 MHz, DMSO-d$_6$) δ: 3.070 (bs, 2H), 3.584 (m, 2H), 7.590 (t, 1H, J = 7.52 Hz), 7.844 (t, 1H, J = 7.52 Hz), 7.959 (bm, 3H), 8.065 (d, 1H, J = 8.01 Hz), 8.111 (d, 1H, J = 8.01 Hz), 8.364 (d, 1H, J = 8.14 Hz), 8.678 (s, 1H), 9.083 (t, 1H, J = 5.32 Hz). ^{13}C NMR (100 MHz, DMSO-d$_6$): ppm 37.75, 39.02, 123.51, 124.98, 128.53, 129.58, 129.68, 130.14, 132.00, 133.51, 135.28, 135.73, 135.77, 139.92, 178.95, 179.02.

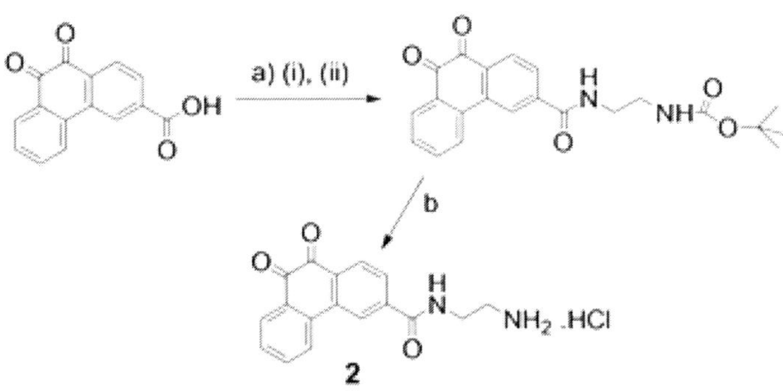

Scheme 3. Synthesis of 2. Reagents and conditions: (a) (i) HOBt, DCC, dioxane, (ii) BocNH(CH$_2$)$_2$NH$_2$, Et$_3$N, CH$_2$Cl$_2$; (b) HCl, CH$_3$COOH.

Synthesis of 5

Compound 5 was synthesized as described by Vlahov et.al.[61] and purified by silica column using 10% EtOAc/Hexane. Yield 72%. ^{1}H NMR (400 MHz, CDCl$_3$) δ: 3.023 (t, 2H, J = 6.13 Hz), 3.523 (m, 2H), 3.634 (m, 2H), 4.385 (t, 2H, J = 6.09 Hz), 7.070 (m, 1H), 7.495 (bm, 1H), 7.513 (bm, 1H), 7.614 (bm, 1H), 7.733 (bm, 1H), 7.841 (d, 1H, J = 8.19 Hz), 8.186 (bm, 3H), 8.375 (d, 1H,

J = 4.73 Hz), 8.563 (s, 1H). ^{13}C NMR (100 MHz, $CDCl_3$): ppm 37.758, 40.44, 42.28, 63.56, 119.93, 121.01, 123.56, 124.48, 127.41, 130.04, 130.66, 130.81, 131.13, 132.55, 135.32 136.29, 137.19, 140.55, 149.60, 157.99, 159.71, 170.00, 179.96,. LC/MS 507.58 (calcd.) 508.57 found $[M+H]^+$.

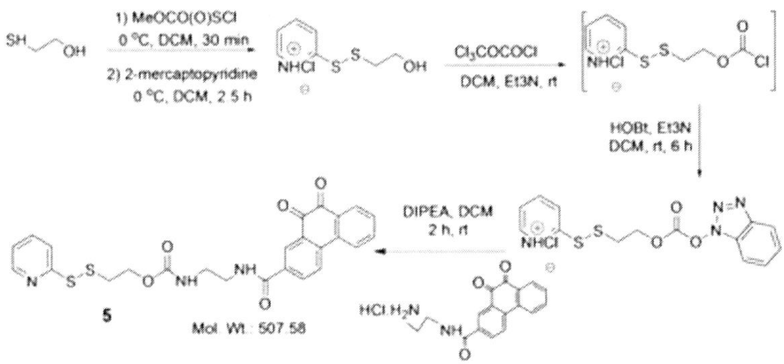

Scheme 4. Synthesis of 5.

Synthesis of 6

1. Fmoc-Cys(4-methoxytrityl)-Wang resin was swelled (1.0 equiv) with DCM (3 mL) using a solid phase peptide synthesis vessel. After decanting, the swelling procedure was repeated with DMF (3 mL). After decanting DMF, 3 mL of 20% piperidine in DMF was added to the resin and bubbled with Ar to promote homogeneous mixing for 5 min. This procedure was repeated 3 times and the resin was then washed with DMF (3 × 3 mL) and i-PrOH (3 × 3 mL). Formation of free amine was assessed by the Kaiser/ninhydrin test.

2. The resin was swelled again in DMF and the solvent decanted. Three mLs of 20% piperidine in DMF was added to the resin and Ar bubbled for 5 min. This procedure was repeated 3 times and the resin washed with DMF (3 × 3 mL) and i-PrOH (3 × 3 mL). Free amine formation was assessed by the Kaiser test.

3. The resin was again swelled in DMF and a solution of δ-protected Fmoc-Glu(O^{t}Bu)-OH (2.5 equiv), PyBOP (2.5 equiv), and DIPEA (4.0 equiv) in DMF was added. Ar was bubbled for 2 h. The resin was then washed with DMF (3 × 3 ml) and i-PrOH (3 × 3 mL). The coupling efficiency was assessed using the Kaiser test. Steps 3 and 4 were repeated to load two units of δ-protected Fmoc-Glu(O^{t}Bu)-OH.

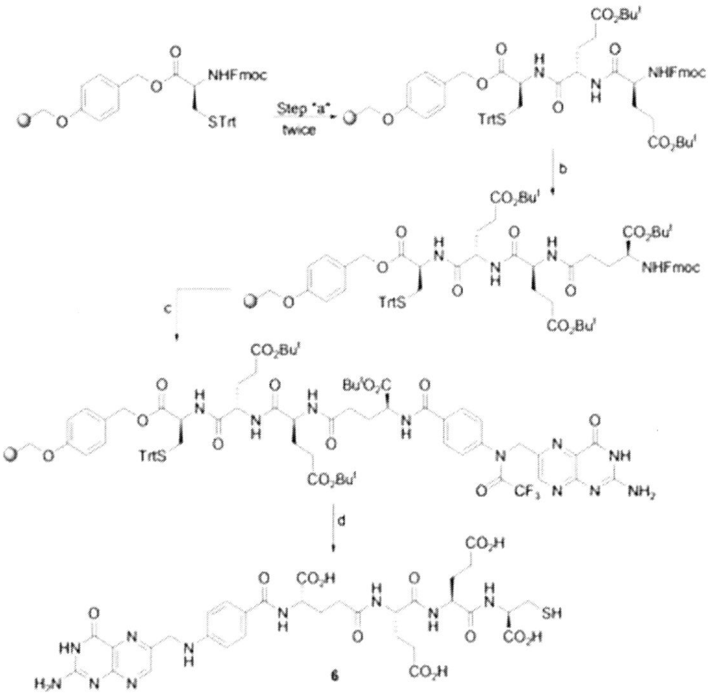

Scheme 5. Synthesis of 6: reagents and conditions: (a) (i) 20% piperidine/DMF, (ii) Fmoc-Glu(O^{t}Bu)-OH, PyBOP, DIPEA/ DMF; (b) (i) 20% piperidine/DMF, (ii) α-protected Fmoc-Glu(O^{t}Bu)-OH, PyBOP, DIPEA/DMF; (c) (i) 20% piperidine/ DMF, (ii) N^{10}TFA-Pteroic acid, PyBOP, DIPEA/DMF; (d) (i) 2% NH_2NH_2/DMF, (ii) TFA:TIPS:Water:EDT (92.5:2.5:2.5:2.5).

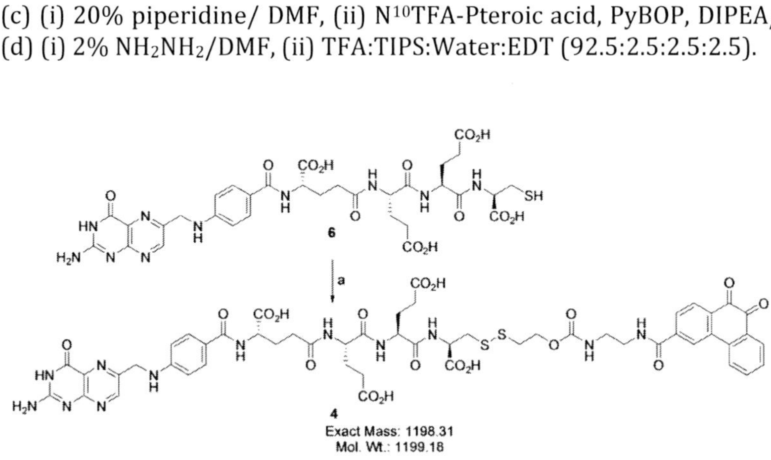

Scheme 6. Synthesis of 4: (a) 5, THF:H_2O (1:1), pH = 7.0, argon bubbling, rt, 45 min.

4. The resin was swelled in DMF and the solvent decanted. Three mLs of 20% piperidine in DMF were then added to the resin and Ar bubbled for 5 min. This procedure was repeated three times and the resin washed with DMF (3 × 3 mL) and i-PrOH (3 × 3 mL). Free amine formation was assessed by the Kaiser Test.

5. The resin was again swelled in DMF and a solution of α-protected Fmoc-Glu(O^{t}Bu)-OH (2.5 equiv), PyBOP (2.5 equiv), and DIPEA (4.0 equiv) in DMF was added. Ar was bubbled for 2 h. The resin was then washed with DMF (3 × 3 ml) and i-PrOH (3 × 3 mL). The coupling efficiency was assessed using the Kaiser Test.

6. The resin was again swelled in DMF. A solution of N^{10}-TFA-pteroic acid (1.25 equiv), PyBOP (2.5 equiv), and DIPEA (4.0 equiv) in DMF was then added. Ar was then bubbled for 8 h and the resin washed with DMF (3 × 3 ml) and i-PrOH (3 × 3 mL). The coupling efficiency was assessed using the Kaiser Test.

7. The resin was then treated with 2% anhydrous hydrazine in DMF for 2-min (3 × 2 mL) and washed with DMF (3 × 3 mL) and i-PrOH (3 × 3 mL). The final compound was cleaved from the resin using 3mL TFA: H_2O: TIPS (95:2.5:2.5) cocktail (30 min × 3) and concentrated under vacuum. The concentrated product 6 was then precipitated in ice old diethyl ether, centrifuged and dried under vacuum.

8. The crude product was purified using preparative RP-HPLC at λ = 285 nm (1%B to 50%B for 30 min, 80%B wash for 5 min; A = 0.1% TFA, pH = 2; B = ACN; column: Waters, xTerra C_{18} 10 μm; 19 x 250 mm, flow rate = 10 mL/min). Fractions were analyzed using analytical RP-HPLC and LC-MS: m/z 802 (calcd.) 803 found [M + H]. The fractions containing pure folate-Cys were combined, can was removed and lyophilized for 36 h to yield final product as yellow solid. LC-MS: m/z 802.2 (calcd.) 803.5 found $[M + H]^+$.

Synthesis of 4

Compound 6 was dissolved in de-oxygenated H_2O under argon bubbling with 0.1 N $NaHCO_3$resulting in a clear yellow solution at pH > 6.5. To this mixture it was added at once, under extensive stirring and bubbling of argon, a solution of 5 in THF. According to the HPLC profile, the reaction was completed in 20 min. HPLC purification gave pure conjugate 4. LC-MS: m/z 1198.3 (calcd.) 1199.2 found $[M + H]^+$.

2.5. Ascorbate Autoxidation

Oxygen consumption rate measurements were performed at 37.0°C ± 0.1°C in 20 mM phosphate buffer (pH 7.4). Stock and sample solutions were prepared with double distilled deionized water and decontaminated from traces of transition metals by exposure to Chelex 100 resin using the batch method [48] . Air-saturated sample solutions were used containing micromolar amounts of quinone and millimolar amounts of ascorbate (simulating parenteral ascorbate concentrations) and 20 mM phosphate buffer (pH 7.4). Oxygen consumption rates were measured with a 5300 Oxygen Biological Monitor (Yellow Springs Instruments Co., USA) using a Clark electrode as a sensor in a YSI 5301 constant temperature bath. Rates were calculated from the initial constant slopes of [O_2] traces. Runs started in the absence of ascorbate followed by ascorbate addition, without interrupting oxygen consumption measurements.

2.6. Quinone Redox Potentials

Since 3 and 4 are not soluble in acetonitrile, half-wave reduction potentials ($E_{1/2}$) were determined in 1:3 (v/v) nitrogen purged DMSO:20 mM phosphate buffer (pH 7.4). In this solvent a single peak corresponding to a 2 electron + 2 proton reduction step is observed for quinones [49] . Solutions contained up to 500 μM quinone and 0.1 M tetra-n-butylammonium perchlorate (TBAP). A BAS CV 50W voltammetric analyzer with a glassy carbon working electrode was used in these determinations. An Ag/AgCl(sat) electrode served as the reference electrode (E' = +0.22 V vs. NHE) and a platinum wire as the counter electrode. Differential pulse voltammograms (DPV) were obtained in the potential range of −2.00 to 0.00 volts, using a 50 mV pulse amplitude and 20 mV/s of scan rate. Reduction potential values were obtained from the DPV peak potential maxima. These are very similar to the half-wave redox potentials, $E_{1/2}$, in normal polarographic measurements [50] . Since compound 2 (see Figure 1 & Table 1) represents the quinone moiety existing in 3 and 4 and 1 has been used in previous quinone + ascorbate cytotoxicity and antitumor activity studies, redox potentials and ascorbate oxidation rates were also determined for these compounds.

Table 1. Ascorbate oxidation rates (Rox)[a] and quinone redox potentials ($E_{1/2}$)[b].

Quinone	Rox/μM·s^{-1}	$E_{1/2}$/mV
2	1.30 ± 0.06	−304
3	1.02 ± 0.05	−316
4	0.87 ± 0.08	−336
1	0.54 ± 0.04	−372

[a]Obtained after addition of 1 mM ascorbate to solutions containing 10 μM quinone in 20 mM, air-saturated, phosphate buffer. [b]Measured in 1:4 (v/v) DMSO:20 mM phosphate buffer (pH 7.4). DPV peak potential maxima are against Ag/AgCl.).

[a]Obtained after addition of 1 mM ascorbate to solutions containing 10 µM quinone in 20 mM, air-saturated, phosphate buffer. [b]Measured in 1:4 (v/v) DMSO:20 mM phosphate buffer (pH 7.4). DPV peak potential maxima are against Ag/AgCl.).

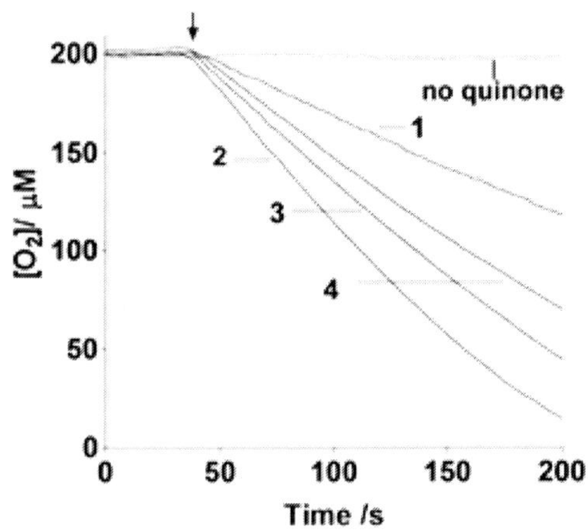

Figure 1. Oxygen consumption traces in air-saturated solutions containing 10 µM quinone and 1.00 mM ascorbate in 20 mM phosphate buffer at pH 7.4 and 37°C. The arrow indicates the time when ascorbate was injected. The numbers labeling the curves identify the quinone used to promote redox cycling.

2.7. In Vitro Cytotoxicity Assays of 2, 3 and 4

FR⁺KB cells and FR⁻ A-549 cells were seeded into 24-well (100,000 cells/well in 500 µL) Falcon plates and allowed to form monolayers over 12 h. Spent medium was replaced with fresh medium (0.5 mL) containing increasing concentrations of 2, 3 or 4 and incubated for 2 h. The unbound test compounds 2, 3 and 4 were washed from the wells with fresh medium (3 × 0.5 ml), followed by addition of 5, 3 and 0 mM (for a control experiments without sodium ascorbate) freshly prepared sodium ascorbate solution (pH = 7.4) in medium (0.5 mL/well). Cells were then incubated for an additional 2, 12, 16 or 68 h at 37°C. Cells were washed with fresh medium (3 × 0.5 ml) and incubated in fresh medium (0.5 mL) for 66, 56 or 0 h at 37°C. Spent medium in each well was replaced with fresh medium (0.5 mL) containing [^{3}H]-thymidine (1 µCi/mL), and cells were incubated for 4 h at 37°C to allow [^{3}H]-thymidine incorporation. Cells were then washed with medium (3 × 0.5 mL) and treated with 5% trichloroacetic acid (0.5 mL) for 10 min at room temperature. Cells were dissolved in 0.25 M NaOH (0.5 mL), transferred into individual vials containing Ecolume scintillation cocktail (3.0 mL), and counted in a scintillation counter (Packard, Packard Instrument Company). IC_{50} values were calculated by plotting % ^3H-thymidine incorporation versus log concentration of compounds 2, 3 and 4 using GraphPad Prism 4. Each % ^3H-thymidine incorporation value represents the average ± standard error of the mean of at least three determinations.

2.8. Mouse Tumor Model

Four to five week-old female nu/nu mice were inoculated subcutaneously with KB or MDA-MB-231 cells (1.0×10^6/mouse in cell growth medium) on their shoulders. Growth of the tumors was measured in two perpendicular directions every 2 days using a caliper (body weights were monitored on the same schedule), and the volumes of the tumors were calculated as $0.5 \times L \times W^2$(L = longest axis and W = axis perpendicular to L in millimeters). Tumor therapy studies were performed with 3 (40 or 80 nmol/mouse) in saline (100 μL) 12 or 20 days after tumor cell implantation, when the tumors reached approximately 75 - 100 mm^3 (for MDA-MB-231 tumor) and 100 - 350 mm^3 (for KB-tumor) in volume.

2.9. In Vivo Potency of 3

Female nude mice bearing KB or MDA-MB-231 xenograft tumors were injected with 3 (2.0 μmol/kg for KB tumors of volume 100 - 350 mm^3or 4.0 μmol/kg for MDA-MB-231 tumors of volume 75 - 100 mm^3) dissolved in 100 μL of saline via lateral tail vein injection. Treatments were conducted 3× per week for two weeks. A single dose of sodium ascorbate in saline (4g/kg or 80 mg/mouse) was injected 30-minutes post 3 injection and also on days when mice were not treated with 3. Tumor volumes and body weights were measured on the same schedule. In vivo efficacy was evaluated by plotting tumor growth % versus days on therapy. Tumor growth % is defined by Equation (1),

$$\text{Tumor Growth \%} = \left[\frac{V - V_0}{V_0}\right] \times 100\% \tag{1}$$

where V and V_0 are the tumor volumes on any given day and the first day of measurement, respectively. Reported tumor growth percents are the averages ± standard errors of the mean of at least three determinations.

3. RESULTS

3.1. Ascorbate Oxidation Rates vs. Redox Potentials

Upon addition of 1mM ascorbate to 10 μM quinone 1, 2, 3 or 4, an increase in the rate of ascorbate oxidation is observed (Figure 1). The reason for including compounds 2 and 1 in this study is that the former mimics the PHQ moiety in the folate conjugate while the latter serves as an appropriate comparison to quinones used by others, since most of the anticancer work with quinones + ascorbate has been performed with 1. It can be concluded from this figure that the ascorbate oxidation enhancement activity of these quinones follows the order 2 > 3 > 4 > 1, which coincides with the ranking of the quinone 2 electrons+2 protons reduction potentials (Table 1).

3.2. Cytotoxicity Assays in Vitro

Quinone 2 displayed cytotoxic activity against KB cells at 3 and 5 mM ascorbate, with IC_{50} values in the 10^{-8}M range, depending on incubation time (Table 2). Once this behavior was detected for 2, we were confident that a similar cytotoxic behavior was to be observed for 3 and 4 and thus were prompted to synthesize the FR-targeted quinones. Incubation of quinone 3 with 3 or 5 mM ascorbate for 2 hours exhibited very low cytotoxic activity in vitro (Table 2, entries 3 - 4), with the greatest cytotoxicity seen when cells were incubated for 12 hours ($IC_{50} < 10^{-9}$ M) in the presence of either 3 or 5 mM ascorbate (Table 2, entries 6 - 7). In contrast, incubation of quinone 4 conjugate with 3 or 5 mM ascorbate for up to 68 hours was only slightly cytotoxic to KB cells (Table 2, entries 8 - 9) and no cytotoxicity was expressed when incubated for 12 hours with ascorbate. In addition, incubation of 3 in the absence of ascorbate with KB cells did not show cytotoxicity if incubated for12 hours with ascorbate (Table 2, entry 5), indicating the need for ascorbate for cytotoxicity. Compound 3 in the presence of 3 mM ascorbate did not show toxicity, while it was mildly toxic when incubated for 12 hours, in the presence of 5 mM ascorbate, with FR negative A549 cells, thus demonstrating the need for cell surface binding to FR for expression of cytotoxic activity (Table 2, entries 10 - 11). The latter was expected, since 3 is extracted from FR negative cells upon rinsing due to its high hydrophilicity. These observations are in contrast to previous studies where KB [51] and A549 cells [21] were incubated for 1 to 2 hours with ascorbate followed by cell rinsing to remove free ascorbate and testing for cytotoxicity after incubation periods of more than one week. Ascorbate cytotoxicity to KB and A549 cells in the absence of quinones was then detected. Thus, interestingly, compound 3 is in fact accelerating the expression of this toxicity in FR+ cells.

Table 2. Cytotoxicity to KB (FR^+) and A549 (FR^-) cells of compounds 2, 3, 4 after incubation for 2 h followed by washing off unbound compounds with fresh medium (3 × 0.5 mL) and different incubation time in the presence of 3 mM or 5 mM ascorbate solution.[b]

S. No.	Compound No.	Cell line	Ascorbate concentration (mM)	Incubation time of cells with ascorbate (h)	IC_{50} nM (Nanomolar), pM (Pico molar)
1.	2	KB (FR^+)	3	16	14.7 nM
2.	2	KB (FR^+)	5	16	12.6 nM
3.	3	KB (FR^+)	3	2	Non-toxic
4.	3	KB (FR^+)	5	2	Non-toxic
5.	3	KB (FR^+)	0	12	Non-toxic
6.	3	KB (FR^+)	3	12	0.14 pM
7.	3	KB (FR^+)	5	12	0.0011 pM
8.	4	KB (FR^+)	3	68	Non-toxic
9.	4	KB (FR^+)	5	68	Non-toxic
10	3	A549 (FR^-)	3	12	Non-toxic
11.	3	A549 (FR^-)	5	12	Non-toxic

3.3. Analysis of Ascorbate plus Folate-Quinone Conjugate Toxicity in Tumor-Bearing Mice

Since the quinone conjugate 3 in combination with ascorbate demonstrated the greatest cytotoxicity in vitro, it was selected for more thorough testing against two FR+ tumor xenografts in vivo. As seen in Figure 2, using the protocol described in Materials and methods, essentially no reduction in tumor size was detected in either xenograft model, independent of whether quinone was injected or not. This result was not anticipated, since, as mentioned above, ascorbate alone, and ascorbate + 1, have been reported to display some anti-tumor activity in murine tumor models, and because FR expression has been found to be similar in murine tumor xenografts and the cultured cancer cells from which they are derived (unpublished observations). Clearly, unknown factors dominate the behavior of ascorbate or folate-PHQ in our murine tumor models.

4. DISCUSSION

4.1. Ascorbate Oxidation Rates vs. Redox Potentials

Although 2 electrons + 2 protons reduction potentials were measured here instead of 1 electron reduction potentials, the values still reflected the relative abilities of the quinones to undergo reduction, since previous work had shown that the rate of ascorbate oxidation in the presence of quinones increased with an increase in the one- electron redox potential of the quinone [9] . The electrochemical reaction, measured in the present work, Equation (2), corresponds to a redox potential ($E_{1/2}$), which is essentially equal to the summation of the first and second electron redox potentials, Equation (3).

$$2H^{+} + Q + 2e \rightarrow QH_2 \tag{2}$$

$$E_{1/2} \approx E\left(Q/Q^{\bullet -}\right) + E\left(Q^{?-}/QH_2\right) \tag{3}$$

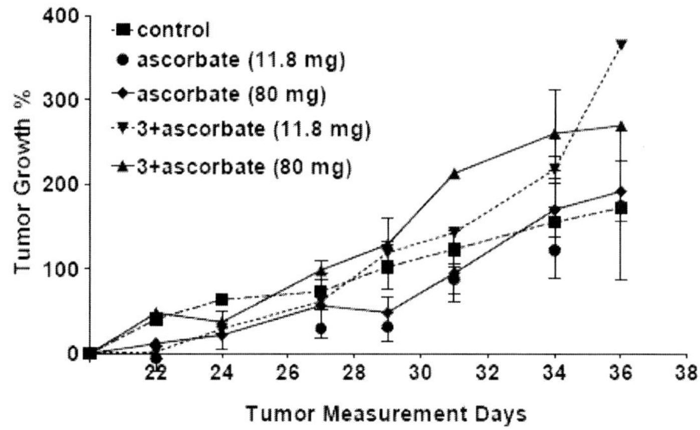

(a)

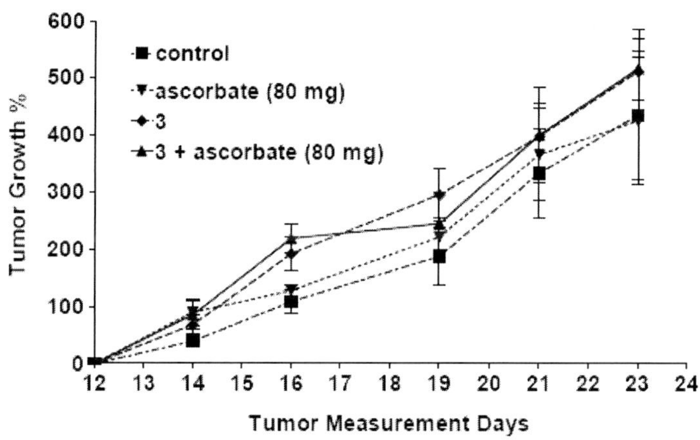

(b)

Figure 2. Effect of different combinations of ascorbate +/− various quinones on growth of FR+ tumor xenografts in athymic nude mice. (a) KB tumor implantation: 1 million KB cells were injected subcutaneously in 0.1 ml FD medium/mouse. Treatment began after the 20th day of tumor implantationwhen the average tumor size was 100 - 350 mm^3. Thecontrols are untreated mice. 3was i.v. injected at a dose of 2.0 mmol/kg, 3 days/week for two weeks. Sodium ascorbate was injected intraperitoneally at a dose of 4 g/kg or 80 mg per mouse 30 min after i.v. injection of 3 or 11.8 mg per mouse every 30 min for 90-min, 30 min after i.v. injection of 3; (b) FR+ MDA-MB-231 tumor implantation: 1 million FR+ MDA-MB-231 cells were injected subcutaneously in 0.1 ml FD medium/mouse. Treatment began after the12th day of tumor implantation when the average tumor size was 75 - 100 mm^3. Thecontrols are untreated mice. 3 was i.v. injected at a dose of 4.0 mmol/kg 3 days/week for two weeks.Sodium ascorbate was injected i.p. (4 g/kg or 80 mg/mouse) 30 min after 3 was injected and also on days when mice were not injected with 3.Tumor growth % values are averages ± standard errors of the mean at least four determinations.

Thus, since in general a quinone with a more positive E(Q/Q•?) will also have a more positive E(Q•?/QH_2) [52] , the trend in $E_{1/2}$ values shown here should roughly follow that in E(Q/Q•?) values. Not surprisingly, the relative order in $E_{1/2}$ values correlates nicely with the rate of ascorbate oxidation in Table 1. The discrepancy in ascorbate oxidation rates and in reduction potentials among compounds 2, 3 and 4 is an interesting subject. First of all, at physiological pH, compound 2 should be mostly protonated at the primary amine moiety. Addition of the folate linkers to the quinone moiety not only will increase the steric hindrance upon approaching the ascorbate ion, but the amide group attached to the quinone moiety has a different pKa value than that of the amine group in compound 2. Those characteristics may be responsible for the smaller redox reactivity of 3 and 4 as compared to 2. Differences in redox reactivity between 3 and 4 may be due to differences in steric or structural conformations when approaching ascorbate, since the disulfide bond is not being reduced by ascorbate. However, the elucidation of the molecular causes for these redox reactivity differences is beyond the scope of this work, which is intended to explore the relative toxicities of releasable and non-releasable FR-targeted quinones.

4.2. Cytotoxicity Assays

Since ascorbate oxidation by quinones generates H_2O_2 and other ROS, it is expected that all PHQ derived compounds would show significant cytotoxic activity when combined with ascorbate. However, neither FA-bound 3 nor 4 showed cytotoxicity to KB cells when those compounds were incubate for only 2 hours in the presence of 3 or 5 mM ascorbate, nor compound 3, the more potent of the two folate conjugates, required 12 h of incubation with ascorbate to achieve its full potency in the subpicomolar range. Although compound 4 exhibits limited toxicity to KB cells after 68 hours of incubation with ascorbate, no cytotoxicity was expressed when incubated with ascorbate for 12 hours.

Previous research on folate-targeted therapeutic agents has suggested that folate-linked cytotoxic warheads must be released from folate to achieve maximum cytotoxicity. To promote this release, self-immolative linkers, generally containing a disulfide bond in close proximity to a labile ester bridge, have been employed to attach folate to its cytotoxic cargo [42] [53] . However, in our studies, the non-releasable FA-bound 3 displayed significantly more potent activity than the releasable FA-bound 4. Incubation of compound 4 with 5 mM ascorbate did not release the quinone even after 24 hours of incubation, as detected from HPLC sample analysis. Furthermore, a decrease in the HPLC peak of 4 was not detected either. Many studies have shown that exogenous dehydroascorbic acid is transported into cells where it is reduced to ascorbic acid by glutathione [54] . The latter is in line with the fact that the ascorbate to dehydroascorbate oxidation potential is +0.10 V, while the oxidation potential of GSH to GSSH is + 0.24 V [55] , indicating that GSH, which is located inside the cell in relatively large concentrations, should efficiently reduce the disulfide moiety in 4, as opposed to ascorbate. In fact, to our best knowledge, all the disulfide-containing folate-targeted drugs reported in the literature are known to release the drugs inside the cells, thus improving their cytotoxic activity as compared to non-releasable FR-targeted analogs [56] .

The lack of toxicity of 4 might have been caused by glucuronide conjugation of the released phenanthraquinone moiety inside the cell, as previously observed for phenanthraquinone [57] . This will decrease the amount of released PHQ available for ROS production. Another cause for the lack of toxicity of 4 could be its smaller redox potential which correlates with its smaller rate of ascorbate oxidation (Figure 1, Table 1). FA-bound 3, in contrast, would be expected to remain cell associated due to the high affinity of folate for FR [56] , forcing the ROS generator to accompany FR wherever the receptor traffics within the cell and at the cell membrane.

4.3. Ascorbate and Folate Conjugate Administration to Tumor-Bearing Mice

No tumor size reduction or decrease in tumor growth rate was detected after i.v. injection of either ascorbate, ascorbate + conjugate 3, or conjugate 3 alone. This observation contrasts with previous results seen following both i.p. administration of ascorbate alone as well as ascorbate + 1 mixture [27] . A possible explanation for the absence of anti-tumor activity in vivo may derive from the fact that neither KB nor MDA-MB-231 tumors are highly vascularized, and since ascorbate/quinone-mediated ROS production requires molecular O_2, the pressure of O_2 in the tissues could have been insufficient to enable adequate ROS generation in the tumors. Whether a more active quinone catalyst can be found that can compensate for the lower oxygen tensions that often exist in tumors will have to await further investigation.

A second possible explanation for the insensitivity of KB and MDA-MB-231 tumors to the ascorbate/quinone combination in vivo could have arisen from an inadequate concentration of the quinone moiety in the tumor.

Thus, Verrax et al. reported antitumor activity of 1 + ascorbate after their i.p. injection in mice [27] . Although less ascorbate was injected in mice in their study (1 g/mouse Kg vs. 4 g/mouse Kg), 15 times more of 1 than the amount of 3 used in our work was injected. Furthermore, previous work from our lab has demonstrated that an average of ~2 μmol/Kg of FR-targeted compound will saturate KB tumors in mice, and any additional folate conjugate injected will simply be excreted [58] . This saturating concentration (i.e. ~2 μmol/Kg) is 30 times lower than the amount of 1 used by Verrax et al. Furthermore, FR-targeted compounds generally saturate all FR in KB xenografts within 5 minutes of tail injection with kinetics that are largely independent of the type of folate linker [56] . In previous work, ascorbate concentrations of up to 8 mM and H_2O_2 concentrations of up to 20 μM were detected in the extracellular fluid at 20 to 30 minutes after tail injection of 0.5 mg ascorbate/g mouse [8] . Furthermore, H_2O_2 concentrations ≥ 25 μM in vitro were found to be toxic to cancer cells [59] . In the current work we are injecting 8 times more ascorbate/g of mouse (4 mg/g). Thus, larger peak concentrations of ascorbate and even higher levels of H_2O_2 (due to the targeted PHQ) in the intracellular fluid must have been produced at the tumor tissues described in the current work. However, taken together the fact that 3 is water-soluble and 1 is hydrophobic, it is conceivable that 3 might be more rapidly excreted than 1, while 1 might partition into cells more efficiently than 3, rendering the amount of 3 available for ascorbate reduction at the tumor site insufficient to generate the cytotoxic levels of ROS induced by 1. While differences in drug metabolism, tumor size, levels of ROS detoxifying enzymes, ascorbate excretion, etc. could have also contributed to the insensitivity of our murine tumor models to the ascorbate/quinone oxidant generating system, an unambiguous description of the actual mechanism is beyond the scope of this study.

In summary, no known eukaryotic cells can survive prolonged exposure to ROS, suggesting that constitutive generators of reactive oxygen species could constitute an evasion-proof method for killing essentially all tumor cells. In an effort to design such a mutation-resistant chemotherapy, the combination of a PHQ-derived non-re- leasable FR-targeted compound with ascorbate was investigated. Our data demonstrate a significant acceleration of the ascorbate cytotoxicity against FR+ KB cells, promoted by the presence of 3, if 3 is incubated with ascorbate in vitro, but not in vivo. Since, as described above, related nontargeted therapies from other labs have shown activity against other types of implanted tumors, our observations suggest that FR-targeting of quinones to KB and MDA-MB-231 tumors may be impaired by either poor vascularization and/or a limited concentration of quinone at the tumor site as compared to nontargeted quinone-ascorbate-based therapies.

ACKNOWLEDGEMENTS

The authors express appreciation for Grants supported by the National Center for Research Resources and the National Institute of General Medical Sciences of the National Institute of Health through Grant number 8P20 GM 103475, S06-GM008216, UPRH-FOPI program and P20 RR-016470 from the National Institutes of Health and a grant from Endocyte, Inc. for financial support of this work.

CITE THIS PAPER

Ajay Kumar, Venkatesh Chelvam,, Mahalingam Sakkarapalayam, Guo Li, Pedro Sanchez-Cruz, Natasha S. Piñero, Philip S. Low*, Antonio E. Alegria* (2016) Synthesis and Evaluation of Folate-Conjugated Phenanthraquinones for Tumor-Targeted Oxidative Chemotherapy. *Open Journal of Medicinal Chemistry*,**06**,1-17. doi: 10.4236/ojmc.2016.61001

REFERENCES

1. Watson, M.B., Lind, M.J. and Cawkwell, L. (2007) Establishment of in-Vitro Models of Chemotherapy Resistance. Anti-Cancer Drugs, 18, 749-754.
2. Apel, K. and Hirt, H. (2004) Reactive Oxygen Species: Metabolism, Oxidative Stress, and Signal Transduction. Annual Review of Plant Biology, 55, 373-399.
3. Slauch, J.M. (2011) How Does the Oxidative Burst of Macrophages Kill Bacteria? Still an Open Question. Molecular Microbiology, 80, 580-583.
4. Sun, Y., Colburn, N.H. and Oberley, L.W. (1993) Depression of Catalase Gene Expression after Immortalization and Transformation of Mouse Liver Cells. Carcinogenesis, 14, 1505-1510.
5. Oberley, T.D. and Oberley, L.W. (1997) Antioxidant Enzyme Levels in Cancer. Histology and Histopathology, 12, 525-535.
6. Yang, J., Lam, E.W., Hammad, H.M., Oberley, T.D. and Oberley, L.W. (2002) Antioxidant Enzyme Levels in Oral Squamous Cell Carcinoma and Normal Human Oral Epithelium. Journal of Oral Pathology & Medicine, 31, 71-77.
7. Trachootham, D., Alexandre, J. and Huang, P. (2009) Targeting Cancer Cells by ROS-Mediated Mechanisms: A Radical Therapeutic Approach? Nat. Rev. Drug Discov, 8, 579-591.
8. Chen, Q., Espey, M.G., Sun, A.Y., Lee, J.H., Krishna, M.C., Shacter, E., Choyke, P.L., Pooput, C., Kirk, K.L., Buettner, G.R. and Levine, M. (2007) Ascorbate in Pharmacologic Concentrations Selectively Generates Ascorbate Radical and Hydrogen Peroxide in Extracellular Fluid in Vivo. Proceedings of the National Academy of Sciences of the United States of America, 104, 8749-8754.
9. Roginsky, V.A., Barsukova, T.K. and Stegmann, H.B. (1999) Kinetics of Redox Interaction between Substituted Quinones and Ascorbate under Aerobic Conditions. Chemico-Biological Interactions, 121, 177-197.

10. Levine, M., Padayatty, S.J. and Espey, M.G. (2011) Vitamin C: A Concentration-Function Approach Yields Pharmacology and Therapeutic Discoveries. Advances in Nutrition, 2, 78-88.
11. Levine, M., Espey, M.G. and Chen, Q. (2009) Losing and Finding a Way at C: New Promise for Pharmacologic Ascorbate in Cancer Treatment. Free Radical Biology and Medicine, 47, 27-29.
12. Fukumura, H., Sato, M., Kezuka, K., Sato, I., Feng, X., Okumura, S., Fujita, T., Yokoyama, U., Eguchi, H., Ishikawa, Y. and Saito, T. (2012) Effect of Ascorbic Acid on Reactive Oxygen Species Production in Chemotherapy and Hyperthermia in Prostate Cancer Cells. The Journal of Physiological Sciences, 62, 251-257.
13. Chen, P., Yu, J., Chalmers, B., Drisko, J., Yang, J., Li, B. and Chen, Q. (2012) Pharmacological Ascorbate Induces Cytotoxicity in Prostate Cancer Cells through ATP Depletion and Induction of Autophagy. Anticancer Drugs, 23, 437-444.
14. Du, J., Martin, S.M., Levine, M., Wagner, B.A., Buettner, G.R., Wang, S.H., Taghiyev, A.F., Du, C., Knudson, C.M. and Cullen, J.J. (2010) Mechanisms of Ascorbate-Induced Cytotoxicity in Pancreatic Cancer. Clinical Cancer Research, 16, 509-520.
15. Verrax, J. and Calderon, P.B. (2009) Pharmacologic Concentrations of Ascorbate Are Achieved by Parenteral Administration and Exhibit Antitumoral Effects. Free Radical Biology and Medicine, 47, 32-40.
16. Deubzer, B., Mayer, F., Kuci, Z., Niewisch, M., Merkel, G., Handgretinger, R. and Bruchelt, G. (2010) H2O2-Mediated Cytotoxicity of Pharmacologic Ascorbate Concentrations to Neuroblastoma Cells: Potential Role of Lactate and Ferritin. Cellular Physiology and Biochemistry, 25, 767-774.
17. Gilloteaux, J., Jamison, J.M., Neal, D.R., Loukas, M., Doberzstyn, T. and Summers, J.L. (2010) Cell Damage and Death by Autoschizis in Human Bladder (RT4) Carcinoma Cells Resulting from Treatment with Ascorbate and Menadione. Ultrastructural Pathology, 34, 140-160.
18. Ranzato, E., Biffo, S. and Burlando, B. (2011) Selective Ascorbate Toxicity in Malignant Mesothelioma: A Redox Trojan Mechanism. American Journal of Respiratory Cell and Molecular Biology, 44, 108-117.
19. Takemura, Y., Satoh, M., Satoh, K., Hamada, H., Sekido, Y. and Kubota, S. (2010) High Dose of Ascorbic Acid Induces Cell Death in Mesothelioma Cells. Biochemical and Biophysical Research Communications, 394, 249-253.
20. Biswas, S., Zhao, X., Mone, A.P., Mo, X., Vargo, M., Jarjoura, D., Byrd, J.C. and Muthusamy, N. (2010) Arsenic Trioxide and Ascorbic Acid Demonstrate Promising Activity against Primary Human CLL Cells in Vitro. Leukemia Research, 34, 925-931.
21. Chen, Q., Espey, M.G., Sun, A.Y., Pooput, C., Kirk, K.L., Krishna, M.C., Khosh, D.B., Drisko, J. and Levine, M. (2008) Pharmacologic Doses of Ascorbate Act as a Prooxidant and Decrease Growth of Aggressive Tumor Xenografts in Mice. Proceedings of the National Academy of Sciences of the United States of America, 105, 11105-11109.
22. Chen, M.F., Yang, C.M., Su, C.M., Liao, J.W. and Hu, M.L. (2011) Inhibitory Effect of Vitamin C in Combination with Vitamin K3 on Tumor Growth and Metastasis of Lewis Lung Carcinoma Xenografted in C57BL/6 Mice. Nutrition and Cancer, 63, 1036-1043.
23. Sawant, R.R., Vaze, O.S., Wang, T., D'Souza, G.G., Rockwell, K., Gada, K., Khaw, B.A. and Torchilin, V.P. (2012) Palmitoyl Ascorbate Liposomes and Free Ascorbic Acid: Comparison of Anticancer Therapeutic Effects upon Parenteral Administration. Pharmaceutical Research, 29, 375-383.
24. Chen, P., Stone, J., Sullivan, G., Drisko, J.A. and Chen, Q. (2011) Anti-Cancer Effect of Pharmacologic Ascorbate and Its Interaction with Supplementary Parenteral Glutathione in Preclinical Cancer Models. Free Radical Biology and Medicine, 51, 681-687.
25. Pollard, H.B., Levine, M.A., Eidelman, O. and Pollard, M. (2010) Pharmacological Ascorbic Acid Suppresses Syngeneic Tumor Growth and Metastases in Hormone-Refractory Prostate Cancer. In Vivo, 24, 249-255.
26. Yeom, C.H., Lee, G., Park, J.H., Yu, J., Park, S., Yi, S.Y., Lee, H.R., Hong, Y., Yang, J. and Lee, S. (2009) High Dose Concentration Administration of Ascorbic Acid Inhibits Tumor Growth

in BALB/C Mice Implanted with Sarcoma 180 Cancer Cells via the Restriction of Angiogenesis. Journal of Translational Medicine, 7, 70.

27. Verrax, J., Stockis, J., Tison, A., Taper, H.S. and Calderon, P.B. (2006) Oxidative Stress by Ascorbate/Menadione Association Kills K562 Human Chronic Myelogenous Leukaemia Cells and Inhibits Its Tumour Growth in Nude Mice. Biochemical Pharmacology, 72, 671-680.
28. Beck, R., Pedrosa, R.C., Dejeans, N., Glorieux, C., Leveque, P., Gallez, B., Taper, H., Eeckhoudt, S., Knoops, L., Calderon, P.B. and Verrax, J. (2011) Ascorbate/Menadione-Induced Oxidative Stress Kills Cancer Cells That Express Normal or Mutated Forms of the Oncogenic Protein Bcr-Abl. An in Vitro and in Vivo Mechanistic Study. Investigational New Drugs, 29, 891-900.
29. Bielski, B.H.J. (1978) Reevaluation of the Spectral and Kinetic Properties of Hydroperoxo and Superoxide Anion Free Radicals. Photochemistry and Photobiology, 28, 645-649.
30. Mladenka, P., Simunek, T., Hubl, M. and Hrdina, R. (2006) The Role of Reactive Oxygen and Nitrogen Species in Cellular Iron Metabolism. Free Radical Research, 40, 263-272.
31. Pardo-Andreu, G.L., Delgado, R., Nunez-Selles, A.J. and Vercesi, A.E. (2006) Dual Mechanism of Mangiferin Protection against Iron-Induced Damage to 2-Deoxyribose and Ascorbate Oxidation. Pharmacological Research, 53, 253-260.
32. Biaglow, J.E., Held, K.D., Manevich, Y., Tuttle, S., Kachur, A. and Uckun, F. (1996) Role of Guanosine Triphosphate in Ferric Ion-Linked Fenton Chemistry. Radiation Research, 145, 554-562.
33. M'Bemba-Meka, P., Lemieux, N. and Chakrabarti, S.K. (2005) Nickel Compound-Induced DNA Single-Strand Breaks in Chromosomal and Nuclear Chromatin in Human Blood Lymphocytes in Vitro: Role of Oxidative Stress and Intracellular Calcium. Mutation Research/Genetic Toxicology and Environmental Mutagenesis, 586, 124-137.
34. Petersen, A.B., Gniadecki, R., Vicanova, J., Thorn, T. and Wulf, H.C. (2000) Hydrogen Peroxide Is Responsible for UVA-Induced DNA Damage Measured by Alkaline Comet Assay in HaCaT Keratinocytes. Journal of Photochemistry and Photobiology B: Biology, 59, 123-131.
35. Roginsky, V.A., Barsukova, T.K., Bruchelt, G. and Stegmann, H.B. (1998) Kinetics of Redox Interaction between Substituted 1,4-Benzoquinones and Ascorbate under Aerobic Conditions: Critical Phenomena. Free Radical Research, 29, 115-125.
36. Verrax, J., Delvaux, M., Beghein, N., Taper, H., Gallez, B. and Buc, C.P. (2005) Enhancement of Quinone Redox Cycling by Ascorbate Induces a Caspase-3 Independent Cell Death in Human Leukaemia Cells. An in Vitro Comparative Study. Free Radical Research, 39, 649-657.
37. Muller, F., Crofts, A.R. and Kramer, D.M. (2002) Multiple Q-Cycle Bypass Reactions at the Qo Site of the Cytochrome bc1 Complex. Biochemistry, 41, 7866-7874.
38. Mason, R.P. (1990) Redox Cycling of Radical Anion Metabolites of Toxic Chemicals and Drugs and the Marcus Theory of Electron Transfer. Environmental Health Perspectives, 87, 237-243.
39. Goldenthal, E.I. (1971) A Compilation of LD50 Values in Newborn and Adult Animals. Toxicology and Applied Pharmacology, 18, 185-207.
40. Low, P.S. and Antony, A.C. (2004) Folate Receptor-Targeted Drugs for Cancer and Inflammatory Diseases. Advanced Drug Delivery Reviews, 56, 1055-1058.
41. Kamen, B.A. and Capdevila, A. (1986) Receptor-Mediated Folate Accumulation Is Regulated by the Cellular Folate Content. Proceedings of the National Academy of Sciences of the United States of America, 83, 5983-5987.
42. Yang, J., Chen, H., Vlahov, I.R., Cheng, J.X. and Low, P.S. (2006) Evaluation of Disulfide Reduction during Receptor-Mediated Endocytosis by Using FRET Imaging. Proceedings of the National Academy of Sciences of the United States of America, 103, 13872-13877.
43. Reddy, J.A., Allagadda, V.M. and Leamon, C.P. (2005) Targeting Therapeutic and Imaging Agents to Folate Receptor Positive Tumors. Current Pharmaceutical Biotechnology, 6, 131-150.
44. Hilgenbrink, A.R. and Low, P.S. (2005) Folate Receptor-Mediated Drug Targeting: From Therapeutics to Diagnostics. Journal of Pharmaceutical Sciences, 94, 2135-2146.

45. Patri, A.K., Kukowska-Latallo, J.F. and Baker Jr., J.R. (2005) Targeted Drug Delivery with Dendrimers: Comparison of the Release Kinetics of Covalently Conjugated Drug and Non-Covalent Drug Inclusion Complex. Advanced Drug Delivery Reviews, 57, 2203-2214.
46. Stephenson, S.M., Low, P.S. and Lee, R.J. (2004) Folate Receptor-Mediated Targeting of Liposomal Drugs to Cancer Cells. Methods in Enzymology, 387, 33-50.
47. Jacobsen, M.H. and Koch, T. (2000) Method of Photochemical Immobilization of Ligands Using Quinines. US Patent No. 6033784.
48. Buettner, G.R. (1988) In the Absence of Catalytic Metals Ascorbate Does Not Autoxidize at pH 7: Ascorbate as a Test for Catalytic Metals. Journal of Biochemical and Biophysical Methods, 16, 27-40.
49. Bouffier, L., Lister, K.E., Higgins, S.J., Nichols, R.J. and Doneux, TH. (2012) Electrochemical Investigations of Dissolved and Surface Immobilised 2-amino-1,4-naphthoquinones in Aqueous Solutions. Journal of Electroanalytical Chemistry, 664, 80-87.
50. Sawyer, D.T. and Roberts, J.L. (1974) Experimental Electrochemistry for Chemists. Wiley and Sons, New York.
51. Noto, V., Taper, H.S., Yi-Hua, J., Janssens, J., Bonte, J. and De Loecker, W. (1989) Effects of Sodium Ascorbate (Vitamin C) and 2-methyl-1,4-naphthoquinone (Vitamin K3) Treatment on Human Tumor Cell Growth in Vitro. I. Synergism of Combined Vitamin C and K3 Action. Cancer, 63, 901-906.
52. Chambers, J.Q. (1974) Electrochemistry of Quinines. In: Patai, S. and Rappoport, Z., Eds., The Chemistry of Quinonoid Compounds, John Wiley & Sons Ltd., New York, 737-791.
53. Leamon, C.P., Reddy, J.A., Vlahov, I.R., Vetzel, M., Parker, N., Nicoson, J.S., Xu, L.-C. and Westrick, E. (2005) Synthesis and Biological Evaluation of EC72: A New Folate-Targeted Chemotherapeutic. Bioconjugate Chemistry, 16, 803-811.
54. Winkler, B.S., Orselli, S.M. and Rex, T.S. (1994) The Redox Couple between Glutathione and Ascorbic Acid: A Chemical and Physiological Perspective. Free Radical Biology and Medicine, 17, 333-349.
55. Foyer, C.H. and Noctor, G. (2005) Redox Homeostasis and Antioxidant Signaling: A Metabolic Interface between Stress Perception and Physiological Responses. Plant Cell, 17, 1866-1875.
56. Low, P.S., Henne, W.A. and Doorneweerd, D.D. (2008) Discovery and Development of Folic-Acid-Based Receptor Targeting for Imaging and Therapy of Cancer and Inflammatory Diseases. Accounts of Chemical Research, 41, 120-129.
57. Taguchi, K., Shimada, M., Fujii, S., Sumi, D., Pan, X., Yamano, S., Nishiyama, T., Hiratsuka, A., Yamamoto, M., Cho, A.K., Froines, J.R. and Kumagai, Y. (2008) Redox Cycling of 9,10-phenanthraquinone to Cause Oxidative Stress Is Terminated through Its Monoglucuronide Conjugation in Human Pulmonary Epithelial A549 Cells. Free Radical Biology and Medicine, 44, 1645-1655.
58. Paulos, C.M., Reddy, J.A., Leamon, C.P., Turk, M.J. and Low, P.S. (2004) Ligand Binding and Kinetics of Folate Receptor Recycling in Vivo: Impact on Receptor-Mediated Drug Delivery. Molecular Pharmacology, 66, 1406-1414.
59. Chen, Q., Espey, M.G., Krishna, M.C., Mitchell, J.B., Corpe, C.P., Buettner, G.R., Shacter, E. and Levine, M. (2005) Pharmacologic Ascorbic Acid Concentrations Selectively Kill Cancer Cells: Action as a Pro-Drug to Deliver Hydrogen Peroxide to Tissues. Proceedings of the National Academy of Sciences of the United States of America, 102, 13604-13609.

CHAPTER 11

Efficient Utilization of 6-Aminouracil to Synthesize Fused and Related Heterocyclic Compounds and Their Evaluation as Prostate Cytotoxic Agents with Cathepsin B Inhibition

Marwa Taha Mostafa Sarg[1*], Shereen Said El-Shaer[2]

[1]*Department of Organic Chemistry, Faculty of Pharmacy (Girls), Al-Azhar University, Cairo, Egypt*

[2]*Department of Biochemistry, Faculty of Pharmacy (Girls), Al-Azhar University, Cairo, Egypt*

ABSTRACT

6-aminouracil 1 was utilized to introduce different heterocyclic rings at C-6 position through various synthetic strategies. The synthesized compounds bear rings that are either directly attached to the uracil back bone as in compounds 6, 12a-c and 15, or attached through an amino bridge as compounds 3a-c, 5a, b, 7a, b, 9 and 10, or through an imino bridge as compound 18. Also, compounds 4, 8, 11a-c, 14, 16 and 17 bearing biologically active side chains were synthesized. In addition to, compounds 13, 19, 20, 21 and 22 bear fused rings to the uracil backbone. All synthesized compounds were evaluated for their anticancer activity against prostate PC3 cell line using in-vitro sulforhodamine-B (SRB) method, from which compounds 3a, c, 4, 5a, b, 6, 7a, b, 11a-c, 12a, b, 17 and 20 were the most active. These active compounds were further evaluated for their ability to inhibit cathepsin B enzyme by using enzyme-linked immunosorbent assay, which revealed that compounds 5a, b, 7a, 11a, 12a and 17 exhibited more than 50% inhibition of cathepsin B. Among which the phenyl thiourea derivative 17 was the most active exhibiting 82.3% inhibition, while the reference doxorubicin exerted 18.7% inhibition.

Keywords: Uracil, Cytotoxicity, PC3 Cell Line, Cathepsin B

1. INTRODUCTION

Cancer is thought to reflect a multistep process, resulting from accumulation of inherited and/or acquired defects in genes involved in the positive or negative regulation of cell proliferation and survival [1] . Among the widespread carcinoma is prostate cancer which is the most common tumor in men and the second cause of cancer-related death among them [2] [3] . However, pathways that play a role in prostate cancer tumorigenesis include chronic inflammation, immunosuppression and angiogenesis [4] .

Angiogenesis could be defined as a biological process involving the division and migration of endothelial cells, leading to the formation of microvasculature [5] [6] . This process facilitates tumor growth by providing oxygenation to the tumor through a series of steps including endothelial cell proliferation, motility of endothelial cells through the extracellular matrix toward the angiogenic stimuli and capillary differentiation [4] . Therefore, changes in lysosomal trafficking and content that support invasion and angiogenesis could account for disorders in the regulation of apoptotic and non-apoptotic cell death, particularly through the apparent cellular release of a class of proteases, cathepsins which are usually sequestered within the lysosomal lumen [7] [8] . Lysosomal cysteine cathepsins belong to a family of 11 human proteolytic enzymes. Some of them correlate with progression in a variety of cancers and therefore are considered as potential therapeutic targets [9] .

In cancer, the cellular localization of lysosomal cysteine cathepsins is often altered. Intracellular, cell surface and secreted cysteine cathepsins are involved in distinct tumorigenic processes in-vivo, such as angiogenesis, invasion through extracellular matrices and metastasis [10] -[12] . The precise role of each cathepsin in carcinogenesis remains unclear [13] . Cathepsin B has been shown to play a dominant role in executing the apoptotic program in several tumor cell lines [14] as it reveals that cathepsin B may play a role in malignancy as an executioner of apoptosis in cytotoxic signaling cascades and as a mediator of tumor invasion [8] . Moreover, it has been observed that increased cathepsin B protein levels, pericellular localization and secretion are thought to contribute to prostate cancer invasion and metastasis with cathepsin B acting both directly and indirectly on extracellular matrix (ECM) remodeling and degradation [15] .

Nucleoside analogues have very well recognized activity as anti-metabolites against several types of neoplastic cells [16] [17] , in which 5-fluorouracil and mercaptopurine are typical pyrimidine and purine analogues, respectively [18] . Recently, uracil derivatives substituted either at C-5 or C-6 positions as well as their nucleosides have proved to exhibit significant status in the field of chemotherapy [19] -[26] . However, significant attention has been paid to nucleoside analogues having a heterocyclic ring as a substituent [27] -[29] . Modifications are mainly introduced on the C-5 atom of the uracil ring among which the introduction of a heterocyclic system to the uracil moiety exhibited increased anticancer activity [29] [30].

In this investigation, construction of several heterocyclic rings either directly attached or attached through an amino or imino bridges to the C-6 atom of the uracil ring were performed using different strategies utilizing 6-aminouracil as a building block. Among the registered active anticancer rings were pyrimidine [31] [32] , furan [33] [34] , pyrrole [35] , thiophene [36] , benzoxazine [37] , quinoxaline [38] , triazole [39] [40] , thiazolidine [41] as well as thienopyrimidine [42] [43] . This active building unit; 6-aminouracil was also used to synthesize some fused rings to the uracil moiety. The synthesized compounds were screened for their anticancer activity against prostate cancer PC3 cell line, from which the most active compounds were subjected to biochemical screening against cathepsin B which is considered as a potential target for prostate cancer therapy.

2. DISCUSSION

2.1. Chemistry

The synthetic strategies adopted for the synthesis of the target compounds are depicted in Schemes 1-4. The starting material; 6-aminouracil was chosen as an active substrate [44] -[47]upon which different reaction conditions were applied utilizing its 6-amino function in the synthesis of various 6-heterocyclic uracil and fused uracil derivatives "(Scheme 1)" comprises the coupling of 6-aminouracil 1 with 6-substituted-4-chloro-2-thioxo-1, 2-dihydropyrimidine-5-carbonitriles 2a-c [48] under basic conditions [49] with subsequent elimination of a hydrochloride molecule to yield 6-substituted-4-(2,6-dioxo-1,2,3,6-tetrahydropyrimidin-4-ylamino)-2-thioxo-1,2-dihydropyrimidine-5-carbonitriles 3a-c. ^{1}H NMR spectra of compounds 3a & 3c exhibited four deuterium oxide exchangeable singlets at δ 6.14 - 6.20, δ 10.01 - 10.25, δ 10.49 - 11.73 and δ 10.61 - 11.92 ppm attributed to uracil-C_6-NH, uracil-N_1-H, uracil-N_3-H and pyrimidine-N_1-H; respectively. However, the ^{1}H NMR spectrum of compound 3b revealed only two deuterium oxide exchangeable singlets each integrated for one proton at δ 6.17 and δ 10.04 ppm corresponding to uracil-C_6-NH and uracil-N_1-H; respectively. In addition to a deuterium oxide exchangeable singlet integrated for two protons at δ 10.56 ppm due to uracil-N_3-H and pyrimidine-N_1-H protons. Furthemore, the amino function of compound 1 was chloroacetylated by using equimolar amount of chloroacetyl chloride under anhydrous conditions at room temperature to yield the useful intermediate 6-chloroacetylaminouracil 4. The ^{1}H NMR spectrum of compound 4 revealed a singlet integrated for two protons at δ 4.85 ppm attributed to CH_2Cl protons. Compound 4 was utilized for the synthesis of several rings attached to the uracil nucleus, among which was the reaction of compound 4 in presence of sodium ethoxide with activated methylene bearing compounds namely; acetyl acetone and diethyl malonate [50] to yield the corresponding substituted 2-furan-2-ylaminouracil derivatives 5a & b; respectively. The preparation of such compounds is expected to proceed via nucleophilic substitution of the chloride atom by the active methylene bearing compounds under the reaction basic conditions followed by tautomerism then subsequent elimination of a water molecule leading to ring closure and formation of the furan ring. The ^{1}H NMR spectrum of compound 5a revealed two singlets at δ 1.62 and δ 4.39 ppm corresponding to furyl-C_5-CH_3protons and $COCH_3$ protons; respectively. In addition to, a singlet at δ 4.28 ppm attributed to furyl-C_3 proton, while the electron impact mass spectrum of compound 5b revealed the molecular ion peak at m/z 309 (2.40) and the base peak at m/z 55 (100).

Moreover, compound 4 was further utilized to synthesize the pyrrolidinone derivative 6 through its treatment with malononitrile in presence of a strong base [51] [52] . IR spectrum of compound 6 exhibited a strong absorption band at 2200 cm^{-1} due to cyano group, while its electron impact mass spectrum showed a peak at m/z 234 (3.84) corresponding to M+1 and a base peak at m/z 64 (100). Fulfilling the aforementioned strategy, the reactive intermediate; 6-chloroacetylaminouracil 4 was refluxed with o-phenylene diamine and o-aminophenol in presence of triethyl amine [50] to yield dihydroquinoxalin-2-ylamino derivative 7a and the benzo[b][1,4]oxazin- 3-ylamino derivative 7b; respectively. The reaction mechanism is believed to proceed through Schiff base formation between the chloroacetamido carbonyl function and the amino group of the selected reagent, followed by intramolecular cyclization via elimination of a hydrochloride molecule to afford the target compounds 7a and 7b. The ^{1}H NMR spectra of both compounds revealed the presence of tautomerism as ^{1}H NMR spectrum of 7a showed two singlets at δ 4.71 and δ 4.85 ppm attributed to quinoxalinyl-C_3-CH_2 and quinoxalinyl-C_3-CH tautomers; respectively. In addition to a deuterium oxide exchangeable singlet at δ 8.14 ppm due to quinoxalinyl-N-H. Also, ^{1}H NMR spectrum of compound 7b revealed a deuterium oxide exchangeable singlet at δ 4.41 ppm due to benzoxazine-N_4-H tautomer. In addition to two singlets at δ 4.68 and δ 6.37 ppm corresponding to benzoxazine-C_2-CH_2 and benzoxazine-C_2-CH tautomers; respectively.

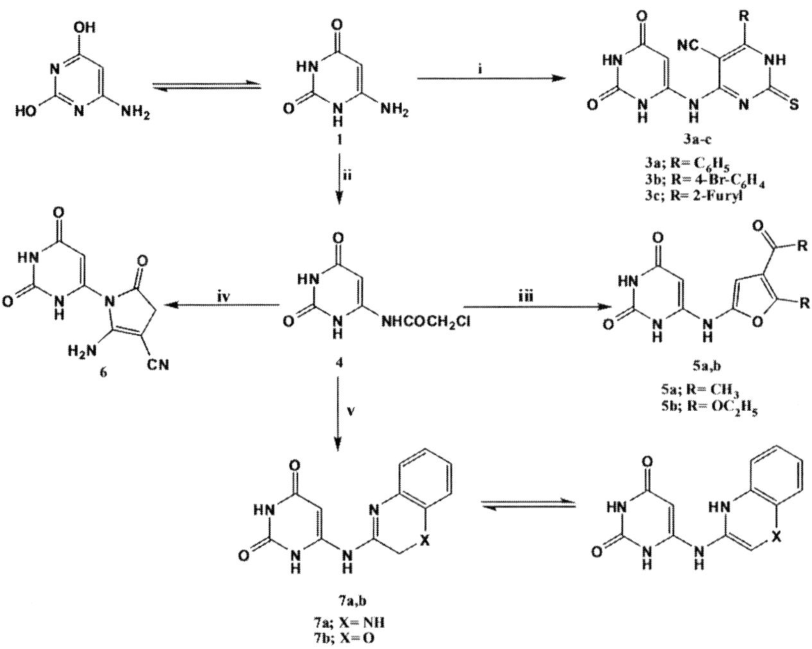

Scheme 1. Synthetic pathways for compounds 3a-c, 4, 5, 6 and 7a, b. Reagents & conditions: i: 4-Chlorodihydropyrimidine derivatives 2a-c/DMF/piperidine/Reflux; ii: $ClCH_2COCl/K_2CO_3$/DMF/R.T.; iii: acetyl acetone or diethyl malonate/sod. ethoxide/Abs.ethanol/R.T.; iv: malononitrile/fusion; v: o-pheneylene diamine or o-aminophenol/Abs. ethanol/TEA/Reflux.

6-Aminouracil 1 was further utilized for the synthesis of thiophene-2-carbonitrile derivative 9 "(Scheme 2)". This was accomplished through the acetylation of compound 1 with acetic anhydride/glacial acetic mixture to yield 6-acetylaminouracil 8 [53] that was further subjected to Gewald reaction [54] [55] via its treatment with sulphur and malononitrile in dry dimethyl formamide which acted as a base and solvent at the same time. The synthesis of 3-aminothiophene-2-carbonitrile derivative 9 is assumed to proceed via Gewald reaction in which malononitrile condenses first with the acetamide ketone yielding a Knoevenagel-Cope condensation product which is then thiolated at the active methyl group with elemental sulphur followed by ring closure [55] (Figure 1). The electron impact mass spectrum of compound 9 showed the molecular ion peak at m/z 249 (10.42) and the base peak at m/z 58 (100). The reactive o-aminonitrile functions in compound 9 were subjected to cyclocondensation through treatment of compound 9 with formamide [55] to afford thienopyrimidine derivative 10. The IR spectrum of compound 10 was devoid of the absorption band due to cyano group of its precursor. However, its ^{1}H NMR spectrum revealed a deuterium oxide exchangeable singlet at δ 5.88 ppm attributed to NH_2 protons. In addition to two singlets at δ 7.95 and δ 8.54 ppm corresponding to thienopyrmidine C_6& C_2 protons; respectively. Moreover, 6-aminouracil 1 was reacted with aromatic aldehydes namely; 4-chlorobenzaldehyde, furan-2-carboxaldehyde and thiophene-2-carboxaldehyde in presence of anhydrous potassium carbonate to give the corresponding Schiff bases 11a-c [56] . It is worth mentioning that, in a previous report [57] the reaction of 6-aminouracil 1 with aromatic aldehydes was suggested to afford the corresponding condensation products 11 or 11' (Figure 2). However, this report revealed that, the formation of the imino compound 11' was not plausible as the imino group is much more nucleophilic than the CH in position 5 of 6-aminouracil. Furthermore, the prepared Schiff bases were used as key precursors for the synthesis of different compounds, in which the azomethine moiety underwent cycloaddition upon treatment with thioglycolic acid in presence of anhydrous zinc chloride [58] -[60] to yield 6-thiazolidinylpyrimidine derivatives 12a-c. The ^{1}H NMR spectrum of compound 12c revealed the presence of two singlets at δ 3.65 and δ 4.37 ppm attributed to thiazolidine-C_4-CH_2 and thiazolidine-C_2-CH protons; respectively. Aligned with the aim of the work, compound 11c was fused with diethyl acetylene dicarboxylate to yield the pyridopyrimidine derivative 13 that was prepared through cycloaddition of compound 11c on the triple bond of the reagent utilizing the methine protons of C_5 uracil and the Schiff base. ^{1}H NMR spectrum of compound 13 revealed the presence of two multiplets at δ 1.00 - 1.18 ppm and δ 3.60 - 3.90 ppm attributed to methyl and methylene protons of the two ethyl ester functions; respectively.

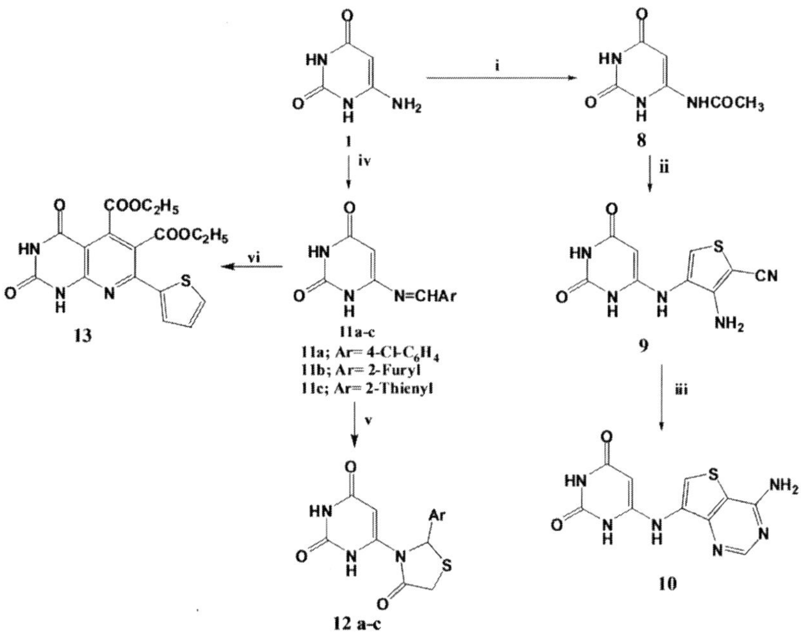

Scheme 2. Synthetic pathways for compounds 8, 9, 10, 11a-c, 12a-c, and 13. Reagents & conditions: i: $(CH_3CO)_2O$/gl.CH_3COOH/steam bath; ii: malononitrile/sulphur/DMF/Reflux; iii: $HCONH_2$/Reflux; iv: ArCHO/ K_2CO_3/DMF/Reflux; v: $HSCH_2COOH$/$ZnCl_2$/Benzene/Reflux; vi: diethyl acetylene dicarboxylate/ Fusion.

Diazotization of 6-aminouracil 1 was accomplished with sodium nitrite and hydrochloric acid at 0°C - 5°C which was followed by coupling with sodium azide to furnish 6-azidouracil 14 [61] "(Scheme 3)". The azide function of compound 14 was coupled with malononitrile in basic medium followed by subsequent nucleophilic cycloaddition on one of the cyano functions [62] to afford the corresponding 5-amino-1,2,3-triazole-4-carbonitrile derivative 15. The IR spectrum of compound 15 lacked the absorption band attributed to azide group and displayed an absorption band at 2204 cm^{-1} corresponding to cyano group. Furthermore, its ^1H NMR spectrum revealed the presence of three deuterium oxide exchangeable singlets at δ 7.28, δ 9.51 and δ 10.76 ppm attributed to triazole-NH_2, uracil-N_1-H and uracil-N_3-H; respectively, while its electron impact mass spectrum exhibited its molecular ion peak at m/z 219 (5.14). Furthermore, the amino group in compound 1 was subjected to acylation with 2,4-dimethoxybenzoyl chloride in presence of piperidine as catalyst which afforded the 2,4-dimethoxybenzamide derivative 16. The ^1H NMR spectrum of compound 16 revealed two singlets at δ 4.42 and δ 6.20 ppm attributed to methoxy protons and phenyl-C_3proton; respectively. In addition to a multiplet at δ 6.30 - 6.38 ppm due to phenyl-$C_{5,6}$ protons.

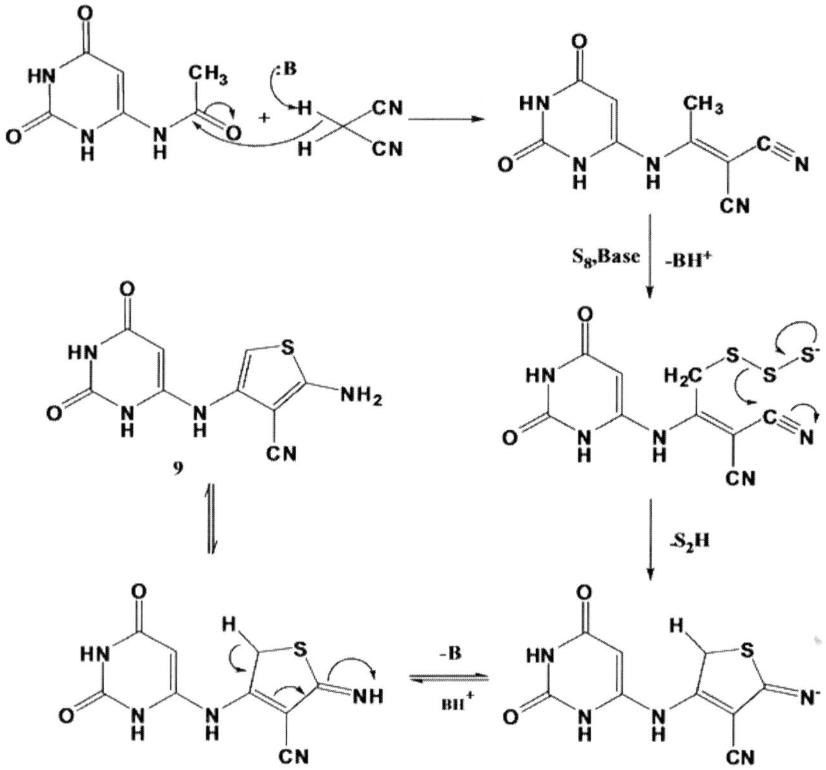

Figure 1. Reaction mechanism for the synthetic route of compound 9.

Moreover, the behavior of compound 1 towards isothiocyanate reagent was investigated to proceed typical to literature [63] [64] . Compound 1 was refluxed with phenyl isothiocyanate in pyridine to yield the acyclic thiourea derivative 17 which its ^1H NMR spectrum revealed the presence of two deuterium oxide exchangeable singlets at δ 8.90 and δ11.63 ppm due to uracil-C_6-NH and NH-phenyl protons; respectively. In addition to, other deuterium oxide exchangeable singlets due to uracil-N_1 and N_3 protons. In an attempt to fulfill the research goal, the thiourea derivative 17 was treated with 4-bromophenacyl bromide in absolute ethanol containing sodium acetate as a base to yield the thiazole derivative 18 that is believed to be formed through nucleophilic substitution reaction involving the reactive SH thiourea tautomer and 4-bromophenacyl bromide with elimination of a hydrobromide molecule followed by tautomerism and cyclocondensation (Figure 3) to yield compound 18 which its structure was established on its elemental and spectral data, as its ^{1}H NMR spectrum revealed the presence of two doublets at δ 7.29 and δ 7.95 ppm due to $C_{2,6}$ and $C_{3,5}$ protons of 4-bromophenyl ring; respectively. In addition to, two singlets at δ 7.58 and δ 7.99 ppm attributed to thiazolidine C_5 and uracil C_5 protons; respectively.

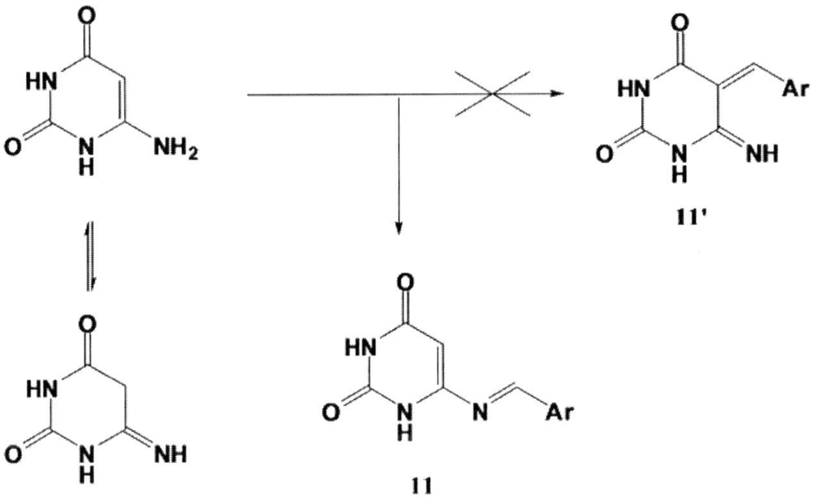

Figure 2. Possible reaction products of compound 11.

Furthermore, compound 1 was further utilized to synthesize different substituted pyrimidopyrimidine derivatives 19, 20 and 21 "(Scheme 4)". This was accomplished through refluxing compound 1 with ethanolamine and paraformaldehyde in absolute ethanol [65] to yield the tetrahydropyrimido[4,5-d]pyrimidinedione derivative 19. The IR spectrum of compound 19 showed strong absorption bands at 2949, 2916 and 2872 cm^{-1} due to aliphatic CH functions, while its ^{1}H NMR spectrum exhibited two singlets at δ 3.96 and δ 4.85 ppm due to tetrahydropyrimidine C_4 and C_2 protons; respectively. In addition to, a deuterium oxide exchangeable singlet at δ 4.42 ppm due to OH proton. Moreover, formation of pyrimido[1,6-a]pyrimidine-2,6,8-trione derivative 20 was achieved through fusion of compound 1 with ethyl acetoacetate that was believed to be formed via Michael type addition of compound 1 on ethyl acetoacetate followed by ethanol elimination to give the acyclic intermediate 20' which then underwent tautomerism and cyclization via elimination of a water molecule. The electron impact mass spectrum of compound 20 showed a peak at m/z 192 (0.34) corresponding to M-1, in addition to the base peak at m/z 60 (100). Also, 2-aminodihydropyrimido[1,6-a]pyrimidine-3-carbonitrile derivative 21 was synthesized through treatment of compound 1 with ethoxymethylene malononitrile [66] in absolute ethanol containing piperidine as catalyst. The reaction is suggested to proceed through Micheal type addition followed by elimination of an ethanol molecule to yield the acyclic intermediate which underwent intramolecular cyclization via nucleophilic addition on one cyano function and subsequent tautomerism leading to formation of compound 21 as demonstrated in (Figure 4). IR spectrum of compound 21 exhibited a strong absorption band at 2200 cm^{-1} due to cyano group. Besides, its ^{1}H NMR spectrum revealed the presence of a deuterium oxide exchangeable singlet at δ 3.75 ppm due to NH_2 protons. In addition to a singlet at δ 7.65 ppm due to pyrimidine C_6 proton. Finally, compound 1 was refluxed with acetyl isothiocyanate in dry acetone [63] to yield the pyrimido[1,6-a][1,3,5] triazine derivative 22 that is assumed to be formed through formation of the thiourea intermediate 22' that underwent tautomerism and subsequent cyclocondensation to yield compound 22. ^{1}H NMR spectrum of compound 22 exhibited a singlet at δ 2.05 ppm due to methyl protons. In addition to, a deuterium oxide exchangeable singlet at δ 6.66 ppm attributed to triazine-N_1 proton.

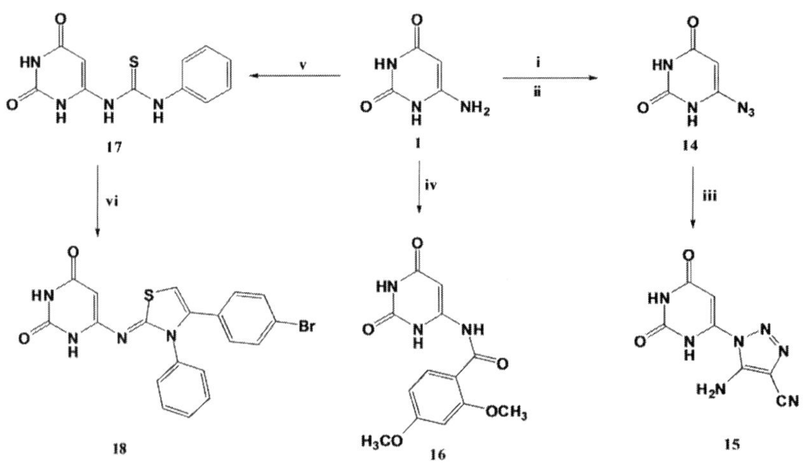

Scheme 3. Synthetic pathways for compounds 14, 15, 16, 17 and 18. Reagents & conditions: i: $NaNO_2$/ HCl/0°C; ii: NaN_3/0°C; iii: malononitrile/sod. ethoxide/Abs. ethanol/Reflux; iv: 2,4-di(OCH_3) $_2C_6H_3COCl$/ piperidine/DMF/Reflux; v: C_6H_5NCS/pyridine/Reflux; vi: 4-bromophenacyl bromide/Na acetate/Abs. ethanol/Reflux.

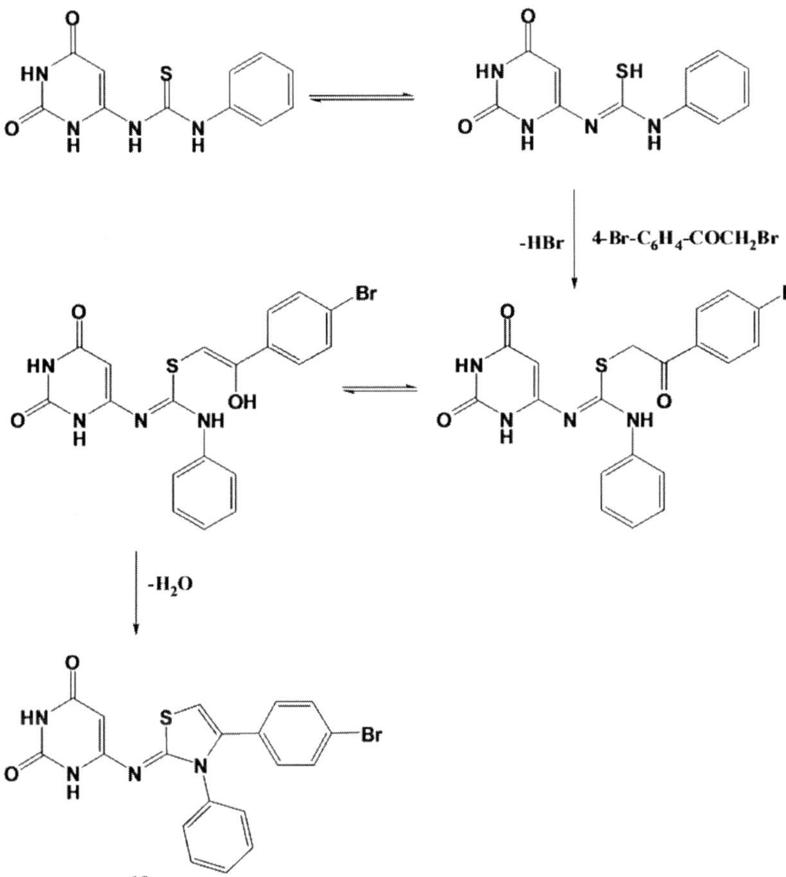

18

Figure 3. Reaction mechanism for the synthetic route for compound 18.

2.2. Biological Screening

2.2.1. Anticancer Screening

Cytotoxicity of all the synthesized compounds was evaluated against prostate cancer; PC3 cell line by using in-vitro sulforhodamine-B (SRB) method. The screening was performed by evaluation of the growth inhibition percent using five different doses from which their IC_{50} values were calculated. Also, we take doxorubicin as a reference drug in which its IC_{50} was 0.93 as presented in Table1 Aligned with the aim of this work; to synthesize and study the effect of various heterocyclic rings upon attachment to C_6 position or fusion to the uracil backbone, the IC_{50} of 6-

aminouracil 1 against PC3 cell line was evaluated which was found to be 362 μM. The attachment of a pyrimidine-2-thione ring through an amino bridge as in compounds 3a and 3c, greatly improved the anticancer activity as they exhibited IC_{50} 43.95 and 79.20 μM; respectively. Also, chloroacetylation of 6-aminouracil as in compound 4 further improved the activity showing IC_{50} 21.21 μM. However, attachment of different substituted furan rings through an amino bridge as in compounds 5a and 5b exhibited moderate anticancer activity with IC_{50} 7.02 and 8.57 μM; respectively, compared to doxorubicin (IC_{50} 0.93 μM), while direct attachment of a pyrrolidinone ring as in compound 6 showed IC_{50} 38.73 μM.

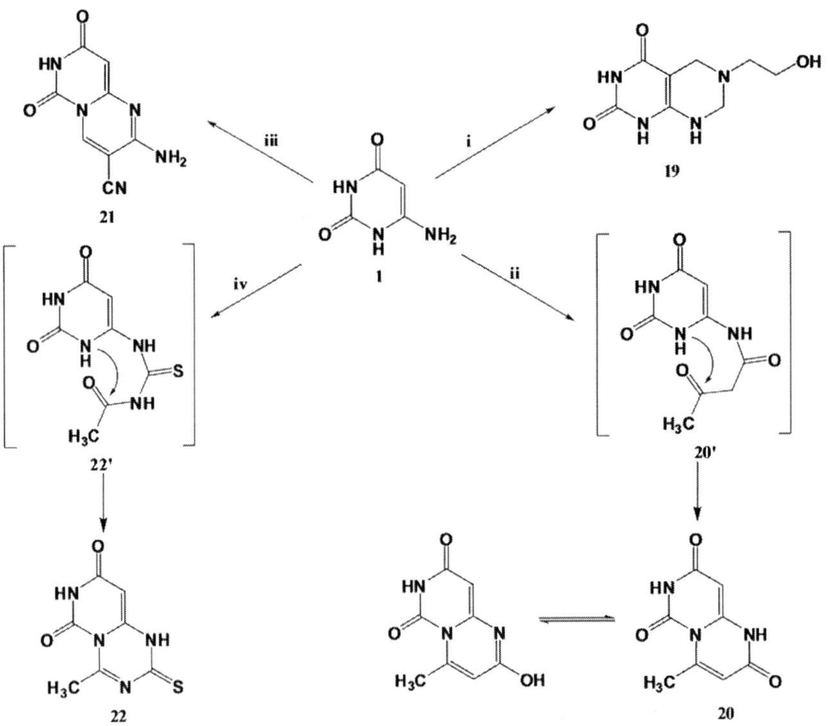

Scheme 4. Synthetic pathways for compounds 19, 20, 21 and 22. Reagents & conditions: i: $H_2NCH_2CH_2OH$/ HCHO/Abs. ethanol/Reflux; ii: $CH_3COCH_2COOC_2H_5$/fusion; iii: Ethoxymethylene malononitrile/piperidine/ Abs. ethanol/Reflux; iv: CH_3COCl/NH_4SCN/Acetone/Reflux.

Furthermore, attachment of a quinoxaline ring through an amino bridge as in compound 7a exhibited strong activity with IC_{50} 2.31 μM, while replacing quinoxaline with 1,4-benzoxazine moiety as in 7b showed IC_{50} 36.61 μM.

Also, it is worth mentioning that attachment of a thiophene ring through an amino bridge as in compound 9 greatly diminished the activity showing IC_{50} 547.40 μM which upon fusion to a pyrimidine ring to yield the thienopyrimidine derivative 10, the activity was slightly improved to exhibit IC_{50} 146 μM. Moreover, the intermediate Schiff bases 11a-c showed moderate to low activity showing IC_{50} 15.66 - 151 μM that upon cyclization to yield compounds 12a-c bearing a thiazolidinone ring directly attached to the uracil ring, showed enhancement in the activity of compounds 12a and 12b with IC_{50} 8.84 and 29.54 μM; respectively, while compound 12c with a thiophene moiety attached to the thiazolidinone ring; exerted very weak activity with IC_{50}272.80 μM. However, cyclization of the Schiff base 11c into the pyridopyrimidine derivative 13 exhibited marked anticancer activity with IC_{50} 1.89 μM which is nearly half potent to that of doxorubicin.

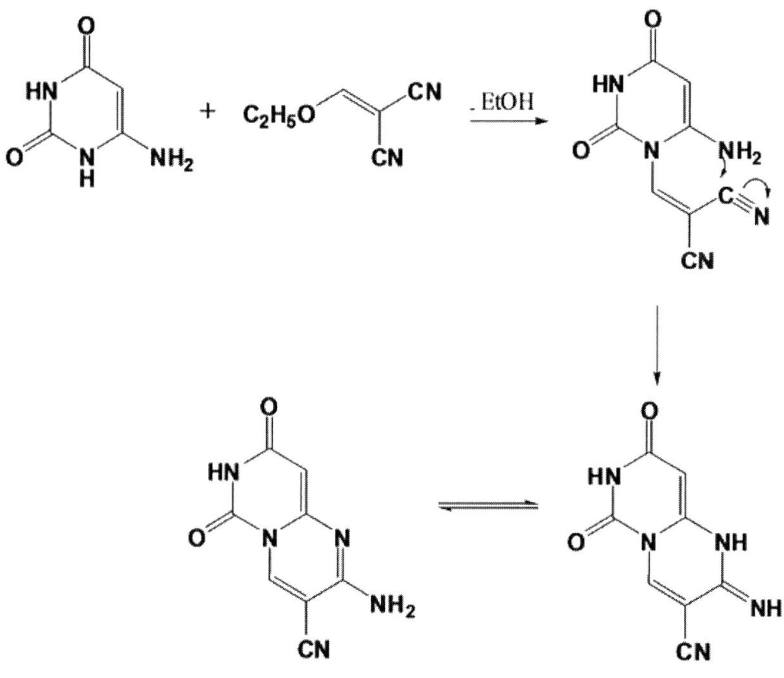

21

Figure 4. Reaction mechanism for the synthetic route for compound 21.

Table 1. Five doses growth inhibition percent and IC_{50} values of the tested compounds against PC3 cell line.

Comp. No. / Conc. (μM)	% Growth Inhibition					IC_{50} (μM)
	0.1	1	10	100	1000	
1	89.40	83.50	86.00	65.80	15.10	362.80
3a	95.10	77.60	75.10	30.10	9.50	43.95
3b	94.10	94.80	89.00	73.60	17.90	432.60
3c	96.20	81.30	76.00	37.80	1.60	79.20
4	92.20	93.20	72.10	14.10	14.10	21.21
5a	93.80	61.00	45.10	26.20	2.20	7.02
5b	83.30	62.00	46.00	23.60	12.30	8.57
6	89.00	81.20	73.20	31.40	15.00	38.73
7a	89.70	58.70	35.70	9.10	9.70	2.31
7b	95.30	99.50	77.40	28.70	6.30	36.61
8	98.20	99.60	83.20	67.10	13.80	233.80
9	87.00	88.30	86.60	71.90	21.10	547.40
10	95.10	96.00	82.30	63.40	25.30	146.00
11a	82.10	72.60	53.10	8.50	7.10	15.66
11b	92.50	93.70	95.20	58.80	18.70	151.00
11c	101.50	102.10	77.60	22.40	1.20	30.72
12a	92.70	80.30	51.30	32.10	12.80	8.84
12b	100.10	101.50	80.20	47.20	28.50	29.54
12c	91.20	83.40	74.60	69.30	36.00	272.80
13	89.30	65.00	42.50	32.90	19.40	1.89
14	92.80	84.40	75.40	61.00	2.40	322.30
15	98.60	89.30	86.80	77.90	33.50	471.50
16	100.00	102.10	90.00	79.80	21.10	487.30
17	28.40	16.90	11.20	6.40	6.20	0.03
18	91.80	87.40	27.40	20.80	8.40	3.67
19	94.20	86.20	78.80	72.50	22.20	522.80
20	87.90	72.70	68.30	39.20	13.80	63.68
21	80.90	69.60	64.10	55.60	39.10	423.4
22	92.10	84.10	81.50	78.80	48.60	975.9
Doxorubicin	92.90	76.80	51.20	23.80	12.00	0.93

It is to be noted that, the presence of an azide function as in compound 14, a directly attached triazole ring as in compound 15 or a substituted benzamide side chain as in compound 16 attached to C_6 position of uracil backbone, resulted in diminished anticancer activity. However, the presence of a phenyl thiourea side chain as in compound 17 exerted highly potent anticancer activity showing IC_{50} 0.03 μM in comparison to doxorubicin IC_{50} 0.93 μM, which upon cyclization to yield substituted thiazole derivative attached to the uracil ring through an imino bridge as in compound 18 resulted in slight decrease of anticancer activity with IC_{50} 3.67 μM. Moreover, fusion of the uracil ring to perhydropyrimidine, 4-aminopyrimidine-5-carbonitrile and 2-thioxo[1,3,5]triazine rings as in compounds 19, 21 and 22; respectively, resulted in diminished anticancer activity, although that, fusion of pyrimidine moiety to the uracil backbone as in compound 20 exerted good anticancer activity with IC_{50} 63.68 μM.

2.2.2. Cathepsin B Assay

The possible link between cathepsins and cancer and the role of cathepsin B in metastatic potential was postulated since many years [67] . Cathepsin B, either the mRNA or the protein, was often detected in higher amounts in malignant tumors than in benign ones or in normal tissues [68] . Moreover, a positive correlation between cathepsin B expression and metastasis of carcinoma cells to lymph nodes has been shown in prostate cancer [69] .

Current prostate cancer specific treatments are limited and development of new therapeutics is needed, specifically therapies against pathways that mediate the aggressiveness of the disease. A clinical approach based on targeting of proteases involved in pathomechanism of given diseases also stimulates the interest as anticancer strategy alternative, or supplementary, to surgical intervention and radiotherapy [70] . In various types of mouse cancer models deletion of cathepsin B led to suppression of the aggressiveness of the respective cancer phenotype. Moreover, delivery of broad spectrum cysteine cathepsin inhibitors in the tumor microenvironment disrupts the permissive ecosystem of the cancer and results in impaired growth or even in regression of the tumor [9] . A logical consequence of these results would be to further pursue selective inhibition of cathepsin B. In this investigation, the synthesized compounds exhibit a good inhibition of cathepsin B in vitro. Thus, we can hypothesize that this inhibition could be due to one or more of the mechanisms by which cathepsin B contribute to metastasis facilitating cell migration and invasiveness.

In general, loss of cell-cell and cell-matrix adhesion and degradation of extracellular matrix components are involved in invasion and migration [71] of cancerous cells. The tumor and metastasis-promoting effect of secreted cysteine cathepsins is thought to be caused by their ability to degrade extracellular matrix molecules, which in turn enables cancer cells to invade into the surrounding tissue and to metastasize [72] . In addition, proteases promote tumor growth by processing growth factors, cytokines, and chemokines or increasing their bioavailability by releasing them from the extracellular matrix [73] -[75] . Furthermore, they may play roles in the regulation of the action of certain growth factors, growth factor-binding proteins and growth factor receptors [13] , all are vital participants in human cancer growth. Therefore, proteases are functionally embedded in the complex proteolytic and cellular network [76] -[78] . Also, the localization of cysteine cathepsins on the tumor cell membrane might be mediated through an association of individual cysteine cathepsins with binding partners in membrane microdomains. The membrane microdomains include lipid rafts. The association on the tumor cell membrane of cathepsin B with caveolae, a subset of lipid rafts that contain the structural protein caveolin, is mediated by a direct interaction of procathepsin B with the light chain of the annexin II heterotetramer (AIIT) [10] .

Previous studies have implicated tumor cell cathepsin B in degradation of ECM proteins including fibronectin, laminin and collagen IV [79] [80] . In accordance to these studies, Cathepsin B has been shown to be implicated in the degradation of basement membrane proteins and invasion of the adjunct stroma in several experimental tumor models, and the over expression of cathepsin B shows a close association with disease progression in various kinds of human malignancy, including prostate cancer [81] . Active cathepsin B is also secreted from tumors, a mechanism likely to be facilitated by lysosomal exocytosis or extracellular processing by surface activators [82] .

Additionally, and in support of our study, inhibition of cathepsin B has been demonstrated to reduce collagen I degradation by prostate carcinoma cell lines [83] . Consistent with cathepsin B role in invasion, [84] demonstrated a critical function of tumor-derived cathepsin B in metastasis in the 4T1.2 model and revealed that preventing ECM degradation by cathepsin B inhibition is likely to reduce metastatic burden.

Therefore, the most active synthesized compounds against prostate PC3 cell line; 3a, 3c, 4, 5a, 5b, 6, 7a, 7b, 11a, 11b, 11c, 12a, 12b, 13, 17, 18 and 20 were further evaluated for their ability to inhibit cathepsin B by using enzyme-linked immunosorbent assay (ELISA) kit (Wuhan ElAab Science Co., Ltd, Wuhan, China), as shown in Table2

The presented results revealed that, all the tested compounds showed better cathepsin B inhibition than the reference drug doxorubicin (18.7%), except the 2-thioxopyrimidine derivative 3c. However, several compounds exhibited more than 50% inhibition of cathepsin B, among which are the furan derivatives 5a and 5b, quinoxaline derivative 7a, 4-chlorophenyl Schiff base derivative 11a and the thiazolidinone derivative 12a. In addition to, the phenyl thiourea derivative 17 which is the most active compound exerting 82.3% inhibition of cathepsin B.

In conclusion, our study reveals that these synthesized compounds have the ability to inhibit cathepsin B which is associated particularly with metastasis of prostate cancer and that inhibition could possibly lead to changes of tumor microenvironment and to decreased cell survival, therefore, to the impairment of metastatic seeding and onset. Further research on prostate cancer will however be crucial to generate therapeutic strategies based on the use of available or novel potential inhibitors of cathepsin B activity.

Table 2. Growth inhibition percent of the selected compounds against cathepsin B enzyme.

Comp. No.	% Inhibition of cathepsin B
3a	38.8
3c	10.0
4	48.8
5a	52.2
5b	57.62
6	40.8
7a	56.3
7b	41.7
11a	54.6
11b	22.2
11c	42.7
12a	59
12b	37
13	22.6
17	82.3
18	26.5
20	18.9
Doxorubicin	18.7

3. EXPERIMENTAL

3.1. Chemistry

All melting points were measured on Electro thermal LA 9000 SERIS, Digital Melting point Apparatus and are uncorrected. IR spectra (KBr) were recorded on FT-IR 200 spectrophotometer (ύ cm^{-1}), pharmaceutical analytical unit, Faculty of Pharmacy, Al-Azhar University. ^{1}H-NMR spectra were recorded in (DMSO-d_6) at 300 MHz on a Varian Gemini NMR spectrometer (δ ppm) using TMS as an internal standard, Main defense chemical laboratory, Cairo. Mass spectra were recorded on GC Ms-QP 5050A mass spectrometer at 70 eV and microanalytical data were performed on Elementar Vario EI III CHN analyzer at the microanalytical unit, in Regional center for Mycology and Biotechnology, Al-Azhar University. Thin layer chromatography was performed on precoated (0.25mm) silica gel GF_{254} plates (E. Merck, Germany). Compounds were detected with 254 nm UV lamp.

3.1.1. General Method for Synthesis of Compounds 3a-c

A mixture of 6-aminouracil 1 (1.27 g, 10 mmol.) and 4-chlorodihydropyrimidine derivative 2a-c [48](5 mmol) was refluxed in dry DMF (10 mL) containing piperidine (2 - 3 drops) for 6 hr. The reaction was triturated with water, the solid obtained was filtered, washed with water, dried then crystallized from DMF/H_2O to yield compounds 3a-c.

1) 4-(2,6-Dioxo-1,2,3,6-tetrahydropyrimidin-4-ylamino)-6-phenyl-2-thioxo-1,2-dihydro-pyrimidine-5-carbonitrile; 3a:

Light brown powder, yield 55%, m.p. < 300°C; IR (KBr) cm^{-1}: 3336, 3174 (NH); 3050 (CH-aromatic); 2210 (CN); 1729 (C=O); 1639 (C=N); 1521, 1386, 1165, 1017 (I, II, III, IV bands N-C=S). ^{1}H NMR (DMSO-d_6) δ ppm: 6.20 (s, 1 H, uracil-C_6-NH, D_2O exchangeable); 6.60 - 6.90 (m, 5 H, C_6H_5); 7.95 (s, 1 H, uracil-C_5-H); 10.25 (s, 1 H, uracil-N_1-H, D_2O exchangeable); 10.49 (s, 1 H, uracil-N_3-H, D_2O exchangeable); 10.61 (s, 1 H, pyrimidine-N_1-H, D_2O exchangeable). Anal. Calcd for $C_{15}H_{10}N_6O_2S$ (338.34): C, 53.25; H, 2.98; N, 24.84. Found: C, 53.31; H, 3.04; N, 24.92.

2) 4-(2,6-Dioxo-1,2,3,6-tetrahydropyrimidin-4-ylamino)-6-(4-bromophenyl)-2-thioxo-1,2-dihydropyrimidine- 5-carbonitrile; 3b:

Dark brown powder, yield, 63%, m.p. < 300°C; IR (KBr) cm^{-1}: 3419 (NH); 2922 (CH-aromatic); 2216 (CN); 1710 (C=O); 1627 (C=N); 1545, 1243, 1163, 1064 (I, II, III, IV bands N-C=S). ^{1}H NMR (DMSO-d_6) δ ppm: 6.17 (s, 1 H, uracil-C_6-NH, D_2O exchangeable); 7.12 - 7.40 (m, 2 H, 4-Br-C_6H_4-$C_{2,6}$-H); 7.71 - 7.79 (m, 2 H, 4-Br-C_6H_4-$C_{3,5}$-H); 8.57 (s, 1 H, uracil-C_5-H); 10.04 (s, 1 H, uracil-N_1-H, D_2O exchangeable); 10.56 (s, 2 H, uracil-N_3-H & pyrimidine-N_1-H, D_2O exchangeable). Anal. Calcd for $C_{15}H_9BrN_6O_2S$ (417.24): C, 43.18; H, 2.17; N, 20.14. Found: C, 43.17; H, 2.21; N, 20.18.

3) 4-(2,6-Dioxo-1,2,3,6-tetrahydropyrimidin-4-ylamino)-6-(2-furyl)-2-thioxo-1,2-dihydro-pyrimidine-5-carbonitrile; 3c:

Light orange crystals, yield, 55%, m.p. < 300°C; IR (KBr) cm^{-1}: 3451, 3152 (NH); 3048 (CH-aromatic); 2219 (CN); 1735, 1670 (two C=O); 1632 (C=N); 1578, 1321, 1133, 1035 (I, II, III, IV bands N-C=S); 1222, 1078 (C-O-C). ^1H NMR (DMSO-d_6) δ ppm: 6.14 (s, 1 H, uracil-C_6-NH, D_2O exchangeable); 6.88 (dd, 1 H, J = 3.3, 1.5 Hz, furyl-C_4-H); 7.43 (d, 1 H, J = 3.3 Hz, furyl-C_5-H); 7.76 (d, 1 H, J = 3.3 Hz, furyl-C_3-H); 8.19 (s, 1 H, uracil-C_5-H); 10.01 (s, 1 H, uracil-N_1-H, D_2O exchangeable); 11.73 (s, 1 H, uracil-N_3-H, D_2O exchangeable); 11.92 (s, 1 H, pyrimidine-N_1-H, D_2O exchangeable). Anal. Calcd for $C_{13}H_8N_6O_3S$ (328.31): C, 47.56; H, 2.46; N, 25.60. Found: C, 47.69; H, 2.52; N, 25.74.

3.1.2. Synthesis of 2-Chloro-N-(2,6-dioxo-1,2,3,6-tetrahydropyrimidin-4-yl)acetamide; 4

A mixture of 6-aminouracil 1 (1.27 g, 10 mmol.), chloroacetyl chloride (1.13 g, 0.8 mL, 10 mmol.) and anhydrous potassium carbonate (1.38 g, 10 mmol.) was stirred at room temperature for 24 hr. The reaction mixture was triturated with water, dried and crystallized from DMF/H_2O.

Canary yellow crystals, yield, 75%, m.p. < 300°C; IR (KBr) cm^{-1}: 3393, 3259 (NH); 3093 (CH-aromatic); 2865 (CH-aliphatic); 1734, 1651 (two C=O); 1528 (C=C); 778 (C-Cl). ^1H NMR (DMSO-d_6) δ ppm: 4.85 (s, 2 H, CH_2Cl); 7.42 (s, 1 H, uracil-C_6-NH, D_2O exchangeable); 7.95 (s, 1 H, uracil-C_5-H); 9.74 (s, 1 H, uracil-N_1-H, D_2O exchangeable); 10.76 (s, 1 H, uracil-N_3-H, D_2O exchangeable). Anal. Calcd for $C_6H_6ClN_3O_3$ (203.58): C, 35.40; H, 2.97; N, 20.64. Found: C, 35.43; H, 3.02; N, 20.85.

3.1.3. General Method for Synthesis of Compounds 5a, b

To an ethanolic solution of sodium ethoxide [prepared from sodium metal (0.05 g, 2.2 mmol.) and 15 mL absolute ethanol] was added (5 mmol.) of the appropriate activated methylene bearing compounds namely; acetyl acetone and dimethyl malonate. After stirring for 15 min. at room temperature, compound 4 (0.2 g, 1 mmol.) was added and stirring was continued for 24 hr. The solvent was evaporated under reduced pressure and the remainder was poured onto ice cold water (25 mL) and neutralized with diluted HCl. The solid product was collected by filtration, washed with water, dried and crystallized from DMF/H_2O.

1) 6-(4-Acetyl-5-methylfuran-2-ylamino)pyrimidine-2,4(1H,3H)dione; 5a
Light green crystals, yield, 86%, m.p. < 300 °C; IR (KBr) cm^{-1}: 3357, 3191 (NH); 2940 (CH-aliphatic); 1710 (C=O); 1561 (C=C); 1286, 1094 (C-O-C). ^1H NMR (DMSO-d_6) δ ppm: 1.62 (s, 3 H, furyl-C_5-CH_3); 4.28 (s, 1 H, furyl-C_3-H); 4.39 (s, 3 H, $COCH_3$); 6.30 (s, 1 H, uracil-C_6-NH, D_2O exchangeable); 8.40 (s, 1

H, uracil-C_5-H); 9.05 (s, 1 H, uracil-N_1-H, D_2O exchangeable); 9.49 (s, 1 H, uracil-N_3-H, D_2O exchangeable). Anal. Calcd for $C_{11}H_{11}N_3O_4$ (249.22): C, 53.01; H, 4.45; N, 16.86. Found: C, 52.98; H, 4.50; N, 16.97.

2) Ethyl 5-(2,6-dioxo-1,2,3,6-tetrahydropyrimidin-4-ylamino)-2-ethoxyfuran-3-carboxylate; 5b Pale yellow crystals, yield, 38%, m.p. < 300°C; IR (KBr) cm^{-1}: 3358 (NH); 3000 (CH-aromatic); 2936 (CH-aliphatic); 1730 (C=O); 1563 (C=C); 1294, 1099 (C-O-C). MS (m/z, %): 309 (M^+, 2.40); 55 (100). Anal. Calcd for $C_{13}H_{15}N_3O_6$ (309.27): C, 50.49; H, 4.89; N, 13.59. Found: C, 50.52; H, 4.91; N, 13.68.

3.1.4. Synthesis of 2-Amino-1-(2,6-dioxo-1,2,3,6-tetrahydropyrimidin-4-yl)-5-oxo-4,5-dihydro-1Hpyrrole-3-carbonitrile; 6

An equimolar mixture of the 6-chloroacetylaminouracil 4 (2.04 g, 10 mmol.) and malononitrile (0.66 g, 10 mmol.) was fused for 20 min. The reaction mixture was triturated with ethanol and the product was collected, washed with ethanol, dried then crystallized from DMF/H_2O.

Light brown powder, yield, 74%, m.p. < 300°C; IR (KBr) cm^{-1}: 3380, 3320, 3180 (NH_2, NH); 3002 (CH-aromatic); 2200 (CN); 1720 (C=O); 1530 (C=C). MS (m/z, %): 234 (M+1, 3.84); 64 (100). Anal. Calcd for $C_9H_7N_5O_3$ (233.18): C, 46.36; H, 3.03; N, 30.03. Found: C, 46.41; H, 3.05; N, 30.14.

3.1.5. General Method for Synthesis of Compounds 7a, b

To a suspension of compound 4 (2.04 g, 10 mmol.) in absolute ethanol (30 mL), an equimolar amount of the appropriate aromatic amine; namely o-phenylene diamine and o-aminophenol, two drops of triethyl amine were added and the suspension was heated under reflux for 12 hr., then left to cool. The solid product was collected by filtration, washed with ethanol, dried and crystallized from DMF/H_2O.

1) 6-(3,4-Dihydroquinoxalin-2-ylamino)pyrimidine-2,4(1H,3H)-dione; 7a Light green crystals, yield, 44%, m.p. < 300°C; IR (KBr) cm^{-1}: 3364, 3178 (NH); 2922 (CH-aromatic); 1707 (C=O); 1618 (C=N); 1506 (C=C). ^1H NMR (DMSO-d_6) δ ppm: 4.71 (s, 1 H, quinoxalinyl-C_3-CH_2 tautomer); 4.85 (s, 1 H, quinoxalinyl-C_3-C=CH tautomer); 7.35 (s, 1 H, uracil-C_6-NH, D_2O exchangeable); 7.62 - 7.64 (m, 2 H, quinoxalinyl-$C_{5,8}$-H); 7.67 (s, 1 H, uracil-C_5-H); 7.75 (t, 1 H, J=7.73 Hz, quinoxalinyl-C_6-H); 7.97 (t, 1 H, J = 7.73 Hz, quinoxalinyl-C_7-H); 8.14 (s, 1 H, quinoxalinyl-N-H, D_2O exchangeable); 10.65 (s, 1 H, uracil-N_1-H, D_2O exchangeable); 10.80 (s, 1 H, uracil-N_3-H, D_2O exchangeable). Anal. Calcd for $C_{12}H_{11}N_5O_2$ (257.25): C, 56.03; H, 4.31; N, 27.22. Found: C, 56.05; H, 4.37; N, 27.29.

2) 6-(2H-Benzo[b][1,4]oxazin-3-ylamino)pyrimidine-2,4(1H,3H)-dione; 7b Dark orange crystals, yield, 63%, m.p. < 300°C; IR (KBr) cm^{-1}: 3408, 3177 (NH); 2921 (CH-aromatic); 1711 (C=O); 1626 (C=N); 1460 (C=C); 1287, 1051 (C-O-C). ^1H NMR (DMSO-d_6) δ ppm: 4.41 (s, 1/2 H, benzoxazine-N_4-H tautomer, D_2O exchangeable); 4.68 (s, 1 H, benzoxazine-C_2-CH_2); 6.22 (s, 1 H, uracil-C_6-NH, D_2O exchangeable); 6.37 (s, 1/2 H, benzoxazine-C_2-C=CH-O tautomer); 7.30 - 7.75 (m, 4 H, benzoxazine-$C_{5,6,7,8}$-H); 8.25 (s, 1 H, uracil-C_5-H); 10.08 (s, 2 H, uracil-N_1-H & N_3-H, D_2O exchangeable). Anal. Calcd for $C_{12}H_{10}N_4O_3$ (258.23): C, 55.81; H, 3.90; N, 21.70. Found: C, 55.87; H, 3.89; N, 21.81.

3.1.6. Synthesis of 3-Amino-4-(2,6-dioxo-1,2,3,6-tetrahydropyrimidin-4-ylamino)thiophene-2- carbonitrile; 9

An equimolar mixture of 6-acetylaminouracil 8 [53] , malononitrile and sulphur (2 mmol.) was refluxed in dry DMF (15 mL) for 6 hr. The reaction mixture was concentrated, allowed to cool then triturated with cold water to yield the target compound which was filtered, washed with water, dried and crystallized from DMF/H_2O.

Chocolate brown crystals, yield, 46%, m.p. < 300°C; IR (KBr) cm^{-1}: 3359, 3186 (NH); 3035 (CH-aromatic); 2210 (CN); 1716 (C=O); 1513 (C=C); 1267, 1052 (C-S-C). MS (m/z, %): 249 ($M^{+\cdot}$, 10.42); 58 (100). Anal. Calcd for $C_9H_7N_5O_2S$ (249.25): C, 43.37; H, 2.83; N, 28.10. Found: C, 43.40; H, 2.87; N, 28.17.

3.1.7. Synthesis of 6-(4-aminothieno[3,2-d]pyrimidin-7-ylamino)pyrimidine-2,4(1H, 3H)-dione; 10

Compound 9 (0.2 g, 5 mmol) was refluxed in excess formamide (10 mL) for 5 hr. The reaction mixture was concentrated under reduced pressure then poured onto crushed ice. The obtained product was filtered, washed with water, dried and crystallized from DMF/H_2O.

Dark brown crystals, yield, 45%, m.p. < 300°C; IR (KBr) cm^{-1}: 3359, 3182 (NH_2, NH); 2921 (CH-aromatic); 1688 (C=O); 1632 (C=N); 1510 (C=C); 1290, 1044 (C-S-C). ^1H NMR (DMSO-d_6) δ ppm: 5.88 (s, 2 H, NH_2, D_2O exchangeable); 7.37 (s, 1 H, uracil-C_6-NH, D_2O exchangeable); 7.83 (s, 1 H, uracil-C_5-H); 7.95 (s, 1 H, thienopyrimidine-C_6-H); 8.54 (s, 1 H, thienopyrimidine-C_2-H); 10.14 (s, 2 H, uracil-N_1-H & N_3-H, D_2O exchangeable). Anal. Calcd for $C_{10}H_8N_6O_2S$ (276.27): C, 43.47; H, 2.92; N, 30.42; S, 11.61. Found: C, 43.53; H, 2.97; N, 30.57; S, 11.69.

3.1.8. General Method for Synthesis of Compounds 11a-c

A mixture of 6-aminouracil 1 (1.27 g, 10 mmol.) and the appropriate aromatic aldehyde (20 mmol.) namely; 4-chlorobenzaldehyde, furan-2-carboxaldehyde, thiophen-2-carboxaldehyde, in presence of anhydrous potassium carbonate (1.38 g, 10 mmol.) was refluxed in dry DMF (25 mL) for 5 hr. The reaction mixture was allowed to cool, then poured onto crushed ice to yield a solid product that was collected by filtration. The obtained product was washed with ice-cold water, dried then crystallized from DMF/H_2O.

1) 6-(4-Chlorobenzylideneamino)pyrimidine-2,4(1H,3H)-dione; 11a Yellow crystals, yield, 75%, m.p. < 300°C [56] ; IR (KBr) cm^{-1}: 3438, 3150 (NH); 3088 (CH-aromatic); 2921, 2855 (CH-aliphatic); 1715 (C=O); 1625 (C=N); 1490 (C=C); 831 (C-Cl). ^1H NMR (DMSO-d_6) δ ppm: 7.10 (s, 1 H, arylidine-CH); 7.19 (d, 2 H, J = 8.4 Hz, 4-Cl-C_6H_4-$C_{2,6}$-H); 7.26 (d, 2 H, J = 8.4 Hz, 4-Cl-C_6H_4-$C_{3,5}$-H); 7.95 (s, 1 H, uracil-C_5-H); 9.13 (s, 1 H, uracil-N_1-H, D_2O exchangeable); 9.89 (s, 1 H, uracil-N_3-H, D_2O exchangeable). Anal. Calcd for $C_{11}H_8ClN_3O_2$ (249.65): C, 52.92; H, 3.23; N, 16.83. Found: C, 52.94; H, 3.23; N, 16.91.

2) 6-(Furan-2-ylmethyleneamino)pyrimidine-2,4(1H,3H)-dione; 11b Dark brown powder, yield, 35%, m.p. < 300°C [56] ; IR (KBr) cm^{-1}: 3413, 3143 (NH); 3050 (CH-aromatic); 2924, 2856 (CH-aliphatic); 1718 (C=O); 1625 (C=N); 1495 (C=C); 1294, 1064 (C-O-C). ^1H NMR (DMSO-d_6) δ ppm: 4.85 (s, 1 H, furyl-C_4-H); 5.89 (s, 1 H, furyl-C_3-H); 6.21 (s, 1 H, furyl-C_5-H); 7.30 (s, 1 H, arylidine-CH); 7.95 (s, 1 H, uracil-C_5-H); 9.86 (s, 1 H, uracil-N_1-H, D_2O exchangeable); 10.22 (s, 1 H, uracil-N_3-H, D_2O exchangeable). Anal. Calcd for $C_9H_7N_3O_3$ (205.17): C, 52.69; H, 3.44; N, 20.48. Found: C, 52.67; H, 3.49; N, 20.54.

3) 6-(Thiophen-2-ylmethyleneamino)pyrimidine-2,4(1H,3H)-dione; 11c Dark yellow powder, yield, 64%, m.p. < 300°C [56] ; IR (KBr) cm^{-1}: 3450, 3150 (NH); 3091 (CH-aromatic); 2923, 2857 (CH-aliphatic); 1717 (C=O); 1636 (C=N); 1543, 1495 (C=C); 1294, 1099 (C-S-C). ^1H NMR (DMSO-d_6) δ ppm: 7.08 (s, 1 H, arylidine-CH); 7.17 - 7.20 (m, 1 H, J = 5.4 Hz, thienyl-C_4-H); 7.50 (d, 1 H, J = 5.4 Hz, thienyl-C_5-H); 7.72 (d, 1 H, J = 5.4 Hz, thienyl-C_3-H); 7.96 (s, 1 H, uracil-C_5-H); 9.03 (s, 1 H, uracilN_1-H, D_2O exchangeable); 9.96 (s, 1 H, uracil-N_3-H, D_2O exchangeable). Anal. Calcd for $C_9H_7N_3O_2S$ (221.24): C, 48.86; H, 3.19; N, 18.99. Found: C, 49.02; H, 3.22; N, 19.12.

3.1.9. General Method for Synthesis of Compounds 12a-c

The appropriate Schiff base 11a-c (2 mmol.) was refluxed with thioglycolic acid (0.37 g, 0.28 mL, 4 mmol.) containing anhydrous zinc chloride (0.27 g, 2 mmol.) in dry benzene (20 mL) for 30 hr. The reaction mixture was cooled and the solid product was collected, washed with benzene and crystallized from DMF/H_2O.

1) 6-[2-(4-Chlorophenyl)-4-oxothiazolidin-3-yl]pyrimidine-2,4(1H,3H)-dione; 12a Dark brown powder, yield, 60%, m.p. 230°C [58] ; IR (KBr) cm^{-1}: 3464 (OH tautomer); 3205 (NH); 3091 (CH-aromatic); 2923 (CH-aliphatic); 1700 (C=O); 1565 (C=C); 1380, 1021 (C-S-C). ^1H NMR (DMSO-d_6) δ ppm: 3.65 (s, 2 H, thiazolidine-C_4-CH_2); 4.93 (s, 1 H, thiazolidine-C_2-CH); 5.11 (s, 1 H, uracil-C_5-H); 7.09 (d, 2 H, J = 8.6 Hz, 4-Cl-C_6H_4-$C_{2,6}$-H); 7.33 (d, 2 H, J = 8.6 Hz, 4-Cl-C_6H_4-$C_{3,5}$-H); 11.10 (s, 1 H, uracil-N_1-H, D_2O exchangeable); 11.80 (s, 1 H, uracil-N_3-H, D_2O exchangeable). Anal. Calcd for $C_{13}H_{10}ClN_3O_3S$ (323.75): C, 48.23; H, 3.11; N, 12.98; S, 9.90. Found: C, 48.29; H, 3.09; N, 13.13; S, 9.97.

2) 6-[2-(Furan-2-yl)-4-oxothiazolidin-3-yl]pyrimidine-2,4(1H,3H)-dione; 12b Dark brown powder, yield, 34%, m.p. < 300°C [58] ; IR (KBr) cm^{-1}: 3200 (NH); 3090 (CH-aromatic); 2919, 2861 (CH-aliphatic); 1701 (C=O); 1534 (C=C); 1395, 1050 (C-S-C); 1262, 1050 (C-O-C). MS. (m/z, %): 279 (M^+, 0.18); 278 (M^+-1, 0.27); 71 (100). Anal. Calcd for $C_{11}H_9N_3O_4S$ (279.27): C, 47.31; H, 3.25; N, 15.05; S, 11.48. Found: C, 47.43; H, 3.29; N, 15.19; S, 11.62.

3) 6-[4-Oxo-2-(thiophen-2-yl)thiazolidin-3-yl]pyrimidine-2,4(1H,3H)-dione; 12c Light brown powder, yield, 36%, m.p. < 300°C; IR (KBr) cm^{-1}: 3163 (NH); 3039 (CH-aromatic); 2843 (CH-aliphatic); 1697 (C=O); 1453 (C=C); 1386, 1071 (C-S-C). ^1H NMR (DMSO-d_6) δ ppm: 3.65 (s, 2 H, thiazolidine-C_4-CH_2); 4.37 (s, 1 H, thiazolidine-C_2-CH); 4.66 (d, 1 H, J = 6 Hz, thienyl-C_3-H); 4.92 - 5.11 (m, 1 H, thienyl-C_4-H); 5.56 (d, 1 H, J = 6 Hz, thienyl-C_5-H); 8.53 (s, 1 H, uracil-C_5-H); 11.53 (s, 1 H, uracil-N_1-H, D_2O exchangeable); 11.92 (s, 1 H, uracil-N_3-H, D_2O exchangeable). Anal. Calcd for $C_{11}H_9N_3O_3S_2$ (295.34): C, 44.73; H, 3.07; N, 14.23; S, 21.71. Found: C, 44.85; H, 3.12; N, 14.51; S, 21.84.

3.1.10. Synthesis of Diethyl 2,4-Dioxo-7-(thiophen-2-yl)-1,2,3,4-tetrahydropyrido[2,3-d] pyrimidine-5,6-dicarboxylate; 13

The Schiff base derivative 11c (0.3 g, 1 mmol.) was fused with excess diethyl acetylene dicarboxylate (1 mL) for 1 hr. The reaction mixture was allowed to cool then triturated with ethanol. The obtained product was filtered, washed with ethanol, left to dry then crystallized from DMF/H_2O.

Light brown powder, yield, 43%, m.p. charrs at 280°C; IR (KBr) cm^{-1}: 3210, 3120 (NH); 3030 (CH-aromatic); 2925, 2812 (CH-aliphatic); 1718 (C=O); 1616 (C=N); 1495 (C=C); 1375, 1067 (C-S-C); 1293, 1261, 1067, 1033 (C-O-C). ^1H NMR (DMSO-d_6) δ ppm: 1.00 - 1.18 (m, 6 H, two CH_3); 3.60 - 3.90 (m, 4 H, two CH_2); 6.70 - 6.90 (m, 1 H, thienyl-C_4-H); 7.09 (d, 1 H, J = 5.4 Hz, thienyl-C_3-H); 7.90 - 8.02 (m, 1 H, thienyl-C_5-H); 10.03 (s, 1 H, uracil-N_1-H, D_2O exchangeable); 10.20 (s, 1 H, uracil-N_3-H, D_2O exchangeable). Anal.

Calcd for $C_{17}H_{15}N_3O_6S$ (389.38): C, 52.44; H, 3.88; N, 10.79. Found: C, 52.58; H, 3.92; N, 11.02.

3.1.11. Synthesis of 6-Azidopyrimidine-2,4(1H, 3H)-dione; 14

To an ice cold mixture of 6-aminouracil 1 (0.5 g, 4 mmol.), water (10 mL) and concentrated hydrochloric acid (3 mL), sodium nitrite solution (0.27 g, 4 mmol. in 1 mL H_2O) was added drop wise; while stirring at 0°C. Stirring was continued for 15 min., then an ice cold aqueous solution of sodium azide (0.26 g, 4 mmol.) was added drop wise. The reaction mixture was stirred for another 30 min. then filtered to yield purple colored powder of compound 13.

Purple powder, yield, 72%, m.p. charrs at 240°C without melting [61] ; IR (KBr) cm^{-1}: 3430 (OH tauotomer); 3292, 3182 (two NH); 3015 (CH-aromatic); 2150 (N_3); 1672 (C=O); 1512, 1428 (N=N); 1543 (C=C). MS. (m/z, %): 153 (M^+, 1.81); 71 (100). Anal. Calcd for $C_4H_3N_5O_2$ (153.10): C, 31.38; H, 1.98; N, 45.74. Found: C, 31.42; H, 2.02; N, 45.81.

3.1.12. Synthesis of 5-Amino-1-(2,6-dioxo-1,2,3,6-tetrahydropyrimidin-4-yl)-1H-1,2,3-triazole- 4-carbonitrile; 15

Compound 14 (0.3 g, 1.96 mmol.) was refluxed with malononitrile (0.26 g, 3.92 mmol.) in an ethanolic solution of sodium ethoxide (0.1 g, 1.96 mmol.) for 12 hr. The reaction mixture was filtered while hot and the filtrate was concentrated and allowed to cool to yield compound 15 which was collected, washed with ethanol and crystallized from ethanol.

Dark orange crystals, yield, 81%, m.p. < 300°C. IR (KBr) cm^{-1}: 3450, 3334 (NH_2, NH); 2204 (CN); 1710 (C=O); 1646 (C=N). ^1H NMR (DMSO-d_6) δ ppm: 7.28 (s, 2 H, NH_2, D_2O exchangeable); 8.50 (s, 1 H, uracil-C_5-H); 9.51 (s, 1 H, uracil-N_1-H, D_2O exchangeable); 10.76 (s, 1 H, uracil-N_3-H, D_2O exchangeable). MS. (m/z, %): 219 (M^+, 5.14); 92 (100). Anal. Calcd for $C_7H_5N_7O_2$ (219.16): C, 38.36; H, 2.30; N, 44.74. Found: C, 38.45; H, 2.36; N, 44.89.

3.1.13. Synthesis of N-(2,6-Dioxo-1,2,3,6-tetrahydropyrimidin-4-yl)-2,4-dimethoxybenzamide; 16

An equimolar mixture of 6-aminouracil 1 (0.2 g, 1.57 mmol.) and 2,4-dimethoxybenzoyl chloride (0.32 g, 1.57 mmol.) was refluxed in dry DMF (10 mL) containing 2 - 3 drops of piperidine for 6 hr. The reaction mixture was concentrated then allowed to cool. The separated product was collected, washed with ethanol then crystallized from DMF/H_2O.

Light brown crystals, yield, 32%, m.p. < 300°C. IR (KBr) cm^{-1}: 3325, 3175 (NH); 3018 (CH-aromatic); 2924, 2850 (CH-aliphatic); 1729 (C=O); 1632

(C=N); 1455 (C=C); 1290, 1088 (C-O-C). ^{1}H NMR (DMSO-d_6) δ ppm: 4.42 (s, 6 H, two CH_3); 6.20 (s, 1 H, C_6H_3-C_3-H); 6.30 - 6.38 (m, 2 H, dimethoxy-C_6H_3-$C_{5,6}$-H); 6.72 (s, 1 H, uracil-C_6-NH, D_2O exchangeable); 7.95 (s, 1 H, uracil-C_5-H); 10.15 (s, 1 H, uracil-N_1-H, D_2O exchangeable); 10.43 (s, 1 H, uracil-N_3-H, D_2O exchangeable). MS. (m/z, %): 219 (M^+, 4.16); 65 (100). Anal. Calcd for $C_{13}H_{13}N_3O_5$ (291.26): C, 53.61; H, 4.50; N, 14.43. Found: C, 54.13; H, 3.91; N, 14.72.

3.1.14. Synthesis of 1-(2,6-Dioxo-1,2,3,6-tetrahydropyrimidin-4-yl)-3-phenylthiourea; 17

An equimolar mixture of 6-aminouracil 1 (0.5 g, 4 mmol.) and phenyl isothiocyanate (0.53 g, 0.4 mL, 4 mmol.) was refluxed in dry pyridine (10 mL) for 60 hr. The reaction mixture was allowed to cool, poured onto crushed ice then neutralized by using 10% hydrochloric acid to yield the desired product. The product was filtered, washed with water, dried and crystallized from DMF/H_2O.

Dark yellow crystals, yield, 55%, m.p. 105°C. IR (KBr) cm^{-1}: 3400, 3195 (NH); 3035 (CH-aromatic); 1680 (C=O); 1538, 1328, 1244, 1050 (I, II, III, IV bands N-C=S). ^{1}H NMR (DMSO-d_6) δ ppm: 7.10 - 7.21 (m, 1 H, C_6H_5-C_4-H); 7.23 - 7.47 (m, 2 H, C_6H_5-$C_{2,6}$-H); 7.53 (d, 2 H, J = 8.1 Hz, C_6H_5-$C_{3,5}$-H); 8.54 (s, 1 H, uracil-C_5-H); 8.90 (s, 1 H, uracil-C_6-NH, D_2O exchangeable); 10.01 (s, 1 H, uracil-N_1-H, D_2O exchangeable); 11.16 (s, 1 H, uracil-N_3-H, D_2O exchangeable); 11.63 (s, 1 H, C_6H_5-NH, D_2O exchangeable). Anal. Calcd for $C_{11}H_{10}N_4O_2S$ (260.27): C, 50.37; H, 3.84; N, 21.36. Found: C, 50.49; H, 3.82; N, 21.52.

3.1.15. Synthesis of 6-[4-(4-Bromophenyl)-3-phenylthiazol-2(3H)-ylideneamino]pyrimidine- 2,4(1H, 3H)-dione; 18

An equimolar mixture of thiourea derivative 17 (0.26 g, 1 mmol.) and 4-bromophenacyl bromide (0.28 g, 1 mmol.) was refluxed in absolute ethanol (30 mL) containing sodium acetate (1 g, 12 mmol.) for 12 hr. The reaction mixture was allowed to cool then filtered. The filtrate was concentrated then diluted with ice cold water. The obtained product was filtered, washed with water, dried and crystallized from DMF/H_2O.

Dark yellow crystals, yield, 40 %, m.p. charrs at 290°C. IR (KBr) cm^{-1}: 3186 (NH); 3058, 3031 (CH-aromatic); 1705 (C=O); 1579 (C=N); 1484 (C=C); 1389, 1068 (C-S-C). ^{1}H NMR (DMSO-d_6) δ ppm: 6.96 - 7.06 (m, 1 H, C_6H_5-C_4-H); 7.29 (d, 2 H, J = 8.4 Hz, 4-Br-C_6H_4-$C_{2,6}$-H); 7.42 - 7.49 (m, 2 H, C_6H_5-$C_{3,5}$-H); 7.58 (s, 1 H, thiazolidine-C_5-H); 7.68 - 7.86 (m, 2 H, C_6H_5-$C_{2,6}$-H); 7.95 (d, 2 H, J = 8.4 Hz, 4-Br-C_6H_4-$C_{3,5}$-H); 7.99 (s, 1 H, uracil-C_5-H); 11.10 (s, 1 H, uracil-N_1-H, D_2O exchangeable); 11.84 (s, 1 H, uracil-N_3-H, D_2O

exchangeable). Anal. Calcd for $C_{19}H_{13}BrN_4O_2S$ (441.30): C, 51.71; H, 2.97; N, 12.70. Found: C, 51.84; H, 3.01; N, 12.84.

3.1.16. Synthesis of 6-(2-Hydroxyethyl)-5,6,7,8-tetrahyropyrimido[4,5-d]pyrimidine-2, 4(1H, 3H)- dione; 19

A mixture of ethanolamine (0.31 g, 0.31 mL, 5 mmol) and paraformaldehyde (0.84 g, 7 mmol.) was added to a suspension of 6-aminouracil 1 (0.39 g, 3 mmol.) in absolute ethanol (20 mL). The reaction mixture was heated under reflux for 12 hr. then allowed to attain room temperature. The obtained precipitate was filtered, washed with ethanol, dried and crystallized from DMF/H_2O.

White powder, yield, 25%, m.p. 280°C. IR (KBr) cm^{-1}: 3350 (OH); 3208, 3126 (NH); 2949, 2916, 2872 (CH-aliphatic); 1704 (C=O); 1585 (C=C). ^1H NMR (DMSO-d_6) δ ppm: 3.40 - 3.62 (m, 4 H, N-CH_2-CH_2-OH); 3.96 (s, 2 H, tetrahydropyrimidine-C_4-H); 4.42 (s, 1 H, OH, D_2O exchangeable); 4.85 (s, 2 H, tetrahydropyrimidine-C_2-H); 6.38 (s, 1 H, tetrahydropyrimidine-N_1-H, D_2O exchangeable); 9.81 (s, 1 H, uracil-N_1-H, D_2O exchangeable); 10.05 (s, 1 H, uracil-N_3-H, D_2O exchangeable). Anal. Calcd for $C_8H_{12}N_4O_3$ (212.21): C, 45.28; H, 5.70; N, 26.40. Found: C, 45.79; H, 4.85; N, 26.83.

3.1.17. Synthesis of 4-Methyl-1H-pyrimido[1,6-a]pyrimidine-2,6,8(7H)-trione; 20

An equimolar mixture of 6-aminouracil 1 (0.50 g, 4 mmol.) and ethyl acetoacetate (0.51 g, 0.5 mL, 4 mmol.) was fused for 6 hr. The reaction mixture was triturated with ethanol, filtered and dried. The obtained product was crystallized from DMF/H_2O.

Pale yellow crystals, yield, 52%, m.p. < 300°C. IR (KBr) cm^{-1}: 3439 (broad OH tautomer); 3280, 3160 (NH); 2950 (CH-aliphatic); 1700 (C=O); 1640 (C=N); 1561 (C=C). MS. (m/z, %): 192 (M-1, 0.34); 60 (100). Anal. Calcd for $C_8H_7N_3O_3$ (193.16): C, 49.74; H, 3.65; N, 21.75. Found: C, 49.88; H, 3.71; N, 21.92.

3.1.18. Synthesis of 2-Amino-6,8-dioxo-7,8-dihydro-6H-pyrimido[1,6-a]pyrimidine-3-carbonitrile; 21

An equimolar mixture of compound 1 (0.50 g, 4 mmol.) and 2-(ethoxymethylene)malononitrile [66](0.48 g, 4 mmol.) was refluxed in absolute ethanol (20 mL) containing a catalytic amount of piperidine (2 - 3 drops) for 12 hr. The reaction mixture was left to cool and the obtained product was filtered, washed with ethanol and crystallized from DMF/H_2O.

Yellow crystals, yield, 45%, m.p. 98°C. IR (KBr) cm^{-1}: 3374, 3197 (NH_2, NH); 3083, 3022 (CH-aromatic); 2220 (CN); 1725 (C=O); 1630 (C=N); 1460

(C=C). ^{1}H NMR (DMSO-d_6) δ ppm: 3.75 (s, 2 H, NH_2, D_2O exchangeable); 7.65 (s, 1 H, pyrimidine-C_6-H); 7.95 (uracil-C_5-H); 10.79 (s, 1 H, uracil-N_3-H, D_2O exchangeable). Anal. Calcd for $C_8H_5N_5O_2$ (203.16): C, 47.30; H, 2.48; N, 34.47. Found: C, 47.46; H, 2.45; N, 34.61.

3.1.19. Synthesis of 4-Methyl-2-thioxo-1H-pyrimido[1,6-a][1,3,5]triazine-6,8(2H,7H)-dione; 22

To a solution of acetyl isothiocyanate (1.01 g, 10 mmol.) [prepared by refluxing acetyl chloride (0.79 g, 0.72 mL, 10 mmol.) with ammonium thiocyanate (0.76 g, 10 mmol.) in dry acetone] in dry acetone (50 mL), compound 1 (1.27 g, 10 mmol.) was added. The reaction was refluxed for 12 hr., allowed to cool then filtered. The filtrate was concentrated and the obtained product was collected then crystallized from acetone.

Orange crystals, yield, 30%, m.p. charrs at 215°C. IR (KBr) cm^{-1}: 3395, 3312 (NH); 3070 (CH-aromatic); 2917, 2850 (CH-aliphatic); 1699 (C=O); 1640 (C=N); 1543 (C=C); 1515, 1367, 1168, 1052 (I, II, III, IV bands N-C=S). ^{1}H NMR (DMSO-d_6) δ ppm: 2.05 (s, 3 H, CH_3); 6.66 (s, 1 H, triazine-N_1-H, D_2O exchangeable); 8.36 (s, 1 H, uracil-C_5-H); 11.06 (s, 1 H, uracil-N_3-H, D_2O exchangeable). Anal. Calcd for $C_7H_6N_4O_2S$ (210.21): C, 40.00; H, 2.88; N, 26.65. Found: C, 40.13; H, 2.94; N, 26.84.

3.2. Biological Screening

3.2.1. Anticancer Screening

1) Cell culture PC3 human prostate cancer cells were grown in RPMI-1640 medium, supplemented with 10% heat inactivated FBS, 100 units/mL of penicillin and 100 mg/mL of streptomycin and maintained at 37° in a humidified atmosphere containing 5% CO_2. The cells were maintained as "monolayer culture" by serial subculturing.

2) SRB cytotoxicity assay Cytotoxicity was determined using SRB method as previously described by Skehan et al. [85] . Exponentially growing cells were collected using 0.25% Trypsin-EDTA and seeded in 96-well plates at 1000 - 2000 cells/well in RPMI-1640 supplemented medium. After 24 h, cells were incubated for 72 h with various concentrations of the tested compounds. Following 72 h treatment, the cells will be fixed with 10% trichloroacetic acid for 1 h at 4°C. Wells were stained for 10 min at room temperature with 0.4% SRB dissolved in 1% acetic acid. The plates were air dried for 24 h and the dye was solubilized with Tris-HCl for 5 min on a shaker at 1600 rpm. The optical density (OD) of each well was measured spectrophotometrically at 564 nm with an ELISA microplate reader (ChroMate-4300, FL, USA). The IC_{50} values were calculated according to the

equation for Boltzman sigmoidal concentration-response curve using the nonlinear regression fitting models (Graph Pad, Prism Version 5).

3.2.2. Cathepsin B Assay

1) Cell culture PC3 human prostate cancer cells were grown in RPMI-1640 medium supplemented with 10% heat inactivated FBS, 100 units/mL of penicillin and 100 mg/mL of streptomycin and maintained at 37° in a humidified atmosphere containing 5% CO_2. Exponentially growing cells were seeded in 6-well plates at (4×10^5) cells/well in RPMI-1640 supplemented medium. After 24 h, cells were incubated for 24 h with the tested compounds at their median inhibitory concentrations (IC_{50}). Then, cells were washed with 1x PBS, scraped, homogenized in ice-cold PBS using Dounce homogenizer and stored overnight at −20°C. After two freeze-thaw cycles, the cells were centrifuged at 5000 × g for 5 minutes. Supernatants were collected and assayed immediately.

2) ELISA assay Cathepsin-B (CTS-B) was quantified in the collected samples using enzyme-linked immunosorbent assay kit (Wuhan ElAab Science Co., Ltd, Wuhan, China) according to the manufacturer instructions. Briefly, the assay was based on the sandwich technique in which specific antibodies to CTS-B were pre-coated onto 96-well plate. The specific detection antibodies were biotinylated. The test samples and biotinylated detection antibodies were added sequentially followed by washing. Avidin-Peroxidase Complex was added and unbound conjugates were washed. A substrate solution was added to the wells to determine the bound enzyme activity. The color development was stopped by sulphuric acid, and the absorbance was read at 450 nm using an ELISA microplate reader (ChroMate-4300, FL, USA). The intensity of the color is directly proportional to the concentration of CTS-B in the sample.

4. CONCLUSION

Marwa Taha Mostafa Sarg, Shereen Said El-Shaer Upon the different prepared synthesized compounds bearing different heterocyclic rings either directly attached or attached through an amino or imino bridge to C-6 position of uracil backbone, the most active anticancer compounds against prostate PC3 cell line were those bearing rings as 2-thioxopyrimidine-5-carbonitrile 3a and 3c, quinoxaline 7a, 1,4-benzoxazine 7b, thiazolidinone 12a, 12b and thiazolidine ring 18. In addition to uracil bearing side chains at C-6 position as the chloroacetylamino derivative 4 and the phenylthiourea derivative 17. However, among fusion of different heterocycles to uracil ring, only the pyridopyrimidine derivative 13 and the pyrimidopyrimidine trione derivative 20 were considerably active. All these compounds exhibited better inhibition of cathepsin B than the reference drug doxorubicin did, except compound 3c, while compounds 5a, 5b, 7a, 11a, 12a and 17 exerted more than 50% inhibition of cathepsin B that is associated with metastasis of prostate cancer.

REFERENCES

1. El-Deeb, I.M. and Lee, S.H. (2010) Design and Synthesis of New Anticancer Pyrimidines with Multiple-Kinase Inhibitory Effect. Bioorganic & Medicinal Chemistry, 18, 3860-3874.
2. Jemal, C., Siegel, R., Xu, J. and Ward, E. (2010) Cancer Statistics. Ca—A Cancer Journal for Clinicians, 60, 277-300.
3. Merino, M., Pinto, A., González, R. and Espinosa, E. (2011) Antiangiogenic Agents and Endothelin Antagonists in Advanced Castration Resistant Prostate Cancer. European Journal of Cancer, 47, 1846-1851.
4. VanCleave, T.T., Moore, J.H., Benford, M.L., Brock, G.N., Kalbfleisch, T., Baumgartner, R.N., Lillard Jr., J.W., Kittles, R.A. and Kidd, L.C.R. (2009) Interaction among Variant Vascular Endothelial Growth Factor (VEGF) and Its Receptor in Relation to Prostate Cancer Risk. The Prostate, 70, 341-352.
5. Folkman, J. (1972) Angiogenesis in Psoriasis: Therapeutic Implications. Journal of Investigative Dermatology, 59, 40-43.
6. Giavazzi, R. and Taraboletti, G. (1999) Angiogenesis and Angiogenesis Inhibitors in Cancer. Forum (Genova), 9, 261-272.
7. Kroemer, G. and Jaattela, M. (2005) Lysosomes and Autography in Cell Death Control. Nature Reviews Cancer, 5, 886-897.
8. Oliveira, C.R., Barbosa, C.M.V., Nascimento, F.D., Lanetzki, C.S., Meneghin, M.B., Pereira, F.E.G., Paredes-Gamero, E.J., Ferreira, A.T., Rodrigues, T., Queiroz, M.L.S., Caires, A.C.F., Tersariol, I.L.S. and Bincoletto, C. (2009) Pre-Clinical Antitumour Evaluation of Biphosphinic Palladacycle Complex in Human Leukaemia Cells. Chemico-Biological Interactions, 177, 181-189.
9. Reinheckel, T., Peters, C., Krüger, A., Turk, B. and Vasiljeva, O. (2012) Differential Impact of Cysteine Cathepsins on Genetic Mouse Models of De Novo Carcinogenesis: Cathepsin B as Emerging Therapeutic Target. Frontiers in Pharmacology, 3, 133.
10. Mohamed, M.M. and Sloane, B.F. (2006) Cysteine Cathepsins: Multifunctional Enzymes in Cancer. Nature Reviews Cancer, 6, 764-775.
11. Gocheva, V. and Joyce, J.A. (2007) Cysteine Cathepsins and the Cutting Edge of Cancer Invasion. Cell Cycle, 6, 60-64.
12. Vasiljeva, O., Reinheckel, T., Peters, C., Turk, D., Turk, V. and Turk, B. (2007) Emerging Roles of Cysteine Cathepsins in Disease and Their Potential as Drug Targets. Current Pharmaceutical Design, 13, 387-403.
13. Doxakis, A., Maria, A., Savvas, P. and Zafiroula, I.K. (2013) Assessment of the Roles of Cathepsins B, H and L in the Progression of Colorectal Cancer. Journal of Cancer Therapy, 4, 1-7.
14. Foghsgaard, L., Wissing, D., Mauch, D., Lademann, U., Bastholm, L., Boes, M., Elling, F., Leist, M. and Jaattela, M. (2001) Cathepsin B Acts as a Dominant Execution Protease in Tumour Cell Apoptosis Induced by Tumour Necrosis Factor. The Journal of Cell Biology, 153, 999-1010.
15. Beckham, T.H., Lu, P., Cheng, J.C., Zhao, D., Turner, L.S., Zhang, X., Hoffman, S., Armeson, K.E., Liu, A., Marrison, T., Hannun, Y.A. and Liu, X. (2012) Acid Ceramidase-Mediated Production of Sphingosine 1-Phosphate Promotes Prostate Cancer Invasion through Upregulation of Cathepsin B. International Journal of Cancer, 131, 2034-2043.
16. El Gemeie, G.H. (2003) Thioguanine, Mercaptopurine: Their Analogs and Nucleosides as Antimetabolites. Current Pharmaceutical Design, 9, 2627-2642.

17. Parker, W.B., Secrist, J. and Waud, W.R. (2004) Purine Nucleoside Antimetabolites in Development for the Treatment of Cancer. Current Opinion in Investigational Drugs, 5, 592-596.
18. Nussbaumer, S., Bonnabry, P., Veuthey, J.L. and Fleury-Souverain, S. (2011) Analysis of Anticancer Drugs: A Review. Talanta, 85, 2265-2289.
19. Heidelberger, C. (1984) Pyrimidine and Pyrimidine Antimetabolites in Cancer Medicine. Lea and Febiger, Philadelphia, 801-824.
20. Kundu, N.G., Das, P., Balzarini, J. and De Clercq, E. (1997) Palladium-Catalyzed Synthesis of [E]-6-(2-Acylvinyl) Uracils and [E]-6-(2-Acylvinyl)-1-[(2-hydroxyethoxy) methyl]uracils—Their Antiviral and Cytotoxic Activities. Bioorganic & Medicinal Chemistry, 5, 2011-2018.
21. Sigmond, J. and Peters, G.J. (2005) Pyrimidine and Purine Analogues, Effects on Cell Cycle Regulation and the Role of Cell Cycle Inhibitors to Enhance Their Cytotoxicity. Nucleosides, Nucleotides and Nucleic Acids, 24, 1997-2022.
22. Prekupec, S., Makuc, D., Plavec, J., Suman, L., Kralj, M., Pavelic, K., Balzarini, J., De Clercq, E., Mintas, M. and Raic-Malic, S. (2007) Novel C-6 Fluorinated Acyclic Side Chain Pyrimidine Derivatives: Synthesis, 1H and 13C NMR Conformational Studies, and Antiviral and Cytostatic Evaluations. Journal of Medicinal Chemistry, 50, 3037-3045.
23. Gazivoda, T., Sokcevic, M., Kralj, M., Suman, L., Pavelic, K., De Clercq, E., Andrei, G., Snoeck, R., Balzarini, J., Mintas, M. and Raic-Malic, S. (2007) Synthesis and Antiviral and Cytostatic Evaluations of the New C-5 Substituted Pyrimidine and Furo[2,3-d]pyrimidine 4',5'-Didehydro-L-ascorbic Acid Derivatives. Journal of Medicinal Chemistry, 50, 4105-4112.http://dx.doi.org/10.1021/jm070324z
24. Gazivoda, T., Raic-Malic, S., Marjanovic, M., Kralj, M., Pavelic, K., Balzarini, J., De Clercq, E. and Mintas, M. (2007) The Novel C-5 Aryl, Alkenyl, and Alkynyl Substituted Uracil Derivatives of L-Ascorbic Acid: Synthesis, Cytostatic, and Antiviral Activity Evaluations. Bioorganic & Medicinal Chemistry, 15, 749-758.
25. Spácilová, L., Dzubák, P., Hajdúch, M., Krupkováa, S., Hradil, P. and Hlavác, J. (2007) Synthesis and Cytotoxic Activity of Various 5-[Alkoxy-(4-nitro-phenyl)-methyl]uracils in Their Racemic Form. Bioorganic & Medicinal Chemistry Letters, 17, 6647-6650.
26. Kraljevic, T.G., Kristafor, S., Suman, L., Kralj, M., Ametamey, S.M., Cetina, M. and Raic-Malic, S. (2010) Synthesis, X-Ray Crystal Structure Study and Antitumoral Evaluations of 5,6-Disubstituted Pyrimidine Derivatives. Bioorganic & Medicinal Chemistry, 18, 2704-2712.
27. Lee, Y.S. and Kim, B.H. (2002) Heterocyclic Nucleoside Analogues: Design and Synthesis of Antiviral, Modified Nucleosides Containing Isoxazole Heterocycles. Bioorganic & Medicinal Chemistry Letters, 12, 1395-1397.
28. Lee, Y.S., Park, S.M. and Kim, B.H. (2009) Synthesis of 5-Isoxazol-5-yl-2'-deoxyuridines Exhibiting Antiviral Activity against HSV and Several RNA Viruses. Bioorganic & Medicinal Chemistry Letters, 19, 1126-1128.
29. Jakubiec, D. and Walczak, K.Z. (2011) A Convenient Synthesis of 5-(1,2,4-Oxadiazol-5-yl)pyrimidine-2,4(1H, 3H)- diones. Tetrahedron Letters, 52, 6890-6891.
30. Lee, Y.S., Park, S.M., Kim, H.M., Park, S.K., Lee, K., Lee, C.W. and Kim, B.H. (2009) C5-Modified Nucleosides Exhibiting Anticancer Activity. Bioorganic & Medicinal Chemistry Letters, 19, 4688-4691.
31. Soliman, R., Habib, N.S., Ismail, K., Moustafa, A., Sarg, M.T. and Fanaki, N.H. (2003) Synthesis of Novel Pyrimidine-2,4-diones and 2-Thioxopyrimidine-4-ones as Potential Anticancer and Antimicrobial Agents. Bollettino Chimico Farmaceutico, 142, 167-174.
32. Habib, N.S., Soliman, R., Ismail, K., Hassan, A.M. and Sarg, M.T. (2003) Synthesis of Novel Pyrimidines, 1,2,4-Triazolo[4,3-a]pyrimidin-7-ones and Pyrimidino[2,1-c][1,2,4]triazin-8-ones for Their Antimicrobial and Anticancer Activities. Bollettino Chimico Farmaceutico, 142, 396-405.
33. Cui, Z., Li, Y., Huang, J., Cui, J., Wang, R. and Yang, X. (2010) New Class of Potent Antitumour Acylhydrazone Derivatives Containing Furan. European Journal of Medicinal Chemistry, 45, 5576-5584.

34. Abdel-Hamid, H.F., Soliman, A., Helaly, F.M. and Ragab, S. (2011) Cytotoxic Potency and Induced Biochemical Parameters in Mice Serum of New Furan Derivatives against Liver Cancer Cell Line. Acta Poloniae Pharmaceutica, 68, 499-505.
35. Idhayadhulla, A., Kumar, R.S., Nasser, A.J.A. and Manilal, A. (2013) Synthesis of Some New Pyrrole and Pyridine Derivatives and Their Antimicrobial, Anticancer Activities. International Journal of Biological Chemistry, 7, 15-26.
36. Alsaid, M.S., El-Gazzar, M.G. and Ghorab, M.M. (2013) Anticancer Activity of Novel Thiophenes Containing a Biological Active Diphenyl Sulfone, Diazepine, Piperidine, Oxazepine, Acrylaldehyde and Sulfonamide Moieties. Drug Research, 63, 263-269.
37. Garg, V., Kumar, A., Chaudhary, A., Agrawal, S., Tomar, P. and Sreenivasan, K.K. (2013) Synthesis, Biological Evaluation and Molecular Docking Studies of 1,3-Benzoxazine Derivative as Potential Anticancer Agents. Medicinal Chemistry Research, 22, 5256-5266.
38. Lee, S.H., Kim, N., Kim, S.J., Song, J., Gong, Y.D. and Kim, S.Y. (2013) Anti-Cancer Effect of Quinoxaline Derivatives GK13 as a Transglutaminase 2 Inhibitor. Journal of Cancer Research and Clinical Oncology, 139, 1279-1294.
39. Bathula, S.N.P. and Vadla, R. (2011) Bioactivity of 1,4-Disubstituted-1,2,3-triazoles as Cytotoxic Agents against the Various Human Cell Lines. Asian Journal of Pharmaceutical and Clinical Research, 4, 66-67.
40. Kurumurthy, C., Rao, P.S., Swamy, B.V., Kumar, G.S., Rao, P.S., Narsaiah, B., Velatooru, L.R., Pamanji, R. and Rao, J.V. (2011) Synthesis of Novel Alkyltriazole Tagged Pyrido[2,3-d]pyrimidine Derivatives and Their Anticancer Activity. European Journal of Medicinal Chemistry, 46, 3462-3468.
41. El-Gaby, M.S.A., Ismail, Z.H., Abdel-Gawad, S.M., Aly, H.M. and Ghorab, M.M. (2009) Synthesis of Thiazolidine and Thiophene Derivatives for Evaluation as Anticancer Agents. Phosphorus, Sulfur and Silicon and the Related Elements, 184, 2645-2654.
42. Abu-Hashem, A.A., El-Shehry, M.F. and Badria, F.A. (2010) Design and Synthesis of Novel Thiophenecarbohydrazide, Thienopyrazole and Thienopyrimidine Derivatives as Antioxidant and Antitumour Agents. Acta Pharmaceutica, 60, 311-323.
43. Zhu, W.F., Zhai, X., Li, S., Cao, Y.Y., Gong, P. and Liu, Y.J. (2012) Synthesis and Cytotoxic Activity of Novel 2,6- Disubstituted-4-morpholinothieno[3,2-d]pyrimidines as Potent Anti-Tumour Agents. Chinese Chemical Letters, 23, 703-706.
44. Bernier, J.L., Henichart, J.P., Warin, V., Trentesaux, C. and Jardillier, J.C. (1985) 5-Cinnamoyl-6-aminouracil Derivatives as Novel Anticancer Agents: Synthesis, Biological Evaluation and Structure-Activity Relationships. Journal of Medicinal Chemistry, 28, 497-502.
45. Dabiri, M., Arvin-Nezhad, H., Khavasi, H.R. and Bazgir, A. (2007) A Novel and Efficient Synthesis of Pyrimido[4, 5-d]pyrimidine-2,4,7-trione and Pyrido[2,3-d:6,5-d]dipyrimidine-2,4,6,8-tetrone Derivatives. Tetrahydron, 63, 1770-1774.
46. Essa, F.B., Bazbouz, A. and Al-Lababidi, S. (2010) Preparation, Chemical and Biological Study of Acyclovir Analogue Starting from 6-Aminouracil. Oriental Journal of Chemistry, 26, 1281-1292.
47. Samai, S., Nandi, G.C., Chowdhury, S. and Singh, M.S. (2011) L-Proline Catalyzed Synthesis of Densely Functionalized Pyrido[2,3-d]pyrimidines via Three-Component One-Pot Domino Knoevenagel Aza-Diels-Alder Reaction. Tetrahydron, 67, 5935-5941.
48. Fargualy, A.M., Habib, N.S., Ismail, K.A., Hassan, A.M.M. and Sarg, M.T.M. (2013) Synthesis, Biological Evaluation and Molecular Docking Studies of Some Pyrimidine Derivatives. European Journal of Medicinal Chemistry, 66, 276-295.
49. Fathalla, O.A., Zeid, I.F., Haiba, M.E., Soliman, A.M., Abd-Elmoez, S.I. and El-Serwy, W.S. (2009) Synthesis, Antibacterial and Anticancer Evaluation of Some Pyrimidine Derivatives. World Journal of Chemistry, 4, 127-132.
50. Bondock, S., Khalifa, W. and Fadda, A.A. (2011) Synthesis and Antimicrobial Activity of Some New 4-Hetarylpyrazole and Furo[2,3-c]pyrazole Derivatives. European Journal of Medicinal Chemistry, 46, 2555-2561.
51. Alqasoumi, S.I., Ghorab, M.M., Ismail, Z.H., Abdel-Gawad, S.M., El-Gaby, M.S.A. and Aly, H.M. (2009) Novel Antitumor Acetamide, Pyrrole, Pyrrolopyrimidine, Thiocyanate, Hydrazone,

Pyrazole, Isothiocyanate and Thiophene Derivatives Containing a Biologically Active Pyrazole Moiety. Arzneimittel Forschung, 59, 666-671.
52. Aly, H.M. (2011) Novel Pyrrolidinone and Pyrazolo[1,5-a][1,3,5]triazine Derivatives Bearing a Biologically Active Sulfamoyl Moiety as a New Class of Antitumor Agents. Monatshefte für Chemie—Chemical Monthly, 142, 935-941.
53. Shim, J.L., Niess, R. and Broom, A.D. (1972) Acylation of Some 6-Aminouracil Derivatives. The Journal of Organic Chemistry, 37, 578-581.
54. Gewald, K.J. (1966) Heterocyclen aus CH-aciden nitrilen. IV. Addition von Schwefelkohlenstoff an Alkyliden Malonitrile. Journal für Praktische Chemie, 31, 205-213.
55. Aly, H.M., Saleh, N.M. and Elhady, H.A. (2011) Design and Synthesis of Some New Thiophene, Thienopyrimidine and Thienothiadiazine Derivatives of Antipyrine as Potential Antimicrobial Agents. European Journal of Medicinal Chemistry, 46, 4566-4572.
56. Shingare, M.S. and Ingle, D.B. (1976) Synthesis of Pyrimidine Schiff Bases as Anticancer Agents. Journal of Indian Chemical Society, LIII, 1036-1037.
57. Mohamed, N.R., El-Saidi, M.M.T., Ali, Y.M. and Elnagdi, M.H. (2007) Utility of 6-Amino-2-thiouracil as a Precursor for the Synthesis of Bioactive Pyrimidine Derivatives. Bioorganic & Medicinal Chemistry, 15, 6227-6235.
58. Shingare, M.S. (1983) Synthesis of New Thiazolidinones from Pyrimidine Schiff Bases and Their Biological Activity. Indian Journal of Chemistry B, 22, 714-715.
59. Archana, Srivastava, V.K. and Kumar, A. (2002) Synthesis of Newer Thiadiazolyl and Thiazolidinoyl Quinazolin-4 (3H)-ones as Potential Anticonvulsant Agents. European Journal of Medicinal Chemistry, 37, 873-882.
60. Patel, N.B. and Patel, V.N. (2010) Synthesis, Antibacterial and Antifungal Activities of Novel Thiazolidinyl Sulfonamides of Quinazolin-4(3H)ones. Pharmaceutical Chemistry Journal, 44, 438-445.
61. Abdou, W.M., Fahmy, A.F.M. and Kamel, A.A. (2002) A Facile Synthesis of Pyrrolo[3,2-d]pyrimidines from 6-Azidouracils and Ylide Phosphoranes. Heteroatom Chemistry, 13, 357-365.
62. Pokhodylo, N.T., Matiychuk, V.S. and Obushak, M.D. (2009) Synthesis of Ethyl 4,5-Disubstituted 2-Azido-3-thiophenecarboxylates and Use in the Synthesis of Thieno[3,2-e][1,2,3]triazolo[1,5-a]pyrimidin-5(4H)-ones. Tetrahydron, 65, 2678-2683.
63. Hafiz, I.S.A., Ramiz, M.M.M., Mahmoud, F.F. and Darwish, E.S. (2008) β-Enaminonitriles in Heterocyclic Synthesis: Synthesis of New Tetrahydropyridinetrione, Pyridopyrimidines, Pyridotriazines and Dihydropyridines. Journal of Chemical Sciences, 120, 339-345.
64. Rostom, S.A.F., El-Ashmawy, I.M., El Razik, H.A., Badr, M.H. and Ashour, H.M.A. (2009) Design and Synthesis of Some Thiazolyl and Thiadiazolyl Derivatives of Antipyrine as Potential Non-Acidic Anti-Inflammatory, Analgesic and Antimicrobial Agents. Bioorganic & Medicinal Chemistry, 17, 882-895.
65. Marugan, J.J., Zheng, W., Motabar, O., Southall, N., Goldin, E., Sidransky, E., Aungst, R.A., Liu, K., Sadhukhan, S.K. and Austin, C.P. (2010) Evaluation of 2-Thioxo-2,3,5,6,7,8-hexahydropyrimido[4,5-d]pyrimidine-4(1H)-one Analogues as GAA Activators. European Journal of Medicinal Chemistry, 45, 1880-1897.
66. Ding, R., He, Y., Xu, J., Liu, H., Wang, X., Feng, M., Qi, C., Zhang, J. and Peng, C. (2012) Preparation and Bioevaluation of 99mTc Nitrido Radiopharmaceuticals with Pyrazolo[1,5-a]pyrimidine as Tumor Imaging Agents. Medicinal Chemistry Research, 21, 523-530.
67. Moin, K., Demchik, L., Mai, J., Duessing, J., Peters, C. and Sloane, B.F. (2000) Observing Proteases in Living Cells. Advances in Experimental Medicine and Biology, 477, 391-401.
68. Frohlich, E., Schlagenhauff, B., Mohrle, M., Weber, E., Klessen, C. and Rassner, G. (2001) Activity, Expression, and Transcription Rate of the Cathepsins B, D, H, and L in Cutaneous Malignant Melanoma. Cancer, 91, 972-982.
69. Sinha, A.A., Quast, B.J., Wilson, M.J., Fernandes, E.T., Reddy, P.K., Ewing, S.L. and Gleason, D.F. (2002) Prediction of Pelvic Lymph Node Metastasis by the Ratio of Cathepsin B to Stefin A in Patients with Prostate Carcinoma. Cancer, 94, 3141-3149.

70. Bialas, A. and Kafarski, P. (2009) Proteases as Anti-Cancer Targets—Molecular and Biological Basis for Development of Inhibitor-Like Drugs against Cancer. Anti-Cancer Agents Medicinal Chemistry, 9, 728-762.
71. Lankelma, J., Voorend, D., Barwari, T., Koetsveld, J., Van der Spek, A., De Porto, A., Van Rooijen, G. and Van Noorden, C. (2010) Cathepsin L, Target in Cancer Treatment? Life Sciences, 86, 225-233.
72. Rothberg, J.M., Sameni, M., Moin, K. and Sloane, B.F. (2012) Live-Cell Imaging of Tumor Proteolysis: Impact of Cellular and Non-Cellular Microenvironment. Biochimica et Biophysica Acta (BBA)—Proteins and Proteomics, 1824, 123-132.
73. Van Damme, J., Struyf, S. and Opdenakker, G. (2004) Chemokine-Protease Interactions in Cancer. Seminars in Cancer Biology, 14, 201-208.
74. Green, K.A. and Lund, L.R. (2005) ECM Degrading Proteases and Tissue Remodeling in the Mammary Gland. BioEssays, 27, 894-903.
75. Joyce, J.A. and Pollard, J.W. (2009) Microenvironmental Regulation of Metastasis. Nature Reviews Cancer, 9, 239-252.
76. López-Otin, C. and Hunter, T. (2010) The Regulatory Crosstalk between Kinases and Proteases in Cancer. Nature Reviews Cancer, 10, 278-292.
77. Mason, S.D. and Joyce, J.A. (2011) Proteolytic Networks in Cancer. Trends in Cell Biology, 21, 228-237.
78. Turk, B., Turk, D.S. and Turk, V. (2012) Protease Signalling: The Cutting Edge. The EMBO Journal, 31, 1630-1643.
79. Buck, M.R., Karustis, D.G., Day, N.A., Honn, K.V. and Sloane, B.F. (1992) Degradation of Extracellular-Matrix Proteins by Human Cathepsin B from Normal and Tumor Tissues. Biochemical Journal, 282, 273-278.
80. Premzl, A., Zavasnik-Bergant, V., Turk, V. and Kos, J. (2003) Intracellular and Extracellular Cathepsin B Facilitate Invasion of MCF-10A neoT Cells through Reconstituted Extracellular Matrix in Vitro. Experimental Cell Research, 283, 206-214.
81. Miyake, H., Hara, I. and Eto, H. (2004) Serum Level of Cathepsin B and Its Density in Men with Prostate Cancer as Novel Markers of Disease Progression. Anticancer Research, 24, 2573-2577.
82. Podgorski, I. and Sloane, B.F. (2003) Cathepsin B and Its Role(s) in Cancer Progression. Biochemical Society Symposia, 70, 263-276.
83. Podgorski, I., Linebaugh, B.E., Sameni, M., Jedeszko, C., Bhagat, S., Cher, M.L. and Solane, B.F. (2005) Bone Micro-environment Modulates Expression and Activity of Cathepsin B in Prostate Cancer. Neoplasia, 7, 207-223.
84. Withana, N.P., Blum, G., Sameni, M., Slaney, C., Anbalagan, A., Olive, M.B., Bidwell, B.N., Edgington, L., Wang, L., Moin, K., Sloane, B.F., Anderson, R.L., Bogyo, M.S. and Parker, B.S. (2012) Cathepsin B Inhibition Limits Bone Metastasis in Breast Cancer. Cancer Research, 72, 1199-1209.
85. Skehan, P., Storeng, R., Scudiero, D., Monks, A., McMahon, J., Vistica, D., Warren, J.T., Bokesch, H., Kenney, S. and Boyd, M.R. (1990) New Colorimetric Cytotoxicity Assay for Anticancer-Drug Screening. Journal of the National Cancer Institute, 82, 1107-1112.

CHAPTER 12

Computer-Aided Drug Design: An Innovative Tool for Modeling

Pranita P. Kore, Madhavi M. Mutha, Rishikesh V. Antre*, Rajesh J. Oswal, Sandip S. Kshirsagar

Medicinal Chemistry Research Laboratory, JSPM's Charak College of Pharmacy & Research, Pune, India

ABSTRACT

Strategies for CADD vary depending on the extent of structural and other information available regarding the target (enzyme/receptor) and the ligands. Computer-aided drug design (CADD) is an exciting and diverse discipline where various aspects of applied and basic research merge and stimulate each other. In the early stage of a drug discovery process, researchers may be faced with little or no structure activity relationship (SAR) information. The process by which a new drug is brought to market stage is referred to by a number of names most commonly as the development chain or "pipeline" and consists of a number of distinct stages. To design a rational drug, we must firstly find out which proteins can be the drug targets in pathogenesis. In present review we reported a brief history of CADD, DNA as target, receptor theory, structure optimization, structure-based drug design, virtual high-throughput screening (vHTS), graph machines.

1. INTRODUCTION

All the world's major pharmaceutical and biotechnology companies use computational design tools. At their lowest level the contributions represent the replacement of crude mechanical models by displays of structure which are a much more accurate reflection of molecular reality capable of demonstrating motion and solvent effects [1-3]. Beyond this, theoretical calculations permit the computation of binding free energies and other relevant molecular properties. The theoretical tools include empirical molecular mechanics, quantum mechanics and, more recently, statistical mechanics. This latest advance has permitted explicit solvent effects to be incorporated. Underpinning all this work is the availability of high quality computer graphics, largely supported on workstations [1-5].

Two distinct categories of research are clearly distinguishable:

1) Crystallography, NMR or homology modelling. A detailed molecular structure of the target macromolecule, the drug receptor, is known from x-ray.

2) Variable activity of otherwise similar molecules.

The target receptor binding site has properties which can only be inferred from a knowledge of the both these types of approach will now be considered and illustrated with some recent examples.

1.1 DNA as Target

The sequencing of the human genome represents one of the major scientific endeavours of this century. A major aspect of the utilization of this information will be the provision of small molecules which will recognize selected sequences, perhaps with the goal of switching off particular genes as in cancer chemotherapy. For some time, antibiotics such as netropsin have been known to bind preferentially to sequences rich in A-T pairs. A variant based on this research has been to try to design a bioreductive ligand based upon netropsin [1]. The idea of bioreductive anti-cancer agents, statts with the fact that tumours receive less blood and hence less oxygen than normal tissue.

A second starting point for sequence selective ligands is an organometallic molecule with chiral properties. The propeller-like ruthenium tris-phenanthroline complexes do show differential binding between A-T and G-C sequences [2] and moreover may exhibit a preference for purine 3', 5' pyrimidine sites in DNA [3]. Perhaps the most intriguin starting point for a molecule upon which to build nucleic acid selectivity is the ubiquitous spermine. It has been proposed [4] that spermine can bind to DNA in a cross-groove manner, with relatively non-specific interactions between the positive nitrogens of the spermine and the negatively-charged phosphate backbone.

The Ras family of protooncogenes (N-Ras, H-Ras and K-Ras) codes for small proteins of 189 amino acids with molecular weight 21 kDa protein [5]. Ras proteins are localized in the inner plasma membrane and are involved in the transduction of external stimuli to effect molecule Raf serine/threonine kinase [6]. These proteins bind GDP/ GTP and possess intrinsic GTPase activity allowing inactivation following signal transduction in the normal cellular environment [7]. Activation of point mutations in the Ras is one of the most frequent genetic alterations associated with human cancers [8]. Approximately 90% of these activating mutations occur in codons 12 and 59, identifying these codons as hot-spot targets [9]. A particular genetic alteration has been identified in a significant percentage of bladder tumors; this mutation changes a single amino acid in the H-Ras protein. Specifically, the mutation replaces the amino acid glycine with the amino acid valine at position 12 (RasG12V) [10]. The mutant form of RasA59T, which is known to undergo autophosphorylation on Thr-59 [11], shows a very strong signal that argues against the occurrence of a covalently bound phosphate [8]. As a result of these mutational changes, the mutated Ras-p21 has a structure that disables its ability to bind with GTPase activating protein (GAP) and creation of an autophosphorylation site [12], thus keeping the Ras-p21 in the GTP-bound, activated state contributing to a malignant cell phenotype . Drug target discovery involves the identification and early validation of disease-associated targets. Mutations occurring in the Ras gene(s) lead to uncontrolled cell growth and proliferation. In general, 30% of human tumor occurs through mutation in Ras gene. In colorectal and pancreatic cancers, the occurrence of mutation in Ras is 50% - 90% [5]. When we consider treatments for cancer, they depend on the types and stages of cancer development. Chemotherapy, targeted therapies, surgery, radiation therapy, biological therapy, and hormonal therapy are the various treatments that currently exist.

2. A BRIEF HISTORY OF CADD [9]

1900: The receptor and lock-and-key concepts P. Ehrich (1909) and E. Fisher (1894);

1970s: Quantitative structure-activity relationships (QSAR), Limitations: 2-Dimensional, retrospective analysis;

1980s: Beginning of CADD Molecular Biology, X-ray crystallography, multi-dimensional NMR Molecular modeling, computer graphics;

1990s: Human genome Bioinformatics, Combinatorial chemistry, High-throughput screening.

2.1. How Does CADD Work? [13]

Target Identification

Genetics

Molecular Biology

Bioinformatics

↓

Structure Determination

X-ray Crystallography

NMR Spectroscopy

↕

Biological Assays

Molecular Modeling

Computer Graphics

↕

Synthetic Chemistry

Peptidomimetics

Combinatorial Chemistry

↓

Clinical Trials

2.2. Software for Molecular Modeling

General purpose molecular modeling (large & small molecules)

—molecular mechanics, dynamics and multifunctional programs;

Quantum Chemistry calculations (small molecules)

—molecular orbital or quantum mechanical calculation;

Database of molecular structures (large & small molecules)

—software for storage and retrieval of molecular structure data;

Molecular graphics (large & small molecules)

—programs to visualize molecules QSAR (small molecules).

2.3. Software for General Purpose Molecular Modeling [13]

For workstations, minicomputers, and supercomputers (SGI, Sun, Cray, etc.)

AMBER—Peter Kollman and coworkers, UCSF.

Computer assisted model building, energy minimization, molecular dynamics, and free energy perturbation calculations.

Midas Plus—UCSF Computer Graphics Laboratory.

CHARMM—Martin Karplus and cowrokers, Harvard.

QUANTA/CHARMm—Molecular Simulations Inc. (MSI) molecular/drug design, QSAR, quantum chemistry.

X-ray & NMR data analysis Insight/DISCOVER— Biosym, Inc. Now MSI and Biosym became Accelrys Inc.

SYBYL—Tripos, Inc.

ECEPP—Harold Scheraga and coworkers, Cornell MM3—Norman Allinger and coworkers, Georgia For personal computers (Apple, Compaq, IBM, etc.)

Alchemy III—Tripos, Inc.

Structure building and manipulation, SYBYL energy minimization, molecular display, conformational searching Chem3D Pro—CambridgeSoft Corp.

Desktop Molecular Modeller—Oxford Elec. Publishing Molecular Modeling Pro—WindowChem Software Energy minimization, QSAR (surface area, volume, logP), etc.

PC MODEL—Serena Software.

2.4. Optimization [14]

The second step of drug discovery involves the modification of the hits in order to improve the biological properties of the compound by changing its pharmacophore. Using QSAR to modify lead compounds would be less tedious then having to physically synthesize the compounds. Moreover, such in silico methods could theoretically help to modify the compounds to exhibit the most potency, most selectivity, best pharmacokinetics and least toxicity. QSAR involves mainly physical chemistry and molecular docking tools that lead to tabulated data and first and second order equations. There are many theories, being the most relevant Hansch's analysis that involves Hammett electronic parameters, Esteric parameters and logP parameters [15].

2.5. Receptor Theory

A receptor [16], in the biochemistry context, is a/are protein molecule(s), found in either the plasma membrane or the cytoplasm of a cell, to which one or more specific kinds of signaling molecules may attach. A molecule which attaches to a receptor is called a ligand, and may be a peptide or other small molecule, such as a neurotransmitter, a hormone, a pharmaceutical drug, or a toxin. Each kind of receptor can bind only certain ligand shapes. Each cell typically has many receptors, of many different kinds.

An agonist is a drug that binds to a receptor of a cell and triggers a response by the cell. An agonist often mimics the action of a naturally occurring substance. An agonist produces an action. An antagonist blocks an action of an agonist. Endogenous (such as hormones and neurotransmitters) or exogenous (such as drugs) agonists and antagonists, either stimulate or inhibit a biological response in receptors [17].

3. CADD STRATEGIES IN THE DRUG DISCOVERY PROCESS

Strategies for CADD vary depending on the extent of structural and other information available regarding the target (enzyme/receptor) and the ligands. "Direct" and "indirect" design are the two major modeling strategies currently used in the drug design process. In the indirect approach the design is based on comparative analysis of the structural features of known active and inactive compounds. In the direct design the three-dimensional features of the target (enzyme/receptor) are directly considered.

3.1. CADD in Lead Generation

3.1.1 D Structure of the Protein Unknown

In the early stage of a drug discovery process, researchers may be faced with little or no structure activity relationship (SAR) information. At this point, assay development and screening should be undertaken immediately by the high-throughput screening (HTS) group [18], and chemists should immediately follow up on any screening leads or other sources of initial information. The compounds screened could be commercially available, natural products, collections of in-house synthesized compounds or emerge from combinatorial libraries. Computational chemists can, however, help in the choice of the compounds to be selected for HTS.

Instead of performing random screening, a set of compounds presenting diversity in their physicochemical properties can be selected to find leads. The aim of these analyses is to select and test fewer compounds, whilst gaining as much information as possible about the dataset [19]. Any reduction of the number of compounds to be tested, while only reducing the amount of redundancy within a database without introducing any voids, should have an important impact on research efficiency and the costs associated [20]. Recently, the use of rational design to maximize the structural diversity of database, for lead findings and refinements, was investigated. Hierarchical clustering and maximum dissimilarity methods were compared to a random approach in order to study their efficiency for the diversity enhancement of three-dimension databases. The investigations were done using twodimensional fingerprints as a validated molecular descriptor and the performance of rational selection methods vs. random approach has been compared [18,19,21-23].

The first step to derive a new lead, also called seconddary lead, will be to study the stereoelectronic properties of the selected primary leads **Table 1**. The primary leads should be selected among a set of compounds showing a large variety in chemical structures, and interact with the same target via the same binding mechanism. By comparison of the stereoelectronic properties of primary leads, a pharmacophore is defined. A pharmacophore model is a spatial arrangement of atoms or functional groups believed to be responsible for biological activity [24]. In this model the rest of the molecule acts as a skeleton to hold the groups in the right place. Typically, the derived pharmacophores consist, generally, of 3 - 5 features, and the distances between them (angles and other geometric measures are sometimes used).

3.2. Structure-Based Drug Design

Within many of the rational drug design projects in the group, computer-aided methods, such as virtual screening and de novo design techniques, play an important role. NMR spectroscopy in conjunction with molecular modeling and other spectroscopic methods allows investtigations to be made into molecular mechanisms of ligand-target recognition at the atomic level [25]. This information is a necessary component in the design of novel therapeutics and in prediction of interactions of drugs with the targets. Also over the years, the group hasstudied details of binding of ligands to the minor groove of DNA, such as Hoechst 33,258, or to tRNA [26]. NMR methods are also used by the group to study interactions of proteins with ligands. There is 300 MHz instrumentation in the school, and the group has shared usage of 500 MHz high-field instruments housed in the Department of Chemistry. The group collaborates extensively with Professor Gareth Morris, inventor and pioneer of many modern NMR techniques, thereby bringing novel techniques to bear on red biological problems [27].

Table 1. Main stereoelectronic properties used in CADD [13].

Steric	L (Substituent length) B5 (Substituent width) MR (Molar refractivity) Volume Surface area
Electronic	σ (Hammet constant) F, R (Field and resonance parameters) pKa (Ionization constants) q (atomic charges) MEP (Molecular Electrostatic Potential)...
Lipophilic	π (Hansch constant) f (Hydrophic fragmental constant) LogP (partition coefficients) Logkw (capacity factor values from RP-HPLC) CLOGP (calculated log P values) MLP (Molecular lipophilic potential)
H-bonding	HA (number of H-bond acceptors) HD (number of H-bond donors) ΔlopP (oct-hex) (H-bond capability)...

3.3. Bioinformatics in Computer-Aided Drug Design

A few years ago, the National Institutes of Health (NIH) created the Biomedical Information Science and Technology Initiative (BISTI) to examine the current state of bioinformatics in the United States. BISTI's working definition of bioinformatics included its use in biomedical research, in particular for drug discovery and development programs. Bioinformatics was seen as an emergin how drugs are found, brought to clinical trials and eventually released to the marketplace [28].

Computer-Aided Drug Design (CADD) is a specialized discipline that uses computational methods to simulate drug-receptor interactions. CADD methods are heavily dependent on bioinformatics tools, application and on the support side of the hub, information technology, information management, software applications, databases and computational resources all provide the infrastructure for bioinformatics. On the scientific side of the hub, bioinformatics methods are used extensively in molecular biology, genomics, proteomics, other emerging areas (i.e. metabolomics, transcriptomics) and in CADD research. There are several key areas where bioinformatics supports CADD research **Figure 1** [29].

3.4. Virtual High-Throughput Screening (vHTS)

Pharmaceutical companies are always searching for new leads to develop into drug compounds. One search method is virtual high-throughput screening. In vHTS, protein targets are screened against databases of small molecule compounds to see which molecules bind strongly to the target. If there is a "hit" with a particular compound, it can be extracted from the database for further testing. With today's computational resources, several million compounds can be screened in a few days on sufficiently large clustered computers. Pursuing a handful of promising leads for further development can save researchers considerable time and expense. ZINC is a good example of a vHTS compound library [30].

4. SERENDIPITY IN DRUG RESEARCH

Accidental discoveries always played an important role in science [31], especially in the search for new drugs [32-36]. Even if we do not count the traditional evaluation of plants, animal toxins and minerals for therapeutic potential, in ancient history, and the more or less systematic screening of synthetic compounds in our century, the number of serendipitous findings in drug history is legion. "Ein glücklicher Zufall hat uns ein Präparat in die

Hand gespielt" (a lucky accident played a new drug in our hands) are the first words of a publication which describes the fortunate discovery of the fever-reducing activity of acetanilide. Erroneously this compound was clinically tested, instead of naphthalene that should have been investigated as an intestinal worm-killing agent [32-36].

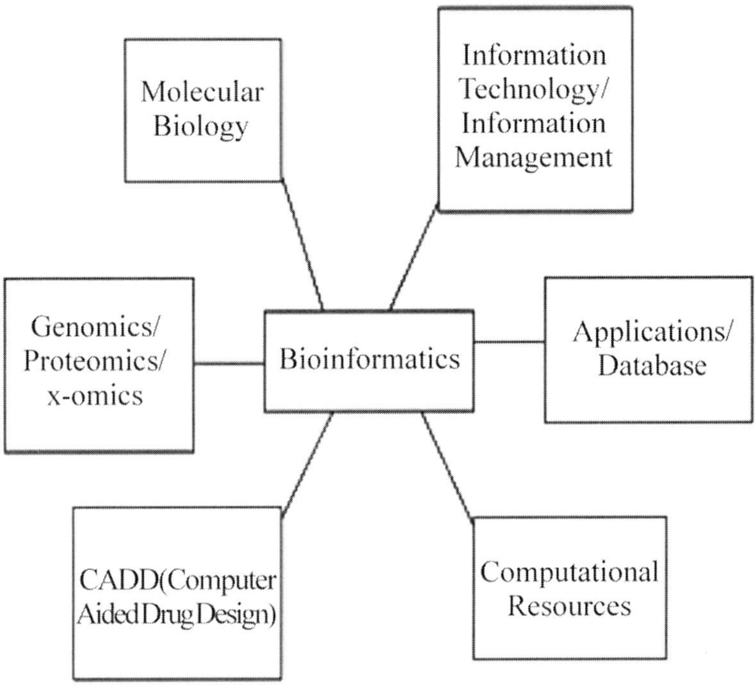

Figure 1. Bioinformatics in computer-aided drug design.

The two best known examples of serendipitous findings are the discovery of the antibiotic effect of a certain Penicillium strain by Sir Alexander Fleming, which led to the development of penicillin and its synthetic derivatives, and the discovery that Chlordiazepoxide, which resulted from an unexpected chemical rearrangement, is a potent tranquillizer. Five years later, he prepared the compound once again; this time he experienced hallucinations, after accidental intake or inhalation of minute amounts of this highly potent compound [31,34,36]. Acetylsalicylic acid was originally designed as a prodrug of salicylic acid to treat headache, fever and rheumatic diseases. Much later it turned out to be an irreversible cyclooxygenase inhibitor, preventing blood coagulation by the inhibition of thrombocyte aggregation. A careful inspection of the reasons for the observed cytotoxic effect led to the surprising result that it was due to the action of ammonium and chloride ions on the platinum electrode, forming Cisplatin in the electrolysis medium [34]. Cyclosporin was developed because of its antifungal activity. Sandoz was already going to stop the program when the compound turned out to be an immunosuppressant, highly valuable to prevent the rejection of organ transplants. The story of the anticoagulants Dicoumarol and Warfarin is full of serendipitous findings. First, cattle bleeded to death after they were fed with moldy hay. The toxicagent Dicoumarol was isolated and introduced into human therapy. Because of its narrow therapeutic range and its frequent side effects it was abandoned after a short period. The Wisconsin Alumni Research Foundation developed the Dicoumarol analog Warfarin as a rat poison. New clinical trials started when a US army cadet unsuccessfully attempted to commit suicide. Warfarin is now the drug of choice to protect against stroke and other acute thrombotic events [37]. Recently it was recognized as a valuable lead for the development of potent HIV protease inhibitors. Application of a water-soluble salt, lithium urate, led to the serendipitous discovery of the beneficial effect of lithium salts [31,34,37]. Phenylbutazone and Valproic acid were designed as solubility enhancers for other drugs. However, both compounds turned out to be valuable drugs on their own. Phenolphthalein was discovered to be a potent laxative when it was tested as a possible marker to label cheap Hungarian wines. The three most important artificial sweeteners, saccharine, cyclamate and aspartame, were also serendipitous discoveries. Chemists experienced the sweet taste when licking their fingers or smoking a cigarette [38]. An important discovery in receptor research was also a case of serendipity. The second messenger cyclic AMP was discovered in 1957, adenylate cyclase in 1958. Fluoride ions activated adenylate cyclase but the mechanism of this surprising stimulation of enzymatic activity could not be explained for the next 24 years. Adenylate cyclase originally consisted of two components, the cyclase and a regulatory unit, the G protein. Fluoride activation of the G protein was observed in disposable glass tubes or in the presence of tap water, but not with distilled water in plastic tubes [39]. Further systematic investigation of these confusing results made clear that fluoride ions activate the G proteins only in the presence of minute amounts of aluminum ions. Whereas a GDP-G protein complex is inactive, the GTPG protein or GDP-fluoroaluminate-G protein complexes activate adenylate cyclase: It was supposed that the fluoroaluminate ion mimics the outer phosphate group of GTP [39].

5. GRAPH MACHINES

Definitions and notations for handling acyclic graphs, and the construction of graph machines from general graphs (possibly cyclic). It is shown that the training and model selection methods developed for vector machines can be extended to graph machines.

5.1. Handling Directed Acyclic Graphs

Definitions: We consider the mapping from a set of acyclic graphs G to a set of realvalued numbers. Graph machines and their applications to computer-aided drug design for each acyclic graph Gi of G, a parameterized function gi Rn R is constructed, which is intended 1) to encode the structure of the graph [40].

5.2. Model Selection

Similarly to vector machines, usual model selection techniques such as hold-out, K-fold cross-validation, leaveone-out, can be applied to recursive networks and to graph machines. In the present section, we show how virtual leave-one-out, a powerful method for estimating the generalization capability of a vector machine, can be extended to graph machines.

5.3. Graph Machines for the Prediction of Properties and/or Activities of Molecules

The prediction of the physico-chemical properties and pharmaceutical activities of molecules is a critical task in the drug industry for shortening the development times and costs. Typically, one synthesized molecule out of 10,000 becomes a commercial drug, and the development time of a new drug is approximately 10 years. Therefore, predicting the activity of a hitherto non-existent molecule may lead to tremendous savings in terms of development time and cost. Hence, over the past few years, QSPR/ QSAR has become a major field of research in the chemical industry. In a typical QSAR/QSPR scenario, a database of measured properties or activities of molecules is available, and it is desired to infer, from those data, the property/activity of molecules that have not yet been synthesized.

5.4. Encoding the Molecules

Molecules are usually described in databases in a representation called SMILES (Simplified Molecular Input Line Entry System), which provides a description of the graph structure of the molecule as a character string. In the applications described here, the functions gi were generated from the SMILES files of the molecules by the following procedure: The molecules, described by these files, were converted into labeled graphs by the association of each non-hydrogen atom to a node, and each bond to an edge. The nodes were also assigned labels describeing the atoms they were related to (e.g. their nature, their degree or stereoisomery). Then the adjacency matrices associated to those labeled graphs were generated. In the subsequent step, the matrices were cast into a canonical form, by an algorithm that ranks the nodes according to criteria such as their degree, the fact that they belong to a cycle [41]. This canonical form allowed the choice of the root nodes, and the conversion of the graphs into directed acyclic graphs.

5.5. Predicting the Boiling Points of Halogenated Hydrocarbons

The volatility of halogenated hydrocarbons is an important property, because those compounds are widely used in the industry, for example as solvents, anaesthetics, blowing agents, and end up in the environment, where they can damage the ozone layer or be greenhouse gases. The volatility of a molecule can be assessed by its boiling point, a property measured only for a small proportion of possible halogenated hydrocarbons. We studied a data set of 543 haloalkanes, whose boiling points were previously predicted by Multi Linear Regression (MLR) [41,42]. This regression required the computation of numerous molecular descriptors, including arithmetic descriptors, topological indices, geometrical indices, and counts of substructures and fragments. The best feature subset was then selected, and generally comprised 6 or 7 descriptors.

To provide a comparison with the results obtained by this method, we used the same training and test sets [43]. They feature 507 and 36 halo alkanes respectively, whose boiling points range from −128°C to 249°C. In order to select the number of neurons required by the complexity of the problem, we first performed 10-fold crossvalidation on the 507 examples of the training set. The results suggested the use of neural networks with 4 hidden neurons. We then trained the selected graph machines, and predicted the boiling points of the test set molecules. The results of this study are shown in **Table 2**, where they are compared to the results obtained by [43] on the same sets, using a 7-regressor MLR model.

5.6. Predicting the Anti-HIV Activity of TIBO Derivatives

TIBO (Tetrahydroimidazobenzodiazepinone) derivatives are a family of chemicals with a potential anti-HIV activeity. They belong to the category of non-nucleoside inhibitors, which block the reverse transcriptase of the retrovirus and prevent its duplication. We studied a data set of 73 of those compounds, whose activity was previously modeled with several QSAR methods, including conventional neural networks [44], multi-linear regression, comparative molecular field analysis (CoMFA) [45], and the substructural molecular fragments (SMF) method. The latter approach is based on the representation of the molecules with graphs, which are split into fragments, whose contribution to the modeled activity is then computed with linear or non-linear regression. Those fragments are either atom-bond sequences, or "augmented atoms", defined as atoms with their nearest neighbors. In order to compare the prediction abilities of graph machines to the performances of the SMF method , the data set was split into a training and a test set of 66 and 7 examples respectively, exactly as in. The activity is expressed as log ($1/IC_{50}$) where IC_{50} is the concentration leading to the inhibition of 50% of the HIV-1 reverse transcriptase enzyme. Since some compounds of the set are stereoisomers, a label that described the enantiomer (R or S) of the atoms was added when necessary. We first performed 6-fold cross-validation on the training set with node functions having up to five hidden neurons to select the complexity of the model [46].

Table 2. Four major cases in CADD also known as "direct" and "indirect design when the structure of the target is respectively known or unknown [18].

Compounds (inhibitor/ligand)	Target (enzyme/receptor)	
	Unknown	Known
Unknown	Similarity Searching Rational Screening	*de novo* design
Known	Analog-Based Drug Design	Structure-Based Drug Design

6. DRUG DISCOVERY PROCESS

The process by which a new drug is brought to market stage is referred to by a number of names most commonly as the development chain or "pipeline" and consists of a number of distinct stages. Broadly it can be grouped under two stages Preclinical and the Clinical. Subsequently, the drug goes through many phases of clinical development in humans. In the clinical phase, the drug is administered to human volunteers to determine:

- The passage of the drug through the body-from the time it is taken to the time it is excreted;
- The effect of the drug on the body;
- Its effectiveness on the disease being targeted;
- Undesirable side effects of the drug.

6.1. Pitfall in Current Drug Discovery Process the Productivity Gap

A recent US Government Accountability Office (GAO) report found that Pharma R & D spending grew by 147% between 1993 and 2004 while the overall number of New drug applications (NDAs) submitted to the FDA increased only 38% and, worse still, the number of NDAs submitted for the presumably more innovative New molecular entities (NMEs) increased by only 7% in that time.

The attrition rate is unacceptably high. Only 1 out of 12 drugs entering clinical trials become a new drug. A particular worry or the pharmaceutical industry is that, despite a variety of approaches being used for R & D, attrition rates remain high during drug development. There are a number of factors attributed to the high attrition rates observed, but the number of active substances with poor pharmacological properties has been cited as a major concern. These are active substances that lack appropriate bioavailability, exhibit poor pharmacokinetics or cause adverse events and will therefore need to be withdrawn from development. It is estimated that these types of failures represent approximately 50% of all failures in drug development.

6.2. Need for an Alternative Tool

From the above facts and figures it is evident that there is an urge for an alternative tool that would not only shorten the R & D time cycle but also reduce the ever increasing cost involved in the drug discovery process.

There is a general perception that applied sciences have not kept pace with the advances of basic sciences [48].

6.3. Impact of Technology

The process of finding a drug molecule that attaches itself to the target protein in the body has now moved from the lab to the computer [49]. The words in silico drug design and computer aided drug design are almost synonymous. In the post genomic era, computer aided drug design (CADD) has considerably extended its range of applications, spanning almost all stages in the drug discovery pipeline, from target identification to lead discovery, from lead optimization to preclinical or clinical trials **Figure 2** [50].

6.4. In Silico Drug Discovery Process Comprises of 3 Stages

Stage 1 It involves Identification of a therapeutic target and building a heterogeneous small molecule library to be tested against it. This is followed by the development of a virtual screening protocol initialized by either docking of small molecules from the library or building these structures in the active site by employing De novo design methods.

Stage 2 These selected hits are checked for specificity by docking at binding sites of other known drug targets.

Stage 3 These selected hits are subjected to detail in silico ADMET profiling studies and those molecules that pass these studies are termed as leads.

6.5. Target Identification and Validation in Silico

Target identification and validation is the first key stage in the drug discovery pipeline. However, identification and validation of drug able targets from among thousands of candidate macromolecules is still a challenging task. Numerous technologies for addressing the targets have been developed recently. Genomic and proteomic approaches are the major tools for target identification. For example, a proteomic approach for identification of binding proteins for a given small molecule involvescomparison of the protein expression profiles for a given cell or tissue in the presence or absence of the given molecule.

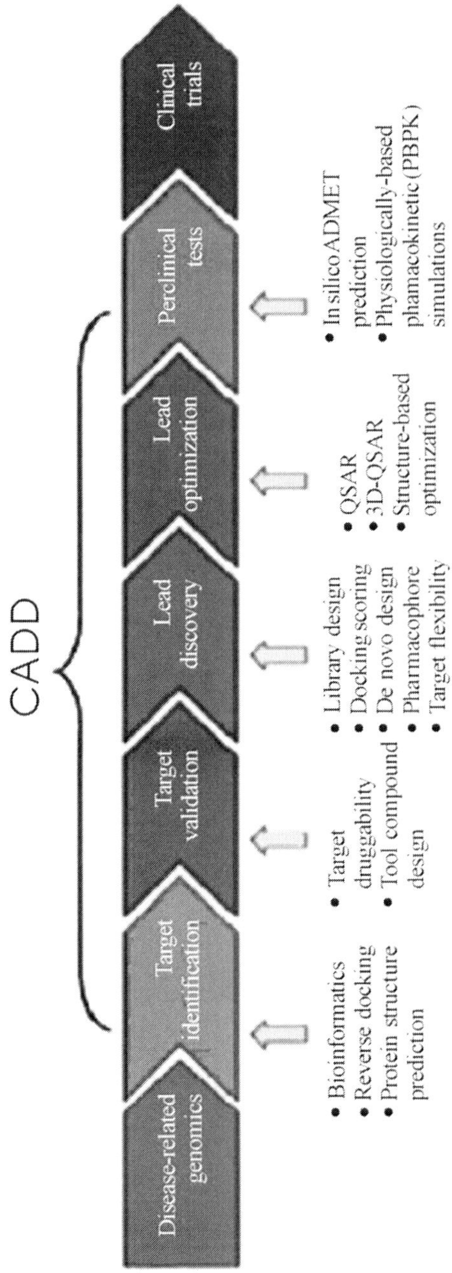

Figure 2. It shows the applications of CADD to the various stages of drug development [47].

6.6. In Silico ADMET (Absorption, Distribution, Metabolism, Excretion, Toxicity) Prediction [51]

Studies indicate that poor pharmacokinetics and toxicity are the most important causes of costly late stage failures in drug development and it has become widely appreciated that these areas should be considered as early as possible in the drug discovery process. Combinatorial chemistry and high throughput screening have significantly increased the number of compounds for which early data on absorption, distribution, metabolism, excretion (ADME) and toxicity (T) are needed. With use of in silico tools it is possible to model the most relevant pharmacokinetic, metabolic and toxicity endpoints, thereby accelerating the drug discovery process.

6.7. In Silico Prediction of Drug Safety [52]

There is considerable interest in computational models to predict drug safety in drug discovery and development. Significant adverse toxicological findings for a drug in late-stage clinical trials or post marketing can cause enormous financial losses and place patients at risk. The earlier such molecules are identified and the drug development process halted the better.

There are tools to predict toxicities like

1) Genotoxicity;

2) Liver toxicity;

3) CYP450 inhibition; and 4) Cardiotoxicity.

6.8. In Silico Prediction of Drug-Drug Interactions [53]

Recently, metabolic drug-drug interactions (MDDI) have raised some high profile problems in drug development resulting in restricted use, withdrawal or non approval by regulatory agencies. The use of in vitro technologies to evaluate the potential for MDDI has become routine in the drug development process. Nevertheless, in the absence of an integrated approach, their interpretation and value remains the subject of debate, and the vital distinction between a useful "simulation" and a precise "prediction" is not often appreciated. Various in silico software are now available for the simulation of MDDI. One such software is SIMCYP.

6.9. Virtual Screening [54]

Virtual screening involves the docking of selected lead molecules against the biological target. This is followed by a scoring pattern. There is a number of software available for this. Some are commercially available and some are free to use.

7. CONCLUSION

Pranita P. Kore, Madhavi M. Mutha, Rishikesh V. Antre, Rajesh J. Oswal, Sandip S. Kshirsagar (QSAR/ QSPR, structure-based design, combinatorial library design, chemoinformatics & bioinformatics); the growing number of chemical and biological databases; and an explosion in currently available software tools are providing a much improved basis for the design of ligands and inhibitors with desired specificity. In future our review will helpful for design of drug with minimal side effect and high potency.

REFERENCES

1. W. G. Richards, "Computer-Aided Drug Design," Pure and Applied Chemistry, Vol. 66, No. 8, 1994, pp. 1589- 1596.
2. S. Haworth, C. Burt, F. Gago, C. A. Reynolds and W. G. Richards, "A Prototype Bioreductive DNA Groove Binding Ligand," Anticancer Drug Design, Vol. 6, No. 1, 1991, pp. 59-61.
3. S. Haworth, A. H. Elcock, A. Rodger and W. G. Richards, "Sequence Selective Binding to the DNA Major Groove: Tris (1,lO-phenanhline) Metal Complexes Binding to Poly(dG-dC) and Poly(dA-dT)," Journal of Biomolecular Structure and Dynamics, Vol. 9, No. 1, 1991, pp. 23-29.
4. S. Haworth, A. H. Elcock, A. Rodger and W. G. Richards, "A Binding Mode of A-[tris(l,lO-phenanthrohe)ruthenium(II)]$^{2+}$ Exhibiting Preference for Purine-3',5'-Pyrimidine Sites of DNA," Journal of Biomolecular Structure and Dynamics, Vol. 9, No. 1, 1991, pp. 553-569.
5. M. Jayakanthan, G. Wadhwa, T. M. Mohan, L. Arul, P. Balasubramanian and D. Sundar, "Computer-Aided Drug Design for Cancer-Causing H-Ras p21 Mutant Protein," Letters in Drug Design and Discovery, Vol. 6, No. 1, 2009, pp. 14-20.
6. D. A. Spandidos and N. M. Wilkie, "Malignant Transformation of Early Passage Rodent Cells by a Single Mutated Human Oncogene," Nature, Vol. 310, 1984, pp. 469- 475.

A. Valencia, P. Chardin, A. Wittinghofer and C. Sander, "The Ras Protein Family: Evolutionary Tree and Role of Conserved Amino Acids," Biochemistry, Vol. 30, No. 19, 1991, pp. 4637-4648.

7. M. Barbacid, "Ras Genes," Annual Review of Biochemistry, Vol. 56, 1987, pp. 779-827.
8. R. Khosravi-Far and C. J. Der, "The Ras Signal Transduction Pathway," Cancer and Metastasis Reviews, Vol. 13, No. 1, 1994, pp. 67-89.
9. Russo, V. Bazan, V. Agnese, V. Rodolico and N. Gebbia, "Prognostic and Predictive Factors in Colorectal Cancer: Kirsten Ras in CRC (RASCAL) and TP53CRC Collaborative Studies," Annals of Oncology, Vol. 4, Suppl 4, 2005, pp. 44-49.
10. G. Oxford and D. Theodorescu, "The Role of Ras Superfamily Proteins in Bladder Cancer Progression," Journal of Urology, Vol. 170, No. 5, 2003, pp. 1987-1993.

11. J. John, M. Frech and A. Wittinghofer, "Biochemical Properties of Haras Encoded p21 Mutants and Mechanism of the Autophosphorylation Reaction," The Journal of Biological Chemistry, Vol. 263, 1988, pp. 11792-11799.
12. H. Kiaris and D. A. Spandidos, "Mutations of Ras Genes in Human Tumors," International Journal of Oncology, Vol. 7, No. 3, 1995, pp. 413-421.
13. L. George and J. H. Ching, "VPython Application To The Computer-Aided Drug Design Problem," The Python Papers Monograph, Vol. 2, 2010, pp. 1-5.
14. D. G. Sprous, J. Zhang, L. Zhang, Z. Wang and M. A. Tepper, "Kinase Inhibitor Recognition by Use of a Multivariable QSAR Model," Journal of Molecular Graphics and Modelling, Vol. 24, No. 4, 2006, pp. 278-295.
15. C. D. Selassie, "Burger's Medicinal Chemistry, Drug Discovery and Development," 6th Edition, John Wiley & Sons, Inc., New York, 2003.
16. Structure-Activity Relationships (QSAR), "A Review, Combinatorial Chemistry & High Throughput Screening," Vol. 9, 2006, pp. 213-228.
17. K. R. Oldenburg, "Annual Report in Medicinal Chemistry," J. A. Bristol, Ed., Academic Press, London, Vol. 33, 1998, pp. 301-307.
18. W. H. Moos, G. D. Green and M. R. Pavia, "Chapter 33. Recent Advances in the Generation of Molecular Diversity," Annual Reports in Medicinal Chemistry, Vol. 28, 1993, pp. 315-324.
19. F. Ooms, "Molecular Modeling and Computer Aided Drug Design. Examples of their Applications in Medicinal Chemistry," Current Medicinal Chemistry, Vol. 7, No. 2, 2000, pp. 141-158.
20. J. Kuhlman, International Journal of Clinical Pharmacology and Therapeutics, Vol. 35, 1997, pp. 541-552.
21. R. G. Halliday, S. R. Walker and C. E. Lumley, Journal of Pharmaceutical Medicine, Vol. 2, 1992, pp. 139-154.
22. K. Ghose and J. J. Wendoloski, "Perspective in Drug Discovery and Design," Kluwer/Escom, Vol. 9-11, 1998, pp. 253-271.
23. D. J. Abraham and G. E. Kellogg, "3D-QSAR in Drug Design," H. Kubinyi, Ed., Escom, Leiden, Vol. 1, 1993, pp. 506-522.
24. D. Scherer, P. Dubois and B. Sherwood, "VPython: 3D Interactive Scientific Graphics for Students," Computing in Science and Engineering, Vol. 2, No. 5, 2000, pp. 56-62.
25. J. Irwin, D. M. Lorber, S. L. McGovern, B. Wei and B. K. Shoichet, "Docking and Drug Discovery," Computational Nanoscience and Nanotechnology, Vol. 2, 2002, pp. 50-51.
26. C. A. Taft, V. B. da Silva and C. H. T de P. da Silva, "Current Topics in Computer-Aided Drug Design," Journal of Pharmaceutical Sciences, Vol. 97, No. 3, 2008, pp. 1089-1098.
27. D. Bernard, A. Coop and A. D. MacKerell Jr., "Computer-Aided Drug Design: Structure-Activity Relationships of Delta Opioid Ligands," Drug Design Reviews, Vol. 2, No. 4, 2005, pp. 277-291.
28. W. G. Richards, "Computer-Aided Drug Design," Pure and Applied Chemistry, Vol. 66, No. 8, 1994, pp. 1589- 1598.
29. D. S. Park, J. M. Kim, Y. B. Lee and C. H. Ahn, "QSID Tool: A New Three-Dimensional QSAR Environmental Tool," Journal of Computer-Aided Drug Design, Vol. 22, No. 12, 2008, pp. 873-883.
30. H. Kubinyi, "Chance Favors the Prepared Mind—From Serendipity to Rational Drug Design," Journal of Receptor and Signal Transduction Research, Vol. 19, No. 1-4, 1999, pp. 15-39.
31. M. von Itzstein, W. Y. Wu, G. B. Kok, M. S. Pegg, J. C. Dyason, B. Jin, T. V. Phan, M. L. Smythe, H. F. White, S. W. Oliver, P. M. Colman, J. N. Varghese, D. M. Ryan, J. M. Woods, R. C. Bethell, V. J. Hotham, J. M. Cameron and C. R. Penn, "Rational Design of Potent SialidaseBased Inhibitors of Influenza Virus Replication," Nature, Vol. 363, No. 6428, 1993, pp. 418-423.
32. J. Greer, J. W. Erickson, J. J. Baldwin and M. D. Varney, "Application of the Three-Dimensional Structures of Protein Target Molecules in Structure-Based Drug Design," Journal of Medicinal Chemistry, Vol. 37, No. 8, 1994, pp. 1035-1054.
33. J. P. Vacca and J. H. Condra, "Clinically Effective HIV-1 Protease Inhibitors," Drug Discovery Today, Vol. 2, No. 1, 1997, pp. 6-18.

34. P. C. Sternweis and A. G. Gilman, "Aluminum: A Requirement for Activation of the Regulatory Component of Adenylate Cyclase by Fluoride," Proceedings of the National Academy of Sciences USA, Vol. 79, No. 16, 1982, pp. 4888-4891.
35. J. Sondek, D. G. Lambright, J. P. Noel, H. E. Hamm and P. B. Sigler, "GTPase Mechanism of Gproteins from the 1.7-Å Crystal Structure of Transducin α.GDP.AlF4-," Nature, Vol. 372, 1994, pp. 276-279.
36. G. De Stevens, "Serendipity and Structured Research in Drug Discovery," Progress in Drug Research, Vol. 30, 1986, pp. 189-203.
37. P. J. Goodford, "Drug Design by the Method of Receptor Fit," Journal of Medicinal Chemistry, Vol. 27, No. 5, 1984, pp. 557-564.
38. J. P. Vacca and J. H. Condra, "Clinically Effective HIV-1 Protease Inhibitors," Drug Discovery Today, Vol. 2, No. 1, 1997, pp. 6-18.
39. Goulon, A. Duprat and G. Dreyfus, "Graph Machines and Their Applications to Computer-Aided Drug Design: A New Approach to Learning from Structured Data," Lecture Notes in Computer Science, Vol. 4135, 2006, pp. 1-19.
40. Jochum and J. Gasteiger, "Canonical Numbering and Constitutional Symmetry," Journal of Chemical Information and Computer Sciences, Vol. 17, No. 2, 1977, pp. 113-117.
41. T. Balaban, S. C. Basak, T. Colburn and G. D. Grunwald, "Correlation between Structureand Normal Boiling Points of Haloalkanes C1-C4 Using Neural Networks," Journal of Chemical Information and Computer Sciences, Vol. 34, No. 5, 1994, pp. 1118-1121.
42. Rucker, M. Meringer and A. Kerber, "QSPR Using MOLGEN-QSPR: The Example of Haloalkane Boiling Points," Journal of Chemical Information and Computer Sciences, Vol. 44, No. 6, 2004, pp. 2070-2076.
43. L. Douali, D. Villemin and D. Cherqaoui, "Exploring QSAR of Non-Nucleoside Reverse Transcriptase Inhibitors by Neural Networks: TIBO Derivatives," International Journal of Molecular Sciences, Vol. 5, No. 2, 2004, pp. 48-55.
44. J. Huuskonen, "QSAR Modeling with the Electrotopological State, TIBO Derivatives," Journal of Chemical Information and Computer Sciences, Vol. 41, No. 2, 2001, pp. 425-429.
45. Z. Zhou and J. D. Madura, "CoMFA 3D-QSAR Analysis of HIV-1 RT Nonnucleoside Inhibitors, TIBO Derivatives Based on Docking Conformation and Alignment," Journal of Chemical Information and Computer Sciences, Vol. 44, No. 6, 2004, pp. 2167-2178.
46. Hammer, "Recurrent Networks for Structured Data— A Unifying Approach and its Properties," Cognitive Systems Research, Vol. 3, No. 2, 2002, pp. 145-165.
47. Jochum and J. Gasteiger, "Canonical Numbering and Constitutional Symmetry," Journal of Chemical Information and Computer Sciences, Vol. 17, No. 2, 1977, pp. 113-117.
48. T. Balaban, S. C. Basak, T. Colburn and G. D. Grunwald, "Correlation between Structure and Normal Boiling Points of Haloalkanes C1-C4 Using Neural Networks," Journal of Chemical Information and Computer Sciences, Vol. 34, No. 5, 1994, pp. 1118-1121.
49. Rucker, M. Meringer and A. Kerber, "QSPR Using MOLGEN-QSPR: The Example of Haloalkane Boiling Points," Journal of Chemical Information and Computer Sciences, Vol. 44, No. 6, 2004, pp. 2070-2076.
50. J. Augen, "The Evolving Role of Information Technology in the Drug Discovery Process," Drug Discovery Today, Vol. 7, Suppl 5, 2002, pp. 275-282.
51. H. Van de waterbeemd and E. Gifford, "ADMET in Silico Modelling, towards Prediction Paradise?" Nature Reviews Drug Discovery, Vol. 2, No. 3, 2003, pp. 192-204.
52. W. J. Egan, G. Zlokarnik and P. D. J. Grootenhuis, "In Silico Prediction of Drug Safety: Despite Progress There Is Abundant Room for Improvement," Drug Discovery Today: Technologies, Vol. 1, Suppl 4, 2004, pp. 381-387.
53. Rostami-Hodjegan and G. Tucker, "'In Silico' Simulations to Assess the 'in Vivo' Consequences of 'in Vitro' Metabolic Drug-Drug Interactions," Drug Discovery Today: Technologies, Vol. 1, Suppl 4, 2004, pp. 441-448.

CHAPTER 13

New acetylenic derivatives of betulin and betulone, synthesis and cytotoxic activity

Ewa Bębenek1, Monika Kadela-Tomanek1, Elwira Chrobak1, Joanna Wietrzyk2, Joanna Sadowska2, Stanisław Boryczka1

Department of Organic ChemistryMedical University of Silesia in Katowice, School of Pharmacy with the Division of Laboratory Medicine in SosnowiecSosnowiecPoland

2.Department of Experimental Oncology, Polish Academy of SciencesLudwik Hirszfeld Institute of Immunology and Experimental TherapyWrocławPoland

ABSTRACT

Betulin **1** and its semisynthetic derivatives exhibit a cytotoxic activity toward various cancer cell lines. These compounds are a promising and potential anticancer candidates. A series of betulin derivatives was prepared and tested for the antiproliferative activity in vitro against T47D breast cancer, CCRF/CEM leukemia, HL-60 promyelocytic leukemia, SW707 colorectal, murine P388 leukemia, as well as BALB3T3 normal fibroblasts cell lines. Cisplatin and betulin **1** were used as a reference compounds. Some derivatives of betulin showed a higher cytotoxic activity than the parent compound **1**. Two derivatives (**5** and **17**) were 24-fold potent than betulin **1** against the human promyelocytic leukemia cell line (HL-60), with an IC_{50} value of 0.3 μg/mL.

Keywords: Betulin Betulone Synthesis Cytotoxic activity Lipophilicity

1. INTRODUCTION

Betulin (lup-20(29)-ene-3β,28-diol) **1** is a pentacyclic triterpene of the lupane type which is isolated from bark of white birch species (Figure 1). The compound **1** has three active positions in its structure, namely the primary hydroxyl group at C-28, the secondary group at C-3 and the isopropenyl side chain at C-19. It's possible to make a chemical modification of these positions to obtain new betulin derivatives with important biological properties such as antitumor, antiviral, antimicrobial, anti-inflammatory, as well as hepatoprotective activities (Alakurtti et al., 2006; Tolstikov et al., 2005).

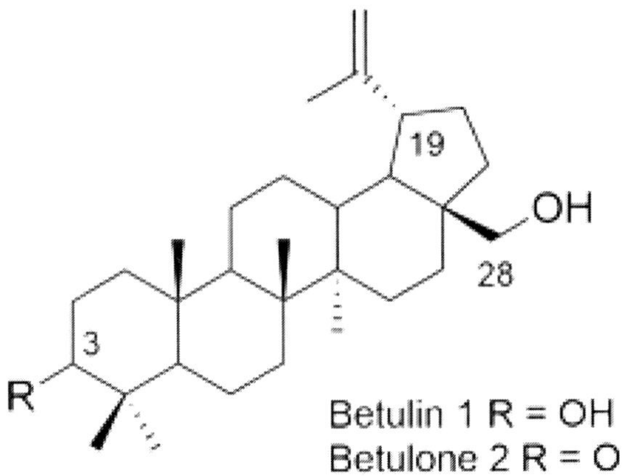

Figure 1. Chemical structure of betulin **1** and betulone **2**

Betulone (lup-20(29)-en-28-ol-3-one) **2** similarly to betulin **1** represents a class of pentacyclic triterpenes which can be isolated from various plants (Figure 1) (Diouf et al., 2009; Liu et al., 2012; Reyes et al., 2006). The content of betulone **2** in the native plant material is very low, therefore the compound **2** was obtained by oxidation or biotransformation of naturally occurring betulin **1** (Grishko et al., 2013; Hase et al., 1981; Mao et al., 2012). Despite the fact that betulone **2** has been isolated from *Betula lenta* in 1991 (Cole et al., 1991), the crystal structure of this compound was determined for the first time in 2013 (Boryczka et al., 2013a).

It has already been reported that betulone **2** possess interesting pharmacological activities such as anti-leishmanial, anti-inflammatory, and aniparasitic against *Plasmodium falciparum* and *Trypanosoma brucei* *rhodesiense* (Alakurtti et al., 2010; Gachet et al., 2011; Reyes et al., 2006). Triterpene **2** exhibited also antifouling activity against cyprid larvae of the barnacle *Balanus albicostatus* with the EC_{50} value 8.73 μg/mL slightly higher than betulin **1** (Chen et al., 2011). The compound **2** demonstrated almost the same protective effects as betulin **1** against the cytotoxicity of cadmium at high concentrations (Hiroya et al., 2002). Betulone **2** with the carbonyl group at C-3 position showed anticancer effect on mouse melanoma (B16 2F2) cell line with the IC_{50} value 29.3 μM (Hata et al., 2002). Additionally, the compound **2** and its derivatives showed in vitro cytotoxic activity against different cancer cell lines like stomach (MGC-803), breast (Bcap-37, MCF-7), prostate (PC3), melanoma (SK-MEL-2, A-375), medulloblastoma (Dayo), glioblastoma (LN-229), ovarian carcinoma (OVCAR-3), and colon carcinoma (HT-29) (Koohang et al., 2009; Liu et al., 2012; Mar et al., 2009). Derivatives of betulone containing 3′-substituted glutaryl groups at C-28 position represent a new class of anti-HIV agents. These compounds exhibited anti-HIV activity with EC_{50} values in the range of 4.3–10.0 μM (Sun et al., 1998a; Sun et al., 1998b).

We have previously described the synthesis and evaluation of cytotoxicity of betulin derivatives containing one or two acetylenic groups at the C-3 and/or C-28 positions. Our studies showed, that the derivative of betulin with a propynoyl group at C-28 position, has strong cytotoxic effects against human leukemia (CCRF/CEM) and murine leukemia (P388) cancer cells. Moreover, 28-*O*-propynoylbetulin induces apoptosis in human melanoma (G-361) cells via caspase-3 activation (Boryczka et al., 2013b; Orchel et al., 2014).

Continuing our research project on the development of anticancer agents, we synthesized new compounds, in order to obtain more information about the influence of alkane, alkene, and alkyne moiety at the C-28 position on antiproliferative activity in the group of betulin and betulone derivatives.

2. MATERIALS AND METHODS

2.1 Chemistry

Melting points of betulin derivatives were obtained in open capillary tubes on a Boetius melting point apparatus without correction. Nuclear magnetic resonance (NMR) (600/150 MHz) spectra were measured in $CDCl_3$ as solvent on a Bruker Avance III 600 spectrometer. The chemical shifts values are reported in ppm (δ) and the coupling constants (J) are presented in Hertz. The spin multiplicities are given as singlet (s), doublet (d), triplet (t), q (quartet), and multiplet (m). Mass spectra were measured under EI conditions on a Finnigan MAT 95 spectrometer. Infrared spectra (KBr, pellet) were recorded using the IRAffinity-1 Shimadzu spectrometer and reported in wave number (cm^{-1}). The progress of all reactions were monitored by thin layer chromatography (TLC) on silica gel 60 254F plates using a mixture of chloroform and ethanol (40:1, v/v) as an eluent. The spots were visualized by spraying with a solution of 5 % sulfuric acid and then heating to 100 °C. Purification of the new compounds was carried out by column chromatography (silica gel 60, <63 µm, Merk) using a mixture of chloroform and ethanol (40:1, v/v) as an eluent. All solvents for reactions were dried and purified prior to use.

The synthesis and spectral data of the compounds **3**, **5**, **7**, **12–14** was described previously by Boryczka et al. (2013b).

2.2 General procedure for the synthesis of derivatives 4, 6, 8–9

To a mixture of betulin **1** (0.44 g, 1 mmol) and 2-propenoic acid (0.08 g, 1.10 mmol), 3-cyclopropyl-2-propynoic acid (0.12 g, 1.10 mmol), 2-butenoic acid (0.09 g, 1.10 mmol) or 2-butynoic acid (0.9 g, 1.10 mmol) in dichloromethane (5 mL) was added slowly a solution of dicyclohexylcarbodiimide (0.23 g, 1.12 mmol), and 4-dimethylaminopyridine (0.01 g, 0.08 mmol) in dichloromethane (1 mL) at −10 °C temperature. The reaction was stirred under argon atmosphere at −10 °C temperature for 5 h, and then was allowed to warm to room temperature and stirred overnight. The progress of the reaction was monitored by TLC until completion and then filtered. The solvent was removed under reduced pressure and the residue was purified by silica gel column chromatography (chloroform/ethanol 40:1, v/v).

28-*O*-(2-Propenoyl)betulin (**4**) Yield 74 %; mp 117–121 °C; R_f 0.44 (chloroform/ethanol, 40:1, v/v); IR (KBr) ν_{max} 3445, 2939, 1723, 1456, 1269 cm^{-1}; ^1H NMR (600 MHz, $CDCl_3$): δ 6.42 (1H, m, $CH{=}CH_2$), 6.15 (1H, m, $CH{=}CH_2$), 5.84 (1H, m, $CH{=}CH_2$), 4.71 (1H, s, H-29), 4.61 (1H, s, H-29), 4.36 (1H, d, J = 10.8 Hz, H-28), 3.95 (1H, d, J = 10.8 Hz, H-28), 3.20 (1H, m, H-3), 2.49 (1H, m, H-19), 1.67 (3H, s, CH_3), 1.06 (3H, s, CH_3), 0.98 (3H, s, CH_3), 0.96 (3H, s, CH_3), 0.84 (3H, s, CH_3), 0.77 (3H, s, CH_3); ^{13}C NMR (150 MHz, $CDCl_3$): δ 166.7 (O–C=O), 150.2 (C-20), 130.5, 128.6, 109.9 (C-29), 79.0 (C-3), 62.8 (C-28), 55.3, 50.4, 48.8, 47.7, 46.5, 42.7, 40.9, 38.9, 38.7, 37.6, 37.1, 34.6, 34.2, 29.8, 29.6, 28.0, 27.4, 27.1, 25.2, 20.8, 19.1, 18.3, 16.1, 16.0, 15.4, 14.8; EIMS m/z 496 $[M]^+$ (14), 189 (100).

28-*O*-(3-Cyclopropyl-2-propynoyl)betulin (**6**) Yield 49 %; mp 108–113 °C; R_f 0.44 (chloroform/ethanol, 40:1, v/v); IR (KBr) ν_{max} 3475, 2941, 2228, 1707, 1454, 1258 cm^{-1}; ^{1}H NMR (600 MHz, $CDCl_3$): δ 4.68 (1H, s, H-29), 4.58 (1H, s, H-29), 4.31 (1H, d, *J* = 10.8 Hz, H-28), 3.93 (1H, d, *J* = 10.8 Hz, H-28), 3.18 (1H, m, H-3), 2.42 (1H, m, H-19), 1.67 (3H, s, CH_3), 1.01 (3H, s, CH_3), 0.97 (3H, s, CH_3), 0.96 (3H, s, CH_3), 0.92–0.91 (5H, m, CH, CH_2), 0.82 (3H, s, CH_3), 0.77 (3H, s, CH_3); ^{13}C NMR (150 MHz, $CDCl_3$): δ 154.5 (O–C=O), 150.1 (C-20), 109.9 (C-29), 93.3, 78.9 (C-3), 68.6 (C-28), 64.1, 55.3, 50.4, 48.8, 47.7, 46.4, 42.7, 40.9, 38.9, 38.7, 37.6, 37.1, 34.5, 34.2, 29.7, 29.5, 28.0, 27.4, 27.0, 25.2, 20.8, 19.1, 18.3, 16.1, 16.0, 15.3, 14.8, 9.2, 1.1, -0.6; EIMS *m/z* 534 $[M]^+$ (18), 189 (100).

28-*O*-(2-Butenoyl)betulin (**8**) Yield 57 %; mp 182–184 °C; R_f 0.45 (chloroform/ethanol, 40:1, v/v); IR (KBr) ν_{max} 3560, 2945, 1707, 1443, 1195 cm^{-1}; ^{1}H NMR (600 MHz, $CDCl_3$): δ 7.00 (1H, m, CH=CHCH_3), 5.88 (1H, m, CH=CHCH_3), 4.72 (1H, s, H-29), 4.61 (1H, s, H-29), 4.34 (1H, d, *J* = 10.8 Hz, H-28), 3.93 (1H, d, *J* = 10.8 Hz, H-28), 3.21 (1H, m, H-3), 2.48 (1H, m, H-19), 1.91 (3H, m, CH=CHCH_3), 1.68 (3H, s, CH_3), 1.06 (3H, s, CH_3), 1.00 (3H, s, CH_3), 0.99 (3H, s, CH_3), 0.85 (3H, s, CH_3), 0.71 (3H, s, CH_3); ^{13}C NMR (150 MHz, $CDCl_3$): δ 167.0 (O–C=O), 150.2 (C-20), 144.4, 122.9, 109.8 (C-29), 78.9 (C-3), 62.4 (C-28), 55.3, 50.4, 48.9, 47.7, 46.5, 42.7, 40.9, 38.9, 38.7, 37.6, 37.2, 34.6, 34.2, 29.9, 29.7, 28.0, 27.4, 27.1, 25.2, 20.8, 19.2, 18.3, 16.1, 16.0, 15.4, 14.8, 3.7; EIMS *m/z* 510 $[M]^+$ (14), 189 (100).

28-*O*-(2-Butynoyl)betulin (**9**) Yield 52 %; m.p. 111–113 °C; R_f 0.44 (chloroform/ethanol, 40:1, v/v); IR (KBr) ν_{max} 3482, 2942, 2245, 1709, 1457, 1248 cm^{-1}; ^{1}H NMR (600 MHz, $CDCl_3$): δ 4.69 (1H, s, H-29), 4.59 (1H, s, H-29), 4.33 (1H, d, *J* = 10.8 Hz, H-28), 3.95 (1H, d, *J* = 10.8 Hz, H-28), 3.16 (1H, m, H-3), 2.42 (1H, m, H-19), 1.99 (3H, s, C≡CCH_3), 1.68 (3H, s, CH_3), 1.02 (3H, s, CH_3), 0.98 (3H, s, CH_3), 0.97 (3H, s, CH_3), 0.82 (3H, s, CH_3), 0.76 (3H, s, CH_3); ^{13}C NMR (150 MHz, $CDCl_3$): δ 154.4 (O–C=O), 150.0 (C-20), 109.9 (C-29), 85.5, 79.0 (C-3), 72.5, 64.2 (C-28), 55.3, 50.4, 48.8, 47.6, 46.4, 42.7, 40.9, 38.9, 38.7, 37.6, 37.1, 34.5, 34.2, 29.7, 29.5, 28.0, 27.4, 27.0, 25.2, 20.8, 19.1, 18.3, 16.1, 16.0, 15.3, 14.8, 3.8; EIMS *m/z* 508 $[M]^+$ (22), 189 (100).

2.3 General procedure for the synthesis of derivatives 10–11

To a mixture of betulin **1** (0.44 g, 1 mmol) and pyridine (2.5 mL) in benzene (6 mL) at 0–5 °C temperature was added solution of propyl chloroformate or allyl chloroformate (3 mmol) in benzene (5 mL). The reaction was stirred at 0–5 °C temperature for 4 h. After this time the reaction was allowed to warm to room temperature and stirred overnight.

The reaction mixture was diluted with 5 mL of chloroform and washed successively with 1 N sulfuric acid and water, then dried and concentrated under reduced pressure. The crude product was purified by silica gel column chromatography (chloroform/ethanol 40:1, v/v).

28-*O*-Propoxycarbonylbetulin (**10**) Yield 51 %. m.p. 92–95 °C; R_f 0.43 (chloroform/ethanol, 40:1, v/v); IR (KBr) ν_{max} 3536, 2941, 1743, 1457, 1267 cm^{-1}; ^{1}H NMR (600 MHz, $CDCl_3$): δ 4.72 (1H, s, H-29), 4.61 (1H, s, H-29), 4.37 (1H, d, J = 10.8 Hz, H-28), 4.12 (2H, t, J = 6.6 Hz, OCH_2), 3.94 (1H, d, J = 10.8 Hz, H-28), 3.20 (1H, m, H-3), 2.46 (1H, m, H-19), 1.73 (2H, m, CH_2CH_3), 1.68 (3H, s, CH_3), 1.07 (3H, s, CH_3), 1.01 (3H, s, CH_3), 1.00 (3H, s, CH_3), 0.99 (3H, t, J = 7.2 Hz, CH_2CH_3), 0.85 (3H, s, CH_3), 0.78 (3H, s, CH_3); ^{13}C NMR (150 MHz, $CDCl_3$): δ 156.0 (O–C=O), 150.1 (C-20), 109.9 (C-29), 79.0 (C-3), 69.6, 66.4 (C-28), 55.3, 50.4, 48.8, 47.7, 46.6, 42.7, 40.9, 38.9, 38.7, 37.6, 37.1, 34.4, 34.2, 29.6, 29.5, 28.0, 27.4, 27.0, 25.2, 22.0, 20.8, 19.1, 18.3, 16.1, 16.0, 15.3, 14.8, 10.2; EIMS m/z 528 [M]$^+$ (19), 189 (100).

28-*O*-Allyloxycarbonylbetulin (**11**) Yield 66 %; m.p. 91–94 °C; R_f 0.47 (chloroform/ethanol, 40:1, v/v); IR (KBr) ν_{max} 3405, 2962, 1737, 1457, 1270 cm^{-1}; ^{1}H NMR (600 MHz, $CDCl_3$): δ 5.98 (1H, m, $CH{=}CH_2$), 5.38 (1H, m, $CH{=}CH_2$), 5.31 (1H, m, $CH{=}CH_2$), 4.71 (1H, s, H-29), 4.66 (2H, m, OCH_2), 4.61 (1H, s, H-29), 4.38 (1H, d, J = 10.8 Hz, H-28), 3.95 (1H, d, J = 10.8 Hz, H-28), 3.21 (1H, m, H-3), 2.46 (1H, m, H-19), 1.68 (3H, s, CH_3), 1.06 (3H, s, CH_3), 1.00 (3H, s, CH_3), 0.99 (3H, s, CH_3), 0.84 (3H, s, CH_3), 0.78 (3H, s, CH_3); ^{13}C NMR (150 MHz, $CDCl_3$): δ 155.6 (O–C=O), 150.1 (C-20), 131.7, 118.9, 109.9 (C-29), 78.9 (C-3), 68.5, 66.7 (C-28), 55.3, 50.4, 48.8, 47.7, 46.6, 42.7, 40.9, 38.9, 38.7, 37.6, 37.2, 34.4, 34.2, 29.7, 29.6, 28.0, 27.4, 27.0, 25.2, 20.8, 19.1, 18.3, 16.1, 16.0, 15.4, 14.8; EIMS m/z 496 [M]$^+$ (14), 189 (100).

2.4 General procedure for the synthesis of derivatives 15–26

To a solution of the appropriate monoester **3–14** (1 mmol) in dry dichloromethane (12 mL) was added pyridinium chloroformate (0.53 g, 2.48 mmol). The reaction was stirred at room temperature for 2 h and then diluted with ether (16 mL) and still was stirred for 10 min. The reaction mixture was filtered off through a layer of silica gel and washed with ether (5 mL). The filtrate was concentrated under reduced pressure and the residue was purified by silica gel column chromatography (chloroform/ethanol 40:1, v/v).

28-*O*-Propanoylbetulone (**15**) Yield 82 %; m.p. 65–67 °C; R_f 0.64 (chloroform/ethanol, 40:1, v/v); IR (KBr) ν_{max} 2945, 1733, 1705, 1462, 1187 cm^{-1}; ^{1}H NMR (600 MHz, $CDCl_3$): δ 4.72 (1H, s, H-29), 4.62 (1H, s, H-

29), 4.30 (1H, d, J = 10.8 Hz, H-28), 3.86 (1H, d, J = 10.8 Hz, H-28), 2.49 (1H, m, H-19), 2.37 (2H, q, J = 7.2 Hz, CH_2CH_3), 1.69 (3H, s, CH_3), 1.18 (3H, t, J = 7.2 Hz, CH_2CH_3), 1.10 (3H, s, CH_3), 1.09 (3H, s, CH_3), 1.05 (3H, s, CH_3), 1.01 (3H, s, CH_3), 0.96 (3H, s, CH_3); ^{13}C NMR (150 MHz, $CDCl_3$): δ 218.0 (C=O), 174.9 (O–C=O), 150.1 (C-20), 109.9 (C-29), 62.5 (C-28), 55.0, 49.8, 48.8, 47.7, 47.4, 46.4, 42.8, 40.8, 39.6, 37.7, 36.9, 34.6, 34.2, 33.5, 29.8, 29.6, 27.7, 27.1, 26.6, 25.2, 21.3, 21.1, 19.6, 19.2, 15.9, 15.8, 14.7, 9.2; EIMS *m/z* 526 $[M]^+$(23), 189 (100).

28-*O*-(2-Propenoyl)betulone (**16**) Yield 62 %; m.p. 69–71 °C; R_f 0.62 (chloroform/ethanol, 40:1, v/v); IR (KBr) ν_{max} 2962, 1723, 1706, 1456, 1261 cm^{-1}; ^1H NMR (600 MHz, $CDCl_3$): δ 6.34 (1H, m, $CH=CH_2$), 6.06 (1H, m, $CH=CH_2$), 5.75 (1H, m, $CH=CH_2$), 4.63 (1H, s, H-29), 4.53 (1H, s, H-29), 4.30 (1H, d, J = 10.8 Hz, H-28), 3.86 (1H, d, J = 10.8 Hz, H-28), 2.40 (1H, m, H-19), 1.65 (3H, s, CH_3), 1.01 (3H, s, CH_3), 1.00 (3H, s, CH_3), 0.96 (3H, s, CH_3), 0.92 (3H, s, CH_3), 0.87 (3H, s, CH_3); ^{13}C NMR (150 MHz, $CDCl_3$): δ 217.0 (C=O), 165.6 (O–C=O), 149.0 (C-20), 129.5, 127.6, 108.9 (C-29), 61.8 (C-28), 53.9, 48.7, 47.7, 46.7, 46.3, 45.5, 41.8, 39.8, 38.6, 36.7, 35.9, 33.5, 33.1, 32.4, 28.7, 28.6, 26.1, 25.5, 24.2, 20.3, 20.0, 18.6, 18.1, 14.9, 14.8, 13.7; EIMS *m/z* 495 $[M]^+$ (27), 203 (100).

28-*O*-Propynoylbetulone (**17**) Yield 78 %; m.p. 93–96 °C; R_f 0.65 (chloroform/ethanol, 40:1, v/v); IR (KBr) ν_{max} 3301, 2948, 2117, 1712, 1708, 1225 cm^{-1}; ^1H NMR (600 MHz, $CDCl_3$): δ 4.69 (1H, s, H-29), 4.60 (1H, s, H-29), 4.38 (1H, d, J = 10.8 Hz, H-28), 3.99 (1H, d, J = 10.8 Hz, H-28), 2.89 (1H, s, C≡CH), 2.48 (1H, m, H-19), 1.68 (3H, s, CH_3), 1.07 (3H, s, CH_3), 1.06 (3H, s, CH_3), 1.03 (3H, s, CH_3), 0.88 (3H, s, CH_3), 0.80 (3H, s, CH_3); ^{13}C NMR (150 MHz, $CDCl_3$): δ 217.9 (C=O), 153.2 (O-C=O), 149.8 (C-20), 110.0 (C-29), 74.8, 74.6, 64.8 (C-28), 55.0, 49.7, 48.8, 47.6, 47.3, 46.4, 42.8, 40.8, 39.6, 37.8, 36.9, 34.4, 34.1, 33.5, 29.6, 29.5, 27.0, 26.6, 25.2, 21.3, 21.0, 19.6, 19.1, 15.9, 15.8, 14.7; EIMS *m/z* 492 $[M]^+$ (28), 203 (100).

28-*O*-(3-Cyclopropyl-2-propynoyl)betulone (**18**) Yield 83 %; m.p. 93–97 °C; R_f 0.65 (chloroform/ethanol, 40:1, v/v); IR (KBr) ν_{max} 2947, 2232, 1711, 1705, 1457, 1249 cm^{-1}; ^1H NMR (600 MHz, $CDCl_3$): δ 4.71 (1H, s, H-29), 4.62 (1H, s, H-29), 4.35 (1H, d, J = 10.8 Hz, H-28), 3.96 (1H, d, J = 10.8 Hz, H-28), 2.43 (1H, m, H-19), 1.68 (3H, s, CH_3), 1.09 (3H, s, CH_3), 1.08 (3H, s, CH_3), 1.05 (3H, s, CH_3), 1.00 (3H, s, CH_3), 0.97–0.96 (5H, m, CH, CH_2), 0.95 (3H, s, CH_3); ^{13}C NMR (150 MHz, $CDCl_3$): δ 218.5 (C=O), 155.0 (O–C=O), 150.5 (C-20), 110.5 (C-29), 93.9, 69.1, 64.6 (C-28), 55.5, 50.3, 49.3, 48.2, 47.9, 46.9, 43.3, 41.4, 40.2, 38.3, 37.4, 35.1, 34.7, 34.0, 30.2, 30.1, 27.6, 27.1, 25.8, 21.9, 21.6, 20.2, 19.7, 16.5, 16.4, 15.2, 9.4 1.3, -0.4; EIMS *m/z* 533 $[M]^+$ (14), 93 (100).

28-*O*-Phenylpropynoylbetulone (**19**) Yield 82 %; m.p. 82–85 °C; R_f 0.64 (chloroform/ethanol, 40:1, v/v); IR (KBr) ν_{max} 2945, 2223, 1722, 1707, 1457, 1187 cm^{-1}; ^1H NMR (600 MHz, $CDCl_3$): δ 7.60–7.37 (5H, m, Ar-H),

4.71 (1H, s, H-29), 4.61 (1H, s, H-29), 4.43 (1H, d, *J* = 10.8 Hz, H-28), 4.04 (1H, d, *J* = 10.8 Hz, H-28), 2.41 (1H, m, H-19), 1.69 (3H, s, CH_3), 1.09 (3H, s, CH_3), 1.07 (3H, s, CH_3), 1.03 (3H, s, CH_3), 1.00 (3H, s, CH_3), 0.94 (3H, s, CH_3); ^{13}C NMR (150 MHz, $CDCl_3$): δ 218.0 (C=O), 154.7 (O–C=O), 149.9 (C-20), 132.9, 130.6, 128.5, 110.0 (C-29), 86.3, 80.7, 64.5 (C-28), 54.9, 49.7, 48.8, 47.6, 47.3, 46.4, 42.8, 40.8, 39.6, 37.8, 36.8, 34.5, 34.1, 33.5, 29.6, 29.5, 27.0, 26.6, 25.2, 21.3, 21.0, 19.6, 19.1, 15.9, 15.8, 14.7; EIMS *m/z* 569 $[M]^+$ (16), 129 (100).

28-*O*-(2-Butenoyl)betulone (**20**) Yield 81 %; m.p. 129–131 °C; R_f 0.63 (chloroform/ethanol, 40:1, v/v); IR (KBr) ν_{max} 2944, 1719, 1708, 1457, 1179 cm^{-1}; ^{1}H NMR (600 MHz, $CDCl_3$): δ 7.00 (1H, m, $CH{=}CHCH_3$), 5.89 (1H, m, $CH{=}CHCH_3$), 4.72 (1H, s, H-29), 4.62 (1H, s, H-29), 4.34 (1H, d, *J* = 10.8 Hz, H-28), 3.93 (1H, d, *J* = 10.8 Hz, H-28), 2.42 (1H, m, H-19), 1.92 (3H, m, $CH{=}CHCH_3$), 1.69 (3H, s, CH_3), 1.10 (3H, s, CH_3), 1.09 (3H, s, CH_3), 1.07 (3H, s, CH_3), 1.01 (3H, s, CH_3), 0.96 (3H, s, CH_3); ^{13}C NMR (150 MHz, $CDCl_3$): δ 218.1 (C=O), 167.0 (O–C=O), 150.1 (C-20), 144.5, 122.8, 109.9 (C-29), 62.4 (C-28), 55.0, 49.8, 48.8, 47.7, 47.4, 46.5, 42.8, 40.8, 39.6, 37.7, 36.8, 34.6, 34.2, 33.5, 29.8, 29.6, 27.1, 26.6, 25.2, 21.3, 21.1, 19.6, 18.0, 15.9, 15.8, 14.7, 3.6; EIMS *m/z* 510 $[M]^+$ (14), 189 (100).

28-*O*-(2-Butynoyl)betulone (**21**) Yield 78 %; m.p. 92–94 °C; R_f 0.66 (chloroform/ethanol, 40:1, v/v); IR (KBr) ν_{max} 2960, 2245, 1742, 1707, 1458, 1260 cm^{-1}; ^{1}H NMR (600 MHz, $CDCl_3$): δ 4.69 (1H, s, H-29), 4.59 (1H, s, H-29), 4.34 (1H, d, *J* = 10.8 Hz, H-28), 3.94 (1H, d, *J* = 10.8 Hz, H-28), 2.44 (1H, m, H-19), 1.99 (3H, s, $C{\equiv}CCH_3$), 1.68 (3H, s, CH_3), 1.07 (3H, s, CH_3), 1.06 (3H, s, CH_3), 1.03 (3H, s, CH_3), 0.99 (3H, s, CH_3), 0.93 (3H, s, CH_3); ^{13}C NMR (150 MHz, $CDCl_3$): δ 218.0 (C=O), 154.4 (O–C=O), 149.9 (C-20), 110.0 (C-29), 85.6, 72.5, 64.2 (C-28), 55.0, 49.7, 48.8, 47.6, 47.3, 46.4, 42.7, 40.8, 39.6, 37.7, 36.8, 34.5, 34.1, 33.5, 29.6, 29.5, 27.0, 26.6, 25.2, 21.3, 21.0, 19.6, 19.1, 15.9, 15.8, 14.7, 3.8; EIMS *m/z* 507 $[M]^+$ (29), 422 (100).

28-*O*-Propoxycarbonylbetulone (**22**) Yield 78 %; m.p. 76–78 °C; R_f 0.63 (chloroform/ethanol, 40:1, v/v); IR (KBr) ν_{max} 2963, 1741, 1700, 1458, 1256 cm^{-1}; ^{1}H NMR (600 MHz, $CDCl_3$): δ 4.72 (1H, s, H-29), 4.62 (1H, s, H-29), 4.37 (1H, d, *J* = 10.8 Hz, H-28), 4.13 (2H, t, *J* = 6.6 Hz, OCH_2), 3.95 (1H, d, *J* = 10.8 Hz, H-28), 2.45 (1H, m, H-19), 1.74 (2H, m, CH_2CH_3), 1.68 (3H, s, CH_3), 1.10 (3H, s, CH_3), 1.09 (3H, s, CH_3), 1.05 (3H, s, CH_3), 1.01 (3H, s, CH_3), 0.99 (3H, t, *J* = 7.2 Hz, CH_2CH_3), 0.96 (s, CH_3, 3H); ^{13}C NMR (150 MHz, $CDCl_3$): δ 218.0 (C=O), 156.0 (O–C=O), 150.0 (C-20), 109.9 (C-29), 69.6, 66.4 (C-28), 55.0, 49.7, 48.8, 47.7, 46.6, 42.8, 40.8, 39.6, 37.7, 36.9, 34.4, 34.2, 33.5, 29.6, 29.5, 27.1, 26.6, 25.6, 25.2, 22.1, 21.3, 21.1, 19.6, 19.1, 15.9, 15.8, 14.7, 10.2; EIMS *m/z* 526 $[M]^+$ (15), 422 (100).

28-*O*-Allyloxycarbonylbetulone (**23**) Yield 72 %; m.p. 123–125 °C; R_f 0.66 (chloroform/ethanol, 40:1, v/v); IR (KBr) ν_{max} 2945, 1741, 1710, 1457, 1248 cm^{-1}; ^{1}H NMR (600 MHz, $CDCl_3$): δ 5.98 (1H, m, $CH{=}CH_2$), 5.40 (1H, m, $CH=CH_2$), 5.31 (1H, m, $CH=CH_2$), 4.72 (1H, s, H-29), 4.66 (2H, m, OCH_2), 4.62 (1H, s, H-29), 4.38 (1H, d, J = 10.8 Hz, H-28), 3.96 (1H, d, J = 10.8 Hz, H-28), 2.44 (1H, m, H-19), 1.68 (3H, s, CH_3), 1.10 (3H, s, CH_3), 1.09 (3H, s, CH_3), 1.08 (3H, s, CH_3), 1.05 (3H, s, CH_3), 0.96 (3H, s, CH_3); ^{13}C NMR (150 MHz, $CDCl_3$): δ 218.0 (C=O), 155.6 (O–C=O), 150.0 (C-20), 131.7, 119.0, 109.9 (C-29), 68.5, 66.6 (C-28), 55.0, 53.4, 49.7, 48.8, 47.7, 47.4, 46.6, 42.8, 40.8, 39.6, 37.7, 36.9, 34.4, 34.2, 33.5, 29.6, 29.5, 27.0, 26.6, 25.2, 21.3, 21.1, 19.6, 15.9, 15.8, 14.7; EIMS m/z 524 $[M]^+$ (19), 422 (100).

28-*O*-Propargyloxycarbonylbetulone (**24**) Yield 73 %; m.p. 173–175 °C; R_f 0.64 (chloroform/ethanol, 40:1, v/v); IR (KBr) ν_{max} 3245, 2944, 2131, 1721, 1706, 1452, 1269 cm^{-1}; ^{1}H NMR (600 MHz, $CDCl_3$): δ 4.74 (2H, d, J = 2.4 Hz, OCH_2), 4.69 (1H, s, H-29), 4.60 (1H, s, H-29), 4.38 (1H, d, J = 10.8 Hz, H-28), 3.96 (1H, d, J = 10.8 Hz, H-28), 2.54 (1H, t, J = 2.4 Hz, C≡CH), 2.41 (1H, m, H-19), 1.68 (3H, s, CH_3), 1.09 (3H, s, CH_3), 1.07 (3H, s, CH_3), 1.03 (3H, s, CH_3), 0.99 (3H, s, CH_3), 0.93 (3H, s, CH_3); ^{13}C NMR (150 MHz, $CDCl_3$): δ 218.0 (C=O), 155.1 (O–C=O), 149.9 (C-20), 110.0 (C-29), 76.8, 75.6, 75.4, 67.2 (C-28), 55.2, 54.9, 49.7, 48.7, 47.6, 46.5, 42.8, 40.8, 39.6, 37.7, 36.8, 34.3, 34.1, 33.4, 29.5, 29.4, 26.9, 26.6, 25.2, 21.3, 21.0, 19.6, 19.1, 15.9, 15.8, 14.7; EIMS m/z 523 $[M]^+$ (24), 189 (100).

28-*O*-(3-Butynyloxycarbonyl)betulone (**25**) Yield 77 %; m.p. 81–84 °C; R_f 0.63 (chloroform/ethanol, 40:1, v/v); IR (KBr) ν_{max} 3310, 2948, 2360, 1744, 1705, 1458, 1248 cm^{-1}; ^{1}H NMR (600 MHz, $CDCl_3$): δ 4.69 (1H, s, H-29), 4.59 (1H, s, H-29), 4.36 (1H, d, J = 10.8 Hz, H-28), 4.25 (2H, t, J = 7.2 Hz, OCH_2), 3.94 (1H, d, J = 10.8 Hz, H-28), 2.60 (2H, m, OCH_2CH_2), 2.41 (1H, m, H-19), 2.03 (1H, t, J = 2.4 Hz, C≡CH), 1.68 (3H, s, CH_3), 1.08 (3H, s, CH_3), 1.07 (3H, s, CH_3), 1.03 (3H, s, CH_3), 0.98 (3H, s, CH_3), 0.93 (3H, s, CH_3); ^{13}C NMR (150 MHz, $CDCl_3$): δ 218.2 (C=O), 155.5 (O–C=O), 149.9 (C-20), 110.0 (C-29), 79.4, 76.8, 70.2, 66.7 (C-28), 65.3, 54.9, 49.6, 48.7, 47.6, 47.3, 46.5, 42.7, 40.8, 39.6, 37.6, 36.8, 34.3, 34.1, 33.4, 29.5, 29.4, 26.9, 26.5, 25.1, 21.2, 21.0, 19.6, 19.0, 15.9, 15.8, 14.6; EIMS m/z 537 $[M]^+$ (14), 189 (100).

28-*O*-(2-Butynyloxycarbonyl)betulone (**26**) Yield 75 %; m.p. 150–152 °C; R_f 0.59 (chloroform/ethanol, 40:1, v/v); IR (KBr) ν_{max} 2956, 2231, 1742, 1704, 1458, 1255 cm^{-1}; ^{1}H NMR (600 MHz, $CDCl_3$): δ 4.71 (2H, q, J = 2.4 Hz, OCH_2), 4.69 (1H, s, H-29), 4.59 (1H, s, H-29), 4.36 (1H, d, J = 10.8 Hz, H-28), 3.93 (1H, d, J = 10.8 Hz, H-28), 2.42 (1H, m, H-19), 1.87 (3H, t, J = 2.4 Hz, C≡CCH_3), 1.68 (3H, s, CH_3), 1.08 (3H, s, CH_3), 1.07 (3H, s, CH_3), 1.03 (3H, s, CH_3), 0.98 (3H, s, CH_3), 0.93 (3H, s, CH_3); ^{13}C NMR (150 MHz, $CDCl_3$): δ 218.2 (C=O), 155.2 (O–C=O), 149.9 (C-20), 110.0 (C-29), 84.1, 76.7, 72.6, 66.9 (C-28), 56.1, 54.9, 49.6, 48.7, 47.6, 47.3, 46.5, 42.7, 40.8, 39.6, 37.7, 36.8, 34.3, 33.4, 29.5, 29.4, 26.9, 26.5, 25.1, 21.2, 21.0, 19.6, 19.1, 15.9, 15.8, 14.6, 3.7; EIMS m/z 537 $[M]^+$ (14), 189 (100).

2.5 Antiproliferative Assay in vitro

2.5.1 Cells

The targeted compounds were evaluated for their cytotoxicity towards the cancer cell lines including T47D (human breast cancer), CCRF/CEM (human leukemia), SW707 (human colorectal), HL-60 (human promyelocytic leukemia), P388 (mouse leukemia), as well as BALB3T3 normal mouse fibroblasts cell line. The tested cell lines were obtained from the American Type Culture Collection (Rockville, Maryland, USA) and maintained at the Cell Culture Collection of the Institute of Immunology and Experimental Therapy (Wrocław, Poland). The cells were seeded in 96-well plates (Sarstedt, USA) at a density of 10^4 cells per well in 100 µL of culture medium overnight. The cancer cell lines CCRF/CEM (human leukemia) and P388 (mouse leukemia) were cultured in RPMI 1640 medium (Gibco, Scotland, UK) supplemented with 2 mM glutamine (Sigma-Aldrich, Chemie GmbH, Steinheim, Germany) and 10 % fetal calf serum FBS (Sigma-Aldrich, Chemie GmbH, Steinheim, Germany). The cancer cell lines SW707 (human colorectal), T47D (human breast cancer) and HL-60 (human promyelocytic leukemia) were cultured in mixture of RPMI 1640 and Opti-MEM (1:1) medium (both from Gibco, Scotland, UK) supplemented with 2 mM glutamine (Sigma-Aldrich, Chemie GmbH, Steinheim, Germany) and 5 % or 20 % fetal calf serum FBS (Sigma-Aldrich, Chemie GmbH, Steinheim, Germany). The normal mouse fibroblasts BALB3T3 was cultured in Dulbecco's modified Eagle's medium (Sigma-Aldrich, Chemie GmbH, Steinheim, Germany) supplemented with 2 mM glutamine (Sigma-Aldrich, Chemie GmbH, Steinheim, Germany) and 10 % fetal calf serum FBS (Sigma-Aldrich, Chemie GmbH, Steinheim, Germany). The all culture media were supplemented with streptomycin (100 µg/mL) and penicillin (100 U/mL) (both antibiotics from Polfa, Tarchomin, Poland). The cell cultures were maintained at 37 °C in humid atmosphere saturated with 5 % CO_2.

2.5.2 SRB assay

This technique as first was described by Skehan et al. in 1990. The SRB assay was performed after 96 h exposure of the cultured cells to varying concentrations (ranging from 1 to 100 µg/mL) of the tested substances. The reported derivatives as well as betulin **1** were dissolved in 10 % dimethyl sulfoxide (DMSO) to concentration of 1 mg/mL, and next diluted in culture medium to reach the required concentrations. DMSO as a solvent did not exert any inhibitory effect on cell proliferation. The cells fastened to the plastic were fixed mildly layering cold 50 % TCA (trichloroacetic acid, Aldrich-Chemie, Germany) on the top of the culture medium in each well. The plates were incubated at 4 °C for 1 h and then washed five times with tap water. The cells fixed with TCA were stained with 0.4 % sulforhodamine B (SRB, Sigma, Germany) dissolved in 1 % acetic acid (POCH, Gliwice, Poland) for 30 min. The unbound dye was removed by rinsing four times with 1 % acetic acid. The protein-bound dye was extracted with 10 mM unbuffered tris base (POCH, Gliwice, Poland) and then determined the optical density at 540 nm in a computer-interfaced, 96-well microtiter plate reader Multiscan RC photometer (Labsystems, Helsinki, Finland).

The compounds in given concentration were examined in triplicates in each experiment which was repeated 3–5 times. The results of cytotoxic activity in vitro were expressed as an IC_{50} in µg/mL.

2.5.3 MTT assay

The MTT assay was used for the cytotoxicity screening against leukemia cells growing in suspension culture. This assay was performed after 96 h of the leukemia cells to varying concentrations (ranging from 1 to 100 µg/mL) of the tested substances. The derivatives of betulin were dissolved in 10 % DMSO to concentration of 1 mg/mL, and next diluted in culture medium to reach the required concentrations. DMSO as a solvent did not exert any inhibitory effect on cell proliferation. During the last 3–4 h of incubation 20 µl of MTT solution were added to each well (MTT: 3-(4,5-dimethylthiazol-2-yl)-2,5-diphenyl tetrazolium bromide; stock solution: 5 mg/mL). The pale yellow MTT is reduced to a navy blue formazan in the mitochondria of living cells. At the end of incubation time, 80 µL of the lysing mixture was added to each well (lysing mixture: 225 mL dimethylformamide, 67.5 g sodium dodecyl sulfate and 275 mL of distilled water). After 24 h are formed the crystals of formazan, which are insoluble in aqueous solutions. The formazan crystals had been dissolved. The optical densities were read on a Multiskan RC photometer at 570 nm wavelength.

The compounds in given concentration were examined in triplicates in each experiment which was repeated 3–5 times. The results of cytotoxic activity in vitro were expressed as an IC_{50} in µg/mL.

2.5.4 In silico study

The physicochemical properties of obtained compounds such as lipophilicity (cLogP), molecular mass (M), topological polar surface area (tPSA), hydrogen bond donors (HBD) and hydrogen bond acceptors (HBA) were calculated using the ACD/Labs software.

3. RESULTS AND DISCUSSION

3.1 Chemistry

The synthesis of betulin derivatives **3–14** was accomplished starting with betulin **1**, which was isolated from birch bark of *Betula verrucosa*. The crude compound **1** was purified by flash-chromatography using a mixture of dichloromethane and ethanol as an eluent. Derivatives **3–14** were obtained according to our published procedures (Boryczka et al., 2013a, b). The esterification reactions of betulin **1** with carboxylic acids or chloroformates were performed with 49–86 % yields. The resulting monoesters **3–14** were oxidized with pyridinium chloroformate in dry dichloromethane to the derivatives of betulone **15–26** in 62–83 % yields. The synthesis of compounds **3–26** was presented via Scheme 1. All compounds were purified by column chromatography using the mixture of chloroform and ethanol. The chemical structures of new derivatives were determined on the basis of their ^{1}H-NMR, ^{13}C-NMR, IR and MS spectra.

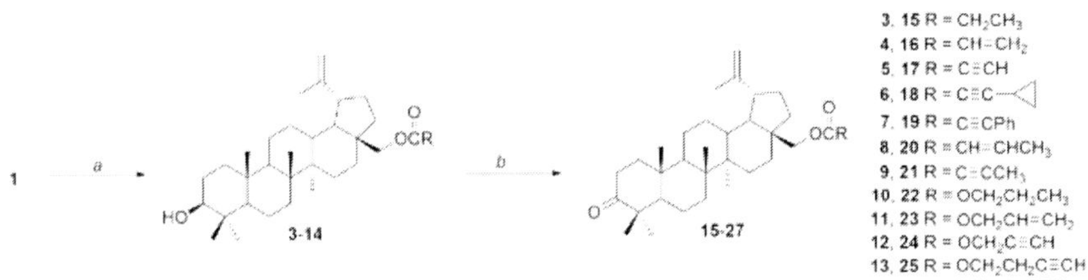

Scheme 1. Synthesis of derivatives **3–26**. *Reagents and conditions*: *a* RCOOH, CH_2Cl_2, DCC, DMAP, rt, 24 h or ROC(O)Cl, benzene, pyridine, rt, 24 h; *b* PCC, CH_2Cl_2, rt, 2 h

3.2 Cytotoxic activity

The newly compounds and betulin **1** were tested in vitro for their antitumor activity towards the following human cancer cell lines: T47D (human breast cancer), CCRF/CEM (human leukemia), SW707 (human colorectal), HL-60 (human promyelocytic leukemia) and murine leukemia P388 as well as BALB3T3 normal mouse fibroblasts cell line. In this study, cisplatin was used as a reference anticancer agent. The IC_{50} values (μg/mL) of the betulin derivatives are reported in Table 1 and Table 2. The compounds **3–9**, containing acyloxy group at the C-28 position, had IC_{50} values in the range of 0.02–49.0 μg/mL against the tested cell lines. In the series of the monoesters **3–9** the cytotoxic activity depends on the type of bond in the substituent at the C-28 position. The structure–activity relationships observed in the monoesters **3–9** indicates that rank order of the antiproliferative activity against the HL-60 cancer cell line, is as follows: propynoyl > propenoyl > 2-butynoyl > propanoyl > 3-cyclopropyl-2-propynoyl. Oxidation of the 3-hydroxyl group in compounds **4**, **6** and **10** to a carbonyl group led to an increase of activity against the HL-60 cancer cell line. The compounds **5** and **17** possessed the most potent activity, with IC_{50} value 0.3 μg/mL towards the human HL-60 cancer cell line. Moreover, these compounds exhibited the same antiproliferative activity as cisplatin. The betulones **24–26** with the acetylenic formate group at the C-28 position showed loss of activity against the human colorectal cell line (SW707) and normal mouse fibroblast cell line (BALB3T3). The rank order of the cytotoxic activity of the compounds **24–26** against T47D and CCRF/CEM cell lines, according to the nature of the formate substituent is as follows: propargyl > 3-butynyl > 2-butynyl.

Table 1. Cytotoxic activity (IC_{50}) of betulin **1**, derivatives of betulin and cisplatin as a reference compound against the four cancer cell lines

Compound	Cytotoxic activity IC_{50} [μg/mL]				
	Human			Murine	
	T47D	CCRF/CEM	SW707	P388	BALB3T3
Betulin 1	32.4 ± 10.7	10.9 ± 5.5	22.9 ± 15.4	5.5 ± 3.3	47.3 ± 7.9
3[a]	12.1 ± 4.4	8.1 ± 0.9	29.2 ± 24.4	3.3 ± 0.8	32.3 ± 23.0
4	7.1 ± 2.7	19.6 ± 9.5	29.6 ± 4.2	9.9 ± 5.9	24.3 ± 6.9
5[a]	9.1 ± 1.9	0.02 ± 0.001	14.9 ± 3.3	0.4 ± 0.1	0.3 ± 0.05
6	33.6 ± 6.3	16.5 ± 8.5	Neg	8.4 ± 9.2	Neg
7[a]	Neg	49.0 ± 9.8	Neg	Neg	Neg
8	29.9 ± 9.3	28.0 ± 21.6	32.0 ± 4.2	5.1 ± 4.9	25.6 ± 1.8
9	16.7 ± 5.5	2.1 ± 0.3	24.3 ± 8.8	2.9 ± 1.4	17.4 ± 1.1
19	Neg	Neg	Neg	Neg	Neg
24	23.3 ± 3.5	18.7 ± 6.6	Neg	30.3 ± 11.8	Neg
25	29.8 ± 10.7	25.2 ± 4.6	Neg	3.8 ± 1.9	Neg
26	79.2 ± 5.2	47.6 ± 5.7	Neg	18.3 ± 12.9	Neg
Cisplatin	3.1 ± 1.0	2.0 ± 0.5	2.2 ± 0.5	0.5 ± 0.3	2.7 ± 0.3

Neg negative in the concentration used

[a] Boryczka et al. (2013a, b)

Table 2. Cytotoxic activity (IC_{50}) of betulin **1**, derivatives of betulin and cisplatin as a reference compound against the HL-60 cancer cell line. The parameters determined by computational methods such as lipophilicity (cLogP), molecular mass (M), topological polar surface area (tPSA), hydrogen bond donors (HBD) and hydrogen bond acceptors (HBA)

Compound	IC_{50} [μg/mL]	cLogP	M	tPSA [Å]	HBD	HBA
Betulin 1	7.2 ± 0.5	6.63	442.71	40.46	2	2
3	29.3 ± 3.4	8.42	498.78	46.53	1	3
4	18.8 ± 4.8	8.23	496.76	46.53	1	3
5	0.3 ± 0.004	7.76	494.75	46.53	1	3
6	44.8 ± 10.9	8.85	534.81	46.53	1	3
9	24.5 ± 1.9	8.31	508.77	46.53	1	3
10	26.8 ± 2.5	8.39	528.80	55.76	1	4
11	14.9 ± 3.9	8.01	526.79	55.76	1	4
12	3.7 ± 0.5	7.64	524.77	55.76	1	4
15	33.7 ± 5.5	8.13	496.76	43.37	0	3
16	5.0 ± 1.3	8.07	494.75	43.37	0	3
17	0.3 ± 0.06	7.66	492.73	43.37	0	3
18	33.9 ± 5.5	8.61	532.79	43.37	0	3
21	29.5 ± 5.4	8.39	506.76	43.37	0	3
22	21.4 ± 2.1	8.23	526.79	52.60	0	4
23	28.4 ± 1.6	8.05	524.77	52.60	0	4
24	11.6 ± 4.9	7.68	522.76	52.60	0	4
Cisplatin	0.3 ± 0.08	—	—	—	—	—

The in silico study of tested compounds was performed by determination of Lipinski's rule of five and tPSA. The number of HBA and HBD significantly modulates the size of the polar surface area of molecule. The increase of the HBA number leads to higher affinity of the derivatives of betulone for P-glycoprotein. The most of betulin derivatives exhibited high values of molecular mass ($M > 500$) and lipophilicity ($cLogP > 5$). However, the tPSA of all compounds is less than 140 Å, what determines high oral bioavailability (Abd El-Karim et al., 2015).

In conclusion, the newly derivatives of betulin and betulone has been synthesized and characterized by spectroscopic analyses. Derivatives of betulin were tested for their antiproliferative activity against the five cancer cell lines. Several compounds exhibited a better cytotoxic effect than betulin **1**. The most active derivatives **5** and **17** were 24-fold potent than betulin **1** against the human promyelocytic leukemia cell line (HL-60).

ACKNOWLEDGMENTS

This work was supported by the Medical University of Silesia in Katowice, Poland. Grant No KNW-1-002/N/5/0.

CONFLICT OF INTEREST

The authors declare that they have no competing interests.

REFERENCES

1. Abd El-Karim SS, Anwar MM, Mohamed NA, Nasr T, Elseginy SA (2015) Design, synthesis, biological evaluation and molecular docking studies of novel benzofuran-pyrazole derivatives as anticancer agents. Bioorg Chem 63:1–12
2. Alakurtti S, Mäkelä T, Koskimies S, Yli-Kauhaluoma J (2006) Pharmacological properties of the ubiquitous natural product betulin. Eur J Pharm Sci 29:1–13
3. Alakurtti S, Bergström P, Sacerdoti-Sierra N, Jaffe CL, Yli-Kauhaluoma J (2010) Anti-leishmanial activity of betulin derivatives. J Antibiot 63:123–126
4. Boryczka S, Michalik E, Kusz J, Nowak M, Chrobak E (2013a) Lup-20(29)-en-28-ol-3-one (betulone). Acta Cryst E69:o795–o796
5. Boryczka S, Bębenek E, Wietrzyk J, Kempińska K, Jastrzębska M, Kusz J, Nowak M (2013b) Synthesis, structure and cytotoxic activity of new acetylenic derivatives of betulin. Molecules 18:4526–4543
6. Chen JD, Yi RZ, Lin YM, Feng DQ, Zhou HC, Wang ZC (2011) Characterization of terpenoids from the root of Ceriops tagal with antifouling activity. Int J Mol Sci 12:6517–6528CrossRefPubMedPubMedCentral
7. Cole BJW, Bentley MD, Hua Y (1991) Triterpenoid extractives in the outer bark of Betula lenta (black birch). Hölzforschung 45:265–268r
8. Diouf PN, Stevanovic T, Boutin Y (2009) The effect of extraction process on polyphenol content, triterpene composition and bioactivity of yellow birch (Betula alleghaniensis Britton) extracts. Ind Crops Prod 30:297–303r
9. Gachet MS, Kunert O, Kaiser M, Brun R, Zehl M, Keller W, Munoz RA, Bauer R, Schuehly W (2011) Antiparasitic compounds from Cupania cinerea with activities against Plasmodium falciparum and Trypanosoma brucei rhodesiense. J Nat Prod 74:559–566
10. Grishko VV, Tarasova EV, Ivshina IB (2013) Biotransformation of betulin to betulone by growing and resting cells of the actinobacterium Rhodococcus rhodochrous IEGM 66. Process Biochem 48:1640–1644r
11. Hase TA, Suokas E, Weckman A (1981) Dehydrogenation of triterpenoid alcohols in alkali melt. A short synthesis of the so called-Jasminol. Synth Commun 11:489–492r

12. Hata K, Hori K, Takahashi S (2002) Differentiation- and apoptosis-inducing activities by pentacyclic triterpenes on a mouse melanoma cell line. J Nat Prod 65:645–648
13. Hiroya K, Takahashi T, Miura N, Naganuma A, Sakamoto T (2002) Synthesis of betulin derivatives and their protective effects against the cytotoxicity of cadmium. Bioorg Med Chem 10:3229–3236
14. Koohang A, Majewski ND, Szotek EL, Mar AA, Eiznhamer DA, Flavin MT, Xu ZQ (2009) Synthesis and cytotoxicity of 2-cyjano-28-hydroxy-lup-1-en-3-ones. Bioorg Med Chem Lett 19:2168–2171
15. Liu M, Yang S, Jin L, Hu D, Wu Z, Yang S (2012) Chemical constituents of the ethyl acetate extract of Belamcanda chinensis (L.) DC roots and their antitumor activities. Molecules 17:6156–6169
16. Mao DB, Feng YQ, Bai YH, Xu CP (2012) Novel biotransformation of betulin to produce betulone by Rhodotorula mucilaginosa. J Taiwan Inst Chem Eng 43:825–829r
17. Mar AA, Koohang A, Majewski ND, Szotek EL, Eiznhamer DA, Flavin MT, Xu ZQ (2009) Synthesis and cytotoxicity of 28-carboxymethoxy lupane triterpenoids. Preference of 28-O-acylation over 28-O-alkylation of betulin by ethyl bromoacetate. Chin Chem Lett 20:1141–1141r
18. Orchel A, Kulczycka A, Chodurek E, Bębenek E, Borkowska P, Boryczka S, Kowalski J, Dzierżewicz Z (2014) Influence of betulin and 28-O-propynoylbetulin on proliferation and apoptosis of human melanoma cells (G-361). Postepy Hig Med Dosw 68:191–197r
19. Reyes CP, Núnez MJ, Jiménez IA, Busserolles J, Alcaraz MJ, Bazzocchi IL (2006) Activity of lupine triterpenoids from Maytenus species as inhibitors of nitric oxide and prostaglandin E2. Bioorg Med Chem 14:1573–1579
20. Skehan P, Storeng R, Scudiero D, Monks A, Mcmachon J, Vistica D, Warren JT, Bokesch H, Kenney S, Boyol MR (1990) New colorimetric cytotoxicity assay for anticancer-drug screening. J Natl Cancer Inst 82:1107–1112
21. Sun IC, Shen JK, Wang HK, Cosentino LM, Lee KH (1998a) Anti-Aids agents. 32 1. Synthesis and anti-HIV activity of betulin derivatives. Bioorg Med Chem Lett 8:1267–1272
22. Sun IC, Wang HK, Kashiwada Y, Shen JK, Cosentino LM, Chen CH, Yang LM, Lee KH (1998b) Anti-Aids agents. 34. Synthesis and structure-activity relationships of betulin derivatives as anti-HIV agents. J Med Chem 41:4648–4657
23. Tolstikov GA, Flekhter OB, Shultz EE, Baltina LA, Tolstikov AG (2005) Betulin and its derivatives. Chemistry and biological activity. Chem Sustain Develop 13:1–29r

CHAPTER 14

Synthesis and anticancer activity of thiosubstituted purines

Alicja Kowalska1 • Małgorzata Latocha2 • Krystian Pluta1

1. *Department of Organic Chemistry, School of Pharmacy with the Division of Laboratory MedicineThe Medical University of SilesiaSosnowiecPoland*

2. *Department of Cell Biology, School of Pharmacy with the Division of Laboratory MedicineThe Medical University of SilesiaSosnowiecPoland*

ABSTRACT

New thiopurines with the propargylthio, pyrrolidinobutynylthio, sulfenamide, and sulfonamide groups in the pyrimidine ring were synthesized. The anticancer activity of these compounds and previously obtained 2- or 6-substituted azathioprine analogs and dialkylaminoalkylthiopurines were tested in vitro against three cell lines: glioblastoma SNB-19, melanoma C-32, and human ductal breast epithelial tumor T47D. 2-Chloro-7-methyl-6-pyrrolidinobutynylthiopurine (**5B**) was the most potent compound against SBN-19 and C-32 cell lines with the activity similar to cisplatin (EC_{50} = 5.00 and 7.58 μg/ml, respectively). The dialkylaminoalkylthio derivatives (**4B**, **4C**, **4E**, and **4F**) showed good activity against SBN-19 cell line (EC_{50} < 10 μg/ml). The azathioprine analogs (**2A**, **2B**, and **3A**) were more active than azathioprine against SBN-19 and C-32 cell lines. The sulfenamide and sulfonamide derivatives of purine were very weak active against tested cell lines. All studied thiopurines were less toxic than cisplatin.

1. INTRODUCTION

The sulfur-containing purines: 6-mercaptopurine, 6-thioguanine, and azathioprine, have been considered as important and effective drugs used in cancer chemotherapy, for immunosuppression in kidney or heart transplantation and autoimmune diseases (Steurer *et al.*, 2006; Hawwa *et al.*, 2008; Relling *et al.*, 2013; Rao *et al.*, 2013). 6-Mercaptopurine is widely used as an antileukemic agent in the lymphoproliferative disorders, including lymphoma, childhood acute lymphoblastic leukemia, and other neoplastic conditions (Hawwa *et al.*, 2008; Prima *et al.*, 2013; Miron *et al.*, 2009). Azathioprine, but also 6-mercaptopurine exert immunosuppressive effects and are the treatments of choice for severe chronic inflammatory diseases such as rheumatoid arthritis, ulcerative colitis, Crohn's disease, inflammatory bowel diseases, systemic lupus erythematosus, and multiple sclerosis (Aldinucci *et al.*, 2010; Colombel *et al.*, 2010; Daehn and Karran, 2009). Various pharmacological effects of synthetic thiopurines including antiviral, antibacterial, antitumor, and antifungal activity (Lech-Maranda *et al.*, 2006; Rao *et al.*, 2013) were developed for treatment of patients with infections of human immunodeficiency virus (HIV) and infections caused by herpes and hepatitis virus (Rao *et al.*, 2013; Schow *et al.*, 1997; Chang-Hyun *et al.*, 2001).

However, the use of thiopurines is limited by their toxicities, which include hepatotoxicity, myelosuppression, pancreatitis, and allergic reactions (Prima *et al.*, 2013). For this reason, many thiopurine derivatives and analogs have been synthesized for evaluation of their biological activities and reduced toxicity. Recently, 9-substituted derivatives of 6-(t-butoxycarbamylaminohexyl)thiopurine were obtained which showed moderate antibacterial activity (Rao *et al.*, 2013); a series of 6-mercaptopurine analogs with 1,2,3-triazole or steroid rings exhibited promising antimalarial and antileishmanial activities (Corrales *et al.*, 2011); and biologically active 6-allyldithiopurine and its riboside inhibited cell proliferation and induced apoptosis (Miron *et al.*, 2009).

Heterocyclic sulfonamides are used as carbonic anhydrase inhibitors, antibacterial, anti-inflammatory, analgesic, hypoglycemic, antifungal, and antiviral agent (Joseph *et al.*, 2010; Guzel *et al.*, 2009; Azab *et al.*, 2013; El-Sayed *et al.*, 2011; Veisi *et al.*, 2011). It is known that aryl and heteroaryl sulfonamides may act as antitumor agents through perturbation of cell cycle in the G_1 phase, distribution of microtubule assembly, or angiogenesis inhibition. They reverse the effect of tumor acidification, consequently inhibit the growth of cancer cell, and suppress tumor invasion mediated by the carbonic anhydrase active side (Zn^{2+}) and so inhibit the catalytic ability of this enzyme (El-Sayed *et al.*, 2011).

Indisulam (E7070) is a novel sulfonamide anticancer agent currently in phase II clinical development for the treatment of several types of cancer. It was shown to act as a nanomolar inhibitor of carbonic anhydrase isoform IX (Sławiński *et al.*, 2013; Mirian *et al.*, 2011).

The sulfamoyl group has been extensively utilized as an activity-modifying substituent in many different classes of drugs. The synthesis of 2-substituted 7- or 9-methylpurines containing a sulfanilamide group at the positions 2, 6, or 8 of the purine ring with moderate antibacterial activity was reported by the reaction of chloropurines with sodium sulfanilamide (Beaman *et al.*, 1966). Structural modification of the amino group at C-6 atom of adenosine leads to selective A_1 receptor agonists, while modification at C-2 atom confers high potency and selectivity on A_{2A} receptors. The synthesis of N-9- and 6-amino monosulfonyladenine and disulfonyladenine derivatives and its sulfonylated nucleoside with high potency and selectivity on A_1 and A_{2A} receptor agonists has been described (Zinic *et al.*, 2001). A number of 6-sulfenamide, sulfinamide, and sulfonamide derivatives of 6-mercaptopurine, 6-thioguanine, and related ribonucleosides have been synthesized and evaluated for antileukemic activity in mice (Revankar *et al.*, 1990). It is quite interesting that a subtle change to an oxidized sulfur atom in the form of sulfonamide resulted in a new group of purine derivatives possessing significant antitumor properties. Majority of the tested compounds exhibited very significant anti-L1210 cell activity (Revankar *et al.*, 1990; Robins *et al.*, 1991).

Oligonucleotides containing 1-alkynyl-6-thiopurine or 6-thioguanine bases demonstrate potent antiviral activity in several assays, including the human immunodeficiency virus reverse transcriptase enzyme assay (Broom *et al.*, 1997). The purine derivatives possessing the alkynyl groups at position 6 are cytokinin analogs with profound plant growth-stimulating effect and have attracted considerable interest as potential anticancer compounds. All 6-phenylacetylenes and certain enyne of the purine are highly toxic against K-562 cells and exhibit cytotoxic activity against a human chronic myelogenous leukemia cell line comparable to well-known anticancer drugs (6-mercaptopurine and fludarabine) (Brathe *et al.*, 2003). Many adenine and adenosine derivatives with 2-, 6-, or 8-alkynyl substituents are selective adenosine subtypes A_1, A_{2A}, A_{2B}, and A_3 receptor antagonists. 2-Alkynyl-8-aryl adenine derivatives bear methyl or an amide moiety at the position 9, as A_{2B} receptor antagonists have hypoglycemic activity and may be used as antidiabetic agents (Harada *et al.*, 2001). Stimulating A_{2A} receptor by 2-adenosine propargyl phenyl ether composition resulted in vasodilating activity and potential of using in coronary artery disease (Zablocki *et al.*, 2001). Introduction of alkynyl chains on C-8 atom of adenosine led to very selective antagonists of the A_3 receptor (Volpini *et al.*, 2001). Some 2-alkynyl and 2,(6-

amino)dialkynyl adenosine derivatives show high affinity and different degrees of selectivity for A1, A_{2A}, and A3 receptor (Volpini *et al.*, 2001).

Our previous research demonstrated the effective synthesis of thiosubstituted purines such as 2- or 6-substituted azathioprine analogs (Kowalska and Pluta, 2008; Kowalska *et al.*, 2009) and dialkylaminoalkylthiopurines (Kowalska and Pluta, 2012). The aim of this study was to elaborate synthesis of new thiopurines with the propargylthio, pyrrolidinobutynylthio, sulfenamide, and sulfonamide groups in the pyrimidine ring, to examine anticancer activity of not only those compounds but also previously synthesized thiopurines and to discuss the structure–activity relationship.

2. RESULT AND DISCUSSION

2.1 Chemistry

All thiosubstituted 7-methylpurines were obtained from 2-substituted 7-methylpurine-6-thiones **1A–C** and 7-methylpurine-2,6-dithiones **1D**. Purinyl sulfides **2–4** with the pharmacophoric 1-methyl-4-nitroimidazol-5-yl and dialkylaminoalkyl groups in positions 2 and 6 were obtained according to the procedures published previously (Kowalska and Pluta, 2008; Kowalska *et al.*, 2009; Kowalska and Pluta, 2012). The alkynylthio derivatives **5** were obtained from 2-chloro derivative **1A** and propargyl bromide and further via Mannich reaction with pyrrolidine (Scheme 1).

The traditional and general method for preparing sulfonamides is based on reaction of sulfonyl chloride with ammonia or amines. The aryl sulfonyl chlorides may be obtained from thiols, sulfides, thioacetals, or thiocarbamines by oxidative chlorination with chlorine gas in aqueous acid (Veisi *et al.*, 2011; Mirian *et al.*, 2011; Ashfaq *et al.*, 2013; Narasaiah *et al.*, 2012; Rehman *et al.*, 2012). 2-Chloro-7-methyl-6-purinethione **1A** was chosen as a model compound to prepare S-benzylthiopurine **6**. It is known that azinesulfonyl chlorides with the chlorosulfonyl group in aza-activated positions readily lose sulfur dioxide to form chloroazines (Beaman and Robins, 1961; Beaman, 1963; Revankar *et al.*, 1990).

Two-step synthesis of purinesulfonamide **8** via oxidation and chlorination of sulfide **6** with chlorine in acetic acid and subsequent transformation of sulfonyl chloride **7** with ammonia, diethylamine, or aniline failed. In this case, the highly instable chlorosulfonyl group lost sulfur dioxide, and resulted chloropurine reacted with ammonia and amines to give 2-chloro-6-amine derivatives **9A–C** with 71–86 % yield (Scheme 2).

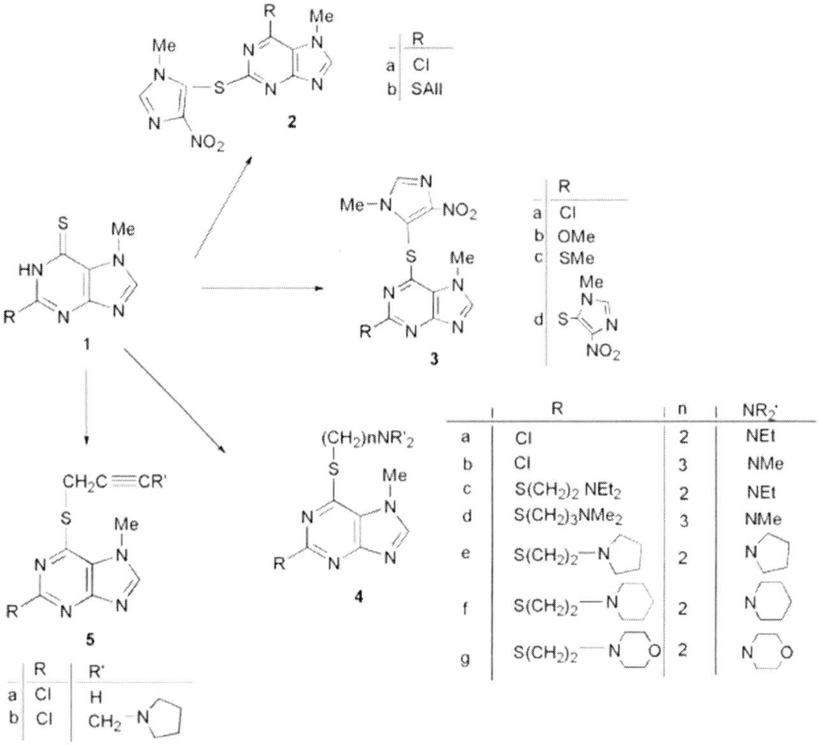

Scheme 1. Synthesis of 2- or 6-substituted azathioprine analogs **2**, **3**, dialkylaminoalkylthiopurines **4**, propargylthio- and pyrrolidinobutynyl-thiopurines **5**

Therefore, the sulfonamide compounds were obtained from purinethione via amination and oxidation. The procedure of S-amination of azinethiones with chloramine solution has been widely used in the preparation of purine-, pyridine-, pyrimidine-, and benzothiazolesulfenamides (Revankar *et al*., 1990; Robins *et al*., 1991; Hurley and Robinson, 1965; Schoenwald *et al*., 1984). For 2-Substituted 7-methyl-6-purinethiones **1A–C**, S-amination with a chloramine solution at 0 °C to sulfenamide **10A–C** in 85–92.5 % yield was carried out. In the case of purinedithione **1D**, two products were obtained: disubstituted compound as the main product, purine-2,6-disulfenamide **10E** (66 % yield) and purine-6-sulfenamide **10D** (17.5 % yield). Treatment of sulfenamides **10** with 4 molar equivalent of 3-chloroperoxybenzoic acid (MCPBA) at room temperature in ethanol gave purinesulfonamides **8A–D** in 66–79 % yield. Lesser equivalents of MCPBA led to sulfenamides (Revankar *et al*., 1990).

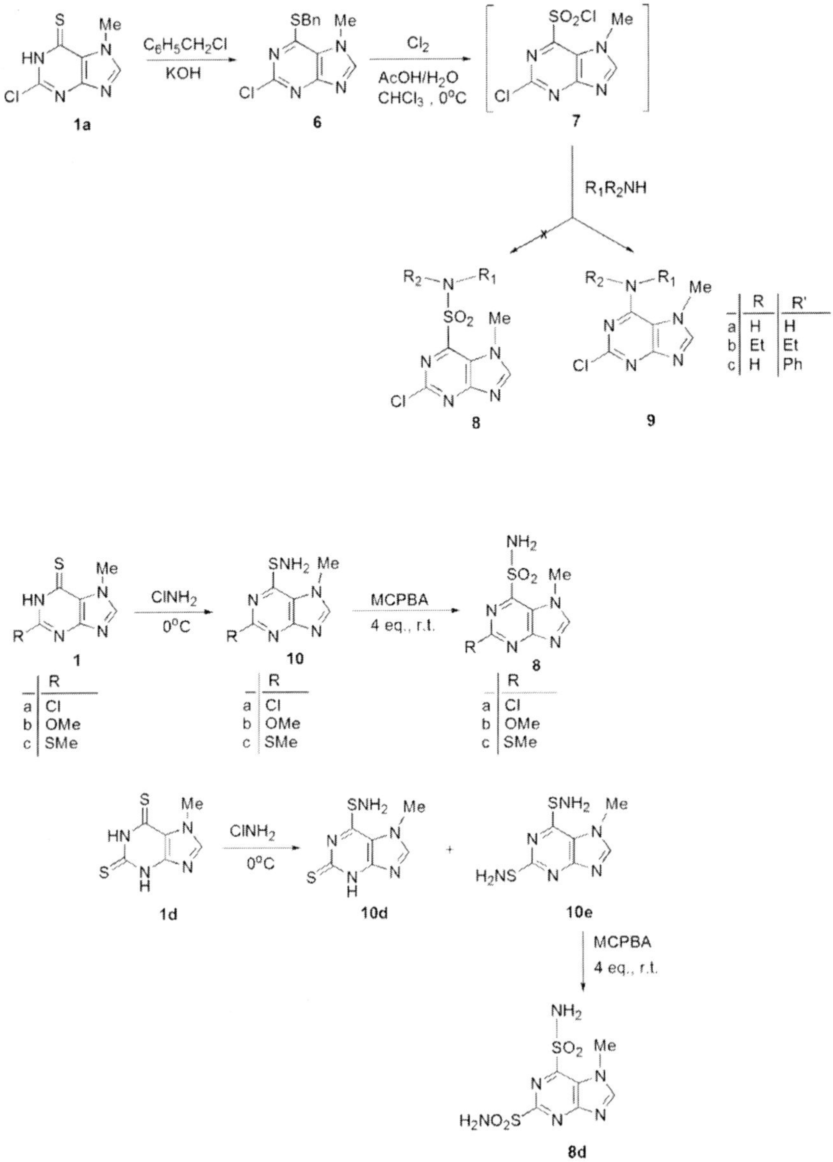

Scheme 2. Synthesis of amine **9**, sulfenamide **10**, and sulfonamide **8** derivatives of purines

The structure of all new compounds was determined by the spectral data analysis: ^{1}H and ^{13}C NMR, EI, CI or FAB MS, and HR MS. ^{1}H NMR spectra of new compounds revealed H-8, NCH$_3$proton signals, and protons from C-2

(OCH_3, SCH_3) and C-6 ($CH_2C_6H_5$, NH_2, $N(C_2H_5)_2$, NHC_6H_5, SCH_2CCH, $SCH_2CCCH_2NC_4H_8$, SNH_2, SO_2NH_2) substituents. ^{13}C NMR spectra showed primary, secondary, tertiary, and quaternary carbon signals. In order to assign all of these signals, 2D NMR techniques (HSQC—the primary, secondary, and tertiary carbon atoms connected with the hydrogen atoms and HMBC—the primary, tertiary, and quaternary carbon atoms connected with the hydrogen atoms via two and mainly three bonds) for selected compounds (**5A**, **10E**) were used. For compounds **5A**, in the HSQC spectrum the H-8 proton at 8.59 ppm correlated with the signal at 151.09 ppm (C-8), the SCH_2 proton at 4.22 ppm correlated with the signal at 18.09 ppm (C, SCH_2), and the proton at 4.04 ppm (NCH_3) correlated with the signal at 34.84 ppm (C, NCH_3). The HMBC spectrum revealed correlation via three bonds between the H-8 proton at 8.59 ppm and the carbon signals at 34.84 ppm (C, NCH_3), at 123.16 ppm (C-5), and at 160.52 ppm (C-4). The proton signal at 4.22 ppm (H, SCH_2) correlated with the signals of carbon atoms via two bonds at 79.55 ppm (C, CCH) and via three bonds at 74.67 ppm (C, CH) and at 153.99 ppm (C-6). The proton signal at 4.04 ppm (H, NCH_3) correlated via three bonds with the signals at 123.16 ppm (C-5) and at 151.09 ppm (C-8). The proton–carbon correlation was presented in Scheme 3.

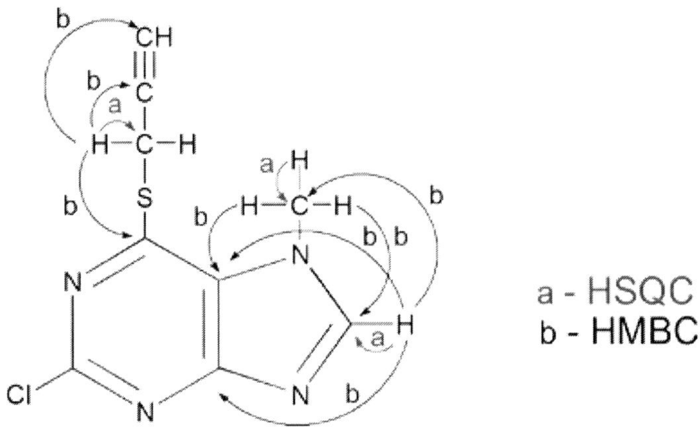

Scheme 3.2D NMR experiments for compound **5A**

2.2 Anticancer activity

For the biological tests, 20 thiopurines of five types were selected:

a) the azathioprine analogs containing the methylnitroimidazolylthio group in position 2 (compounds **2A**, **2B**), in position 6 (**3A–C**), and in both positions (**3D**),

b) the dialkylaminoalkylthio derivatives containing that substituent with an open-chain amine group in positions 6 (**4A, 4B**), 2 and 6 (**4C**, **4D**) and with a cyclic amine group (pyrrolidine, piperidine, and morpholine) in positions 2 and 6 (**4E–G**),

c) the alkynylthio derivatives containing the propynylthio (**5A**) and pyrrolidinobutynylthio (**5B**) groups,

d) the sulfenamide derivative (**10A**),

e) the sulfonamide derivatives with that group in positions 6 (**8A–C**), and 2 and 6 (**8D**).

The anticancer activity of the obtained compounds was investigated in vitro using cultured glioblastoma SNB-19, melanoma C-32, and human ductal breast epithelial tumor T47D cell lines. Normal human fibroblasts (HFF-1) were used as a control, and azathioprine and cisplatin as reference drugs. Table 1 contains the activity of thiopurines as the EC_{50} values.

The most active was 2-chloro-7-methyl-6-pyrrolidinobutynylthiopurine (**5B**) with the EC_{50}values of 5.00 and 7.58 μg/ml against SNB-19 and C-32 cell lines, comparable to the cisplatin effect. This compound exhibited good selectivity and low toxicity, being weak active against T47D cell line and normal fibroblasts HFF-1. The introduction of the pyrrolidine ring to the alkynyl chain enhanced the activity (compound **5B** vs. **5A**).

Good activity against SNB-19 cell line was observed for the dialkylaminoalkylthio derivatives (**4B**, **4C**, **4E**, and **4F**) with the EC_{50} values below 10 μg/ml. The compounds with the ethyl chain were more active than those with the propyl chain. The most active was compound **4E** with the two pyrrolidinoethylthio groups. The activity against other cell lines was lesser as well as their toxicity.

The azathioprine analogs with the imidazolylthio group in position 2 were more active than those with that group in position 6 with an exception of compound (**3A**) which exhibited good activity against C-32 cell line. The most active azathioprine analogs (**2A**, **3A**) possessed the chlorine atom in position 2 or 6. Both compounds were more active than azathioprine (against first two cell lines) and compounds containing the methoxyl and thioalkyl groups (**2B**, **3B**, **3C**). Compound **3D** with two imidazolylthio groups turned out to be unexpectedly inactive. In our opinion, such an effect was the result of the unusual spatial arrangement of those groups directed to N-1 atom with the donor–acceptor interaction between the imidazole rings, as it was observed in X-ray analysis of the monocrystal **3D** (Kowalska *et al.*, 2009).

Table 1. Antiproliferative activity of thiopurines

Compound	Antiproliferative activity EC_{50} (μg/ml)			
	SBN-19	C-32	T47-D	HFF-1
2a	17.92	21.77	41.11	>50
2b	33.26	35.33	49.60	>50
3a	30.38	12.79	>50	34.93
3b	>50	>50	>50	>50
3c	>50	>50	48.63	>50
3d	>50	>50	>50	>50
4a	28.77	22.47	41.27	>50
4b	8.45	>50	>50	>50
4c	9.82	33.55	24.00	40.00
4d	36.28	>50	>50	>50
4e	7.95	27.05	38.40	35.41
4f	9.18	21.80	32.25	45.50
4g	>50	>50	>50	>50
5a	30.95	35.35	44.30	48.95
5b	5.00	7.58	46.00	43.12
8a	>50	>50	>50	>50
8b	35.02	>50	>50	>50
8c	>50	>50	>50	>50
8d	>50	>50	>50	>50
10a	>50	>50	>50	>50
Azathioprine	37.20	35.62	39.00	55.88
Cisplatin	4.45	6.28	8.79	30.00

The sulfenamide and sulfonamide derivatives are very weak active or inactive. The glioblastoma SNB-19 was most sensitive and breast tumor T47D least sensitive cell line for thiopurines. All studied compounds were less toxic than cisplatin.

3. CONCLUSION

In search for novel thiopurine derivatives, we described synthesis of 2-chloro-6-alkynylthio-7-methylpurines **5A**, **5B** from 2-chloro derivative and propargyl bromide and further via Mannich reaction with pyrrolidine. Other types of thiopurines bearing the sulfenamide **10A–E** and sulfonamide **8A–D** groups were obtained through S-amination of purinethiones **1A–D** and oxidation with 3-chloroperoxybenzoic acid.

New thiopurines **5**, **8**, and **10**, previously synthesized azathioprine analogs **2** and **3**, and dialkylaminoalkylthiopurines **4** were investigated as antitumor agents. The most potent compound against SBN-19 and C-32 cell lines was 2-chloro-7-methyl-6-pyrrolidinobutynylthiopurine **5B** with the activity similar to cisplatin. The dialkylaminoalkylthio derivatives **4B**, **4C**, **4E**, and **4F** showed good activity against SBN-19 cell line. The azathioprine analogs **2A**, **2B**, and **3A** were more active than azathioprine against SBN-19 and C-32 cell lines. The sulfenamide and sulfonamide derivatives of thiopurine were very weak active against tested cell lines. All studied thiopurines were less toxic than cisplatin.

4. EXPERIMENTAL

6-Substituted derivatives of 7-methyl-2-(1-methyl-4-nitroimidazol-5-ylthio)purines **2A–B** were prepared from 7-methyl-2,6-di(1-methyl-4-nitroimidazol-5-ylthio)purine **3D** exploiting different reactivity of the imidazolylthio groups toward nucleophilic reagents (Kowalska *et al.*, 2009). 2-Substituted derivatives of 7-methyl-6-(1-methyl-4-nitroimidazol-5-ylthio)purines **3A–D** were obtained in the reaction of 6-purinethiones with 5-chloro-1-methyl-4-nitroimidazole in 70 % ethanol (Kowalska and Pluta, 2008). 2-Substituted 6-(dialkylaminoalkylthio) **4A–B** and 2,6-bis(dialkylaminoalkylthio) derivatives **4C–G** were obtained via the direct S-dialkylamination of the appropriate purinethiones (Kowalska and Pluta, 2012). All commercially available organic solvents and reagents were from Sigma-Aldrich and Chempur and were used without further purification.

The melting points were determined in open capillary tubes on a Boetius melting point apparatus and were uncorrected. ^1H NMR spectra were recorded on a Bruker AVANS 300 spectrometer operating at 300 MHz and 75 MHz for ^1H and ^{13}C nuclei, respectively, in deuterochloroform and dimethyl sulphoxide-d_6 with tetramethylsilane as internal standard. Shifts were given in ppm, coupling constant (J) values were presented in hertz (Hz), and the abbreviations were as follows: s (singlet), d (doublet), t (triplet), and m (multiplet). Electron impact (EI MS), chemical ionization (CI MS), fast atom bombardment (FAB MS), and high-resolution (HR MS) (in the *m*-nitrobenzyl alcohol and glycerol matrix) mass spectra were run on a Finnigan MAT 95 spectrometer at 70 eV. The reactions were monitored by thin-layer chromatography (TLC) using aluminum sheets coated with silica gel 60_{F254} (Merck) and chloroform–ethanol (9:1) as the solvents. Purity of the synthesized compounds was confirmed by TLC in the same way. Spots were detected by their absorption under UV light (λ = 254 nm), and the chromatograms were further visualized by iodine vapor. Column chromatography separations were carried out with Merck Kieselgel 60 or aluminum oxide 90 (Merck) using a mixture of chloroform–ethanol (99:1, v/v) as an eluent.

Synthesis of 2-chloro-6-(prop-2-ynylthio)-7-methylpurine

5A Propargyl bromide (0.17 g, 1.44 mmol) was added to a solution generated from the reaction of 2-chloro-7-methyl-6-thiopurine (0.2 g, 1 mmol) with t-BuOK (0.16 g, 1.44 mmol) in 10 ml of DMF at room temperature for 0.5 h. The reaction mixture was stirred for an additional 24 h at room temperature and then added to 25 ml of water. The resulted solid was filtered off and washed with water to give compound **5A**. It was obtained as a pale yellow solid (0.22 g, 92 %); mp 183–184 °C (EtOH); ^{1}H NMR (DMSO-d$_6$), δ: 3.26 (t, J = 2.2 Hz, 1H, CH), 4.04 (s, 3H, NCH_3), 4.22 (d, J = 2.2 Hz, 2H, SCH_2), 8.59 (s, 1H, H-8); ^{13}C NMR (DMSO-d$_6$), δ: 18.09 (CH_2, SCH_2CCH), 34.84 (CH_3, NCH_3), 74.67 (CH, SCH_2CCH), 79.55 (C, SCH_2CCH), 123.00 (C, C-5), 151.09 (CH, C-8), 152.28 (C, C-2), 153.99 (C, C-6), 160.52 (C, C-4); HSQC NMR, δ: 8.59 (H-8) correlated with 151.09 (C-8), 4.22 (H, SCH_2) correlated with 18.09 (C, SCH_2), 4.04 (H, NCH_3) correlated with 34.84 (C, NCH_3); HMBC NMR, δ: 8.59 (H-8) correlated with 34.84 (C, NCH_3), 123.00 (C-5), and 160.52 (C-4), 4.22 (H, SCH_2) correlated with 79.55 (C, CCH), 74.67 (C, CH), and 153.99 (C-6), 4.04 (H, NCH_3) correlated with 123.00 (C-5) and 151.09 (C-8); EIMS m/z 238 $[M]^+$ (34.5), 240 [M+2] (12), 203 [M-Cl] (39), 133 [M-Cl, -SCH_2CCH] (55), 39 [CH_2CCH] (100); HREIMS m/z $[M]^+$ calcd. for $C_9H_7ClN_4S$ 238.0079, found 238.0076.

Synthesis of 2-chloro-6-(4-N-pyrrolidinylbut-2-ynylthio)-7-methylpurine

5B To a mixture of propargyl derivative **5A** (0.24 g, 1 mmol) and paraformaldehyde (0.06 g, 2 mmol) in 5 ml of dry dioxane, pyrrolidine (0.14 g, 2 mmol) and CuCl (0.01 g) were added. The reaction mixture was stirred at temperature 70 °C for 2 h. After cooling, the resulting solid was filtered off and purified by column chromatography (silica gel, $CHCl_3$, $CHCl_3$–EtOH, 99:1 v/v) to give compound **5B**. It was obtained as a pale yellow solid (0.31 g, 96 %); mp 129–130 °C (EtOH); ^{1}H NMR ($CDCl_3$), δ: 2.13 (m, 8H, $4CH_2$), 3.12 (t, J = 2.2 Hz, 2H, SCH_2), 3.62 (t, J = 2.2 Hz, 2H, CCH_2), 4.19 (s, 3H, NCH_3), 8.25 (s, 1H, H-8); ^{13}C NMR (DMSO-d$_6$), δ: 18.77 (CH_2, SCH_2CC), 23.68 ($2CH_2$), 34.86 (CH_3, NCH_3), 42.58 (CH_2, CCH_2), 51.86 ($2CH_2$, NCH_2), 79.15 (C, SCH_2CC), 80.01 (C, SCH_2CC), 123.16 (C, C-5), 151.08 (CH, C-8), 152.35 (C, C-

2), 154.24 (C, C-6), 160.50 (C, C-4); CIMS m/z 322 $[M+1]^+$ (6.5), 324 [M+1+2] (2.5), 167 $[M+1-SCH_2CCCH_2NC_4H_8]$ (9.5), 72 $[C_4H_8NH_2]^+$ (100); HRFABMS m/z $[M+H]^+$ calcd. for $C_{14}H_{17}ClN_5S$ 322.0893 found 322.0893.

***Synthesis of 2-chloro-6-benzylthio-7-methylpurine**6*

2-Chloro-6-benzylthio-7-methylpurine **6** was prepared from 2-chloro-7-methyl-6-purinethione **1A** (0.2 g, 1 mmol) by alkylation with benzyl chloride (0.25 g, 2 mmol) at room temperature in 4 % aqueous KOH solution (5 ml). After 30 min, the crude product was precipitated, filtered off, washed with water, and crystallized from benzene. It was obtained as a pale yellow solid (0.265 g, 91.5 %); mp 131–132 °C (benzene); ^{1}H NMR ($CDCl_3$), δ: 4.07, (s, 3H, NCH_3), 4.64 (s, 2H, CH_2), 7.38 (m, $5H_{arom}$), 7.98 (s, 1H, H-8); ^{13}C NMR (DMSO-d_6), δ: 33.39 (CH_2, SCH_2), 34.84 (CH_3, NCH_3), 122.72 (C, C-5), 127.98 (CH, p-CH), 128.99 (2CH, m-CH), 129.84 (2CH, o-CH), 137.12 (C, CCH_2), 150.79 (CH, C-8), 152.26 (C, C-2), 155.46 (C, C-6), 160.33 (C, C-4); EIMS m/z 290 $[M]^+$ (100), 292 [M+2] (28.5), 199 $[M-C_6H_5CH_2]^+$ (87); HREIMS m/z $[M^+]$ calcd. for $C_{13}H_{11}ClN_4S$ 290.0392 found 290.0385.

5. CHLORINATION AND AMINATION OF 2-CHLORO-6-BENZYLTHIO-7-METHYLPURINE 6

The chlorinolysis was carried out by passing chlorine gas into a stirred mixture composed of 2-chloro-6-benzylthio-7-methylpurine **6** (0.29 g, 1 mmol), 3 ml of chloroform, and 3 ml of 80 % acetic acid cooled at 5 °C. The passage of chlorine gas was continued for 30 min, and then the mixture was poured into 10 ml of ice water. The chloroform layer was separated, and aqueous layer was extracted with chloroform (2 × 5 ml). The chloroform extracts were combined, washed with water, dried over anhydrous sodium sulfate, and evaporated *in vacuo* to give an oil residue. In order to remove benzyl acetate and benzyl chloride, the residue was triturated with ice-cold dry ether and the ether layer was separated. The mixture of obtained oil (0.26 g, 1 mmol), and 25 % aqueous ammonia (2.5 ml) or diethylamine (0.22 g, 3 mmol) in 10 % NaOH solution or aniline (0.19 g, 2 mmol) in 2.5 ml of benzene was stirred at 40–45 °C for 2–6 h. The resulted solid was filtered off (in the case of the reaction with ammonia, the excess of ammonia was removed in vacuo and the residue was diluted with water up to volume 2.5 ml) to give compounds **9A–C**. Products were purified by a column chromatography (aluminum oxide, $CHCl_3$–EtOH, 99:1 v/v) to give:

2-Chloro-6-amine-7-methylpurine

9A was obtained as a white solid (0.15 g, 82 %); mp 248–249 °C (EtOH), lit. mp 250 °C (Adams and Whitmore, 1945); ^{1}H NMR (DMSO-d$_6$), δ: 3.97 (s, 3H, NCH_3), 7.46 (s, 2H, NH_2), 8.20 (s, 1H, H-8); ^{13}C NMR (DMSO-d$_6$), δ: 34.29 (CH_3, NCH_3), 111.12 (C, C-5), 147.27 (CH, C-8), 153.03 (C, C-2), 153.17 (C, C-6), 161.50 (C, C-4); EIMS m/z 183 $[M]^+$ (100), 185 [M+2] (29); HREIMS m/z $[M]^+$ calcd. for $C_6H_6ClN_5$ 183.0311 found 183.0307.

2-Chloro-6-diethylamine-7-methylpurine

9B was obtained as a white solid (0.20 g, 84 %); mp 106–107 °C (EtOH), lit. mp 107 °C (Adams and Whitmore, 1945); ^{1}H NMR (DMSO-d$_6$), δ: 1.17 (t, J = 7.7 Hz, 6H, 2 × CH_3CH_2N), 3.52 (q, J = 7.7 Hz, 4H, 2 × CH_3CH_2N), 3 0.94 (s, 3H, NCH_3), 8.37 (s, 1H, H-8); ^{13}C NMR (DMSO-d$_6$), δ: 12.93 (2CH_3, NCH_2CH_3), 35.63 (CH_3, NCH_3), 44.32 (2CH_2, NCH_2CH_3), 114.45 (C, C-5), 149.30 (CH, C-8), 151.81 (C, C-2), 154.70 (C, C-6), 163.03 (C, C-4); EIMS m/z 239 $[M]^+$ (19), 241 [M+2] (6); 210 [M-C_2H_5], (100); HREIMS m/z $[M]^+$ calcd. for $C_{10}H_{14}ClN_5$ 239.0937 found 239.0935.

2-Chloro-6-phenylamine-7-methylpurine

9C was obtained as a white solid (0.18 g, 69 %); mp 202–203 °C. (EtOH); ^{1}H NMR ($CDCl_3$), δ: 3.96 (s, 3H, NCH_3), 6.84 (s, 1H, NH), 7.23 (t, J = 7.20 Hz, 1H, p-C_6H_5), 7.40 (t, J = 7.20 Hz, 2H, m-C_6H_5), 7.52 (d, J = 7.20 Hz, 2H, o-C_6H_5), 7.90 (s, 1H, H-8); ^{13}C NMR (DMSO-d$_6$), δ: 34.63 (CH_3, NCH_3), 112.40 (C, C-5), 123.17 (2CH, m-CH), 124.61 (CH, p-CH), 129.06 (2CH, o-CH), 138.97 (C, CNH), 148.47 (CH, C-8), 149.75 (C, C-2), 152.29 (C, C-6), 162.07 (C, C-4); FABMS m/z 260 $[M+1]^+$ (100), 262 [M+1+2] (31.5); HREIMS m/z $[M]^+$ calcd. for $C_{12}H_{10}ClN_5$ 259.0625 found 259.0626.

6. GENERAL SYNTHESIS OF 2-SUBSTITUTED 7-METHYL-6-PURINESULFENAMIDE 10A–E

A solution of commercial NaOCl (14.5 %, 1.7 ml) and 1.7 ml of water was cooled to 0 °C in an ice–NaCl bath. Ammonium hydroxide (25 %, 2 ml) was similarly cooled and added with stirring to the bleach solution. The mixture was stirred at −5 to 0 °C for 15 min and then allowed to cool to 0 °C solution of 2-substituted 7-methyl-6-purinethione **1A–D** (2 mmol) in 10 % KOH (5 ml) was dropped. The mixture was stirred in stoppered flask for 30 min at temperature −5 to 0 °C. The reaction mixture was initially clear, light yellow solution, but after 30 min a solid began separating. After allowing the reaction mixture to stand at room temperature for an additional 1 h, the precipitated product was filtered off and washed with a small amount of cold water, followed by cold ethanol to yield purine 6-sulfenamide **10A–C, 10E**. The crude products were purified by a column chromatography (silica gel, $CHCl_3$, $CHCl_3$–EtOH, 99:1 v/v). The aqueous filtrate separated after precipitation of product **10E** was evaporated to dryness and extracted with absolute ethanol (3 × 5 ml). The ethanolic solvent was evaporated, and the residue was purified by column chromatography (aluminum oxide, $CHCl_3$–EtOH, 10:1 v/v) to give compound **10D**.

2-Chloro-7-methyl-6-purinesulfenamide

10A was obtained as a yellow solid (0.35 g, 87.5 %); mp 149–150 °C (EtOH); ^{1}H NMR (DMSO-d_6), δ: 3.98 (s, 3H, NCH_3), 4.41 (s, 2H, NH_2), 8.51 (s, 1H, H-8); ^{13}C NMR (DMSO-d_6), δ: 35.13 (CH_3, NCH_3), 120.85 (C, C-5), 150.41 (CH, C-8), 152.99 (C, C-4), 159.88, 164.43 (C, C-2, and C-6); FABMS m/z 216 $[M+1]^+$ (38), 218 [M+1+2] (12), 185 (100); HREIMS m/z $[M]^+$ calcd. for $C_6H_6ClN_5S$ 215.0032 found 215.0029.

2-Methoxy-7-methyl-6-purinesulfenamide

10B was obtained as a yellow solid (0.39 g, 92.5 %); mp 158–159 °C (EtOH); ^{1}H NMR (DMSO-d_6), δ: 3.93 (s, 6H, OCH_3, NCH_3), 4.20 (s, 2H, NH_2), 8.30 (s, 1H, H-8); ^{13}C NMR (DMSO-d_6), δ: 34.80 (CH_3, NCH_3), 54.81 (CH_3, OCH_3), 117.79 (C, C-5), 148.81 (CH, C-8), 160.55 (C, C-4), 161.72, 162.98 (C, C-2, and C-6); FABMS m/z 212 $[M+1]^+$ (100); HREIMS m/z $[M]^+$ calcd. for $C_7H_9N_5OS$ 211.0527 found 211.0523.

2-Methylthio-7-methyl-6-purinesulfenamide

10C was obtained as a yellow solid (0.385 g, 85 %); mp 164–165 °C (EtOH); ^{1}H NMR (DMSO-d_6), δ: 2.57 (s, 3H, SCH_3), 3.94 (s, 3H, NCH_3), 4.25 (s, 2H, NH_2), 8.35 (s, 1H, H-8); ^{13}C NMR (DMSO-d_6), δ: 14.32 (CH_3, SCH_3), 34.86 (CH_3, NCH_3), 119.06 (C, C-5), 151.57 (CH, C-8), 159.32 (C, C-4), 161.52, 164.01 (C, C-2, and C-6); FABMS m/z 228 $[M+1]^+$ (100); HREIMS m/z $[M]^+$ calcd. for $C_7H_9N_5S_2$ 227.0299 found 227.0299.

7-Methyl-6-sulfenamide-2-purinethione

10D was obtained as a yellow solid (0.07 g, 16.5 %); mp 254–256 °C (EtOH); ^{1}H NMR (DMSO-d_6), δ: 4.20(s, 3H, NCH_3), 4.37 (s, 2H, NH_2), 8.18 (s, 1H, H-8), 10.83 (s, 1H, NH); ^{13}C NMR (DMSO-d_6), δ: 34.68 (CH_3, NCH_3), 121.30 (C, C-5), 147.96 (CH, C-8), 157.40 (C, C-4), 169.22, 173.95 (C, C-2, and C-6); FABMS m/z 214 [M+1] (100); HREIMS m/z $[M]^+$ calcd. for $C_6H_7N_5S_2$ 213.0143 found 213.0141.

7-Methyl-2,6-purinedisulfenamide

10E was obtained as a yellow solid (0.315 g, 69 %); mp 185–186 °C (EtOH); ^{1}H NMR (DMSO-d_6), δ: 3.94(s, 3H, NCH_3), 4.03 (s, 2H, NH_2), 4.18 (s, 2H, NH_2), 8.35 (s, 1H, H-8); ^{13}C NMR (DMSO-d_6), δ: 34.82(CH_3, NCH_3), 119.09 (C, C-5), 148.72 (CH, C-8), 159.45 (C, C-4), 161.26, 169.60 (C-2 and C-6); HSQC NMR, δ: 8.35 (H-8) correlated with 148.72 (C-8), 3.94 (H, NCH_3) correlated with 34.82 (C, NCH_3); HMBC NMR, δ: 8.35 (H-8) correlated with 34.82 (C, NCH_3), 119.09 (C-5), and 159.45 (C-4), 3.94 (H, NCH_3) correlated with 119.09 (C-5) and with 148.72 (C-8); FABMS m/z 229 [M+1] (100); HREIMS m/z $[M]^+$ calcd. for $C_6H_8N_6S_2$ 228.0252 found 228.0255.

7. GENERAL SYNTHESIS OF 2-SUBSTITUTED 7-METHYL-6-PURINESULFONAMIDE 8A–C, 8E

A solution of 3-chloroperbenzoic acid (MCPBA) (77 %, 0.86 g, 4 mmol) in 10 ml of ethanol was added dropwise for 10 min to a solution of sulfonamides **10A–C** and **10E** (1 mmol) in 79 ml of absolute ethanol. The mixture was stirred at room temperature for 1.5 h, and a precipitated solid was filtered off to give crude products **8A–C** and **8E**. The ethanolic filtrate was evaporated to dryness and extracted with ether to remove 3-chlorobenzoic acid. The solid residue together with crude product was purified by a column chromatography (aluminum oxide, $CHCl_3$, $CHCl_3$–EtOH, 9:1 v/v) to give also compounds **8A–C**, **8E**.

2-Chloro-7-methyl-6-purinesulfonamide

8A was obtained as a white solid (0.18 g, 73 %); mp 174–175 °C (EtOH); ^{1}H NMR (DMSO-d_6), δ: 4.07 (s, 3H, NCH_3), 6.91 (s, 2H, NH_2), 8.82 (s, 1H, H-8); ^{13}C NMR (DMSO-d_6), δ: 36.46 (CH_3, NCH_3), 121.55 (C, C-5), 151.93 (CH, C-8), 153.88 (C, C-4), 160.39, 164.62 (C, C-2, and C-6); EIMS m/z 247 $[M]^+$ (1.5), 249 [M+2] (0.5); 183 [M-SO_2] (14), 168 [M-SO_2NH_2] (100); HREIMS m/z $[M]^+$ calcd. for $C_6H_6ClN_5O_2S$ 246.9930 found 246.9927.

2-Methoxy-7-methyl-6-purinesulfonamide

8B was obtained as a white solid (0.185 g, 76 %); mp 203–204 °C. (EtOH); ^{1}H NMR (DMSO-d_6), δ: 3.88 (s, 3H, OCH_3), 4.04 (s, 3H, NCH_3), 7.99 (s, 2H, NH_2), 8.70 (s, 1H, H-8); ^{13}C NMR (DMSO-d_6), δ: 35.86 (CH_3, NCH_3), 55.59 (CH_3, OCH_3), 117.32 (C, C-5), 133.80 (CH, C-8), 153.60 (C, C-4), 157.55, 166.54 (C, C-2, and C-6); EIMS m/z 243 $[M]^+$ (26), 180 [M-SO_2] (55.5), 162 [M-SO_2NH_2] (8), 148 (100); HREIMS m/z $[M]^+$ calcd. for $C_7H_9N_5O_3S$ 243.0426 found 243.0419.

2-Methylthio-7-methyl-6-purinesulfonamide

8C was obtained as a white solid (0.17 g, 66 %); mp 198–199 °C (EtOH); ^{1}H NMR (DMSO-d_6), δ: 2.56 (s, 3H, SCH_3), 4.03 (s, 3H, NCH_3), 7.86 (s, 2H, NH_2), 8.75 (s, 1H, H-8); ^{13}C NMR (DMSO-d_6), δ: 15.59 (CH_3, SCH_3), 34.04 (CH_3, NCH_3), 116.45 (C, C-5), 129.28 (CH, C-8), 133.20 (C, C-4), 147.01, 166.54 (C, C-2, and C-6); CIMS m/z 260 [M+1] (22), 196 [M+1-SO_2] (64), 180 (M+1-SO_2NH_2] (100); HREIMS m/z $[M]^+$ calcd. for $C_7H_9N_5O_2S_2$ 259.0198 found 259.0195.

7-Methyl-2,6-purinedisulfonamide

8D was obtained as a white solid (0.23 g, 79 %); mp 214–215 °C (EtOH); ^{1}H NMR (DMSO-d_6), δ: 4.17(s, 3H, NCH_3), 7.01 (s, 2H, NH_2), 7.69 (s, 2H, NH_2), 8.97 (s, 1H, H-8); ^{13}C NMR (DMSO-d_6), δ: 36.30 (CH_3, NCH_3), 122.52 (C, C-5), 154.76 (CH, C-8), 158.11 (C, C-4), 159.98, 162.99 (C, C-2, and C-6); EIMS m/z 292 $[M]^+$ (9), 229 (M-SO_2] (7), 149 [M-SO_2 NH_2] (22), 57 (100); HREIMS m/z $[M]^+$ calcd. for $C_6H_8N_6O_4S_2$ 292.0048 found 292.0044.

8. ANTIPROLIFERATIVE ASSAY IN VITRO

8.1 Cell culture

Compounds were evaluated for their anticancer activity using three cultured cell lines: SNB-19 (human glioblastoma, DSMZ—German Collection of Microorganisms and Cell Cultures, Braunschweig, Germany), C-32 (human amelanotic melanoma, ATCC—American Type Culture Collection, Manassas, VA, USA), T47D (human ductal breast epithelial tumor cell line, ATCC, Manassas, VA, USA), and HFF-1 (human fibroblast cell line, ATCC, Manassas, VA, USA). The cultured cells were kept at 37 °C and 5 % CO_2. The cells were seeded (1×10^4 cells/well/100 µl DMEM supplemented with 10 % FCS and streptomycin and penicillin) using 96-well plates (Corning).

8.2 Proliferation

In recent years, tetrazolium salts have been described which can be used for the measurement of cell proliferation and viability. The tetrazolium salts are cleaved to formazan by cellular enzymes. An expansion in the number of viable cells results in an increase in the overall activity of mitochondrial dehydrogenases in the sample. This augmentation in enzyme activity leads to an increase in the amount of formazan dye formed, which directly correlates with the number of metabolically active cells in the culture. The formazan dye produced by metabolically active cells is quantified by a scanning ELISA reader by measuring the absorbance of the dye solution at appropriate wavelengths (λ = 420–480 nm with a reference wavelength λ = 600 nm).

8.3 WST-1 assay

Antiproliferative effect of compounds was determined using the Cell Proliferation Reagent WST-1 assay (Roche Diagnostics, Mannheim, Germany). After exposure to tested compounds (at concentrations between 0 and 100 μg/ml) for 72 h, cells were incubated with WST-1 (10 μl) for 1 h, and the absorbance of the samples against a background control was read at 450 nm using with a reference wavelength λ = 600 nm a microplate reader. Results are expressed as means of at least two independent experiments performed in triplicate.

ACKNOWLEDGMENTS

The synthesis and the structure elucidation are supported by the Medical University of Silesia (Grant KNW-1-004/K/4/0).

REFERENCES

1. Adams RR, Whitmore FC (1945) Heterocyclic basic compounds. Dialkylaminoalkylaminopurines. J Am Chem Soc 67:1271–1273
2. Aldinucci A, Biagioli T, Manuelli C, Repice AM, Massacesi L, Ballerini C (2010) Modulating dendritic cells (DC) from immunogenic to tolerogenic responses. J Neuroimmunol 218:28–35
3. Ashfaq M, Shah SSA, Najjam T, Shaheen S, Riviera G (2013) Synthetic routes of sulfonamide derivatives; a brief review. Mini Rev Org Chem 10:160–170
4. Azab ME, Youssef MM, El-Bordany EA (2013) Synthesis and antibacterial evaluation of novel heterocyclic compounds containing a sulfonamide moiety. Molecules 18:832–844
5. Beaman AG (1963) Heterocyclic nitrogen compounds containing sulfonyl fluoride groups and method of producing some. USA Patent 3(074):948Google Scholar
6. Beaman AG, Robins RK (1961) Potential purine antagonists. Synthesis and reactions of some purinesulfonyl fluorides. J Am Chem Soc 83:4038–4044
7. Beaman AG, Tantz W, Duschinsky R, Grunberg E (1966) Purine sulfonamides. J Med Chem 9:373–378
8. Brathe A, Gundersen LL, Nissen-Meyer J, Rise F, Spilsberg B (2003) Cytotoxic activity of 6-alkynyl- and 6-alkenylpurines. Bioorg Med Chem Lett 13:877–880Pub MedCross Ref Google Scholar
9. Broom AD, Gall AA, Meyer RB (1997) Oligonucleotides containing n-methyl thiolated bases having antiviral activity. US Patent 5652359 A

10. Chang-Hyun O, Hee-Kwon K, Su-Chul L, Changsok O, Boem-Seok Y, Seok RJ (2001) Synthesis and biological propertis of C-2, C-8, N-9 substituted 6-(3-chloroanilino)-purine derivatives cyclin—dependent kinase inhibitors. Arch Pharm Med Chem 334:345–350
11. Colombel JF, Sandborn WJ, Reinisch W, Mantzaris GJ, Kornbluth A, Rachmilewitz D, Lichtiger S, D'Haens G, Diamond RH, Broussard DL, Tang KI, van der Wande CJ, Rutgeerts P (2010) Infliximab, azathioprine, or combination therapy for Crohn's disease. N Engl J Med 362:1383–1395
12. Corrales RCNR, de Souza NB, Pinheiro LS, Abramo C, Coimbra ES, da Silva AD (2011) Thiopurine derivatives containing triazole and steroid: synthesis, antimalarial and antileishmanial activities. Biomed Pharmacother 65:198–203
13. Daehn EI, Karran P (2009) Immune effector cells produce lethal DNA damage in cells treated with a thiopurine. Cancer Res 69:2393–2399
14. El-Sayed NS, El-Bendary ER, El-Ashry SM, El-Kerdawy MM (2011) Synthesis and antitumor activity of new sulfonamide derivatives of thiadiazolo[3,2-α]pyrimidines. Eur J Med Chem 46:3714–3720
15. Guzel O, Innocenti A, Scozzafava A, Salman A, Supuran CT (2009) Carbonic anhydrase inhibitors. Phenacetyl-, pyridylacetyl- and thienylacetyl substituted aromatic sulfonamides act as potent and selective isoform VII inhibitors. Bioorg Med Chem Lett 19:3170–3173
16. Harada H, Asano O, Kawata T, Inoue T, Horizoe T, Yasuda N, Nagata K, Murakami M, Nagaoka J, Kobayashi S, Tanaka I, Abe S (2001) 2-Alkynyl-8-aryladenines possessing an amide moiety: their synthesis and structure–activity relationships of effects on hepatic glucose production induced via agonism of the A_{2B} adenosine receptor. Bioorg Med Chem 9:2709–2726
17. Hawwa AF, Collier PS, Millership JS, McCarthy A, Dempsey S, Cairns C, McElnay JC (2008) Population pharmacokinetic and pharmacogenetic analysis of 6-mercaptopurine in paediatric patients with acute lymphoblastic leukaemia. Br J Clin Pharmacol 66:826–837
18. Hurley TJ, Robinson MA (1965) The reaction of chloramine with mercaptopyrimidine and mercaptopyrimidine derivatives. J Med Chem 8:888–889
19. Joseph P, Turtant F, Ouahrani-Bettache S, Montero JL, Nishimori I, Minakuchi T, Vullo D, Scozzafava A, Kohler S, Winum JY, Supuran CT (2010) Cloning, characterization, and inhibition studies of a β-carbonic anhydrase from Brucelle suis. J Med Chem 53:2277–2285
20. Kowalska A, Pluta K (2008) Synthesis of the azathioprine analogs. Heterocycles 75:555–569
21. Kowalska A, Pluta K (2012) Synthesis of 2- and 6-(dialkylaminoalkylthio)- and 2,6-bis(dialkylaminoalkylthio)-7-methylpurines. Heterocycles 85:2473–2483
22. Kowalska A, Pluta K, Suwińska K (2009) Synthesis of isomeric analogs of azathioprine. Heterocycles 78:2455–2466
23. Lech-Maranda E, Robak E, Korycka A, Robak T (2006) Pharmacological and clinical studies on purine nucleoside analogs—new anticancer agents. Mini Rev Med Chem 6:575–581
24. Mirian M, Zarghi A, Sadeghi S, Tabaraki P, Tavallaee M, Dadrass O, Sadeghi-aliabadi H (2011) Synthesis and cytotoxic evaluation of some novel sulfonamide derivatives against a few human cancer cells. Iran J Pharm Res 10:741–748
25. Miron T, Arditti F, Konstantinovski L, Rabinkov A, Mirelman D, Berrebi A, Wilchek M (2009) Novel derivatives of 6-mercaptopurine: synthesis, characterization and antiproliferative activities of S-allylmercaptopurines. Eur J Med Chem 44:541–550
26. Narasaiah T, Rao DS, Ramana KV, Adam S, Raju CN (2012) Synthesis of new sulfonamide derivatives of tryptamine and their antimicrobial activity. Der Pharma Chem 4:1582–1590
27. Prima FD, Shenoy A, Moses SR, Shabaraya AR (2013) Antitumor activity of mercaptopurine in combination with trikatu and gomutra on 20-methylcholantrene induced carcinogenesis. J App Pharm Sci 3:20–23
28. Rao PV, Ravindhranath K, Kumar KR (2013) Antibacterial activity of novel substituted mercaptopurine derivatives. Inst J Pharm Biomed Res 4:127–131

29. Rehman A, Afroz S, Abbasi MA, Tanveer W, Khan KM, Ashraf M, Ahmad I, Afzal I, Ambreen N (2012) Synthesis, characterization and biological screening of sulfonamides derived form 2-phenylethylamine. Pak J Pharm Sci 25:809–814
30. Relling MV, Gardner EE, Sandborn WJ, Schmiegelow K, Pui CH, Yee SW, Stein CM, Carrillo M, Evans WE, Hicks JK, Schwab M, Klein TE (2013) Clinical pharmacogenetics implementation consortium guidelines for thiopurine methyltransferase genotype and thiopurine dosing. Clin Pharmacol Ther 93:324–325
31. Revankar GR, Hanna NB, Imamura N, Lewis AF, Larson SB, Finch RA, Avery TL, Robins RK (1990) Synthesis and in vivo antitumor activity of 2-amino-9H-purine-6-sulfenamide, -sulfinamide, and sulfonamide and related purine ribonucleosides. J Med Chem 33:121–128
32. Robins RK, Revankar GR, Hanna NB (1991) 6-Sulfenamide, 6-sulfinamide and 6-sulfonamide purines, purine nucleosides, purine nucleotides, pharmaceutical compositions, and processes of making. USA Patent 5(026):836
33. Schoenwald RD, Eller MG, Dixon JA, Barfknecht CF (1984) Topical carbonic anhydrase inhibitors. J Med Chem 27:810–812
34. Schow SR, Mackman RL, Blum CL, Brooks E, Horsma AG, Joly A, Kerwar SS, Lee G, Shiffman D, Nelson MG, Wang X, Wick MM, Zhang X, Lum RT (1997) Synthesis and activity 2,6,9-trisubstituted purines. Bioorg Med Chem Lett 7:2697–2702
35. Sławiński J, Szafrański K, Vullo D, Supuran CT (2013) Carbonic anhydrase inhibitors. Synthesis of heterocyclic 4-substituted pyridine-3-sulfonamide derivatives and their human cytosolic isozymes I and II and transmembrane tumor-associated isozymes IX and XII. Eur J Med Chem 69:701–710
36. Steurer M, Pall G, Richards S, Schwarzer G, Bohlius J, Greil R (2006) Single-agent purine analogues for the treatment of chronic lymphocytic leukaemia: a systematic review and meta-analysis. Cancer Treat Rev 32:377–389
37. Veisi H, Ghorbani-Vaghei R, Mahmoodi J (2011) Poly(*N*, *N*′-dichloro-*N*-ethyl-benzene-1,3-disulfonamide) and *N*, *N*, *N*′, *N*′-tetrachlorobenzene-1,3-disulfonamide as efficient reagents to direct oxidative conversion of thiols and disulfide to sulfonyl chlorides. Bull Korean Chem Soc 32:3692–3695
38. Volpini R, Costanzi S, Lambertucci C, Vittori S, Klotz KN, Lorenzen A, Cristalli G (2001) Introduction of alkynyl chains on C-8 of adenosine led to very selective antagonists of the A_3adenosine receptor. Bioorg Med Chem Lett 11:1931–1934
39. Zablocki AJ, Palle VP, Elzein EO, Nudelman G (2001) Propargyl phenyl ether A_{2A} receptor agonists. US Patent 6180615 B1
40. Zinic B, Krizmanic I, Vikic-Topic D, Srzic D, Zinic M (2001) Synthesis, NMR and MS study of novel N-sulfonylated purine derivatives. Croat Chem Acta 74:399–414

CHAPTER 15

Synthesis of quinoline/ naphthalene-containing azaphenothiazines and their potent in vitro antioxidant properties

Małgorzata Jeleń, Eugenia I. Bavavea, Maria Pappa ·Angeliki P. Kourounakis, Beata Morak-Młodawska, Krystian Pluta

1.Department of Organic Chemistry, School of Pharmacy with the Division of Laboratory MedicineThe Medical University of SilesiaSosnowiecPoland

2.Department of Medicinal Chemistry, Faculty of PharmacyUniversity of AthensAthensGreece

ABSTRACT

New tetracyclic and pentacyclic azaphenothiazines containing one or two quinoline rings instead of benzene rings were obtained in the original reactions of isomeric diquinodithiins, dichlorodiquinolinyl sulfides, and disulfide with aromatic amines. The type of ring fusion in the azaphenothiazine system was concluded from the ^{1}H NMR spectra. The obtained azaphenothiazines were evaluated in vitro for their antioxidant activity on rat hepatic microsomal membranes for protection of non-enzymatic lipid peroxidation promoted by the Fe^{2+}/ascorbic acid redox system. Most compounds exhibited a very significant antioxidant activity with IC_{50} values between 1 and 23 μM. The degree of antioxidant activity depends on the lipophilicity and molecular size as well as the (non)substitution of the thiazine nitrogen atom and type of ring system fusion. It is the first time to our knowledge that azaphenothiazines are shown to exhibit such potent antioxidant activity.

Keywords: Phenothiazine derivatives, NH-azaphenothiazines, Quino-naphthothiazines, Diquinothiazines, Lipid, peroxidation, Lipophilicity

1. INTRODUCTION

Phenothiazines are an important class of drugs exhibiting antipsychotic, antihistaminic, antitussive, and anti-emetic activities (Gupta and Kumar, 1988). The most significant modifications of the phenothiazine structure are the introduction of new pharmacophoric substituents at the thiazine nitrogen atom and the substitution of the benzene rings with other homoaromatic or heteroaromatic ones. Recently studied phenothiazines exhibit promising antibacterial, antifungal, anticancer, antiviral, anti-inflammatory, antimalarial, antifilarial, trypanocidal, anticonvulsant, analgesic, immunosuppressive, and multidrug resistance reversal properties (Aaron *et al.*, 2009; Dasgupta *et al.*, 2008; Motohashi *et al.*, 2006; Pluta *et al.*, 2011). In our study of new azaphenothiazines, we elaborated the synthesis of new types of phenothiazines containing the heterocyclic rings of pyridine or quinoline. Some of those azaphenothiazines exhibited promising immunosuppressive and anticancer activities against cell lines of ten types of human cancer in vitro: leukemia, non-small cell lung cancer, melanoma, as well as colon, CNS, ovarian, renal, prostate, breast, and skin cancer (Jeleń *et al.*, 2013; Pluta *et al.*, 2010; Zimecki *et al.*, 2009).

Free radicals, generated in many redox processes, may induce oxidative damage of proteins, lipids, and DNA. They affect living cells and mediate the pathogenesis of many chronic diseases, such as atherosclerosis, Parkinson's and Alzheimer's diseases, stroke, and arthritis, acting by various mechanisms. A recent trend in the field of antioxidant development focuses on multipotent antioxidant agents that not only can prevent biological substrates from radical induced oxidative damage but also possess additional pharmacological properties (Zhang *et al.*, 2006). The study of antioxidant activity among N-heterocycles has attracted attention. One such heterocyclic structural scaffold is the 1,4-thiazine ring present in the multi-target phenothiazines. Therefore, recent reports on promising antioxidant compounds deal with classical and new phenothiazines (Asghar *et al.*, 2012; Borges *et al.*, 2010; Liu *et al.*, 2009;

Naik *et al.*, 2012;) and their derivatives, benzothiazines (Matralis *et al.*, 2011), and azaphenothiazines (Kumar *et al.*, 2010; Morak-Młodawska *et al.*, 2010).

Our previous work (Morak-Młodawska *et al.*, 2010) revealed that tricyclic azaphenothiazines being dipyridothiazines have a variable degree of antioxidant activity depending on substitution at the thiazine nitrogen atom, with the unsubstituted compound being the most active. In this study, we obtained eleven tetracyclic and pentacyclic (linearly and angularly fused) azaphenothiazines containing one or two quinoline rings instead of the benzene rings and determined their antioxidant properties to find an influence of the number of rings, their type of fusion, and their substituents.

2. MATERIALS AND METHODS

2.1 General techniques

Melting points were determined in open capillary tubes on a Boetius melting point apparatus and were uncorrected. The ^{1}H NMR spectra were recorded on a Bruker Fourier 300 and a Bruker DRX spectrometer at 500 MHz in $CDCl_3$ and DMSO-d_6 with tetramethylsilane as the internal standard. The ^{13}C NMR spectra were recorded at 75 MHz. Electron impact (EI MS) mass spectra were run on a Finnigan MAT 95 spectrometer at 70 eV. The thin-layer chromatography was performed on aluminum oxide 60 F_{254} neutral (type E, Merck 1.05581) with CH_2Cl_2 and on silica gel 60 F_{254} (Merck 1.05735) with $CHCl_3$-EtOH (10:1 v/v) as eluents.

2.2 Synthesis of substrates 1, 2, 7, 8, 10, and 11

The substrates for the title compounds, i.e., diquinodithiins **1**, **7**, **10**, sulfides **8**, **11**, and disulfide **2**, were obtained as described previously (Nowak *et al.*, 2002, 2003, 2007; Pluta, 1994).

2.3 Quino[3,2-b]benzo[1,4]thiazines (3A–C)

2.3.1 From diquino-1,4-dithiin 1

A mixture of diquino-1,4-dithiin **1** (0.16 g, 0.5 mmol) and hydrochloride of aniline, or *p*-chloroaniline or *p*-methoxyaniline (2.5 mmol) was finely powdered together and then heated on an oil bath at 200–205 °C for 4 h and after cooling water was added (10 ml) and the insoluble solid was filtered off. The filtrate was alkalized with 5 % aqueous sodium hydroxide to pH 10, and the resulting solid was filtered off and washed with water. The combined solids were purified by column chromatography (silica gel, $CHCl_3$) to give quinobenzothiazines **3A–C**.

2.3.2 6*H*-Quinobenzothiazine (3A)

0.06 g (24 %), yellow, mp 169–170 °C (mp 169–170 °C, Jeleń and Pluta, 2009). ^{1}H NMR ($CDCl_3$) δ: 6.62 (m, 1H, H-7), 6.87 (m, 1H, H-9), 7.03 (m, 2H, H-8, H-10), 7.26 (t, 1H, H-2), 7.47 (m, 2H, H-1, H-3), 7.53 (s, 1H, H-12), 7.56 (d, 1H, H-4). ^{13}C NMR ($CDCl_3$) δ: 115.57 (C-7), 116.49 and 116.69 (C-10a, C-11a), 122.95 (C-9), 124.19 (C-2), 125.86 (C-10), 126.04 and 126.45 (C-1, C-8), 126.56 (C-12a), 127.57 (C-4), 129.52 (C-3), 131.69 (C-12), 138.45 (C-6a), 145.40 (C-4a), 150.98 (C-5a).

2.3.3 6H-9-Chloroquinobenzothiazine (3B)

0.08 g (28 %), yellow, mp 224–225 °C (mp 224–225 °C, Jeleń and Pluta, 2009). ^{1}H NMR ($CDCl_3$) δ: 6.63 (d, 1H, H-7), 6.99 (s, 1H, H-10), 7.01 (d, 1H, H-8), 7.33 (t, 1H, H-2), 7.51 (d, 1H, H-1), 7.52 (t, 1H, H-3), 7.59 (d, 1H, H-4), 7.60 (s, 1H, H-12). ^{13}C NMR ($CDCl_3$) δ: 115.80 (C-11a), 116.71 (C-7), 118.19 (C-10a), 124.84 and 124.91 (C-8, C-10), 125.65 (C-2), 126.13 (C-12a), 126.61 (C-1), 127.59 (C-4), 128.56 (C-9), 130.31 (C-3), 132.35 (C-12), 136.29 (C-6a), 143.81 (C-4a), 150.04 (C-5a),

2.3.4 6*H*-9-Methoxyquinobenzothiazine (3C)

0.09 g (32 %), orange, mp 159–160 °C.

^{1}H NMR ($CDCl_3$) δ 3.76 (s, 3H, CH_3), 6.54 (d, 1H, H-7), 6.63 (d, 1H, H-10), 6.76 (d, 1H, H-8), 7.29 (t, 1H, H-2), 7.46 (d, 1H, H-1), 7.52 (t, 1H, H-3), 7.55 (s, 1H, H-12), 7.57 (d, 1H, H-4). ^{13}C NMR ($CDCl_3$) δ: 111.59 (C-10), 113.22 (C-8), 116.41 (C-11a), 116.82 (C-7), 117.39 (C-10a), 124.36 and 124.49 (C-1, C-2), 125.80 (C-12a), 126.55 (C-4), 130.10 (C-3), 130.60 (C-6a), 132.07 (C-12), 143.40 (C-4a), 150.36 (C-5a), 156.12 (C-9). EIMS *m/z*: 280 (M^+, 100), 265 (M-CH_3, 90). Anal. Calcd. for $C_{16}H_{12}N_2OS$: C, 68.55; H, 4.31; N, 9.99. Found: C, 68.45; H, 4.36; N, 9.82.

2.3.5 From 2,2′-dichloro-3,3′-diquinolinyl disulfide (2)

A solution of disulfide **2** (0.20 g, 0.5 mmol) and *p*-methoxyaniline (0.25 g, 2 mmol) in monomethyl ether of diethylene glycol (MEDG) (5 ml) was refluxed for 3 h. After cooling, the solution was poured into water (20 ml) and alkalized with 5 % aqueous sodium hydroxide to pH 10. The resulting solid was filtered off, washed with water, and purified by column chromatography (silica gel, $CHCl_3$) to give 0.18 g (64 %) of 6*H*-9-methoxyquinobenzothiazine (**3C**).

2.4 Quino[3,2-b]naphtho[1′,2′-e][1,4]thiazine (4)

Diquinodithiin **1** (0.16 g, 0.5 mmol) was finely powdered together with 1-naphthylamine hydrochloride (0.45 g, 2.5 mmol) on an oil bath at 200–205 °C for 4 h. After cooling, the solution was poured into water (10 ml) and alkalized with 5 % aqueous sodium hydroxide to pH 10. The resulting solid was filtered off, washed with water, and purified by column chromatography (Al_2O_3, $CHCl_3$) to give 0.08 g (27 %) of 14*H*-quinonaphthothiazine (**4**), orange, mp 147-148 °C.

^1H NMR ($CDCl_3$) δ: 7.01 (d,1H, H-6), 7.30 (t, 1H, H-10), 7.47 (m, 4H, H-3, H-4, H-5, H-9), 7.52 (t, 1H, H-2), 7.56 (s, 1H, H-8), 7.60 (t, 1H, H-11), 7.64 (d, 1H, H-12), 7.75 (d, 1H, H-1). ^{13}C NMR ($CDCl_3$) δ: 110.98 (C-6a), 116.91 (C-7a), 118.43 (C-1), 121.89 (C-14b), 122.87 (C-6), 123.70 (C-5), 124.49 (C-10), 125.93, 126.45 and 126.83 (C-2, C-3, C-9), 126.90 (C-8a), 128.92 and 129.65 (C-4, C-12), 131.54 (C-11), 132.55 (C-4a), 133.04 (C-8), 135.07 (C-14a), 145.23 (C-12a), 150.98 (C-13a). EIMS *m/z*: 300 (M^+, 100), 268 (M-S, 45). Anal. Calcd. for $C_{19}H_{12}N_2S$: C, 75.97; H, 4.03; N, 9.33. Found: C, 75.82; H, 4.07; N, 9.21.

2.5 Quino[3,2-b]naphtho[2′,1′-e][1,4]thiazine (5)

Diquinodithiin **1** (0.16 g, 0.5 mmol) was finely powdered together with 2-naphthylamine hydrochloride (0.45 g, 2.5 mmol) on an oil bath at 200–205 °C for 4 h. After cooling, the solution was poured into water (10 ml) and alkalized with 5 % aqueous sodium hydroxide to pH 10. The resulting solid was filtered off, washed with water, and purified by column chromatography (Al_2O_3, $CHCl_3$) to give 0.12 g (40 %) of 7*H*-quinonaphthothiazine (**5**), greenish, mp 244-245 °C.

^1H NMR ($CDCl_3$) δ: 7.06 (d, 1H, H-6), 7.37 (t, 1H, H-11), 7.47 (t, 1H, H-3), 7,57 (m, 3H, H-2, H-10, H-12), 7.65 (d, 1H, H-5), 7.66 (d, 1H, H-4), 7.72 (s, 1H, H-13), 7.80 (m, 2H, H-9, H-1). ^{13}C NMR ($CDCl_3$) δ: 107.94 (C-14a), 115.77 (C-13a), 116.04 (C-6), 121.32 (C-1), 123.33, 123.66 and 123.89 (C-3, C-9, C-11), 125.23 (C-12a), 125.62 (C-2), 126.36, 126.99 and 127.56 (C-4, C-5, C-12), 128.73 (C-4a), 129.22 (C-10), 129.62 (C-14b), 131.51 (C-13), 133.54 (C-6a), 142.13 (C-8a), 149.64 (C-7a). EIMS *m/z*: 300 (M^+, 100), 268 (M-S, 50). Anal. Calcd. for $C_{19}H_{12}N_2S$: C, 75.97; H, 4.03; N, 9.33. Found: C, 75.88; H, 4.05; N, 9.19.

2.6 Diquino[3,2-b;6′,5′-e][1,4]thiazine (6)

Diquinodithiin **1** (0.16 g, 0.5 mmol) was finely powdered together with 6-aminoquinoline hydrochloride (0.46 g, 2.5 mmol) on an oil bath at 200–205 °C for 4 h. After cooling, the solution was poured into water (10 ml) and alkalized with 5 % aqueous sodium hydroxide to pH 10. The resulting solid was filtered off, washed with water, and purified by column chromatography (Al_2O_3, $CHCl_3$) to give 0.10 g (33 %) of 7*H*-diquinothiazine (**6**), brown, mp 260–261 °C.

^1H NMR ($CDCl_3$) δ: 7.44 (t, 1H, H-11), 7.49 (d, 1H, H-6), 7.57 (m, 2H, H-2, H-12), 7.64 (t, 1H, H-10), 7.70 (d, 1H, H-9), 7.75 (s, 1H, H-13), 8.10 (d, 1H, H-5), 8.19 (d, 1H, H-1), 8.90 (d, 1H, H-3). ^{13}C NMR ($CDCl_3$) δ: 107.62 (C-14a), 114.59 (C-13a), 119.33 (C-6), 120.76 (C-2), 124.05 (C-11), 124.37 and 125.45 (C-12a, C-14b), 125.65 (C-12), 128.27, 129.24, 129.62 and 129.64 (C-1, C-5, C-9, C-10), 131.80 (C-13), 134.54 (C-6a), 144.53 (C-7a), 147.55 (C-3), 149.49 and 149.55 (C-4a, C-8a). EIMS *m/z*: 301 (M^+, 100), 269 (M-S, 45). Anal. Calcd. for $C_{18}H_{11}N_3S$: C, 71.74; H, 3.68; N, 13.94. Found: C, 71.59; H, 3.71; N, 13.72.

2.7 Diquino[3,2-b;2′,3′-e][1,4]thiazines (9)

2.7.1 6*H*-Diquinothiazine 9A

This compound was obtained in the reaction of diquinodithiin **7** with acetamide (Nowak *et al.*, 2007), orange, mp > 300 °C (mp > 300 °C, Nowak *et al.*, 2007). ^1H NMR ($CDCl_3$) δ: 7.42 (t, 2H, H-2, H-10), 7.55 (d, 2H, H-1, H-11), 7.62 (t, 2H, H-3, H-9), 7.72 (s, 2H, H-12, H-14), 7.86 (d, 2H, H-4, H-8). ^{13}C NMR (DMSO-d_6) δ: 124.83 (C-12a, C-13a), 127.29 (C-2, C-10), 128.00 (C-11a, C-14a), 128.16 and 128.28 (C-1, C-11 and C-4, C-8), 131.29 (C-3, C-9), 135.26 (C-12, C-14), 146.58 (C-4a, C-7a), 156.22 (C-5a, C-6a).

2.7.2 6-(*p*-Fluorophenyl)diquinothiazine (9B)

From diquinodithiin 7

Diquinodithiin **7** (0.16 g, 0.5 mmol) was finely powdered together with *p*-fluoroaniline hydrochloride (0.37 g, 2.5 mmol), and the mixture was heated on an oil bath at 200–205 °C for 3 h. After cooling, water (10 ml) was added to the reaction mixture and the resulting solid was filtered off, washed with water, air-dried, and purified by column chromatography (Al_2O_3, CH_2Cl_2) to give 0.14 g (35 %) of 6-(*p*-fluorophenyl)diquinothiazine (**9B**), yellow, mp 248–249 °C.

From 2,2′-dichloro-3,3′-diquinolinyl sulfide 8

A solution of sulfide **8** (0.18 g, 0.5 mmol) and *p*-fluoroaniline (0.17 g, 1.5 mmol) in MEDG (5 ml) was refluxed for 3 h. After cooling, the solution was poured into water (20 ml) and alkalized with 5 % aqueous sodium hydroxide to pH = 10. The resulting solid was filtered off, washed with water, and purified by column chromatography (Al_2O_3, CH_2Cl_2) to give 0.16 g (81 %) 6-(*p*-fluorophenyldiquinothiazine (**9B**), yellow, mp 248–249 °C.

^{1}H NMR ($CDCl_3$) δ: 7.31 (m, 4H, H-2, H-10, C_6H_2), 7.47 (m, 4H, H-3, H-9, C_6H_2), 7.56 (d, 2H, H-1, H-11), 7.67 (d, 2H, H-4, H-8), 7.83 (s, 2H, H-12, H-14). ^{13}C NMR ($CDCl_3$) δ: 115.85 (J = 22.6 Hz, *m*-C of C_6H_4F), 115.98 (C-12a, C-13a), 125.16 (C-2, C-10), 125.78 (C-11a, C-14a), 125.96 (C-1, C-11), 128.07 (C-4, C-8), 129.37 (C-3, C-9), 132.07 (C-12, C-14), 132.40 (J = 7.5 Hz, *o*-C of C_6H_4F), 135.59 (J = 2.5 Hz, *ipso*-C of C_6H_4F), 145.13 (C-4a, C-7a), 150.98 (C-5a, C-6a), 161.83 (J = 244.6 Hz, *p*–C of C_6H_4F).

EIMS *m/z*: 395 (M^+, 75), 394 (M-1, 100), 363 (M-S, 5). Anal. Calcd. for $C_{24}H_{14}FN_3S$: C, 72.89; H, 3.57; N, 10.63. Found: C, 72.80; H, 3.55; N, 10.41.

2.8 Diquino[3,4-b;4′,3′-e][1,4]thiazines (12A–C)

6*H*-Diquinothiazine (**12A**) and 6-methyldiquinothiazine (**12B**) were obtained from the reaction of sulfide **11** with ammonia and methylamine in hot phenol (Pluta, 1997).

2.8.1 6*H*-Diquinothiazine (12A)

Beige, mp 200–201 °C (mp 200–201 °C, Pluta, 1997). ^{1}H NMR ($CDCl_3$) δ: 7.64 (t, 2H, H-2, H-12), 7.71 (t, 2H, H-3, H-11), 7.81 (d, 2H, H-4, H-10), 8.04 (d, 2H, H-1, H-13), 8.40 (s, 2H, H-6, H-8). ^{13}C NMR ($CDCl_3$) δ: 109.10 (C-6a, C-7a), 117.18 (C-13a, C-14b), 117.41 (C-1, C-13), 127.25 (C-2, C-12), 129.49 (C-3, C-11), 130.78 (C-4, C-10), 142.21 (C-4a, C-9a), 147.94 (C-6, C-8), 148.07 (C-13b, C-14a).

2.8.2 6-Methyldiquinothiazine (12B)

Yellow, mp 156–157 °C (mp 156–157 °C, Pluta, 1997). ^{1}H NMR ($CDCl_3$) δ: 3.54 (s, 3H, CH_3), 7.66 (t, 2H, H-2, H-12), 7.72 (t, 2H, H-3, H-11), 8.11 (d, 2H, H-4, H-10), 8.34 (d, 2H, H-1, H-13), 8.66 (s, 2H, H-6, H-8). ^{13}C NMR ($CDCl_3$) δ: 43.63 (CH_3), 122.09 (C-1, C-13), 124.17 and 124.42 (C-6a, C-7a and C-13a, C-14b), 127.46 (C-2, C-12), 129.44 (C-3, C-11), 130.11 (C-4, C-10), 148.33 (C-6, C-8), 148.76 and 148.85 (C-4a, C-9a and C-13b, C-14a).

2.8.3 14-(*p*-Fluorophenyl)diquinothiazine (12C)

From diquinodithiin 10

Diquinodithiin **10** (0.16 g, 0.5 mmol) was finely powdered together with *p*-fluoroaniline hydrochloride (0.37 g, 2.5 mmol), and the mixture was heated on an oil bath at 200–205 °C for 3 h. After cooling, water (10 ml) was added to the reaction mixture and the resulting solid was filtered off, washed with water, air-dried, and purified by column chromatography (Al_2O_3, CH_2Cl_2) to give 0.12 g (30 %) of 14-(*p*-fluorophenyl)diquinothiazine (**12C**), beige, mp 315-316 °C.

From 4,4′-dichloro-3,3′-diquinolinyl sulfide (11)

A solution of sulfide **11** (0.18 g, 0.5 mmol) and *p*-fluoroaniline (0.17 g, 1.5 mmol) in MEDG (5 mL) was refluxed for 3 h. After cooling, the solution was poured into water (20 ml) and alkalized with 5 % aqueous sodium hydroxide to pH 10. The resulting solid was filtered off, washed with water and purified by column chromatography (Al_2O_3, $CHCl_3$) to give 0.17 g (86 %) of 14-(*p*-fluorophenyl)diquinothiazine (**12C**), beige, mp 315–316 °C.

^{1}H NMR ($CDCl_3$) δ: 6.43 (dd, 2H, C_6H_2), 6.77 (m, 2H, C_6H_2), 7.75 (t, 2H, H-2, H-12), 7.85 (t, 2H, H-3, H-11), 8.34 (d, 2H, H-4, H-10), 8.39 (d, 2H, H-1, H-13), 9,06 (s, 2H, H-6, H-8). ^{13}C NMR ($CDCl_3$) δ: 115.75 (J = 22.5 Hz, *m*-C of C_6H_4F), 116.30 (J = 7.5 Hz, *o*-C of C_6H_4F), 122.87 (C-1, C-13), 126.82 (C-13a, C-14b), 128.51 (C-2, C-12), 129.89 (C-6a, C-7a), 130.13 (C-3, C-11), 130.25 (C-4, C-10), 140.57 (J = 2.5 Hz, *ipso*-C of C_6H_4F), 145.54 (C-13b, C-14a), 147.98 (C-4a, C-9a), 149.49 (C-6, C-8), 158.07 (J = 238.5 Hz, *p*–C of C_6H_4F). EIMS *m/z*: 395 (M^+, 100), 363 (M-S,20), 300 (M-C_6H_4F, 17). Anal. Calcd. for $C_{24}H_{14}FN_3S$: C, 72.89; H, 3.57; N, 10.63. Found: C, 72.77; H, 3.59; N, 10.46.

2.9 In vitro lipid peroxidation

Heat-inactivated hepatic microsomes from untreated rats were prepared as described (Rekka *et al.*, 1989). The incubation mixture contained microsomal fraction (corresponding to 2.5 mg of hepatic protein per ml or 4 mM fatty acid residues), ascorbic acid (0.2 mM) in Tris–HCl/KCl buffer (50 mM/150 mM, pH 7.4), and the studied compounds (50–1 μM) dissolved in DMSO. The reaction was initiated by addition of a freshly prepared $FeSO_4$ solution (10 μM), and the mixture was incubated at 37 °C for 45 min. Lipid peroxidation of aliquots was assessed spectrophotometrically (535 against 600 nm) as TBAR. Both compounds and solvents were found not to interfere with the assay. Each assay was performed in duplicate, and IC_{50} values represent the mean concentration of compounds that inhibit the peroxidation of control microsomes by 50 % after 45 min of incubation. All standard errors are within 10 % of the respective reported values.

2.10 Calculation of lipophilicity, molecular mass, surface area, and molecular volume

Lipophilicity (as cLogP), molecular mass (M), surface area (S), and molecular volume (V_M) were calculated using CS Chem 3D Ultra 7.0 (CambridgeSoft) and Spartan'04 (Wavefunction, Inc. Irvine, CA).

3. RESULTS AND DISCUSSION

3.1 Synthesis

The synthesis of the title azaphenothiazines was based on the reactions of isomeric diquinodithiins, dichlorodiquinolinyl sulfides, and disulfide with amines, ammonia, and acetamide. The fusion reactions of linearly condensed diquinodithiin **1** with hydrochlorides of aniline and its *p*-substituted derivatives such as *p*-chloroaniline and *p*-methoxyaniline led to tetracyclic 9-substituted 6*H*-quinobenzothiazines **3A–C** (Scheme 1). 9-Methoxy-6*H*-quinobenzothiazine **3C** was obtained in better yield in the reaction of 2,2′-dichloro-3,3′-diquinolinyl disulfide **2** with *p*-methoxyaniline in monomethyl ether of diethylene glycol. The similar reaction of diquinodithiin **1** with hydrochlorides of 1-naphthylamine, 2-naphthylamine, and 6-aminoquinoline gave pentacyclic 7*H*-quinonaphthothiazine **4**, 14*H*-quinonaphthothiazine **5**, and 7*H*-diquinothiazine **6**. The reaction of isomeric diquinodithiin **7** with acetamide and *p*-fluoroaniline hydrochloride gave linearly condensed pentacyclic 6*H*-diquinothiazines **9A** and 6-(*p*-fluorophenyl)diquinothiazine **9B** (Scheme 2). Analogous reaction of another isomeric diquinodithiin **10** with *p*-fluoroaniline hydrochloride led to angularly condensed diquinothiazine **12C**. Better yields of the fluoroaniline products **9B** and **12C** were achieved when x,x'-dichloro-3,3′-diquinolinyl sulfides **8** and **11** (*x* = 2 and 4) were used. Sulfide **11** reacted also with ammonia or methylamine in hot phenol to give diquinothiazines **12A**, **B**.

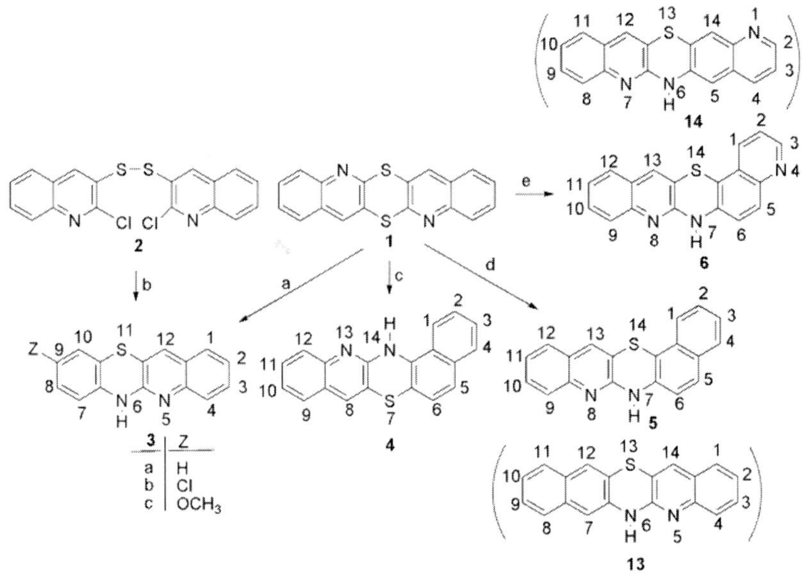

Scheme 1. Reactans: **A** $C_6H_5NH_2 \cdot HCl$ (*p*-$ClC_6H_4NH_2 \cdot HCl$, *p*-$CH_3OC_6H_4NH_2 \cdot HCl$), 200–205 °C, 4 h; **B***p*-$CH_3OC_6H_4NH_2$, MEDG, reflux, 3 h; **C** 1-naphthylamine·HCl, 200–205 °C, 4 h; **D** 2-naphthylamine·HCl, 200–205 °C, 4 h; **E** 6-aminoquinoline·HCl, 200–205 °C, 4 h

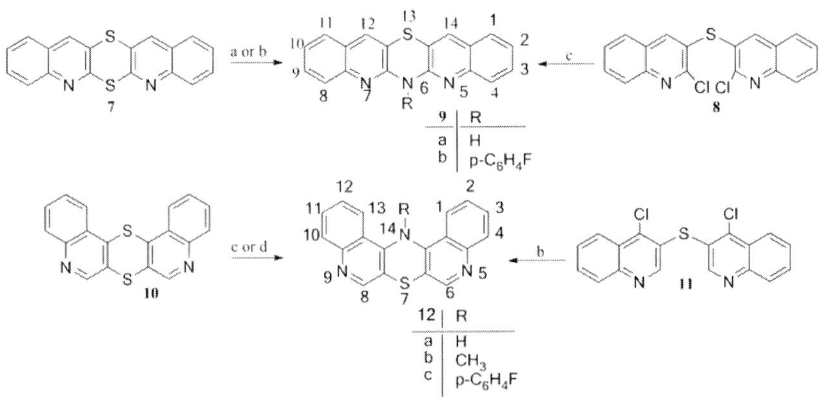

Scheme 2. Reactans: **A** CH_3CONH_2, K_2CO_3, 180 °C, 0.5 h; **B***p*F-$C_6H_4NH_2 \cdot HCl$), 200–205 °C, 3 h; **C***p*-$FC_6H_4NH_2$, MEDG, reflux, 3 h; **D** NH_3 (CH_3NH_2), phenol, 180 °C, 1 h

The described syntheses were monitored by TLC analysis. All chromatograms of new compounds showed characteristic for azaphenothiazines (Jeleń *et al.*, 2011) color changing during irradiation with UV light from blue to yellow (**4**, **9B**), from yellow to green (**5**, **6**), from orange to yellow (**12C**), and from yellow to orange (**7C**).

3.2 Structure

It is well known that the synthesis of phenothiazines can proceed via the Smiles rearrangement of the S–N type of the appropriate sulfide (Pluta *et al.*, 2009). The identification of the product structures was based on the spectroscopic ^{1}H NMR and MS analysis. In the case of the reactions of sulfides **7** and **11**, the products **9** and **12** possessed the C_{2v} symmetry (the left part was a mirror image of the right one) what excluded the stage of rearrangement. The reactions of diquinodithiin **1** and disulfide **2** with anilines proceeded similarly without the stage of rearrangement to give tetracyclic quinobenzothiazines **3A–C** (Jeleń and Pluta, 2009). The reaction with 1-naphthylamine gave pentacyclic quinonaphthothiazine **4**. On the contrary, the reactions with 2-naphthylamine and 6-aminoquinoline were more complex as there were two possibilities of the thiazine ring formation. The ^{1}H NMR analysis of the reaction products pointed at compounds **5** and **6** excluding compounds **13** and **14**, as evidenced from coupling constants; the H-5 and H-6 protons in compounds **5** and **6** showed a coupling constant J_{ortho}, whereas analogous protons in compounds **13** and **14** (H-7/H-12 and H-5/H-14, respectively) would have shown a coupling constant J_{para}, which is very small (i.e., $J_{1,4}$ = 0.6-0.8 Hz in naphthalene (Hamm and von Philipsborn, 1971; Lucchini and Wells, 1976) and $J_{5,8}$ = 0.5-0.8 Hz in quinoline (Hamm and von Philipsborn, 1971; Jones, 1977). We did not observe such small values of coupling constants in the reaction products **5** and **6**.

3.3 Antioxidant activity

The effect of the new derivatives on non-enzymatic lipid peroxidation of rat hepatic microsomal membrane lipids was investigated in vitro. Most of the studied derivatives demonstrated significant antioxidant activity, with IC_{50} values between 1 and 23 μM (Table 1). It is worthwhile to mention that under the same experimental conditions known potent antioxidants, trolox ((S)-(-)-6-hydroxy-2,5,7,8-tetramethylchromane-2-carboxylic acid) and probucol (4,4′-[(1-methylethylidene)bis(thio)]bis[2,6-bis(1,1-dimethylethyl)phenol]), exhibited IC_{50} values of 25 μM and >1 mM, respectively (Kourounakis *et al.*, 2008). Further, all of the active new derivatives were significantly much more potent than previously studied tricyclic dipyridothiazines (IC_{50} of most active compounds was between

64 and 470 μM) (Morak-Młodawska *et al.*, 2010). The time course of lipid peroxidation, as affected by various concentrations of representative compounds, is depicted in Figure 1.

Table 1.IC_{50} values for in vitro lipid peroxidation (LP), LogP, molecular volume (V_M), and molecular mass (M) as well as surface area (S) of the tested compounds

Compound	LP IC_{50} (μM)	LogP	M	S (Å^2)	V_M (Å^3)
3a	23	3.37	250.06	253.13	246.02
3b	3	3.93	284.02	268.84	259.50
3c	2	3.25	280.07	283	273.38
4	2	4.37	300.07	297.74	296.96
5	6	4.37	300.07	297.68	296.87
6	16	3.46	301.07	293.28	291.10
9a	>1000	4.20	301.07	295.91	291.54
9b	>1000	6.00	395.09	374.91	379.66
12a	1	2.71	301.07	291.11	290.87
12b	500	4.77	315.08	317.08	321.82
12c	>1000	4.51	395.09	359.77	375.69

Tetracyclic NH-azaphenothiazines **3A–C** exhibited significant activity dependent on the substitution (H, Cl, and OCH_3) on the benzene ring (Table 1). From the pentacyclic compounds, the angularly fused with unsubstituted, the thiazine nitrogen atom (**4–6** and **12A**) exhibited very significant activity with most active compound **12A**, which showed an IC_{50} of 1 μM. The change of the quinoline moiety into naphthalene (compare compounds **4** and **5** with **6**) marginally increased activity. However, compounds with a linearly fused ring system (**9A** and **9B**) and/or a large aryl substituent at the thiazine nitrogen atom (**9B** and **12C**) did not show any antioxidant activity, while compound **12B**, with a small substituent, exhibited very weak activity.

Considering three isomers (**6**, **9A**, and **12A**), one can find that their antioxidant activity increased with decreasing lipophilic character represented by the logP values. On the other hand, the least active compounds (**9B**, **12B**, and **12C**) exhibited high values of molecular descriptors such as molecular mass (M > 315), molecular volume (V_M > 321), and surface area (S > 317, Table 1). However, attempts to the correlate the activity with those properties turned out to be unsatisfactory.

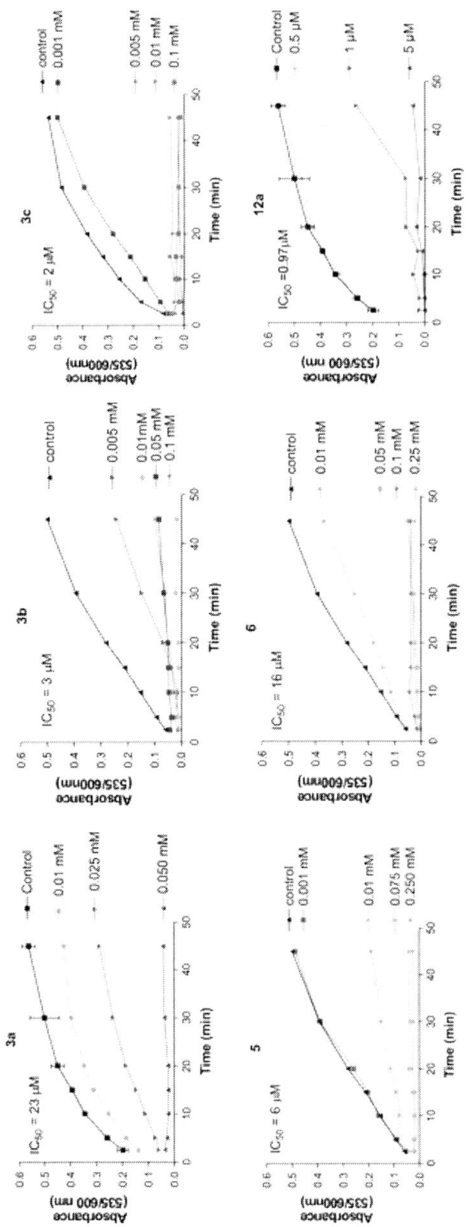

Figure 1. Representative *graphs* of the time course of lipid peroxidation as affected by various concentrations of compounds **3A–C**, **5**, **6**, and **12A**. IC_{50} values are calculated according to these results as the concentration showing 50 % inhibition of the lipid peroxidation reaction at 45 min incubation time

In conclusion, eleven tetracyclic and pentacyclic (linearly or angularly condensed) azaphenothiazines were synthesized, and structure–(antioxidant)activity relationships were investigated. The type of the ring fusion was concluded from the ^{1}H NMR spectra. The degree of antioxidant activity of these derivatives seems to depend on their lipophilicity and molecular mass. The non-substitution of the thiazine nitrogen atom, the type of ring system fusion, and the nature of substituents promote activity. Finally, it is the first time to our knowledge that azaphenothiazines are shown to exhibit such potent antioxidant activity.

ACKNOWLEDGMENTS

The synthesis and the structure elucidation are supported by the Medical University of Silesia (Grant KNW-1-032/K/3/0.

CONFLICT OF INTEREST

Authors have no financial/commercial conflicts of interest.

REFERENCES

1. Aaron JJ, Gaye Seye MD, Trajkovska S, Motohashi N (2009) Bioactive phenothiazines and benzo[a]phenothiazines: spectroscopic studies and biological and biomedical properties and applications. Top Heterocycl Chem 16:153–231
2. Asghar MN, Alam Q, Augusten S (2012) Fluphenazine hydrochloride radical cation assay: a new, rapid and precise method to determine in vitro total antioxidant capacity of fruit extracts. Chin Chem Lett 23:1271–1274
3. Borges MBD, Dos Santos CG, Yokomizo CH, Sood RR, Vitovic PP, Kinnunen KJ, Rodrigues T, Nantes IL (2010) Characterization of hydrophobic interaction and antioxidant properties of the phenothiazine nucleus in mitochondrial and model membranes. Free Radical Res 44:1054–1063
4. Dasgupta A, Dastridara SG, Shirataki Y, Motohashi Y (2008) Antibacterial activity of artificial phenothiazines and isoflavones from plants. Top Heterocycl Chem 15:67–132
5. Gupta RR, Kumar M (1988) Synthesis, properties and reactions of phenothiazines. In: Gupta RR (ed) Phenothiazines and 1,4-benzothiazines—chemical and biological aspects. Elsevier, Amsterdam, pp 1–161
6. Hamm P, von Philipsborn W (1971) Protonenresonanzspektren von aromatischen N-Oxiden Berechnung der chemischen Verschiebungen, verursacht durch die Feldeffekte der N-O-gruppe. Helv Chim Acta 54:2363–2401
7. Jeleń M, Pluta K (2009) Synthesis of quinobenzo-1,4-thiazines from diquino-1,4-dithiin and 2,2′-dichloro-3,3′-diquinolinyl disulfide. Heterocycles 78:2325–2336
8. Jeleń M, Morak-Młodawska B, Pluta K (2011) Thin-layer chromatographic detection of new azaphenothiazines. J Pharm Biomed Anal 55:466–471

9. Jeleń M, Pluta K, Zimecki M, Morak-Młodawska B, Artym J, Kocięba M (2013) Synthesis and selected immunological properties of substituted quino[3,2-b]benzo[1,4]thiazines. Eur J Med Chem 63:444–456
10. Jones G (1977) In: Jones G (ed), The chemistry of heterocyclic compounds, vol 32, Quinolines, Part 1. Wiley, London, p 11–12
11. Kourounakis AP, Charitos C, Rekka EA, Kourounakis PN (2008) Lipid-lowering (hetero)aromatic tetrahydro-1,4-oxazine derivatives with antioxidant and squalene synthase inhibitory activity. J Med Chem 51:5861–5865
12. Kumar M, Sharma K, Samarth RM, Kumar A (2010) Synthesis and antioxidant activity of quinobenzothiazinones. Eur J Med Chem 45:4467–4472
13. Liu Z-Q, Tang Y-Z, Wu D (2009) Antioxidant effects of phenothiazine, phenoxazine, and iminostilbene on free-radical-induced oxidation of linoleic acid and DNA. J Phys Org Chem 22:1009–1014
14. Lucchini V, Wells PR (1976) Proton magnetic resonance spectra of monosubstituted naphthalenes. Org Magn Reson 8:137–140
15. Matralis AN, Katselou MG, Nikitakis A, Kourounakis AP (2011) Novel benzoxazine and benzothiazine derivatives as multifunctional antihyperlipidemic agents. J Med Chem 54:5583–5591
16. Morak-Młodawska B, Pluta K, Matralis AN, Kourounakis AP (2010) Antioxidant activity of newly synthesized 2,7-diazaphenothiazines. Arch Pharm Chem Life Sci 343:268–273
17. Motohashi N, Kawase M, Satoh K, Sakagami H (2006) Cytotoxic potential of phenothiazines. Curr Drug Targets 7:1055–1066
18. Naik N, Kumar HV, Veena V (2012) Novel phenothiazine analogous: synthesis and a new perceptivity into their antioxidant potential. Pharmacia Lett 4:786–794
19. Nowak M, Pluta K, Suwińska K (2002) Synthesis of novel heteropentacenes containing nitrogen, sulfur and oxygen or selenium. New J Chem 26:1216–1220
20. Nowak M, Pluta K, Kloc K, Siegriest T (2003) Synthesis and X-ray analysis of isomeric diazadithiapentacenes. Heterocycles 60:2045–2056
21. Nowak M, Pluta K, Suwińska K, Straver L (2007) Synthesis of new pentacyclic diquinothiazines. J Heterocycl Chem 44:543–550
22. Pluta K (1994) Synthesis and NMR assignment of 1,4-oxathiino[3,2-c;5,6-c']diquinoline. J Heterocycl Chem 31:557–560
23. Pluta K (1997) Synthesis and properties of 14-substituted 1,4-thiazinodiquinolines. Phosphorus Sulfur Silicon 126:145–156
24. Pluta K, Morak-Młodawska B, Jeleń M (2009) Synthesis and properties of diaza-, triaza- and tetraazaphenothiazines. J Heterocycl Chem 46:355–391
25. Pluta K, Jeleń M, Morak-Młodawska B, Zimecki M, Artym J, Kocięba M (2010) Anticancer activity of newly synthesized azaphenothiazines in NCI's anticancer screening. Pharmacol Rep 62:319–332
26. Pluta K, Morak-Młodawska B, Jeleń M (2011) Recent progress in biological activities of synthesized phenothiazines. Eur J Med Chem 46:3179–3189
27. Rekka E, Kolstee J, Timmerman H, Bast A (1989) The effect of some H2-receptor antagonists on rat hepatic microsomal cytochrome P-450 and lipid peroxidation in vitro. Eur J Med Chem 24:43–54
28. Zhang H-Y, Yang D-P, Tang G-Y (2006) Multipotent antioxidants: from screening to design. Drug Discov Today 11:749–754
29. Zimecki M, Artym J, Kocięba M, Pluta K, Morak-Młodawska B, Jeleń M (2009) Immunosupressive activities of newly synthesized azaphenothiazines in human and mouse models. Cell Mol Biol Lett 14:622–635

CHAPTER 16

Homology Modeling of Human Alpha-Glucosidase Catalytic Domains and SAR Study of Salacinol Derivatives

Shinya Nakamura*, Kazunori Takahira, Genzoh Tanabe, Osamu Muraoka, Isao Nakanishi

Department of Pharmaceutical Sciences, Kinki University, Kinki, Japan

ABSTRACT

Maltase-glucoamylase (MGAM) and sucrase-isomaltase (SI) belong to human intestinal alpha-glucosidase and their N-terminal side catalytic domains are called NtMGAM and NtSI, and their C-terminal side catalytic domains are called CtMGAM and CtSI. As an antidiabetic, alpha-glucosidase inhibitor is required to bind to all of these domains to inhibit disaccharides hydrolysis. Salacinol and kotalanol isolated from Salacia reticulata are novel seed compounds for alpha-glucosidase inhibitor. Even though the complex structures of NtMGAM or NtSI have been determined experimentally, those of CtMGAM and CtSI have not been revealed. Thus, homology modeling for CtMGAM and CtSI has been performed to predict the binding mode of salacinol and its derivatives for each domain. The binding affinities for these compounds were also calculated to explain the experimental structure-activity relationships (SARs). After a docking study of the derivatives to each catalytic domain, the MM/PBSA method has been applied to predict the binding affinities. The predicted binding affinities were almost consistent with the experimental SARs. The comparison of the complex structures and binding affinities provided insights for designing novel compounds, which inhibit all catalytic domains.

Keywords: Homology Modeling; Docking Study; MM/PBSA; Alpha-Glucosidase; Salacinol; Kotalanol

1. INTRODUCTION

Starch is one of the most important malnutrition sources. Saccharides after ingestion per oral are digested into oligosaccharides by amylase in saliva and pancreas, and reach the small intestine. Disaccharides such as maltose with alpha-1,4-glycosidic bond, isomaltose with alpha-1,6- glycosidic bond, and sucrose with alpha-1,2-glycosidic bond are digested into monosaccharides by alpha-glucosidase located on the small intestinal brush border membrane.

Two kinds of human intestinal alpha-glucosidase, maltase-glucoamylase (MGAM) and sucrase-isomaltase (SI), have two catalytic domains on the N-terminal side

(NtMGAM, NtSI) and the C-terminal side (CtMGAM, CtSI) [1], as shown in **Figure 1**. These four catalytic domains are homologous, and every domain catalyzes the maltose hydrolysis at the alpha-1,4-glycosidic bond. In addition, NtSI and CtSI catalyze isomaltose and sucrose hydrolysis, respectively [2].

Acarbose (**Figure 2**(a)) [3] and voglibose (**Figure 2**(b)) [4] are well-known alpha-glucosidase inhibitors that are used in diabetes therapy. However, abdominal discomfort and hepatotoxicity [5] have been reported for these drugs, and therefore, novel alpha-glucosidase inhibitors without these side effects, especially hepatotoxicity, are required to improve the patients' quality of life.

On the other hand, salacia, a genus of plants, is used to treat diabetes in Ayurveda, the traditional medicine of India and Sri Lanka. Furthermore, kotalanol (**Figure 2**(c)) [6] and salacinol (1 of **Figure 2**(d)) [7], isolated from Salacia reticulata have almost the same inhibition acti ity to alpha-glucosidase as acarbose and voglibose [6]. Therefore, several drug discovery studies based on these natural compounds have been carried out [8-13].

Recently, the X-ray crystal structure analysis of NtMGAM or NtSI complexes of these compounds has been reported [8,9]. The structure-activity relationships (SARs) for salacinol derivatives are also vigorously studied

[10, 11]. In addition, docking simulations between salacinol derivatives and NtMGAM have been performed [12] and provided the guidelines for optimization studies, such as the replacement of the sulfate group of salacinol and kotalanol with the hydrophobic group. Indeed, several compounds that enhanced the inhibitory activities have been discovered [13].

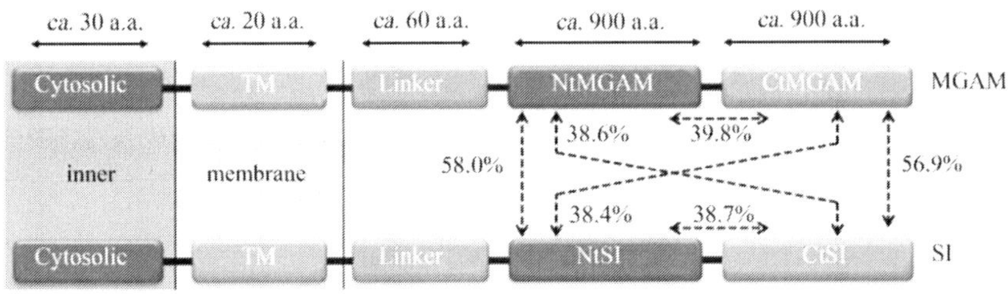

Figure 1. Two kinds of human alpha-glucosidase, Maltase-Glucoamylase (MGAM) and Sucrase-Isomaltase (SI), located on intestinal brush border membrane. Percentage values show the amino acid sequence homology between the two domains.

Though current studies have focused on NtMGAM, they need to take into account other catalytic domains such as NtSI, CtMGAM, and CtSI. The ideal alphaglucosidase inhibitors are expected to simultaneously bind these four domains. The rational design of ideal inhibitors requires clarifying the structural differences among the four catalytic domains.

The structures of the two C-terminal catalytic domains have not been experimentally determined; however, it is possible to construct model structures of both domains by the homology modeling method if the structure of a homologous protein is known. As shown in **Figure 1**, the sequence homologies of amino acid residues for the four catalytic domains is almost 40%, which is over the threshold value of 30% that the homology modeling method is applicable [14]. Therefore, the model structures of CtMGAM and CtSI are constructed using the crystal structures of NtMGAM and NtSI.

Binding mode analysis is an essential process for structure-based drug design, and typically, the binding modes of the ligands are experimentally determined by X-ray crystallography. However, recent progress in the computational docking method enabled us to predict the ligand-binding mode with a positional error less than 2 Å. In this study, the binding modes and binding affinities of the salacinol derivatives (compounds 2-4, shown in Figure 2(d)) to the four catalytic domains have been computationally predicted and the structure-activity relationships (SARs) were analyzed in detail.

The enzyme inhibition activity of the ligand can be computationally predicted by estimating the binding affinity as the binding free energy between the enzyme and the ligand. The binding affinity can be calculated by various computational methods such as the linear interaction energy method [15], the molecular mechanics/ Poisson-Boltzmann surface area (MM/PBSA) method [16]. In this study, the MM/PBSA method was applied to predict the binding affinities of the salacinol derivatives for each catalytic domain. Then, we examined if the predicted binding affinity values could explain the experimental SARs for salacinol derivatives against alphaglucosidase inhibition activity.

2. MATERIALS AND METHODS

2.1. Homology Modeling

Homology modeling is a method to construct a threedimensional (3D) structure of a target protein from the experimental 3D structure of a homologous protein as a template. To obtain a reliable model structure, a homology greater than the 30% of the amino acid sequences between a target protein and a template protein is required [14]. As the discrepancy of one amino acid residue pair on the sequence alignment causes a 4 Å displacement of the position [14], the alignment should be performed with care to minimize this error, and a multiple sequence alignment using more than three amino acid sequences is desirable for this purpose.

The amino acid sequences of human CtMGAM and CtSI were retrieved from NCBI-GeneID: 8972 and 6476 [17], respectively. First, the multiple sequence alignment for the four catalytic domains has been performed using the BLOSUM62 score [18]. Then, homology modeling of the two C-terminal catalytic domains has been performed using MODELLER [19].

Prior to the 3D structure construction of the two domains, the calculation conditions of MODELLER were examined by the prediction of the structure of NtSI from the experimental structure of NtSI itself (PDB-ID: 3LPP). Ramachandran plot [20] and the Profile-3D [21] score validated the reliabilities of the built structure models. The calculation conditions that reproduced the experimental structure of NtSI were most appropriately adopted for modeling the unknown structures.

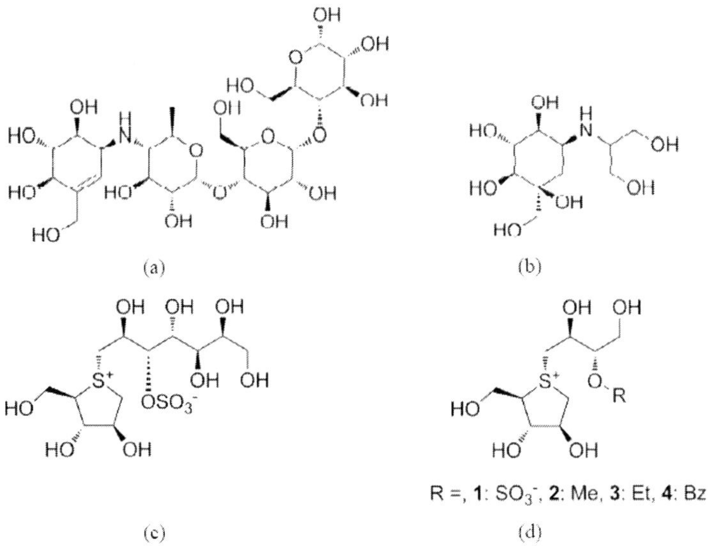

Figure 2. Chemical structures of the known alpha-glucosidase inhibitors: (a) Acarbose; (b) Voglibose; (c) Kotalanol; and (d) Salacinol (1) and its derivatives (2-4).

Model structures of CtMGAM and CtSI were built using the experimental structure of the NtMGAM complex with kotalanol (PDB-ID: 3L4V) as a template by MODELLER according to the alignment data and the verified calculation conditions. The positions of the three disulfide bonds were taken from the NtMGAM domain. To consider the hydrogen bond network between kotalanol and the target protein during model building, kotalanol and several water molecules within 10 Å from kotalanol on the binding pocket were also included as a template. Model structures were built two times and loop structures were created three times in each simulation. Therefore, six model structures were proposed. The CHARMM22 force field parameter [22] was used in the model construction.

After the model construction by MODELLER, the hydrogen atoms were assigned to all models using MOE [23]. The geometries of the models were gradually optimized using CHARMM [24]. First, the positions of the hydrogen atoms were optimized, and then, the side chain atoms were relaxed, and finally, all atoms were optimized. At each step, energy minimizations up to 20,000 cycles using the steepest descent and the conjugate gradient method was performed until the RMS gradient was below 0.01 kcal/mol/Å. The model with the most stable potential energy was selected as the representative model structure from the six candidates.

2.2. Binding Mode Prediction

Experimental binding modes for salacinol and/or kotalanol to the N-terminal domains have been already reported [8,9]. In the narrow binding pockets, salacinol and kotalanol formed many hydrogen bonds and a salt bridge with the protein. As the binding modes of the salacinol derivatives may vary according to the shape of each domain mentioned above, docking studies for salacinol and its derivatives (**Figure 1**(d)) to four catalytic domains have been performed.

For the protein structure to dock, the experimental structures from PDB-ID, 3L4Z and 3LPP, were used for Nt-MGAM and NtSI, respectively, and the predicted model structures were used for CtMGAM and CtSI. The ligand structures were also created based on salacinol from the 3L4Z coordinates. The positions of hydrogen atoms were appropriately assigned using MOE.

Docking studies for each domain have been performed using the MOE-Dock module. The ligand binding site definition was done using the MOE-Alpha SiteFinder module. Before the pose prediction, the conformation generation for the ligands was performed under fixed bond lengths and bond angles. The top 50 poses by the London dG [23] score were chosen from all generated poses for each ligand. These poses were further optimized by the MMFF94x parameter [25] with the generalized born/volume integral solvation model [26] (MM/ GBVI). The pose with the best MM/GBMV score was adopted as its binding mode. The most appropriate docking scheme was explored to reproduce the binding mode observed in the experimental structure, using the NtMGAM complex with salacinol (PDB-ID: 3L4Z).

Recently, the experimental structure of CtMGAM in complex with acarbose has been determined (PDB-ID: 3TOP) [27]. Therefore, the experimental structure of CtMGAM was also employed in the binding mode prediction for comparison purposes. On the other hand, for the modeling of the CtSI structure, using the experimenttal structure of CtMGAM as a template may be more appropriate than using that of NtMGAM, because the sequence homology of CtMGAM to CtSI is much higher than that of NtMGAM (**Figure 1**). In this study, however, the experimental structure of NtMGAM in complex with kotalanol was adopted to consider an induced-fit of the enzyme to the thiosugar moiety in kotalanol and salacinol derivatives.

2.3. Binding Affinity Prediction

To analyze the ligand affinities considering the thermodynamic behavior of molecules, molecular dynamics (MD) simulations have been performed for each complex. For the MD simulations in a solvent, a spherical cluster of water molecules was generated around 30 Å from the ligand coordinate center using InsightII [28]. Before the MD simulation, the structure optimization was carried out until the RMS gradient was below 0.01 kcal/mol/Å. To relax the molecule system, heating from 0 K to 300 K was performed in 50 ps. After 2 ns in the equilibration stage, 100 snapshots were sampled from a 500 ps simulation. The following MD conditions were used: a time step of 1 fs; a dielectric constant of 1; and a cutoff distance for a nonbonded interaction of 15 Å. SHAKE [29] was applied to fix the bond length in all hydrogen atoms. The surface of the spherical water cluster was constrained by 100 kcal/mol/Å to prevent the water molecules from evaporating. The amino acid residues outside the water cluster were also harmonically constrained to the original position by 100 kcal/mol/$Å^2$. All dynamics simulations were performed using CHARMM [24] with the CHARMm.cfrc parameters [28].

Binding affinities for salacinol and its derivatives were estimated using 100 snapshots for each MD simulation by the MM/PBSA method. The van der Waals and electrostatic interaction energies between the ligand and protein for every snapshot were calculated by CHARMM and their averages were used as the MM interaction term. Desolvation energies for the solvent effect were calculated by DelPhi [30] as the PB term for the polar solvent effect, and by MSMS [31] as the SA term for the nonpolar solvent effect. The dielectric constants of 4 and 80 were used for the solute and solvent, respectively. The ion strength was set to 0.145 M. The PARSE parameters [32] for the atomic radius in the PB and SA calculations were used. To calculate the nonpolar SA term, the following parameters were adopted for the PARSE radius surface: the surface-scale factor was 0.00542 and the constant was 0.92 [32].

3. RESULTS AND DISCUSSION

3.1. Homology Modeling

The sequence alignment of the amino acids in the four catalytic domains is shown in **Figure 3**. Homology modeling has been performed using these alignment pairs. The remodeling structure of NtSI and its crystal structure (PDB-ID: 3LPP) were compared and shown in **Figure 4**. The RMSD value of the main chain atoms between the model and the crystal structure was 1.8 Å and that of all heavy atoms within 5 Å from kotalanol was 0.7 Å. These values are small enough to consider that the model structure reproduces the experimental structure. However, the directions of the side chain atoms of several amino acid residues were not reproduced (**Figure 4**(b)). Most of these amino acid residues are hydrophobic and do not affect the hydrogen bond network in ligand binding. Such positional differences in these residues are also observed between the experimental structures of NtMGAM and NtSI, the most homologous pairs among the four domains (**Figure 1**).

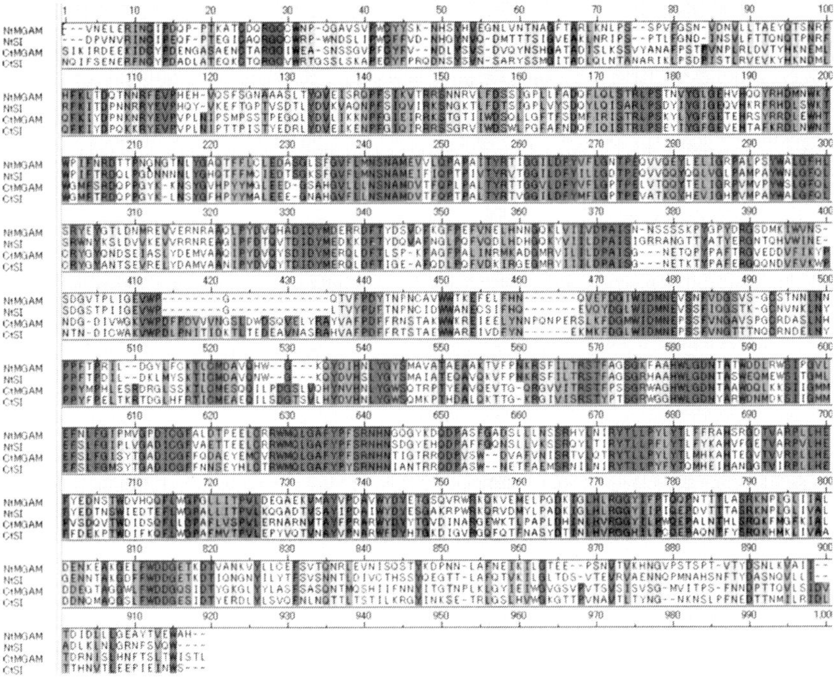

Figure 3. Multiple sequence alignment of the amino acid residues in the four human alpha-glucosidase catalytic domains. Homologous pairs in the alignment are shown in blue background. Blue color depth stands for the characteristic similarity of the amino acid residues.

Also, the predicted CtMGAM structure was compared to the experimental CtMGAM structure (PDB-ID: 3TOP) [27]. The RMSD value of the main chain atoms between the two structures was 3.6 Å, and this large displacement was originated in the long insertion loop near the substrate binding site (**Figure 3**). Predicting the structure of such a long insertion loop by MODELLER is extremely difficult without reference coordinates in the template structure. However, the RMSD value of 1.2 Å for all heavy atoms within 5 Å from kotalanol indicates that the model structure around the ligand binding site reproduces the experimental structure well.

The structural validity of the model was also examined by using the Ramachandran plot and the Profile-3D method. The φ-ψ angle distribution was normal except for several amino acid residues located at the loop region, which is distant from the active site. The qualitative validation of the model structure was confirmed by Profile-3D [21]. Although the threshold score for NtSI was 17 - 8.22, the calculated score of 369.80 for the NtSI model indicated that the model had no significant structural problem. As for the modeling scheme adopted here, it reproduced the experimental structure well and the structures of the two C-terminal domains were predicted with the same scheme.

The predicted CtMGAM and CtSI models were evaluated by the Ramachandran plot as well as the Profile-3D method. No abnormal φ-ψ angle distributions were observed in the Ramachandran plot of both models. The Profile-3D score for the CtMGAM and CtSI models was 366.45 and 371.46, respectively, and these values exceeded the corresponding threshold values of 185.50 and 184.46. Therefore, the model structures are considered valid for use in the comparison study that investigated the shape of the active site.

The detailed structural comparisons among the four catalytic domains are shown in **Figure 5**. The amino acid residues, which form hydrogen bonds or a salt bridge with kotalanol, were also compared (**Figure 5**(a)). These residues were conserved among the four catalytic domains and no significant structural difference was observed. That is, hydrogen bonds and salt-bridge interacttion that are essential for kotalanol binding are common among all the catalytic domains. This means that such interactions would be conserved in the binding of salacinol and its derivatives. Then, the mutated amino acid residues around the kotalanol binding site were compared (**Figure 5**(b)). Subsequently, we focus our discussion on the five amino acid residues: The205, Tyr299, Trp406, Ala576, and Gly602 of NtMGAM. The residues corresponding to Thr205 in the NtMGAM and located near the sulfate group were deemed distant enough not to affect the ligand interaction; however, the large hydrophobic residue Leu233 in NtSI would make the pocket narrow and may show a different ligand recognition and specificity. Tyr-299 in NtMGAM mutates to Trp327 in NtSI and is conserved in the other domains. As the hydroxy groups of the conserved tyrosine residues interact with the other residues, there are no polar interactions with the ligand. The larger tryptophan residue in NtSI would form a smaller binding pocket. The position of Trp406 at the sulfate group-binding pocket in NtMGAM was occupied with the proline residue in CtSI. Even though the tryptophan residue interacted through the CH-O interaction with the sulfate group of kotalanol, the proline residue cannot interact with the same manner and it is supposed to interact weakly. Ala576 in NtMGAM corresponded to valine in NtSI, phenylalanine in CtMGAM and CtSI. Although this residue is slightly apart from kotalanol, the binding pockets of both the C-terminal domains are narrower than the N-terminal domains. Gly602 in NtMGAM varied among the domains. For NtSI and CtMGAM, this residue is mutated by the serine and threonine residues, respectively. As for CtSI, the isoleucine residue is located in this position. These residues do not significantly contribute to the ligand-binding interaction because this residue is located behind Tyr299 and Phe575, and is not exposed to the active site surface.

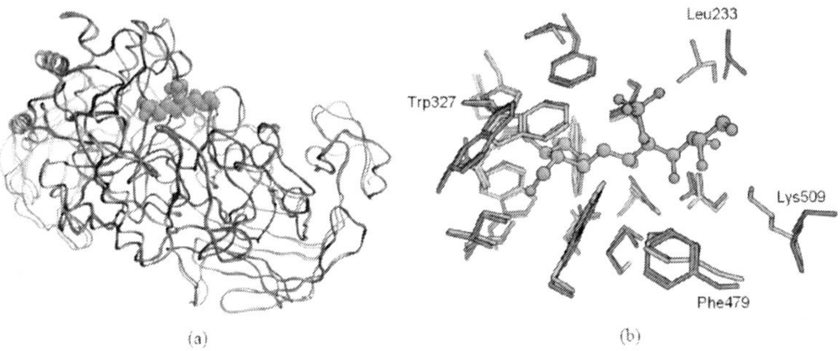

Figure 4. Comparison of the NtSI structure (magenta: homology model, green: experiment), and kotalanol in 3LPP is shown in orange. (a) Overall folding of the main chain and (b) Active site, amino acid residues with relatively large displacement are labeled.

To compare the shape of the binding pockets, the molecular surfaces of the kotalanol binding site (around 5 Å from kotalanol in NtMGAM) are shown in **Figure 6**. It was predicted that the pockets were deeper in the order of NtMGAM, NtSI, CtMGAM, and CtSI. In particular, in CtSI, the overlaid sulfate group of kotalanol seemed to be entirely covered by the pocket and might contact with the wall (**Figure 6**(d)). This means that the van der Waals interaction between the sulfate group of salacinol and each domain might be stronger in the same order of the pocket depth except for CtSI.

3.2. Binding Mode Prediction of Salacinols

The structure of the calculated NtMGAM complex with salacinol was compared with that of the experiments to validate the docking condition. The RMSD value for non- hydrogen atoms was 0.4 Å, which encouraged us to predict the binding modes of other ligand complexes by this condition.

The docked structure for salacinol and its derivatives to NtMGAM is shown in **Figure 7**(a). Interestingly, the methyl group of 2 is located on a similar position of the sulfate group of salacinol (1), and the ethyl group of 3 and the benzyl group of 4 are located where the C4' hydroxy group exists. In the following discussion, the sulfate group-binding site and the C4' hydroxy groupbinding site are referred to as the SG site and HG site, respectively.

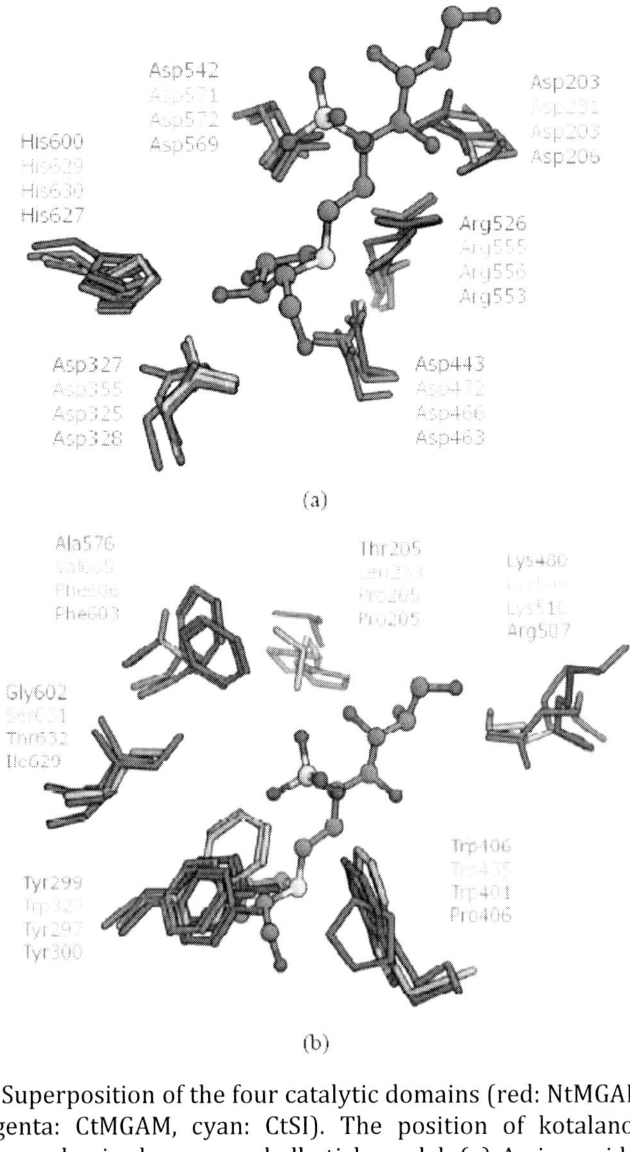

Figure 5. Superposition of the four catalytic domains (red: NtMGAM, green: NtSI, magenta: CtMGAM, cyan: CtSI). The position of kotalanol in the NtMGAM complex is shown as a ball-stick model. (a) Amino acid residues related to hydrogen bonds and a salt bridge with kotalanol; (b) Neighboring amino acid residues with mutation.

In NtMGAM, the SG site consists of many aromatic residues such as Phe575, Tyr299, and Trp406; the HG site consists of various kinds of residues such as Phe450, Asp203, and Lys480. This indicates that the hydrophobic group is likely to bind at the SG site, but it conflicts with the docking study that the hydrophobic ethyl and phenyl group of compound 2 and 3 bind at the HG site. However, the HG site is formed by the aromatic ring and the hydrophobic methylene chains of the hydrophilic amino acid residues, and the environment may be considerably hydrophobic. It is assumed that shape complementarity is important for ligand recognition. For example, the size of the phenyl ring of 4 is sufficient to bind either to the SG site or the HG site. However, the whole phenyl ring interacts with the HG site, whereas part of the ring can interact with the SG site.

The predicted binding mode of the salacinol derivatives to NtSI is shown in **Figure 7**(b). The NtSI complex structure with salacinol (1) has not been reported but that with kotalanol was known [9]. The hydrogen bond network and the salt-bridge formation were nearly common between both ligands on the NtMGAM complex [8]. Thus, the binding mode of salacinol on NtSI was supposed to be the same as that of kotalanol. Similar to the NtMGAM case, the methyl group of 2 was bound to the SG site. The SG site of NtSI consists of many aromatic amino acid residues similar to NtMGAM, but the difference from Tyr299 in NtMGAM to Trp327 in NtSI made the SG site a little smaller as previously described. Furthermore, Ser448 around the HG site in NtMGAM changed into the basic and more hydrophilic Lys509 in NtSI, and the HG site became narrow as shown in **Figure 7**(b). Therefore, the hydrophobic substituent of 3 and 4 bind to the SG site and gain more van der Waals interacttions with the hydrophobic SG site than those with the HG site, whose hydrophobicity decreased. These results suggest that the introduction of the acidic group at the C4' position to interact with the basic Lys509 residue would enhance the binding affinity to NtSI.

The docking of salacinol derivatives to the CtMGAM model is shown in **Figure 7**(c). The binding mode of 1, 2, and 3 in CtMGAM were similar to those in NtMGAM and NtSI. On the other hand, the interaction between the phenyl ring of 4 and the HG site wall seen in NtMGAM was lost. The phenyl ring of 4 was bound to the SG site benefiting from the many van der Waals interactions to the hydrophobic amino acid residues (Tyr297, Trp401, and Ile633) because the HG site became shallower by Phe473, which moved to the HG site that was being affected by the long insertion loop in CtMGAM.

The docking simulation to the experimental CtMGAM structure was also performed. The structural difference of the long insertion loop near the ligand binding region between the experimental and the model structures did not affected the binding mode of salacinol derivatives. That is, the binding mode of each derivative to the both structures was quite similar.

However, attention would be required in docking much larger compounds such as acarbose.

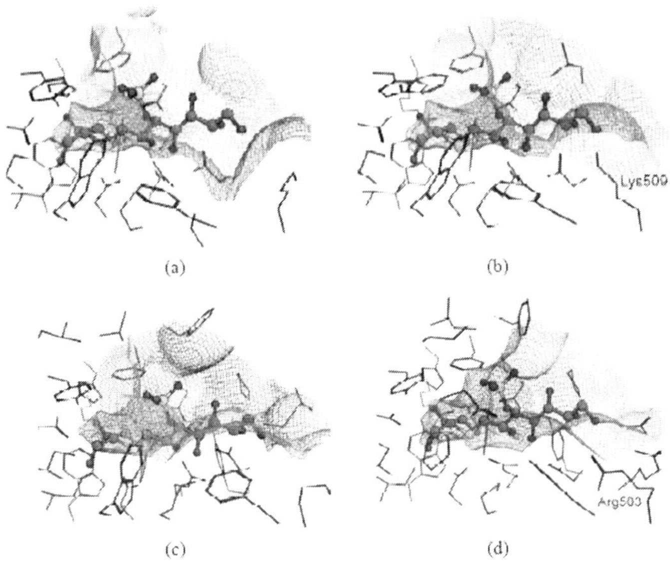

Figure 6. Molecular surfaces of the active sites in the four domains. The green mesh shows the van der Waals contact surface. The position of kotalanol in the NtMGAM complex is overlaid as a ball-stick model: (a) NtMGAM; (b) NtSI; (c) CtMGAM; and (d) CtSI.

Finally, the binding modes for CtSI are shown in Figure 7(d). Interestingly, the sulfate group of 1 was predicted to bind to the HG site like the phenyl ring of 4 in NtMGAM as mentioned above because (a) the SG site of CtSI surrounded by Phe603, Ile629, and Leu407 were much narrower than the other domains (Figures 5 and 7) and (b) positively charged Arg503 was located at the HG site. Therefore, the negatively charged sulfate group was attracted by Arg503 in the HG site. As the size of the SG site was fitted to the hydroxy group at the C4' position, the methyl group of 2 and the ethyl group of 3 were located toward the HG site. The HG site also became narrow because of the existence of Arg503, and then, the phenyl ring of 4 was caught in a narrow groove formed by Arg503 and Pro208.

Thus, the binding mode of salacinol and its derivatives for the four domains has been deduced by the docking study. The substituent groups were not uniformly bound and their location varied according to the shape and properties of the active site in each catalytic domain.

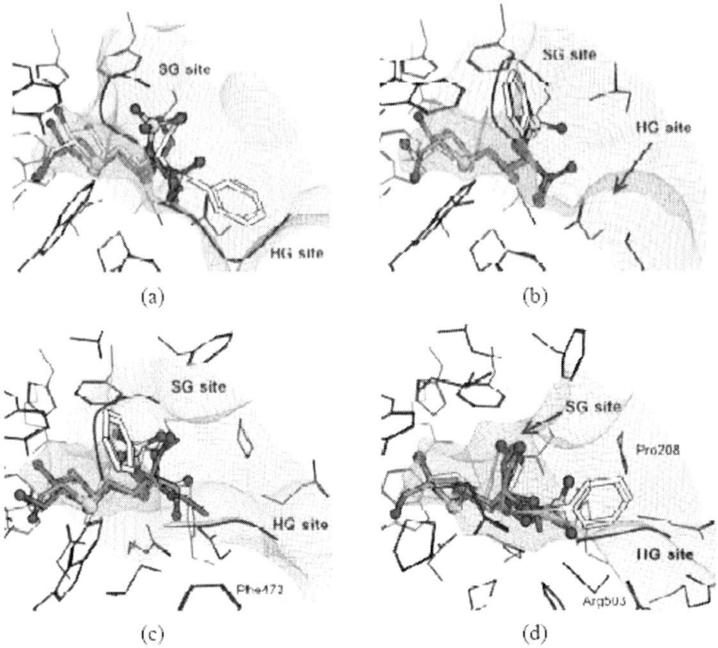

Figure 7. Predicted binding modes of salacinol and its derivatives for each catalytic domain: (a) NtMGAM, (b) NtSI, (c) CtMGAM, and (d) CtSI. Compounds 1, 2, 3, and 4 are shown in orange, magenta, blue, and white, respectively, by the stick model with a partially ball model for noncarbon atoms. The green mesh shows van der Waals contact surface (same as **Figure 6**). Blue curves or arrows depict the sulfate group-binding site (SG site) and the C4' hydroxy group-binding site (HG site).

3.3. Binding Affinity Prediction

The experimental inhibition activities of the three substrates [7,13] and calculated binding affinities of salacinol derivatives for the four domains are shown in **Table 1**. As for the calculated binding affinities for NtMGAM, three compounds (1, 2, and 3) have almost the same strengths and 4 was much stronger than others in all four domains. For CtMGAM(model) and NtSI, 4 was the strongest but 1 was weaker than 2 and 3. In CtSI, however, 2 was much weaker than 1.

The binding affinities of the compounds in CtMGAM (X-ray) were relatively weaker than those in CtMGAM (model). The induced-fit of the enzyme to the thiosugar moiety would have arisen this difference and affected the affinity order. However, it is noted that 4 was still strongest.

It is known that maltose hydrolysis, which is most important for diabetes, is related to all domains, whereas isomaltose and sucrose hydrolysis are related to NtSI and CtSI, respectively [9,33]. The calculated affinities for NtSI agreed quite well with the isomaltase inhibition activity. For maltase, the inhibitions by 3 and 4 are three times and ten times as strong as those by 1 and 2, respectively. It is difficult to evaluate the maltase inhibition simply due to the effect of the four domains. That is, the maltase inhibitory activity should be considered for all catalytic domains. For example, the weak affinity of 2 for NtMGAM and CtSI could be compensated by the strong affinities for NtSI and CtMGAM. Thus, maltose hydrolysis can be explained by the calculated binding affinities.

As described above, the inhibition activities of maltose and isomaltose hydrolysis were consistent with the predicted binding affinities. However, for sucrose, the calculated values for 3 were inconsistent with the experiments. This means that the current MM/PBSA scheme, including the model structure construction of the C-terminal domains, leaves room for improvement. For example, the recent alpha-glucosidase study showed the existence of splicing variants on CtMGAM [34]. As some amino acid residues of the active site might mutate, we have to consider such effects.

Table 1. Experimental inhibitory activity and estimated binding affinities of salacinol and its derivatives.

	Experimental alpha-glucosidase inhibitory activity (pIC_{50})[a]			Predicted binding affinity by MM PBSA (kcal mol)				
Compound	Maltase	Isomaltase	Sucrase	NtNGAM (X-ray)	CtMGAM (model)	CtMGAM (X-ray)	NtSI (X-ray)	CtSI (model)
1	5.28	5.89	5.80	−20.8	−22.6	−21.5	−21.9	−25.2
2	5.28	6.41	6.34	−19.9	−23.5	−21.4	−23.8	−20.9
3	5.77	6.57	6.92	−20.1	−23.7	−19.4	−24.4	−27.2
4	6.36	6.85	6.49	−27.8	−25.9	−25.8	−27.8	−29.5

[a]Reported values [12,13]

4. CONCLUSIONS

In this study, the inhibition of alpha-glucosidase, a target protein in diabetic therapy, has been computationally investigated. As the four catalytic domains of human alpha-glucosidase have been known, the shapes and the characters of their binding pockets were analyzed. Furthermore, the binding modes and the affinities of the salacinol derivatives have been compared.

Because the structures of CtMGAM and CtSI were unknown at the time we started this study, they have been predicted using the homology modeling method. The comparison of the ligand-binding pockets among the four catalytic domains indicated few differences on the salt bridge and hydrogen bond network for the ligand binding observed in the NtMGAM complex with salacinol. However, several amino acid residues mutated, resulting in differences in the shape and physicochemical properties of the SG site and HG site of each catalytic domain.

The binding modes of salacinol and its derivatives to the four domains have been predicted by the docking study. The locations of the substituent groups in the active site varied according to the shape features of the active site in each catalytic domain.

To predict the binding affinity quantitatively, the binding affinities of salacinol and its derivatives for each domain were estimated by the MM/PBSA method. The inhibition activities of the maltose and isomaltose hydrolysis were consistent with the experimentally determined binding affinities, whereas that of sucrose hydrolysis was partly inconsistent.

Though more accuracy is required for the model construction and the affinity prediction, several designed compounds based on the model structures and SARs have improved the activities of all alpha-glucosidase [35].

ACKNOWLEDGEMENTS

Shinya Nakamura, Kazunori Takahira, Genzoh Tanabe, Osamu Muraoka, Isao Nakanishi (Ministry of Education, Culture, Sports, Science and Technology), 2007-2011.

REFERENCES

1. H. A. Ernst, L. L. Leggio, M. Willemoës, G. Leonard, P. Blum and S. Larsen, "Structure of the Sulfolobus Solfataricus Alpha-Glucosidase: Implications for Domain Conservation and Substrate Recognition in GH31," Journal of Molecular Biology, Vol. 358, No. 4, 2006, pp. 1106-1124.
2. G. M. Gray, B. C. Lally and K. A. Conklin, "Action of Intestinal Sucrase-Isomaltase and Its Free Monomers on an α-Limit Dextrin," The Journal of Biological Chemistry, Vol. 254, No. 13, 1979, pp. 6038-6043.

3. J. L. Chiasson, R. G. Josse, R. Gomis, M. Hanefeld, A. Karasik and M. Laakso, "Acarbose for Prevention of Type 2 Diabetes Mellitus: The STOP-NIDDM Randomised Trial," Lancet, Vol. 359, No. 9323, 2002, pp. 2072-2077.
4. B. Göke, H. Fuder, G. Wieckhorst, U. Theiss, E. Stridde, T. Littke, P. Kleist, R. Arnold and P. W. Lücker, "Voglibose (AO-128) Is an Efficient α-Glucosidase Inhibitor and Mobilizes the Endogenous GLP-1 Reserve," Digestion, Vol. 56, No. 6, 1995, pp. 493-501.
5. R. J. Andrade, M. Lucena, J. L. Vega, M. Torres, F. J. Salmeron, V. Bellot, M. D. Garcia-Escano and P. Moreno, "Acarbose-Associated Hepatotoxicity," Diabetes Care, Vol. 21, No. 11, 1998, pp. 2029-2030.
6. M. Yoshikawa, T. Murakami, K. Yashiro and H. Matsuda, "Kotalanol, a Potent α-Glucosidase Inhibitor with Thiosugar Sulfonium Sulfate Structure, from Antidiabetic Ayurvedic Medicine Salacia reticulata," Chemical and Pharmaceutical Bulletin, Vol. 46, No. 8, 1998, pp. 1339- 1340.
7. M. Yoshikawa, T. Murakami, H. Shimada, H. Matsuda, J. Yamahara, G. Tanabe and O. Muraoka, "Salacinol, Potent Antidiabetic Principle with Unique Thiosugar Sulfonium Sulfate Structure from the Ayurvedic Traditional Medicine Salacia reticulata in Sri Lanka and India," Tetrahedron Letters, Vol. 38, No. 48, 1997, pp. 8367-8370.
8. L. Sim, K. Jayakanthan, S. Mohan, R. Nasi, B. D. Johnston, B. M. Pinto and D. R. Rose, "New Glucosidase Inhibitors from an Ayurvedic Herbal Treatment for Type 2 Diabetes: Structures and Inhibition of Human Intestinal Maltase-Glucoamylase with Compounds from Salacia reticulata," Biochemistry, Vol. 49, No. 3, 2010, pp. 443- 451.
9. L. Sim, C. Willesma, S. Mohan, Y. H. Naim, B. M. Pinto and D. R. Rose, "Structural Basis for Substrate Selectivity in Human Maltase-Glucoamylase and Sucrase-Isomaltase N-Terminal Domains," The Journal of Biological Chemistry, Vol. 285, No. 23, 2010, pp. 17763-17770.
10. R. Eskandari, K. Jones, D. R. Rose and B. M. Pinto, "Probing the Active-Site Requirements of Human Intestinal N-Terminal Maltase Glucoamylase: The Effect of Replacing the Sulfate Moiety by a Methyl Ether in Ponkoranol, a Naturally Occurring α-Glucosidase Inhibitor," Bioorganic & Medicinal Chemistry Letters, Vol. 20, No. 19, 2010, pp. 5686-5689.
11. M. Yoshikawa, F. Xu, S. Nakamura, T. Wang, H. Matsuda, G. Tanabe and O. Muraoka, "Salaprionol and Ponkoranol with Thiosugar Sulfonium Sulfate Structure from Salacia prinoides and α-Glucosidase Inhibitory Activity of Ponkoranol and Kotalanol Desulfate," Heterocycles, Vol. 75, No. 6, 2008, pp. 1397-1405.
12. S. Nakamura, K. Takahira, G. Tanabe, T. Morikawa, M. Sakano, K. Ninomiya, M. Yoshikawa, O. Muraoka and I. Nakanishi, "Docking and SAR Studies of Salacinol Derivatives as α-Glucosidase Inhibitors," Bioorganic & Medicinal Chemistry Letters, Vol. 20, No. 15, 2010, pp. 4420- 4423.
13. G. Tanabe, T. Otani, W. Cong, T. Minematsu, K. Ninomiya, M. Yoshikawa and O. Muraoka, "Biological Evaluation of 3'-O-Alkylated Analogs of Salacinol, the Role of Hydrophobic Alkyl Group at 3' Position in the Side Chain on the α-Glucosidase Inhibitory Activity," Bioorganic & Medicinal Chemistry Letters, Vol. 21, No. 10, 2011, pp. 3159-3162.
14. N. Eswar, B. Webb, M. A. Marti-Renom, M. S. Madhusudhan, D. Eramian, M. Shen, U. Pieper and A. Šali "Comparative Protein Structure Modeling Using MODELLER," Current Protocols in Protein Science, 2007, pp. 2.9.1-2.9.31. doi:10.1002/0471140864.ps0209s50
15. J. Aqvist, C. Medina and J. E. Samuelsson, "New Method for Predicting Binding-Affinity in Computer-Aided Drug Design," Protein Engineering Design and Selection, Vol. 7, No. 3, 1994, pp. 385-391.
16. P. A. Kollman, I. Massova, C. Reyes, B. Kuhn, S. Huo, L. Chong, M. Lee, T. Lee, Y. Duan, W. Wang, O. Donini, P. Cieplak, J. Srinivasan, D. A. Case and T. E. Cheatham III, "Calculating Structures and Free Energies of Complex Molecules: Combining Molecular Mechanics and Continuum Models," Accounts of Chemical Research, Vol. 33, No. 12, 2000, pp. 889-897.
17. B. L. Nichols, S. Avery, P. Sen, D. M. Swallow, D. Hahn and E. Sterchi, "The Maltase-Glucoamylase Gene: Common Ancestry to Sucrase-Isomaltase with Complementary Starch Digestion Activities," Proceedings of the National Academy of Sciences, Vol. 100, No. 3, 2003, pp. 1432-1437.

18. S. R. Eddy, "Where Did the BLOSUM62 Alignment Score Matrix Come From?" Nature Biotechnology, Vol. 22, No. 8, 2004, pp. 1035-1036.
19. Šali and T. L. Blundell, "Comparative Protein Modelling by Satisfaction of Spatial Restraints," Journal of Molecular Biology, Vol. 234, No. 3, 1993, pp. 779-815.
20. S. S. Sheik, P. Sundararajan, A. S. Z. Hussain and K. Sekar, "Ramachandran Plot on the Web," Bioinformatics, Vol. 18, No. 11, 2002, pp. 1548-1549.
21. R Lüthy, J. U. Bowie and D. Eisenberg, "Assessment of Protein Models with Three-Dimensional Profiles," Nature, Vol. 356, No. 6364, 1992, pp. 83-85.
22. D. MacKerell Jr., D. Bashford, M. Bellott, R. L. Dunbrack Jr., J. D. Evanseck, M. J. Field, S. Fischer, J. Gao, H. Guo, S. Ha, D. Joseph-McCarthy, L. Kuchnir, K. Kuczera, F. T. K. Lau, C. Mattos, S. Michnick, T. Ngo, D. T. Nguyen, B. Prodhom, W. E. Reiher III, B. Roux, M. Schlenkrich, J. C. Smith, R. Stote, J. Straub, M. Watanabe, J. Wiórkiewicz-Kuczera, D. Yin and M. Karplus, "All-Atom Empirical Potential for Molecular Modeling and Dynamics Studies of Proteins," The Journal of Physical Chemistry B, Vol. 102, No. 18, 1998, pp. 3586-3616.
23. MOE ver.2010, Chemical Computing Group Inc., Montreal. http://www.chemcomp.com/
24. B. R. Brooks, R. E. Bruccoleri, B. D. Olafson, D. J. States, S. Swaminathan and M. Karplus, "CHARMM: A Program for Macromolecular Energy, Minimization, and Dynamics Calculations," Journal of Computational Chemistry, Vol. 4, No. 2, 1983, pp. 187-217.
25. T. A. Halgren and R. B. Nachbar, "Merck Molecular Force Field. IV. Conformational Energies and Geometries for MMFF94," Journal of Computational Chemistry, Vol. 17, No. 5-6, 1996, pp. 587-615.
26. P. Labute, "The Generalized Born/Volume Integral Implicit Solvent Model: Estimation of the Free Energy of Hydration Using London Dispersion Instead of Atomic Surface Area," Journal of Computational Chemistry, Vol. 29, No. 10, 2008, pp. 1693-1698.
27. L. Ren, X. Qin, X. Cao, L. Wang, F. Bai, G. Bai and Y. Shen, "Structural Insight into Substrate Specificity of Human Intestinal Maltase-Glucoamylase," Protein Cell, Vol. 2, No. 10, 2011, pp. 827-836.
28. InsightII ver.2000, Accelrys Inc. San Diego. http://accelrys.com/
29. J. P. Ryckaert, G. Ciccotti and H. J. C. Berendsen, "Numerical Integration of the Cartesian Equations of Motion of a System with Constraints: Molecular Dynamics of n-Alkanes," Journal of Computational Physics, Vol. 23, No. 3, 1977, pp. 327-341.
30. M. K. Gilson, K. Sharp and B. Honig, "Calculating the Electrostatic Potential of Molecules in Solution: Method and Error Assessment," Journal of Computational Chemistry, Vol. 9, No. 4, 1988, pp. 327-335.
31. M. Sanner, A. J. Olson and J. C. Spehner, "Reduced Surface: An Efficient Way to Compute Molecular Surfaces," Biopolymers, Vol. 38, No. 3, 1996, pp. 305-320.
32. D. Sitkoff, K. A. Sharp and B. Honig, "Accurate Calculation of Hydration Free Energies Using Macroscopic Solvent Models," The Journal of Physical Chemistry, Vol. 98, No. 7, 1994, pp. 1978-1988.
33. L. Ren, X. Cao, P. Geng, F. Bai and G. Bai, "Study of the Inhibition of Two Human Maltase-Glucoamylases Catalytic Domains by Different α-Glucosidase Inhibitors," Carbohydrate Research, Vol. 346, No. 17, 2011, pp. 2688- 2692.
34. K. Jones, L. Sim, S. Mohan, J. Kumarasamy, H. Liu, S. Avery, H. Y. Naim, R. Quezada-Calvillo, B. L. Nichols, B. M. Pinto and D. R. Rose, "Mapping the Intestinal Alpha-Glucogenic Enzyme Specificities of Starch Digesting Maltase-Glucoamylase and Sucrase-Isomaltase," Bioorganic & Medicinal Chemistry, Vol. 19, No. 13, 2011, pp. 3929-3934.
35. G. Tanabe, S. Nakamura, N. Tsutsui,G. Balakishan, W, Xie, S. Tsuchiya, J. Akaki, T. Morikawa, K. Ninomiya, I. Nakanishi, M. Yoshikawa and O. Muraoka, "In Silico Design, Synthesis and Evaluation of 3'-O-Benzylated Analogs of Salacinol, a Potent-Glucosidase Inhibitor from Ayurvedic Traditional Medicine 'Salacia'," Chemical Communications, 2012, in Press.

INDEX